Laser-Plasma Interactions and Applications

Scottish Graduate Series

The Scottish Graduate Series is a long-standing series of graduate level texts proceeding from the Scottish Universities Summer Schools in Physics (SUSSP). SUSSP was established in 1960 to contribute to the dissemination of advanced knowledge in physics, and the formation of contacts among scientists from different countries through the setting up of a series of annual summer schools of the highest international standard. Each school is organized by its own committee which is responsible for inviting lecturers of international standing to contribute an in-depth lecture series on one aspect of the area being studied.

For further volumes:
http://www.springer.com/series/11662

Paul McKenna • David Neely • Robert Bingham
Dino A. Jaroszynski
Editors

Laser-Plasma Interactions and Applications

Springer

Editors
Paul McKenna
University of Strathclyde
Glasgow, UK

David Neely
Rutherford Appleton Laboratory
Oxfordshire, UK

Robert Bingham
Rutherford Appleton Laboratory
Oxfordshire, UK

Dino A. Jaroszynski
University of Strathclyde
Glasgow, UK

ISBN 978-3-319-00037-4 ISBN 978-3-319-00038-1 (eBook)
DOI 10.1007/978-3-319-00038-1
Springer Switzerland Heidelberg New York Dordrecht London

Library of Congress Control Number: 2013932230

Printed on acid-free paper

Springer is part of Springer Science+Business Media (www.springer.com)

Previous SUSSP Schools

1	1960	Dispersion Relations
2	1961	Fluctuation, Relaxation and Resonance in Magnetic Systems
3	1962	Polarons and Excitons
4	1963	Strong Interactions and High Energy Physics
5	1964	Nuclear Structure and Electromagnetic Interactions
6	1965	Phonons in Perfect Lattices and in Lattices with Point Imperfections
7	1966	Particles Interactions at High Energy
8	1967	Methods in Solid State and Superfluid Theory
9	1968	Physics of Hot Plasmas
10	1969	Quantum Optics
11	1970	Hadronic Interactions of Photons and Electrons
12	1971	Atoms and Molecules in Astrophysics
13	1972	Electronic and Structural Properties of Amorphous Semiconductors
14	1973	Phenomenology of Particles at High Energy
15	1974	The Helium Liquids
16	1975	Nonlinear Optics
17	1976	Fundamentals of Quark Models
18	1977	Nuclear Structure Physics
19	1978	Metal Non-metal Transitions in Disordered Solids
20	1979	Laser-Plasma Interactions: 1
21	1980	Gauge Theories and Experiments at High Energy
22	1981	Magnetism in Solids
23	1982	Lasers: Physics, Systems and Techniques
24	1982	Laser-Plasma Interactions: 2
25	1983	Quantitative Electron Microscopy
26	1983	Statistical and Particle Physics
27	1984	Fundamental Forces
28	1985	Superstrings and Supergravity

29	1985	Laser-Plasma Interactions: 3
30	1985	Synchrotron Radiation
31	1986	Localisation and Interaction
32	1987	Computational Physics
33	1987	Astrophysical and Laboratory Spectroscopy
34	1988	Optical Computing
35	1988	Laser-Plasma Interactions: 4
36	1989	Physics of the Early Universe
37	1990	Pattern Recognition and Image Processing
38	1991	Physics of Nanostructures
39	1991	High Temperature Superconductivity
40	1992	Quantitative Microbeam Analysis
41	1992	Spatial Complexity in Optical Systems
42	1993	High Energy Phenomenology
43	1994	Determination of Geophysical Parameters from Space
44	1994	Simple Quantum Systems
45	1994	Laser-Plasma Interactions 5: Inertial Confinement Fusion
46	1995	General Relativity
47	1995	Laser Sources and Applications
48	1996	Generation and Application of High Power Microwaves
49	1997	Physical Processes in the Coastal Zone
50	1998	Semiconductor Quantum Opto-Electronics
51	1998	Muon Science
52	1998	Advances in Lasers and Applications
53	1999	Soft and Fragile Matter: Nonequilibrium Dynamics, Metastability and Flow
54	2000	The Restless Universe: Applications of Gravitational N-body Dynamics to Planetary, Stellar and Galactic Systems
55	2001	Heavy Flavour Physics
56	2002	Ultrafast Photonics
57	2003	Large Hadron Collider Phenomenology
58	2004	Hadron Physics
59	2004	Soft Condensed Matter in Molecular and Cell Biology
60	2005	Laser Plasma Interactions
61	2006	Neutrino Physics
62	2007	Extrasolar Planets
63	2008	High Pressure Physics
64	2008	Advanced Techniques in Electron Spin Resonance
65	2009	LHC Physics
66	2010	Ultrafast Nonlinear Optics
67	2011	Quantum Coherence and Quantum Information
68	2011	Laser Plasma Interactions and Applications

Preface

The *Scottish Universities Summer School in Physics, SUSSP*, was established in 1960 to contribute to the dissemination of advanced knowledge in physics and the formation of contacts among scientists from different countries through a series of annual summer schools of the highest international standard. The 68th SUSSP was on the topic of *Laser-Plasma Interactions and Applications* (the 7th on this topic) and was run as a combined SUSSP and NATO Advanced Study Institute. The academic programme was designed to provide students with a thorough grounding in the core foundation physics of laser-plasma interactions, to follow this up with advanced topics in the field and finally to provide details on potential applications. The format was a mixture of lectures designed to introduce and advance the student's understanding of core topics, and guest talks and discussion sessions to provide insight to the challenges and opportunities in the field. The chapters in this proceedings text are a record of the lectures given and reflect progress made in the field at the time of the school. The text is organised such that the theoretical foundations of the subject are discussed first, in Part I. In Part II topics in the area of High Energy Density Physics are covered. Parts III and IV deal with the applications to Inertial Confinement Fusion and as a driver of particle and radiation sources, respectively. Finally, Part V describes the principle diagnostic, targetry and computational approaches used in the field.

In a break with tradition, this latest SUSSP on the topic of laser-plasma interactions was held at the University of Strathclyde – the previous schools on this topic were held at the University of St. Andrews. There were 104 registered PhD students and post-doctoral researchers, and 23 lecturers and guest speakers. The School also included a public lecture and outreach event, on *Prospects for Inertial Fusion Energy*, which attracted additional University academics and members of the public. The lecture programme consisted of 38 lectures/tutorials, given by international experts in high power laser-plasma interactions and applications. Two poster sessions were also held to enable the student participants to discuss their research with the lecturers and their peers. There was also a social programme, which included a Civic Reception in Glasgow City Chambers, excursions to Edinburgh, Stirling Castle and Loch Katrine, and a banquet with whisky tasting event. These activities enabled participants to get to know each other, to establish new networks and collaborations, and provide a relaxed environment for discussion between lecturers and students.

The SUSSP-68 School was directed by Prof. P. McKenna (University of Strathclyde). Professors D. Neely and R. Bingham (Rutherford Appleton Laboratory and University of Strathclyde) were the Treasurer and Scientific Secretary, respectively, and Prof. D.A. Jaroszynski (University of Strathclyde) was the School Bursar. Additional programme direction and advice was provided by Profs. A.A. Andreev (Vavilov State Optical Institute, Russia) and W. Kruer (University of California Davis, USA). Ms M. King (Rutherford Appleton Laboratory) was the School Secretary and helped to coordinate the social programme. Dr. D.C. Carroll, Mr. R. Gray, Mr. O. Tresca and Mr. D. MacLellan (University of Strathclyde) acted as Stewards and helped to manage the School website, teaching materials and the social programme, and Dr. M.N. Quinn (University of Strathclyde) assisted in compiling this proceedings text.

The SUSSP-68 Organising Committee acknowledge the assistance of staff in the Department of Physics, Conference and Catering Services and Financial Services at the University of Strathclyde. We also gratefully acknowledge the helpful support, advice and governance provided by Alan Walker (SUSSP Secretary/Treasurer) and Profs. Tony Doyle and Ken Bowler (present and former SUSSP Chair, respectively).

Financial support for SUSSP-68 from NATO, the EPSRC, SUPA, the STFC-Central Laser facility, the HiPER project, AWE plc., LLE-University of Rochester, LLNL, the IOP-Plasma Group, Andor Technology, Amplitude, Thales, Coherent Scotland and Laser Support Services is gratefully acknowledged.

Professor Paul McKenna
SUSSP-68 Director

List of Participants

Lecturers and Guest Speakers (alphabetically)

- Prof. A.A. Andreev, Vavilov State Optical Institute, Russia
- Prof. S. Atzeni, University of Rome, Italy
- Prof. R. Bingham, Rutherford Appleton Laboratory and University of Strathclyde, UK
- Prof. J. Collier, Rutherford Appleton Laboratory, UK
- Mr. C. Edwards, Rutherford Appleton laboratory, UK
- Prof. S. Eliezer, Soreq Nuclear Research Centre, Israel
- Dr. T. Goldsack, AWE plc, UK
- Dr. V. Goncharov, Laboratory for Laser Energetics, University of Rochester, USA
- Prof. D.A. Jaroszynski, University of Strathclyde, UK
- Prof. B. Kruer, University of California Davis, USA
- Prof. V. Malka, Laboratoire d'Optique Appliquée, Ecole Polytechnique, France
- Prof. P. McKenna, University of Strathclyde, UK
- Prof. D. Neely, Rutherford Appleton Laboratory and University of Strathclyde, UK
- Dr. C. O'Reilly, Engineering and Physical Sciences Research Council, UK
- Dr. J. Pasley, University of York, UK
- Dr. A.P.L. Robinson, Rutherford Appleton Laboratory, UK
- Prof. S. Rose, Imperial College London, UK
- Prof. M. Rosen, Lawrence Livermore National Laboratory, USA
- Prof. M. Roth, Technical University of Darmstadt, Germany
- Prof. G. Shvets, University of Texas at Austin, USA
- Prof. L. Silva, Instituto Superior Tecnico, Portugal
- Mr. M. Tolley, Rutherford Appleton Laboratory, UK
- Prof. M. Zepf, Queens University Belfast, UK

Registered Students and Postdoctoral Researchers

- Diaw Abdourahmane, Ecole Polytechnique, France
- Florian Abicht, Max-Planck-Institute, Germany
- Wameedh Adress, Queens University Belfast, UK
- Hamad Ahmed, Queens University Belfast, UK
- Constantin Aniculaesei, University of Strathclyde, UK
- Lior Bakshi, Soreq Nuclear Research Center, Israel
- Romeo Banici, National Institute for Laser, Plasma and Radiation Physics, Romania
- Victor Baranov, Joint Institute for High Temperatures, Russian Academy of Sciences, Russia
- Alexei Bashinov, Institute of Applied Physics, Russian Academy of Sciences, Russia
- Stefan Bedacht, Technical University of Darmstadt, Germany
- Jianhui Bin, Max-Planck-Institut fr Quantenoptik, Germany
- Michael Bloom, Imperial College London, UK
- Sergey Bochkarev, Lebedev Physics Institute of Russian Academy of Sciences, Russia
- Cristian Bontoiu, Imperial College London, UK
- Ceri Brenner, University of Strathclyde, UK
- Remi Capdessus, University of Bordeaux, France
- Anthony Carr, University of Lancaster, UK
- David Carroll, University of Strathclyde, UK
- Witold Cayzac, CELIA – Centre Lasers Intenses et Applications, France
- Shao-wei Chou, Max-Planck-Institut fr Quantenoptik, Germany
- Cristian Ciocarlan, National Institute for Laser, Plasma and Radiation Physics, Romania
- Fabrizio Consoli, ENEA, Italy
- Mireille Coury, University of Strathclyde, UK
- Ozgur Culfa, University of York, UK
- Paul Cummings, University of Michigan, USA
- Rachel Dance, University of York, UK
- Christopher Davie, Imperial College London, UK
- Olivier Delmas, Laboratoire de Physique des Gaz et des Plasmas, France
- Jan Dostal, Institute of Plasma Physics ASCR, Czech Republic
- Steffen Faik, Goethe University Frankfurt am Main, Germany
- John Farmer, University of Strathclyde, UK
- Javier Fernandez, University of Madrid, Spain
- Daniel Fletcher, University of Warwick, UK
- Stephen Flood, University of Lancaster, UK
- Peta Foster, Central Laser Facility, UK
- Thomas Fox, University of York, UK
- Yechiel Frank, Soreq Nuclear Research Center, Israel

- Tom Goffrey, University of Warwick, UK
- Evgeniy Govras, Lebedev Physics Institute of Russian Academy of Sciences, Russia
- David Grant, University of Strathclyde, UK
- Ross Gray, University of Strathclyde, UK
- James Green, Central Laser Facility, UK
- Chris Harvey, Umea University, Sweden
- Steve Hawkes, Central Laser Facility, UK
- Amol Holkundkar, Umea University, Sweden
- Jaroslav Huynh, Institute of Plasma Physics ASCR, Czech Republic
- Konstantin Ivanov, M.V. Lomonosov Moscow State University, Russia
- Robert Jaeger, Technical University of Darmstadt, Germany
- Axel Jochmann, Helmholtz-Zentrum-Dresden-Rossendorf, Germany
- Jinchuan Ju, Universite Paris-Sud XI, France
- Ajay Kawshik, Friedrich Schiller University Jena, Germany
- Yevgen Kravets, University of Strathclyde, UK
- Miroslav Krus, Czech Technical University in Prague, Czech Republic
- Bjorn Landgraf, Friedrich Schiller University Jena, Germany
- Utaz Lawi, University of Nouakchott, Jordan
- Maxim Legkov, National Research Nuclear University, Russia
- Remi Lehe, Ecole Normale Suprieure (ENS), France
- Sebastien Lemaire, CEA, France
- David MacLellan, University of Strathclyde, UK
- Yury Malkov, Institute of Applied Physics of the Russian Academy of Sciences, Russia
- Grace Manahan, University of Strathclyde, UK
- David Martin, University of Strathclyde, UK
- Joana Martins, Instituto Superior Tcnico, Portugal
- Matthew McCormack, University of Lancaster, UK
- Sebastian Meuren, Max Planck Institute for Nuclear Physics, Germany
- Rohan Mittal, Lawrence Berkeley National Laboratory, USA
- Iani-Octavian Mitu, National Institute for Laser, Plasma and Radiation Physics, Romania
- Chris Murphy, University of Oxford, UK
- Liviu Neagu, National Institute for Laser, Plasma and Radiation Physics, Romania
- Norman Neitz, Max Planck Institute for Nuclear Physics, Germany
- Alex Ortner, Technical University of Darmstadt, Germany
- Evgeny Papeer, Hebrew University, Israel
- Sergey Pikuz, Joint Institute for High Temperatures, Russian Academy of Sciences, Russia
- Haydn Powell, University of Strathclyde, UK
- Chris Price, Imperial College London, UK
- Peter Racz, Research Institute for Solid State Physics and Optics, Hungary
- Martin Ramsay, AWE plc., UK

- Ayesha Rehman, Imperial College London, UK
- Aakash Sahai, Duke University, USA
- Nasrulla Satlikov, Arifov Institute of Electronics, Uzbekistan Academy of Science, Uzbekistan
- Tiberiu Sava, National Institute for Laser, Plasma and Radiation Physics, Romania
- Gabriel Schaumann, Technical University of Darmstadt, Germany
- Elad Schleifer, Hebrew University, Israel
- William Schumaker, University of Michigan, USA
- Graeme Scott, University of Strathclyde, UK
- Hanan Sheinfux, Hebrew University, Israel
- Andrew Simpson, Imperial College London, UK
- Boussaidi Slah, University Tunis El Manar, Tunisia
- Peter Smorenburg, Eindhoven University of Technology, Netherlands
- Chris Spindloe, Central Laser Facility, UK
- Martin Stefan, Umea University, Sweden
- Anna Subiel, University of Strathclyde, UK
- Steinke Sven, Max Born Instituite for Nonlinear Optics, Germany
- Elena Svystun, V.N. Karazin Kharkiv National University, Ukraine
- Olivier Tresca, University of Strathclyde, UK
- Marija Vranic, Instituto Superior Tcnico, Portugal
- Florian Wagner, Technical University of Darmstadt, Germany
- Thomas White, University of Oxford, UK
- Steven White, Queens University Belfast, UK
- Ben Williams, AWE plc., UK
- Lucy Wilson, University of York, UK
- Mark Yeung, Queens University Belfast, UK
- Jens Zamanian, Umea University, Sweden

Contents

Nomenclature

AFI	Advanced Fast Ignition
ALD	Atomic Layer Deposition
ALE	Arbitrary-Lagrangian-Eulerian
ARP	Apparent Reflection Point
ASE	Amplified Sponanteous Emission
AWE	Atomic Weapons Establishment
BOA	Break-Out Afterburner
CBET	Cross-Beam Energy Transfer
CCD	Charge-Coupled Device
CFD	Computational Fluid Dynamics
CNC	Computer Numerical Control
CPS	Charged-Particle Spectrometer
CVD	Chemical Vapour Deposition
DCA	Detailed Configuration Accounting
DS	Decaying-Shock
DSR	Down Scatter Ratio
DT	Deuterium-Tritium
EM	ElectroMagnetic
EOS	Equation Of State
FABS	Full-Aperture Backscattering Stations
FI	Fast Ignition
FIBS	Field Ionization by Barrier Suppression
FLW	Forced Laser Wakefield
FST	Free Standing Target
GPK	Photon Kinetics theory
HFM	High Flux Model
HHG	High Harmonic Generation
HOPG	Highly-Oriented Pyrolytic Graphite
HPL	High Power Laser
HPT	HiPER Baseline Target
IFAR	In-Flight Aspect Ratio

IFE	Inertial Fusion Energy
ISI	Induced Spatial Incoherence
ITF	Ignition Threshold Factor
LEH	Laser Entrance Holes
LLNL	Lawrence Livermore National Laboratory
LPI	Laser Plasma Instabilities
LTE	Local Thermodynamic Equilibrium
LWFA	Laser WakeField Acceleration
MCF	Magnetic Confinement Fusion
MEMS	Micro-Electro-Mechanical Systems
MRS	Magnetic Recoil Spectrometer
MVSS	MultiVariable Sensitivity Study
NIC	National Ignition Campaign
NIF	National Ignition Facility
NLTE	Non-Local-Thermodynamic-Equilibrium
NTD	Neutron Temporal Diagnostics
ORVIS	Optically Recording Velocity Interferometer System
PECVD	Plasma Enhanced Chemical Vapour Deposition
PET	Positron Emission Tomography
PFI	Proton Fast Ignition
PGI	Particle-Grid Interpolation
PVD	Physical Vapour Deposition
QPM	Quasi-Phase Matching
RCF	RadioChromic Film
RF	Radio Frequency
RHU	Radiation Hohlraum Units
RM	Richtmyer-Meshkov
ROM	Relativistically Oscillating Mirror
RT	Rayleigh-Taylor
RTI	Rayleigh-Taylor Instability
RW	Rarefaction Wave
SBS	Stimulated Brillouin Scattering
SEM	Scanning Electron Microscopy
SMWF	Self-Modulated laser WakeField
SRS	Stimulated Raman Scattering
SSD	Smoothing by Spectral Dispersion
TDP	Technology Development Plan
TNSA	Target Normal Sheet Acceleration
TOF	Time Of Flight
TPD	Two-Plasmon-Decay
TRLs	Technology Readiness Levels
URLLE	University of Rochester's Laboratory for Laser Energetics
VISAR	Velocity Interferometry System for Any Reflector
WDM	Warm Dense Matter

Part I
Theoretical Foundations

Chapter 1
Theory of Underdense Laser-Plasma Interactions with Photon Kinetic Theory

Luis O. Silva and Robert Bingham

Abstract We review recent developments in the theory of laser-plasma interactions, with a focus on generalisations of the theory of parametric instabilities driven by lasers in underdense plasmas to include the effects of broadband or partially incoherent radiation via generalised photon kinetic theory. After an introduction addressing the fundamental concepts underlying parametric instabilities, the key concepts and techniques of photon kinetic theory are presented, along with the steps required to obtain the generalised dispersion relations for the different parametric instabilities. The main details of generalised photon kinetic theory are presented such that this chapter can also be used as reference for future work on generalised photon kinetic theory. As a particular example of the application of this theoretical approach, focus will be given to the derivation of the dispersion relation for stimulated Brillouin scattering (SBS). The main results for stimulated Raman scattering (SRS) by a broadband or partially coherent radiation pump field will also be reviewed.

1.1 Introduction

All material substances interact nonlinearly with intense electromagnetic radiation and plasma is no exception. Such nonlinearity leads to so-called parametric excitation or parametric instabilities. Parametric excitation may be defined as an

L.O. Silva (✉)
GoLP/Instituto de Plasmas e Fusao Nuclear, Instituto Superior Tecnico, Lisboa 1096, Portugal
e-mail: luis.silva@ist.utl.pt

R. Bingham
STFC, Rutherford Appleton Laboratory, Harwell Oxford, Didcot, Oxfordshire 0X11 OQX, UK

SUPA, Department of Physics, University of Strathclyde, Glasgow, Scotland G4 ONG, UK
e-mail: r.bingham@stfc.ac.uk

P. McKenna et al. (eds.), *Laser-Plasma Interactions and Applications*,
Scottish Graduate Series, DOI 10.1007/978-3-319-00038-1_1,

amplification of an oscillation due to a periodic modulation of a parameter that characterises the oscillation. Physically, parametric excitation can be looked upon as a nonlinear instability of two waves (an idler and a signal) by a modulating wave (a pump) due to a mode coupling or wave-wave interaction. The simplest example is the three-wave interaction subject to the frequency and wavenumber matching conditions known as the Manley-Rowe relations

$$\omega_0 = \omega_1 + \omega_2$$
$$\mathbf{k}_0 = \mathbf{k}_1 + \mathbf{k}_2 \tag{1.1}$$

where ω_0 is the pump frequency, $4\omega_{1,2}$ are the decay waves.

The concept of parametric excitation dates back to Lord Rayleigh and has subsequently found extensive application in electronic devices and nonlinear optics. Its more recent application to laser produced plasmas has led to the prediction of a large number of possible plasma instabilities. Some lead to anomalous electron and ion heating, others to scattering of electromagnetic energy out of the plasma. These collective phenomena are therefore of paramount importance in laser driven fusion and laser plasma accelerators. Radiation intensities of the order of 10^{14} W/cm^2 and greater are involved here. Other applications are the heating of magnetically confined plasmas by intense radio frequency radiation, the heating of the ionosphere by intense radar and acceleration of particles by the intense electromagnetic fields surrounding a pulsar. The description of nonlinear effects is therefore one of the central problems in modern plasma physics. Nonlinear theory of waves in plasmas, or of any other nontrivial phenomena, does not consist of a sweeping treatment of the subject which contains linear theory as a simple case; rather, it constitutes a number of cautious excursions from the familiar and comparatively safe territory of linear theory into the regime of nonlinearity. Computer simulations of the problems are helpful, but this can be expensive if complicated situations are to be modelled realistically, and it is easy for the physics to be obscured. What is required is a formalism which will simplify the analysis to the greatest possible degree.

Plasmas can support a number of normal modes of collective excitations, which can co-exist independently for sufficiently weak perturbations from the equilibrium. If the characteristic parameters of the plasma, like density, are modulated periodically in time and space by some large amplitude (pump) wave propagating as $\exp i(\mathbf{k}_0 \cdot \mathbf{x} - \omega_0 t)$, the dielectric properties of the plasma are likewise changed. As a result beat waves (or forced oscillations), with frequency $\omega_0 \underline{+} \omega_1$ and wavenumber $\mathbf{k}_0 \underline{+} \mathbf{k}_1$, can be excited in the plasma, where $\omega_1, \mathbf{k}$ represents the frequency and wavenumber of one of the normal modes. If the following resonance conditions are satisfied $\omega_0 \underline{+} \omega_1 = \omega_2, \mathbf{k}_0 \underline{+} \mathbf{k}_1 = \mathbf{k}_2$ where $\omega_2, \mathbf{k}_2$ are the frequency and wavenumber of another normal mode of the plasma, the amplitudes of the two normal modes $(\omega_1, \mathbf{k},)$ and $(\omega_2, \mathbf{k}_2)$ can grow in time leading to instability, and an exchange of energy and momentum will take place among the three waves. This is not the only significant nonlinear process; the beat wave can also interact with charged particles moving at its phase velocity such that $\omega_0 - \omega_1 = (\mathbf{k}_0 - \mathbf{k}_1).\underline{v}$ where $\underline{v}$ is the velocity of a bunch of particles. Nonlinear wave-particle interaction then occurs, sometimes referred to as nonlinear Landau damping or stimulated Compton scattering. In the presence of a magnetic field the number of normal modes increases and hence the number of possible coupling processes increases. Parametric excitation in a magnetic field is of interest to fusion scientists because of the possibility of heating magnetically confined plasmas by high frequency electric fields. In laser fusion stimulated scattering, filamentation and modulational instabilities are seen as detrimental to the coupling of the laser energy to the plasma. Processes such as stimulated Raman and Brillouin scattering can result in a large function of the laser energy being scattered back out of the plasma while filamentation of the laser beams creates hot spots in the plasma. Stimulated Brillouin scattering (SBS) is a serious instability since it can scatter most of the incident laser light out of the plasma before it reaches the critical surface. Stimulated Brillouin scattering describes the decay of an incident electromagnetic wave into a scattered electromagnetic wave and an ion acoustic wave. Although seen to be an instability to be avoided in direct drive laser fusion indirect drive experiments using hohlraums have taken advantage of SBS by using it to transfer energy between beams. This has been demonstrated at NIF where energy transfer between outer and inner cones of laser beams can be controlled. This is not the case in direct drive where potentially a large fraction of the laser energy can be scattered out of the plasma. SBS arises when an incident laser beam with electric field E_0 couples to a low frequency ion acoustic density perturbation δn producing a transverse current $\propto \delta n E_0$ producing a scattered wave with field E_s. The ponderomotive force $<E_o E_s>$, where $<>$ denotes time average, set up by the heating of the incident and scattered electromagnetic wave enhances the original density perturbation δn, thus producing a feedback mechanism that results in exponential growth of stimulated Brillouin scattering provided that the frequency and wavenumber of the three waves satisfy the conditions

$$\omega_0 = \omega_s + \omega_{ia}, \mathbf{k}_0 = \mathbf{k}_s + \mathbf{k}_{ia} \tag{1.2}$$

where subscripts $0, s$ represent pump and scattered waves and ia is the ion acoustic wave.

For underdense plasmas where $\omega_o \gg \omega_{pe}(\omega_{pe} = n_o e^2/m_e E_o$ is the plasma frequency) and $|\mathbf{k}_0| \simeq |\mathbf{k}_s|$, the matching conditions imply

$$\mathbf{k}_s = -\mathbf{k}_0 \tag{1.3}$$

$$\mathbf{k}_{ia} = 2\mathbf{k}_0 \simeq 2\omega_0/c \tag{1.4}$$

The growth rate and threshold of stimulated Brillouin scattering can easily be derived from the three wave equations that describe stimulated Brillouin scattering. For a comprehensive treatment of the linear phase of the instability [1]. SBS can be responsible for significant loss of photon energy that is scattered out of the plasma. The nonlinear evolution of SBS at high laser intensities where radiation pressure is greater than thermal pressure, $2I_L/c > n_e T_e$, where I_L is laser intensity, c is the speed of light, n_e is plasma density and T_e is the electron temperature, momentum coupling steepens the density profile. This causes the effectiveness of SBS to decrease and reduce the reflectivity. For the opposite case with the radiation pressure $2I_L/c < n_eT_e$ and in a long underdense plasma SBS growth rapidly producing a strong instability, with a large amount of radiation backscattered out of the plasma. In this case the ion acoustic wave can reach large amplitudes and the SBS saturates by a number of processes such as particle trapping, wave-breaking or shock formation, nonlinear ion heating and wave mixing due to light reflected from critical. A discussion of these saturation processes can be found in [2]. Saturation limits the amplitude of the ion acoustic wave density perturbation $\delta n_i/n_o$. Wave breaking and particle trapping occur at high laser intensities and can lead to ion acoustic wave amplitude levels of up to 20 %. The fraction of the incident to laser beam that goes into the ion wave during SBS is given by $\omega_{ia}/\omega_0 = 2c_s/c$, where c_s is the ion acoustic speed. The ion wave can produce a tail on the ion distribution function forming high energy ions. In reality, laser plasmas are usually far from being homogeneous and gradients in density and velocity controls the threshold and growth rate [2]. In a plasma with a velocity gradient dv/dx the frequency of the ion acoustic wave $\omega_{ia} = k_{ia}(c_s - v(x))$ changes with position and this limits the region that satisfies the frequency and wavenumber matching conditions to a small region around their resonant position. SBS can be controlled by several processes such as broad laser bandwidth where the spectral width of the laser light is larger than the effective SBS gain width. Density and velocity of expansion irregularities can reduce stimulated Brillouin scattering significantly. Irregularities destroy the coherence of the waves involved and thus reduce convective amplification significantly.

Stimulated Raman scattering results in the incident laser beam decaying into an electron plasma wave and a scattered electromagnetic wave. Since the frequency of the electron plasma wave sometimes referred to as a Langmuir wave is given by,

$$\omega_{pw}^2 = \omega_{pe}^2 + 3k_{pw}^2 v_{Te}^2$$

the wavenumber and frequency matching conditions can only be satisfied for densities less than quarter critical i.e $n_e < n_c/4$. Stimulated Raman scattering therefore occurs for densities up to $n_c/4$. At the quarter critical another instability namely the two plasmon decay suitability competes with the Raman instability. In two plasmon decay the incident laser beam decays onto two Langmuir waves, the wavenumber matching conditions require that the Langmuir waves propagate in almost opposite directions at angles near 45° to the incident laser wavenumber. Raman backscatter at quarter critical is an absolute instability, the backscatter radiation group speed is zero, at the same time the two plasmon decay instability can also be an absolute instability, with the result that the Raman and two plasmon instabilities thresholds due to inhomogeneity are relatively low near quarter critical. These instabilities have therefore strong non-linear effects near $n_c/4$. The ponderomotive force due to the different plasma waves is strong enough to create density structures and in some cases the wave can be trapped in the density cavities. At the same time, large ion density fluctuations are formed that propagate down the density gradient. Profile steepening can also be responsible for increasing the inhomogeneity threshold switching off the instabilities. A consequence of Raman and two plasmon instability is the generation of a high energy electron tail. These heated electrons are a major concern since they preheat the fuel in laser fusion capsules. At intermediate densities $n_e < 0.2n_c$ stimulated Raman scattering is less strong but it can still result in the backscatter of a large fraction of the incident laser energy and contribute to the formation of high energy electron tails.

As with SBS, a broad laser bandwidth can reduce the growth rate of stimulated Raman scattering. At the critical surface it is possible for the incident laser beam to excite the parametric decay instability where the resultant waves are a Langmuir wave and an ion acoustic wave. The parametric decay instability has maximum growth rate for the decay waves to propagate almost parallel to the laser electric field. The parametric decay instability results in wave absorption of the laser energy at the critical density. A variation of the parametric decay instability is the oscillating two stream which is basically a four wave instability where the ion acoustic mode is purely growing. The oscillating two stream instability also occurs at the critical density and results in the laser beam absorption. It can also ripple the critical density surface resulting in non-uniform absorption.

An instability that is also four wave is the filamentation instability. The filamentation instability occurs for densities less than the critical density where the laser beam couples to an ion acoustic perturbation that is purely growing producing density ripples. Filamentation produces an intensity modulation across the laser beam. This modulation grows and results in the laser beam breaking up into filaments that become more pronounced as the beam propagates through the plasma towards the critical density. Filamentation is caused by variations in intensity across the beam regions of initially higher intensity push the plasma aside due to the ponderomotive force. As a consequence this reduces the density locally and increases the index of refraction of the plasma in the higher intensity region bending the wave fronts in such a way that the curvature of the wave fronts produces a focusing effect increasing the intensity still further. Filamentation is a convective instability amplifying any intensity variation initially present in the beam or plasma. Since the density fluctuation in filamentation is purely growing and does not correspond to a resonant mode, the instability is not as sensitive to plasma inhomogeneity as the three wave instabilities. In addition to ponderomotive driven filamentation and self-focusing they can also be driven by thermal or relativistic effects. In laser fusion parametric instabilities are normally seen as being harmful to successful coupling into fusion pellet. However, there is a great deal of research into controlling both simulated Brillouin and Raman scattering with a view to applications. In particular SBS is employed to control the energy distribution between the inner and outer laser cones in hohlraum targets and SRS in being studied as a possible future amplification technique where energy is transferred from a long pump beam to a much shorter probe beam [3]. This has the potential to reach laser intensities of $10^{25}\ \mathrm{W/cm^2}$.

This overview demonstrates the wide range of scenarios and phenomenology associated with parametric instabilities in plasmas and their roles in many applications. There is a tremendous body of theoretical and numerical work on these instabilities, with a good starting point being Refs. [1, 2]. Here we will focus on novel theoretical approaches to study parametric instabilities in the presence of broadband radiation fields of arbitrary intensity.

1.2 Motivation for a Generalised Photon Kinetic Theory

The study of parametric instabilities is important in many fields of science [4–7]. Standard methods use a coherent wave description to study this problem, but the externally induced incoherence or the partial coherence of most systems render this method incomplete. A plan to describe the instabilities of broadband radiation must therefore include an alternative (but formally equivalent) representation of the full nonlinear wave equation for electromagnetic waves in plasmas. A statistical description of the photons in a phase space $(\mathbf{r}, \mathbf{k})$, with the corresponding distribution function of photons in this phase space to represent the radiation field, would therefore meet the requirements for a fully self-consistent description of parametric instabilities driven by broadband radiation of arbitrary intensity.

The Wigner-Moyal statistical theory provides the toolbox to study parametric instabilities, as first explored in nonlinear optics. With a derivation of a statistical description of a partially incoherent electromagnetic wave propagating in a nonlinear medium [8], it became clear that a stabilisation of the modulational instability is possible as a result of an effect similar to Landau damping and caused by random phase fluctuations of the propagating wave, equivalent to the broadening of the Wigner spectrum. Similar studies [9, 10] focused on the onset of the transverse instability in nonlinear media in the presence of a partially incoherent light. The Wigner-Moyal theory applied to electromagnetic waves in plasmas or photon kinetics also provides an alternative approach to numerical modelling of laser propagation in plasmas via the photon-in-cell paradigm [11]. In nonlinear optics, the standard Wigner-Moyal theory is perfectly adjusted, without any required generalisations, since the paraxial wave equation is usually sufficient to describe the main physical processes, thus justifying a forward propagating *ansatz* for the evolution of electromagnetic waves in dispersive nonlinear media. In this context the standard Wigner-Moyal formalism, which is formally equivalent to the Schrödinger equation, can be used directly. In plasma physics, this is clearly a limitation, as many critical aspects in ICF, fast ignition and several applications in laser-plasma and astrophysical scenarios demand a detailed analysis and the inclusion of the backscattered radiation.

The inclusion of bandwidth or incoherence effects in laser driven parametric instabilities has also been studied extensively. The addition of small random deflections to the phase of a plane wave was shown to significantly suppress the three-wave decay instability [12], which was one of the first mechanisms where the introduction of some degree of incoherence in the laser was proposed as a way to avoid the deleterious effects of the instability. The threshold values for some electrostatic instabilities can also be effectively increased either by applying a random amplitude modulation to the laser or by the inclusion of a finite bandwidth of the pump wave [13, 14]. A new method for the inclusion of finite bandwidth effects on parametric instabilities, allowing arbitrary fluctuations of any group velocity, has also been developed in [15, 16]. As far as the stimulated Raman scattering instability is concerned, it became clear that, although it may seriously decollimate a coherent laser beam, the increase of the laser bandwidth is an effective way to suppress the instability [17].

A statistical description of light can be achieved through the Wigner-Moyal formalism of quantum mechanics, which provides, in its original formulation, a one-mode description of systems ruled by Schrödinger-like equations. In order to address other processes where side or backscattering can be important, a generalisation of the Photon Kinetics theory (GPK) was recently developed by J.E. Santos and L.O. Silva [18]. This new formulation is completely equivalent to the full Klein-Gordon equation and was employed to derive a general dispersion relation for stimulated Raman scattering driven by white light [19]. This is the basis for the discussion in the next chapters.

1.3 Generalised Photon Kinetic Theory

Let us first consider the propagation of a linearly polarised electromagnetic wave (polarised along the y direction) in a plasma. The wave equation describing the evolution of the vector potential of the electromagnetic wave A_y can be written as

$$\frac{1}{c^2}\partial_t^2 A_y - \partial_x^2 A_y \simeq -\frac{\omega_{p0}^2}{c^2}\left(1+\frac{\delta n}{n_0}-\frac{1}{2}\frac{e^2A_y^2}{m^2c^4}\right)A_y \tag{1.5}$$

For now we will not discuss the particular plasma response *i.e.* the dynamics of the plasma in the presence of the light wave, since this will be different depending on the particular regime of interest (coupling with electron plasma waves as in stimulated Raman scattering, or coupling with ion acoustic waves as in stimulated Brillouin scattering). Using normalised units, where length is normalised to c/ω_{p0}, with c the velocity of light in vacuum and $\omega_{p0} = (4\pi e^2 n_{e0}/m_e c^2)^{1/2}$ the electron plasma frequency, time to $1/\omega_{p0}$, mass and absolute charge to those of the electron, respectively, m_e and e, with $e > 0$, and the vector potential A_y is normalised to $m_e c^2$, and neglecting the relativistic mass correction term associated with A_y^2, Eq. 1.5 reduces to

$$\left(\partial_t^2 - \partial_x^2\right) a_y + (1+\delta n)a_y = 0 \tag{1.6}$$

Using the standard methods [2, 20] it is not feasible to consider a broadband or partially incoherent radiation field associated with a_y, and determine the properties of the parametric instabilities from Eqs. (1.5 and 1.6). To achieve this it is critical to provide a statistical description of the field. This is the main goal of generalised photon kinetic theory (GPK) [18, 19].

As detailed in Ref. [19], instead of performing the calculations with respect to linear polarisation we will focus our discussion in circularly polarised light, being straightforward the modification for linearly polarised radiation. We will use $\mathbf{a}_p(\mathbf{r},t) = 2^{-1/2}(\hat{z}+i\hat{y})a_0 \int d\mathbf{k}A(\mathbf{k})\exp[i(\mathbf{k}.\mathbf{r}-(\mathbf{k}^2+1)^{1/2}t)]$ as the normalised vector potential of the circularly polarised pump field, $\mathbf{a}_p = e\mathbf{A}_p/m_ec^2$, where $(\mathbf{k}^2+1)^{1/2} \equiv \omega(\mathbf{k})$ is the monochromatic dispersion relation in a uniform plasma, where n_{e0} and n_{i0} are the equilibrium (zeroth order) particle densities of the electrons and ions, respectively, and the densities are normalised to the equilibrium electron density, such that $n_{e0} = 1$ and $n_{i0} = 1/Z$, where Z is the electric charge of the ions in units of e. We also allow for a stochastic component in the phase of the vector potential $A(\mathbf{k}) = \hat{A}(\mathbf{k})\exp[i\psi(\mathbf{r},t)]$ such that $\langle \mathbf{a}_p^*(\mathbf{r}+\mathbf{y}/2,t).\mathbf{a}_p(\mathbf{r}-\mathbf{y}/2)\rangle = a_0^2 m(\mathbf{y})$ is independent of $\mathbf{r}$ with $m(0) = 1$ and $|m(\mathbf{y})|$ is bounded between 0 and 1, which means that the field is spatially stationary.

Instead of using the field $\mathbf{a}$, GPK replaces the radiation field $\mathbf{a}$ by two auxiliary fields, ϕ and χ, such that

$$\phi,\chi = (\mathbf{a} \pm i\partial_t \mathbf{a})/2 \tag{1.7}$$

With these fields it is possible to easily demonstrate that the full wave equation (e.g. Eq. 1.5) is formally equivalent to two coupled Schrödinger equations for the auxiliary fields [18]. This prescription is due to Feschbach and Villars [21].

With the introduction of four real phase-space densities:

$$W_0 = W_{\phi\phi} - W_{\chi\chi} \tag{1.8}$$

$$W_1 = 2\mathrm{Re}[W_{\phi\chi}] \tag{1.9}$$

$$W_2 = 2\mathrm{Im}[W_{\phi\chi}] \tag{1.10}$$

$$W_3 = W_{\phi\phi} + W_{\chi\chi} \tag{1.11}$$

with the usual definition for the Wigner transform

$$W_{\mathbf{f}.\mathbf{g}}(\mathbf{k},\mathbf{r},t) = \left(\frac{1}{2\pi}\right)^3 \int e^{i\mathbf{k}.\mathbf{y}} \mathbf{f}^* \left(\mathbf{r}+\frac{\mathbf{y}}{2}\right).\mathbf{g}\left(\mathbf{r}-\frac{\mathbf{y}}{2}\right) d\mathbf{y} \tag{1.12}$$

as in Refs. [22–25], the coupled equations for ϕ,χ (and, therefore, the complete Klein-Gordon equation corresponding to Eqs. (1.5 and 1.6)) are shown [18] to be equivalent to the following set of transport equations for the W_i, $i = 0,\ldots,3$

$$\partial_t W_0 + \mathscr{L}(W_2 + W_3) = 0 \tag{1.13}$$

$$\partial_t W_1 - \hat{\mathscr{G}}(W_2 + W_3) - 2W_2 = 0 \tag{1.14}$$

$$\partial_t W_2 - \hat{\mathscr{L}} W_0 + \hat{\mathscr{G}} W_1 + 2W_1 = 0 \tag{1.15}$$

$$\partial_t W_3 + \hat{\mathscr{L}} W_0 - \hat{\mathscr{G}} W_1 = 0 \tag{1.16}$$

with the following definition for the operators $\hat{\mathscr{L}}$ and $\hat{\mathscr{G}}$

$$\hat{\mathscr{L}} \equiv \mathbf{k}.\nabla_{\mathbf{r}} - n\sin\left(\frac{1}{2}\overleftarrow{\nabla}_{\mathbf{r}}.\overrightarrow{\nabla}_{\mathbf{k}}\right) \tag{1.17}$$

$$\hat{\mathscr{G}} \equiv \left(\mathbf{k}^2 - \frac{\nabla_{\mathbf{r}}^2}{4}\right) + n\cos\left(\frac{1}{2}\overleftarrow{\nabla}_{\mathbf{r}}.\overrightarrow{\nabla}_{\mathbf{k}}\right) \tag{1.18}$$

where the arrows denote the direction of the operator and the trigonometric functions represent the equivalent series expansion of the operators.

The transport equations (1.13,1.14,1.15, and 1.16) are formally equivalent to the full wave equation that describes the propagation of an arbitrarily intense electromagnetic wave in a plasma (e.g. Eq. 1.5) and describe the evolution of the radiation field and therefore are the field equations in GPK . Even if formally more complex, it is now possible to describe arbitrary distributions of photons in the phase space $(\mathbf{r},\mathbf{k})$ and the perturbation techniques over distribution functions, common in plasma physics, can also be used over the transport equations of GPK.

For instance it is illustrative to evaluate the zeroth order terms of each W_i, $i = 0,\ldots,3$, so we use $\mathbf{a} = \mathbf{a}_p$. It can be easily shown that

$$W_{\phi\phi}^{(0)} = \tfrac{\rho_0(\mathbf{k})}{4}[1+\omega^2(\mathbf{k})+2\omega(\mathbf{k})] \tag{1.19}$$

$$W_{\chi\chi}^{(0)} = \tfrac{\rho_0(\mathbf{k})}{4}[1+\omega^2(\mathbf{k})-2\omega(\mathbf{k})] \tag{1.20}$$

$$W_{\phi\chi}^{(0)} = \tfrac{\rho_0(\mathbf{k})}{4}[1-\omega^2(\mathbf{k})] = -\tfrac{\rho_0(\mathbf{k})}{4}\mathbf{k}^2 \tag{1.21}$$

where $\rho_0(\mathbf{k}) \equiv W_{\mathbf{a}_p.\mathbf{a}_p}$ can be interpreted as the equilibrium distribution function of the photons. We can also immediately write

$$W_0^{(0)} = W_{\phi\phi}^{(0)} - W_{\chi\chi}^{(0)} = \rho_0(\mathbf{k})\omega(\mathbf{k}) \tag{1.22}$$

$$W_1^{(0)} = 2\mathrm{Im}\left[W_{\phi\chi}^{(0)}\right] = 0 \tag{1.23}$$

$$W_2^{(0)} = 2\mathrm{Re}\left[W_{\phi\chi}^{(0)}\right] = -\tfrac{\rho_0(\mathbf{k})}{2}\mathbf{k}^2 \tag{1.24}$$

$$W_3^{(0)} = W_{\phi\phi}^{(0)} + W_{\chi\chi}^{(0)} = \rho_0(\mathbf{k})\left(1+\tfrac{\mathbf{k}^2}{2}\right) \tag{1.25}$$

where we have taken into account the Wigner function can take only real values [22–25].

The first order perturbative term of the transport equations are critical to understand coupling with the plasma density perturbation. From the first transport equation (1.13) we obtain, in first order,

$$\partial_t \tilde{W}_0 + \mathbf{k}.\nabla_{\mathbf{r}}(\tilde{W}_2 + \tilde{W}_3) - \tilde{n}\sin\left(\frac{1}{2}\overleftarrow{\nabla}_{\mathbf{r}}.\overrightarrow{\nabla}_{\mathbf{k}}\right)\rho_0(\mathbf{k}) = 0 \tag{1.26}$$

where $\tilde{}$ describes the first order perturbed quantities. Performing time and space Fourier transforms ($\partial t \rightarrow i\omega_L, \nabla_{\mathbf{r}} \rightarrow -i\mathbf{k}_L$), leads to

$$i\omega_L \tilde{W}_0 - i\mathbf{k}.\mathbf{k}_L(\tilde{W}_2 + \tilde{W}_3) + \tilde{n}\sin\left(\frac{i}{2}\mathbf{k}_L.\nabla_k\right)\rho_0(\mathbf{k}) = 0 \tag{1.27}$$

We note that we can write $\sin\hat{\mathscr{A}} = \frac{e^{i\hat{\mathscr{A}}} - e^{-i\hat{\mathscr{A}}}}{2i}$, for any operator $\hat{\mathscr{A}}$. Similarly, $\cos\hat{\mathscr{A}} = \frac{e^{i\hat{\mathscr{A}}} + e^{-i\hat{\mathscr{A}}}}{2}$. Making use of these relations, we have

$$e^{\mathbf{A}.\nabla_k} f(\mathbf{k}) = \sum_{n=0}^{\infty} \frac{(\mathbf{A}.\nabla_k)^n}{n!} f(\mathbf{k}) = f(\mathbf{k} + \mathbf{A}) \tag{1.28}$$

The first transport equation can then be reduced to

$$\omega_L \tilde{W}_0 - \mathbf{k}.\mathbf{k}_L(\tilde{W}_2 + \tilde{W}_3) - \tilde{n}\frac{\rho_0\left(\mathbf{k} - \frac{\mathbf{k}_L}{2}\right) - \rho_0\left(\mathbf{k} + \frac{\mathbf{k}_L}{2}\right)}{2} = 0 \tag{1.29}$$

We proceed analogously with the other three transport equations, leading to a system of four independent first order equations for the four variables $\tilde{W}_i$. We also note that

$$W_2 + W_3 = W_{\phi\phi} + W_{\chi\chi} + 2\mathrm{Re}[W_{\phi\chi}] = W_{\mathbf{a}.\mathbf{a}} \tag{1.30}$$

In zeroth order, as expected,

$$W_2^{(0)} + W_3^{(0)} = W_{\mathbf{a}_p.\mathbf{a}_p} = \rho_0(\mathbf{k}) \tag{1.31}$$

while in first order

$$\tilde{W}_2 + \tilde{W}_3 = W_{\mathbf{a}_p.\tilde{\mathbf{a}}} + W_{\tilde{\mathbf{a}}.\mathbf{a}_p} = 2W_{\mathbf{a}_p.\tilde{\mathbf{a}}} \tag{1.32}$$

where we have used the symmetry property of the Wigner distribution function that can be immediately derived from its realness ($W_{\mathbf{f}.\mathbf{g}} = W_{\mathbf{g}.\mathbf{f}}$).

Since the plasma response is proportional to the beating of the pump wave $\mathbf{a}_p$ and the scattered wave $\tilde{\mathbf{a}}$ (and the real part of this beating) it is important to obtain an equation for $W_{\mathrm{Re}[\mathbf{a}_p.\tilde{\mathbf{a}}]}$. Taking the real part of Eq. 1.32 and solving this equation together with the four independent equations for each $\tilde{W}_i$ yields, after some lengthy but straightforward calculations,

$$W_{\mathrm{Re}[\mathbf{a}_p.\tilde{\mathbf{a}}]} = \frac{1}{2}\tilde{n}\left[\frac{\rho_0\left(\mathbf{k} + \frac{\mathbf{k}_L}{2}\right)}{D_s^-} + \frac{\rho_0\left(\mathbf{k} - \frac{\mathbf{k}_L}{2}\right)}{D_s^+}\right] \tag{1.33}$$

with

$$\frac{1}{D_s^{\mp}} = \frac{1 \pm \frac{2\mathbf{k}.\mathbf{k}_L}{\omega_L^2} \pm \frac{2\omega\left(\mathbf{k}+\frac{\mathbf{k}_L}{2}\right)}{\omega_L}}{\omega_L^2 - 4\mathbf{k}_L^2 - \mathbf{k}_L^2 + 4\frac{(\mathbf{k}.\mathbf{k}_L)^2}{\omega_L^2} - 4} \tag{1.34}$$

The expression for $D_s^{\mp}$ can simplified to

$$D_s^{\pm} = \frac{(\omega_L^2 \mp 2\mathbf{k}\mathbf{k}_L)^2 - \left[2\omega_L\omega\left(\mathbf{k} \mp \frac{\mathbf{k}_L}{2}\right)\right]^2}{\omega_L^2 \mp 2\mathbf{k}.\mathbf{k}_L \mp 2\omega_L\omega\left(\mathbf{k}+\frac{\mathbf{k}_L}{2}\right)}, \tag{1.35}$$

providing the driving term of the parametric instability as

$$W_{\mathrm{Re}\left[\mathbf{a}_p.\tilde{\mathbf{a}}\right]} = \frac{1}{2}\tilde{n}\left[\frac{\rho_0\left(\mathbf{k}+\frac{\mathbf{k}_L}{2}\right)}{D^-} + \frac{\rho_0\left(\mathbf{k}-\frac{\mathbf{k}_L}{2}\right)}{D^+}\right], \tag{1.36}$$

where $D^{\pm} = \omega_L^2 \mp \left[\mathbf{k}.\mathbf{k}_L - \omega_L\omega\left(\mathbf{k} \mp \frac{\mathbf{k}_L}{2}\right)\right]$ and ω_L ($\mathbf{k}_L$) represents the instability frequency (wave vector).

As can be observed from Eq. 1.36, an arbitrary distribution function of photons ρ_0 can be considered for the pump field, thus allowing for the inclusion of a broadband or a partially coherent radiation pump field. Eq. 1.36 connects the propagation of the pump and the scattered fields with the plasma response (associated with $\tilde{n}$).

1.4 Derivation of the Dispersion Relation for Stimulated Brillouin Scattering

We now consider the coupling of the intense radiation field with the plasma, when the radiation field is described by the transport equations derived in the previous section. In our plan to describe parametric instabilities, we should now analyse the plasma response to the presence of the pump and scattered fields. For illustration purposes, we will analyse coupling with ion acoustic waves *i.e.* stimulated Brillouin scattering) [26–30]. We consider the plasma as an interpenetrating fluid of both electrons and ions, with n_{e0} and n_{i0} their equilibrium (zeroth order) particle densities, respectively.

To obtain a dispersion relation for SBS we must couple the typical plasma response to the independently derived driving term, obtained within the GPK framework, and given by Eq. 1.36.

Combining the continuity and force equations for each species and closing the system with an isothermal equation of state, we can readily obtain the plasma response to the propagation of a light wave $\mathbf{a}_p$, beating with its scattered component $\tilde{\mathbf{a}}$, to produce the ponderomotive force of the laser [2]

$$\left(\frac{\partial^2}{\partial t^2} - 2\tilde{\nu}\partial t - c_S^2\nabla^2\right)\tilde{n} = \frac{Z}{M}\nabla^2\mathrm{Re}[\mathbf{a}_p.\tilde{\mathbf{a}}], \tag{1.37}$$

with $c_S \equiv \sqrt{\frac{Z\theta_e}{M}}$ being the ion-sound velocity, M the mass of the ions, θ_e the electron temperature and $\tilde{\nu}$ an integral operator whose Fourier transform is $\nu|\mathbf{k}_L|c_S$, accounting for the damping of the ion acoustic waves (e.g. via Landau damping).

Performing time and space Fourier transforms ($\partial t \rightarrow i\omega_L, \nabla_{\mathbf{r}} \rightarrow -i\mathbf{k}_L$) on the plasma response (1.37) gives

$$\mathscr{F}[\tilde{n}] = \frac{Z}{M}\frac{k_L^2}{\omega_L^2 + 2i\nu\omega_L|\mathbf{k}_L|c_S - c_S^2\mathbf{k}_L^2}\mathscr{F}\left[\mathrm{Re}[\mathbf{a}_p.\tilde{\mathbf{a}}]\right], \tag{1.38}$$

and the same operations on the driving term Eq. 1.36 yield

$$\mathscr{F}\left[W_{\mathrm{Re}\left[\mathbf{a}_p.\tilde{\mathbf{a}}\right]}\right] = \frac{1}{2}\mathscr{F}[\tilde{n}]\left[\frac{\rho_0\left(\mathbf{k}+\frac{\mathbf{k}_L}{2}\right)}{D^-} + \frac{\rho_0\left(\mathbf{k}-\frac{\mathbf{k}_L}{2}\right)}{D^+}\right], \tag{1.39}$$

with $D^{\pm} = \omega_L^2 \mp \left[\mathbf{k}.\mathbf{k}_L - \omega_L\omega\left(\mathbf{k}\mp\frac{\mathbf{k}_L}{2}\right)\right]$, as before, and $c_S \equiv \sqrt{\frac{Z\theta_e}{M}}$.

The general dispersion relation can now be obtained using one of the properties of the Wigner function [22–25]

$$\int W_{f.g}d\mathbf{k} = f^*g \Rightarrow \int \frac{W_{\mathrm{Re}\left[\mathbf{a}_p.\tilde{\mathbf{a}}\right]}}{\mathrm{Re}\left[\mathbf{a}_p.\tilde{\mathbf{a}}\right]}d\mathbf{k} = 1 \tag{1.40}$$

as

$$1 = \frac{\omega_{pi}^2}{2}\frac{\mathbf{k}_L^2}{\omega_L^2 + 2i\nu\omega_L|\mathbf{k}_L|c_S - c_S^2\mathbf{k}_L^2}\int\left[\frac{\rho_0\left(\mathbf{k}+\frac{\mathbf{k}_L}{2}\right)}{D^-} + \frac{\rho_0\left(\mathbf{k}-\frac{\mathbf{k}_L}{2}\right)}{D^+}\right]d\mathbf{k}, \tag{1.41}$$

with $\omega_{pi} = \sqrt{Z/M}$ the ion plasma frequency in normalised units.

By making an appropriate change of variables, our general dispersion relation can then be written in a more compact way

$$1 = \frac{\omega_{pi}^2}{2}\frac{\mathbf{k}_L^2}{\omega_L^2 + 2i\nu\omega_L|\mathbf{k}_L|c_S - c_S^2\mathbf{k}_L^2}\int \rho_0(\mathbf{k})\left(\frac{1}{D^+} + \frac{1}{D^-}\right)d\mathbf{k}, \tag{1.42}$$

with $D^{\pm}(\mathbf{k}) = [\omega(\mathbf{k}) \pm \omega_L]^2 - (\mathbf{k}\pm\mathbf{k}_L)^2 - 1$.

We first apply our general dispersion relation (1.42) to the simple and common case of a pump plane wave of wave vector $\mathbf{k}_0$, which means that $\rho_0(\mathbf{k}) = a_0^2\delta(\mathbf{k}-\mathbf{k}_0)$, and we drop the Landau damping contribution. The dispersion relation then becomes

$$1 = \frac{\omega_{pi}^2}{2}\frac{\mathbf{k}_L^2}{\omega_L^2 - c_S^2\mathbf{k}_L^2}a_0^2\left\{\frac{1}{[\omega(\mathbf{k}_0)+\omega_L]^2-(\mathbf{k}_0+\mathbf{k}_L)^2-1}+\right.$$

$$\left.+\frac{1}{[\omega(\mathbf{k}_0)-\omega_L]^2-(\mathbf{k}_0-\mathbf{k}_L)^2-1}\right\}. \tag{1.43}$$

This result recovers that of Ref. [2], which studies the case of a pump wave $\mathbf{A}_L = \mathbf{A}_{L0}\cos(\mathbf{k}_0.\mathbf{r}-\omega_0 t)$, if we account for the difference in polarisation and use $\omega_0 = \omega(\mathbf{k}_0)$. All the conclusions derived in Ref. [2], based on this dispersion relation, are then consistent with the predictions of GPK.

A more in depth analysis of the dispersion relation is beyond the scope of this chapter. The interested reader can find additional details on the consequences of Eq. 1.42 in [29, 30]. We observe that to obtain the dispersion relation for stimulated Raman scattering [19] it would suffice to replace the equation for the plasma response (1.37) that in the case discussed here describes coupling with the (low frequency) ion acoustic waves, with the plasma response corresponding to the (high frequency) electron plasma waves viz.

$$\left(\partial_t^2 + \frac{\omega_{p0}^2}{\gamma_0}\right)\tilde{n} = \frac{n_0c^2}{\gamma_0^2}\nabla_{\mathbf{r}}{}^2\mathrm{Re}\left[\mathbf{a}_p\cdot\tilde{\mathbf{a}}^*\right] \tag{1.44}$$

A summary of the results obtained with generalised photon kinetic theory for stimulated Raman scattering, stimulated Brillouin scattering, and the (relativistic) modulation instability is presented in the next section.

1.5 Main Results Obtained with Generalised Photon Kinetic Theory

Generalised photon kinetic theory has been applied to the study of several parametric instabilities, in particular SRS [19], SBS [29,30], and the modulational instability (L.O. Silva and R. Bingham, 2009, "unpublished"). Photon kinetic simulations of modulational type instabilities has been applied in not only laser plasmas but also in tokamaks and space plasmas [31–36].

For the monochromatic pump field described in the previous section $\rho_0(\mathbf{k}) = a_0^2\delta(\mathbf{k}-\mathbf{k}_0)$ the standard results were recovered for all these instabilities, as expected, but the introduction of broadband effects have demonstrated novel features. The main theoretical results have been obtained for a waterbag distribution function of photons, such that analytical results could be obtained. The key parameters in this distribution are the width of the distribution Δ, and the central wavenumber of the photon distribution function $\mathbf{k}_0$.

The analysis of the effects of stimulated Raman scattering has shown two important results [19]. First of all, and for Raman backscattering, the dependence of the growth rate on the bandwidth was found to scale as $\propto 1/\Delta$ as previously works have demonstrated and as expected from a three-wave resonant process. On the other hand, for Raman forward scattering the growth rate scales as $\propto 1/\sqrt{\Delta}$, thus decaying much slower than what was expected and predicted in previous works. This is associated with the fact that Raman forward scattering is, in general, a four-wave non-resonant process, which means that it is less sensitive to broadband effects. This slow decay with Δ also indicates that Raman forward scattering will be significantly more difficult to shutdown by increasing the bandwidth of the pump laser field, unlike what happens in

The role of the bandwidth of the pump radiation in Raman backscattering and stimulated Brillouin scattering bears some resemblance with stimulated Raman backscattering since it is also a three-wave resonant process, showing a growth rate dependence with the bandwidth scaling as $\propto 1/\Delta$. The general dispersion relation for stimulated Brillouin scattering in Eq. 1.42 has been compared with other models for the effects of the bandwidth on the instability and the analysis in Refs. [29, 30] has revealed that the range of validity of GPK is significantly larger than previous models. This work has relied not only on the analysis of the more academic waterbag distribution function but also on the exploration of realistic broadband distribution functions such as those relevant for ICF experiments.

The dispersion relation for the modulational instability can be solved for a Lorentzian distribution function of the transverse wave numbers k_z, with a width $\Delta (f(k_z) = \frac{\Delta}{\pi}\frac{1}{k_z^2+\Delta^2})$ (L.O. Silva and R. Bingham, 2009). It was found that the range of unstable wave numbers has an upper bound at

$$k_{\max} = \frac{\omega_{p0}}{c}\sqrt{2\frac{a_0^2}{\gamma_0^3} - 4\Delta^2} \tag{1.45}$$

which corresponds to an instability threshold given by

$$\frac{a_0^2}{\Delta^2} > 2\gamma_0^3 \tag{1.46}$$

This threshold presents a similar dependence as the threshold for filamentation/self-focusing of a Gaussian laser pulse ($a_0^2 W_0^2 \omega_{p0}^2/c^2 > 32$), which is even more clear if we consider that the spread of the transverse wave-numbers of a Gaussian laser pulse is $\Delta \approx 1/W_0$ where W_0 is the laser spot size. Moreover, the presence of broadband radiation can shutdown the modulational instability even at relativistic intensities in the long wavelength limit.

1.6 Summary

In this chapter we have presented the fundamentals aspects of generalised photon kinetic theory. Particular emphasis was given to the material that is not generally present in the literature and that will allow the reader to use this technique in a broader range of physical conditions. The focus of GPK has been on the parametric instabilities driven by intense lasers in plasmas but the GPK toolbox can be easily used in other nonlinear systems, with the most impact in scenarios where the backscattered radiation/waves play an important role on the overall dynamics of the parametric instabilities. Further generalisations of GPK should address the coupling of the different parametric instabilities, and the spatio-temporal theory of the parametric instabilities driven by broadband or partially coherent radiation.

Acknowledgements Work partially supported by the European Research Council through grant Accelerates ERC-2010-AdG 267841. We would like to thank Jorge Santos and Bruno Brandão for discussions and for their key contributions to the theory of GPK.

References

1. K. Nishikawa, C.S. Liu, *General Formalism of Parametric Excitation*, in Advances in Plasma Physics, ed. by A. Simon, W.B. Thomson (Wiley, New York, 1976)
2. W.L. Kruer, *The Physics of Laser Plasma Interactions* (Redwood City, Addison-Wesley, 1988)
3. G. Shvets, N. Fisch, A. Pukhov, J. Meyer-ter-Vehn, Phys. Rev. Lett. **81**, 4879 (1998)
4. C. Thompson et al., Astrophys. J. **422**, 304 (1994)
5. T.M. Antonsen, P. Mora, Phys. Fluids B **5**, 1440 (1993)
6. P. Mora, T.M. Antonsen, Phys. Plasmas **4**, 217 (1997)
7. R. Trines et al., Nature Phys. **7**, 87 (2011)
8. B. Hall et al., Phys. Rev. E **65**, 035602(R) (2002)
9. D. Anderson et al., Phys. Rev. E **69**, 025601 (2004)
10. D. Anderson et al., Phys. Rev. E **70**, 026603 (2004)
11. L.O. Silva et al., IEEE Trans. Plasma Sci. **28**, 1202 (2000)
12. E.J. Valeo, C.R. Oberman, Phys. Rev. Lett. **30**, 1035 (1973)
13. J.J. Thomson et al., Phys. Fluids **17**, 849 (1974)
14. J.J. Thomson, J.I. Karush, Phys. Fluids **17**, 1608 (1974)
15. G. Laval, R. Pellat, D. Pesme, Phys. Rev. Lett. **36**, 192 (1976)
16. G. Laval et al., Phys. Fluids **20**, 2049 (1977)
17. J.J. Thomson, Phys. Fluids **21**, 2082 (1978)
18. J.P. Santos, L.O. Silva, J. Math. Phys. **46**, 102901 (2005)
19. J.E. Santos, L.O. Silva, R. Bingham, Phys. Rev. Lett. **98**, 235001 (2007)
20. C.D. Decker et al., Phys. Plasmas **3**, 2047 (1996)
21. H. Feshbach, F. Villars, Rev. Mod. Phys. **30**, 24 (1958)
22. M.J. Bastiaans, J. Opt. Soc. Am. **69**, 12 (1979)
23. M.J. Bastiaans, Optica Acta **28**, 9 1215 (1981)
24. M.J. Bastiaans, J. Opt. Soc. Am. **3**, 8 (1986)
25. T.A.C.M. Claasen, W.F.G. Mecklenbrauker, Philips J. Res. **35**, 372–389 (1980)
26. P.K. Kaw, G. Schmidt, T. Wilcox, Phys. Fluids **16**, 1522 (1973)
27. A.J. Schmitt, Phys. Fluids **31**, 3079 (1988)
28. D.H. Froula et al., Phys. Rev. Lett. **101**, 115002 (2008)

29. B. Brandão, *White Light Parametric Instabilities in Plasmas*, MSc Thesis, IST (Portugal), 2008
30. B. Brandão, J.E. Santos, R. Bingham, L.O Silva, *Stimulated Brillouin Scattering of White Light*, Submitted for publication (2011)
31. R. Trines et al., Phys. Rev. Lett. **94**, 165002 (2005)
32. R. Trines et al., Phys. Rev. Lett. **99**, 205006 (2007)
33. R. Bingham et al., J. Plasma Phys. **71**, 899–904 (2005)
34. R. Trines et al., Phys. Plasmas **16**, 055904 (2009)
35. R. Trines et al., Phys. Scripta **T116**, 75 (2005)
36. R. Trines et al., Plasma Phys. Contr. F. **50**, 124048 (2008)

Chapter 2
Theory of Laser-Overdense Plasma Interactions

Alexander Andreev

Abstract This review chapter discusses a rapidly developing area of physics: the interaction of a high intensity laser pulse with a solid target. The aim is to describe the phenomena of absorption and reflection of a short laser pulse in the interaction with over-dense plasma. In particular, a model of laser energy absorption for planar radiation of steep, high density plasma surfaces is presented. It is valid for arbitrary intensity and incidence angle. The model's convenient closed form makes it widely applicable to laser-driven ion acceleration schemes, hard x-ray sources, and the fast igniter fusion concept.

2.1 Introduction

Generation and direct amplification of short laser pulses with high output energies requires the use of amplifiers and other large-aperture components, which greatly increases the difficulties encountered in construction of suitable systems and, in the final analysis, their cost [1]. The progress made in the generation of maximum pulse intensities shows that the application of enhancement and compression has raised the intensity of chirped pulses by 6 orders of magnitude over the past decade [2]. In the interaction of ultra-short pulses with solid targets, the contrast of such pulses should be sufficiently high to ensure that plasma does not form before the arrival of the main pulse on a target (see Fig. 2.1).

The contrast may be improved by several factors, which manifest themselves in different ways at different time intervals. On the microsecond scale, the contrast may be reduced by super-luminescence of laser amplifiers, which can be eliminated

A. Andreev (✉)
Max-Born str.2a, Max Born Institute, Division B, Berlin 12489, Germany
e-mail: alexanderandreev72@yahoo.com

P. McKenna et al. (eds.), *Laser-Plasma Interactions and Applications*, Scottish Graduate Series, DOI 10.1007/978-3-319-00038-1_2,

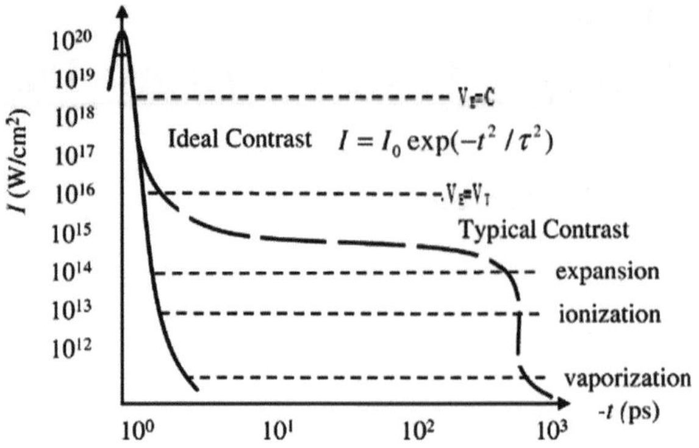

Fig. 2.1 Laser pulse shape and different regimes of laser target interaction

effectively by a variety of methods such as optical switches, spatial filters, etc. A different type of noise grows in a few nanoseconds before the main pulse as a result of multi-pass amplification in a regenerative amplifier. Such pre-pulses can be removed by fast optical switches. Finally, the third group of factors that reduce the contrast (in ps scale) is associated with distortion of the spectrum during the propagation of a beam through the optical components of an amplifying system. The basic approach to improving the contrast of a pulse involves the use of some non-linear processes that depend strongly on the intensity and are characterised by a shorter time.

The use of super-intense laser radiation provides new opportunities for investigating the interaction of ultra-strong laser fields with matter and opens up new avenues in this branch of physics (see Fig. 2.2) [1–3]. This applies particularly to generate electric fields which have intensities considerably greater than the intra-atomic intensity. There is major interest in the creation of ions with a high ionic charge. At laser radiation intensities $I > 10^{17}$ W cm^{-2}, when the acquired velocity of the electron oscillations, υ_E, is higher than the thermal velocity, υ_T, a new physical object can be created: this object is a high-temperature over dense laser plasma, subjected to high-contrast laser pulses, in which electrons do not manage to transfer any significant energy to ions during the plasma lifetime. Such an over dense plasma is of interest as a source of ultra-short pulses of fast particles and of x-ray radiation. A further increase in the intensity above 10^{19} W cm^{-2} makes it possible to reach the next physical threshold when the energy of electron oscillations in the field of an electromagnetic wave becomes equal to the electron rest energy. This situation corresponds to the physics of relativistic laser plasma when the electron energy acquired from the laser beam exceeds 1 MeV and the laser field can influence directly the state of the nuclei. At still higher laser intensities ($I > 10^{20}$ W cm^{-2}) the processes of excitation of nuclei and of nuclear reactions by the direct action of a strong field become probable, so that a considerable number of excited nuclei can be created.

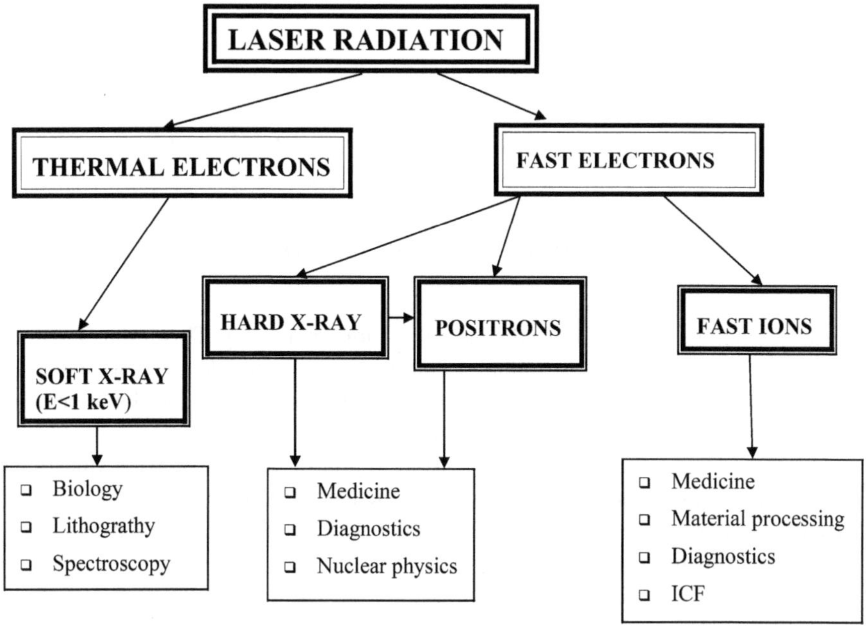

Fig. 2.2 The laser plasma interaction channels and its applications

2.2 Laser Plasma Formation

Our focus will be on the physical processes occurring in the laser plasma produced on solid targets in vacuum. The interest in these processes is due to the fact that the heating of a solid target under these conditions provides the highest plasma density and temperature. Let us consider the basic stages (see Fig. 2.3) in the process of laser radiation interaction with opaque targets and the plasma produced on them [3]. At first, the radiation interacts with a solid, heating its surface to a temperature, at which a low density transparent vapor consisting of neutral atoms of the target substance is formed. The surface heating to a high temperature takes some time varying with the rates of energy delivery to the surface during the radiation absorption and heat transport into the target bulk due to heat conduction. The formation of a transparent vapor is followed by the development of an electron avalanche leading to the vapor breakdown. Seed electrons that may be produced by, say, multi-photon absorption oscillate under the action of the electric field of a laser light wave and collide with atoms, gaining random motion energy. As soon as the energy of an electron becomes equal to the ionisation potential of an atom, a collision between them gives rise to a new electron. New electrons produce the next generation of electrons in a similar way, and the cycle is repeated. The increase in the electron concentration obeys the exponential law. Due to this process, a plasma is produced on the target

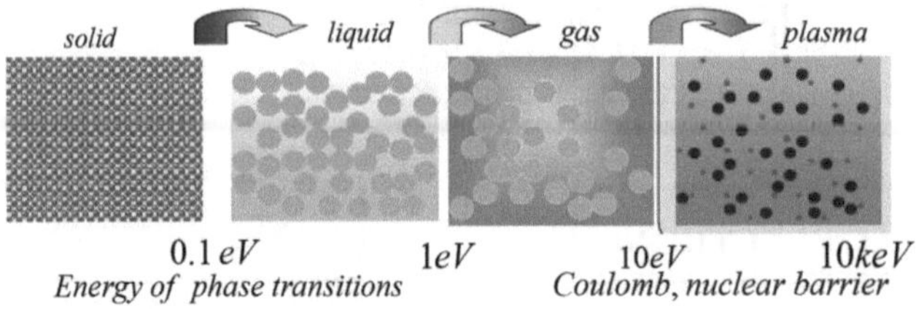

Fig. 2.3 The different matter states and its phase transition energies

surface, which begins to absorb the laser radiation intensively. This moment can be considered to be the onset of radiation interaction with the plasma. The nature of the interaction depends on the density of the heating light beam and the laser pulse duration. In the limiting case of very high beam densities, the plasma formation can be considered to occur at the pulse front. If a laser pulse is short, the plasma is assumed to be quiescent and heated only via the heat conduction mechanism. In the case of a long pulse, one usually analyses the steady-state process of plasma flow off the target surface.

2.2.1 Target Heating and Evaporation by Laser Radiation

Let us consider a plane opaque infinitely thick target and an incident laser beam of intensity I. Suppose, the target surface absorbs the portion P_1 of the incident energy flow. Then the radiation absorbed by the target surface will be $P_1 I$. This radiation will heat the surface and increase the temperature of the deeper target layers. Generally, a solid placed in vacuum begins to evaporate at any surface temperature different from absolute zero. The evaporation rate increases as the temperature rises. In the first approximation, the initial time of intensive evaporation can be considered to be the moment of time when the surface temperature has reached the boiling temperature T_b of the material at atmospheric pressure. The energy flow absorbed by the target surface is transformed to a heat flow q_n which is classically described as $\mathbf{q}_n = -\kappa_T \nabla T$ [4], where κ_T is the heat conductivity of the target material and T is the target temperature. Let us re-write this expression in terms of finite differences for a one-dimensional case. Let T be the surface temperature and x_{fr} the coordinate of the thermal wave front. This means that the temperature is approximately equal to zero at a distance x_{fr} from the surface. Then this expression can be re-written as $q_n \approx -\kappa_T \frac{T}{x_{fr}}$. In addition to the expressions for a heat flow, we will need the heat conduction equation $\rho c_p \frac{\partial T}{\partial t} = -\mathrm{div}\mathbf{q}_n$, where ρ is the density of the target material and c_p is the specific heat at constant pressure. Having re-written this equation as finite differences, we get an expression similar to the case above: $\rho c_p \frac{T}{t} \cong -\frac{q_n}{x_{fr}}$. By combining these equations, we obtain $q_n = \sqrt{\kappa_T T^2 \rho c_p / t}$. Besides, account must be taken of $q_n = \eta_a I$, where η_a is the absorption by the target and I is the intensity of the incident light beam. Then we have:

$$\eta_a I \approx \sqrt{\frac{\kappa_T T^2 \rho c_p}{t}} \tag{2.1}$$

If we substitute into this expression the boiling temperature of the target material, T_b, and replace the time t with the laser pulse duration t_L, we will get the threshold density of the heating laser radiation, I_{th}, at the initial moment of evaporation:

$$I_{th} = \frac{T_b}{\eta_a}\sqrt{\frac{\kappa_T \rho c_p}{t_L}} \tag{2.2}$$

If the radiation density exceeds this value, $I > I_{th}$, an intensive evaporation of matter from the target surface will begin. By solving this equation with respect to t, we will get the time necessary for the development of the evaporation process [3]:

$$\tau_v = \frac{\kappa_T \rho c_p}{I^2}\frac{T_b^2}{\eta_a^2} \tag{2.3}$$

2.2.2 Optical Vapor Breakdown and Target Screening

The vapor produced at the target surface consists, at first, of neutral particles and is transparent to laser radiation. For this reason, the radiation is able to evaporate the target material continuously. Although the vapor is made up of neutral particles, some seed electrons are usually present in the target bulk. These electrons oscillate in the light wave field. Seed electrons may arise from multi-photon ionisation, ionisation by cosmic rays, etc., but this is of no interest to us. When colliding with a heavy neutral particle, an electron transforms the oscillation energy to the energy of random motion [5]. Let us discuss this process in some detail.

The coefficient of collision absorption of an electromagnetic wave by a plasma is given by standard expression [4]. In the case of a transparent vapor, the electron concentration is extremely low, so that $\omega \gg \omega_{pe}$, and the collisions primarily involve neutral particles, rather than ions, and occur with frequency ν_{en}. Then the expression for collision absorption coefficient is

$$K_r \approx \frac{\omega_{pe}^2}{\omega^2 c}\nu_{en} \tag{2.4}$$

To write the equation for the energy gained by an electron from the light wave field, one should invoke Buger's law, $dI = -K_r I dz$. Since this energy contributes to that of random electron motion, the energy released per electron per second will be $I_r q/n_e$. This is the rate of the energy gain by an electron: $K_r I/n_e$. This equation, however, does not allow for the fact that some of the electron oscillation energy is given off to a heavy ion in a collision. This energy can be found by solving the set of equations for the momentum and energy conservation:

$$m_e v_e = m_i v_i + m_e v'_e \tag{2.5}$$

$$m_e v_e^2 = m_i v_i^2 + m_e {v'}_e^2 \tag{2.6}$$

It is assumed here that an electron having a velocity v_e collides with an immobile ion and that the velocities of the electron and the ion after the collision are v'_e and v_i, respectively. After simple algebraic operations, we find that the energy gained by an ion in one collision event is $\Delta E_i = \frac{m_i v_i^2}{2} \sim \frac{m_e}{m_i} E_e$. Thus, the energy lost by an electron due to its transfer to an ion, per unit time, will be $\frac{dE_e}{dt} = -\frac{m_e}{m_i} E_e \nu_{en}$. Having combined the equations, which account for the energy gain and loss by an electron, we eventually get:

$$\frac{dE_e}{dt} = \left(\frac{4\pi e^2}{m_e c \omega^2} I - \frac{m_e}{m_i} E_e \right) \nu_{en} \tag{2.7}$$

The solution to this equation provides the time dependence of the energy of random motion:

$$E_e = \frac{4\pi e^2 I m_i}{c \omega^2 m_e^2} \left[1 - \exp\left(-\frac{m_e}{m_i} \nu_{en} t \right) \right] \tag{2.8}$$

It is clear from this formula that the electron energy increases in time, reaching the maximum value $4\pi e^2 I m_i / (c \omega^2 m_e^2)$. At high radiation densities I, however, the moment when the electron energy becomes equal to the ionisation potential of a neutral atom, $E_e = J$, comes earlier. In that case, an electron ionises an atom in another collision. This produces two electrons which gain energy until they ionise two other atoms. This process is referred to as *avalanche ionisation*. Using the above formula and the equality $E_e = J$, we find the time necessary for a new electron to be produced:

$$\tau_e = -\frac{m_i}{m_e \nu_{en}} \ln\left(1 - \frac{J}{I} \frac{c \omega^2 m_e^2}{2\pi c^2 m_i} \right) \tag{2.9}$$

The growth of the number of electrons in 1 cm^3 in an avalanche occurs exponentially. Indeed, suppose dn_e electrons are produced per unit volume for the time dt. It is then clear that dn_e must be proportional to dt/τ_e, i.e., the gain in the electron concentration, dn_e, is the larger, the larger is the time excess over that necessary for the production of an electron, τ_e. Besides, the concentration gain appears to be proportional to the actual number of electrons, n_e. Hence, we can write $dn_e = n_e \frac{dt}{\tau_e}$. The solution to this equation is:

$$n_e = n_{e0} \exp(t/\tau_e) \tag{2.10}$$

The breakdown time τ_b is arbitrarily taken to be the time necessary for the production of 40 generations of electrons, or $\exp(\tau_b/\tau_e) = \exp 40$. Then one seed electron per 1 cm^3 produces plasma with a concentration $n_e \cong 10^{17}$ cm^{-3}. As a result, we have:

$$\tau_b = -40\frac{m_i}{m_e \nu_{en}} \ln\left(1 - \frac{J}{I}\frac{c\omega^2 m_e^2}{2\pi e^2 m_i}\right) \tag{2.11}$$

The practical application of this formula to describe the breakdown of vapor produced by laser radiation at the target surface requires the knowledge of the variation in ν_{en} with laser light parameters. The frequency of collisions between electrons and neutral particles can be estimated as: $\nu_{en} \sim \sigma_{2n} v_e n_n$. Here, the collision cross section for electrons and neutral particles is practically a constant value, and in estimations, v_e can be substituted by the mean electron velocity during the process of energy gain from $E_e = 0$ to a value equal to the ionisation potential $E_e = J$. Clearly, the only quantity related to laser radiation is the concentration of neutral particles, which can be found in the following way. Suppose a target surface evaporates matter of mass M over the time τ. This mass can be found from the energy conservation law: $\alpha M = \eta_v IS\tau$, where α is the specific heat (per unit mass) of target material evaporation, measured in J/g; η_v is the target absorption during evaporation, and S is the size of the irradiated spot on the target. It follows from the above equation that the energy absorbed by the surface is utilised for the evaporation. The evaporated substance has a temperature equal to the material boiling temperature T_b. The vapor produced is scattered from the target with sound velocity corresponding to this temperature, c_0. If the scattered vapor is assumed to have a uniform density equal to $\rho = n_n m_i$, the mass evaporated for the time τ can be presented as $M = c_0 \tau S n_n m_i$, where $c_0 \tau S$ is the vapor volume. By equating the mass M in the above expressions, we get $n_n = \eta_v I/\alpha c_0 m_i$. As a result, we have $n_n \sim I$ and, hence, $\nu_{en} \sim I$. Then the expression for the breakdown time of a transparent vapor can be re-written as [3] $\tau_b = -\frac{\alpha_1}{I}\ln\left(1 - \frac{\alpha_2}{\lambda^2 I}\right)$, where α_1 and α_2 are numerical factors. An important feature of the process described is that τ_b turns to infinity at a certain value of the radiation intensity. Physically, this means that the electron energy gain in the light wave field at low radiation density does not compensate for the energy loss during collisions with ions.

For high density, optically thick plasma, the radiative and absorptive processes in the constituent plasma ions are balanced, so that a Local Thermal Equilibrium (LTE) is reached, and the ionisation state can be determined. Under these circumstances, the relative ion populations are related by the Saha-Boltzmann equation [6]:

$$\frac{n_e n_{Z+1}}{n_Z} = \frac{g_{Z+1}}{g_Z}\frac{2m_e^3}{h}\left(\frac{2\pi T_e}{m_e}\right)^{3/2} \exp\left(-\frac{\Delta E_Z}{T_e}\right) \tag{2.12}$$

where n_Z, n_{Z+1} is the ion density corresponding to ionisation states Z and $Z+1$; g_Z, g_{Z+1} are the respective statistical weights of this levels and ΔE_Z is the energy difference between two states. This equation is subject to the constraints: $\sum n_Z = n_0$, $\sum Z n_Z = n_e$. For a given element at atomic density n_0, the Saha equation gives the relative proportions of ions – from singly charged to fully stripped (H-like) - plus the net electron density as function of temperature T_e. For a foil target the ion charges can be defined self-consistently by using the formula for the electric field at the rear side of the target

$$E \approx \left(\eta_e \sqrt{1+a^2}\right)^{1/2} m_e c\omega/e \tag{2.13}$$

and the tunnel ionisation probability [5]:

$$\nu_{fi}(Z) = \nu_z \left(\frac{E_z}{E}\right)^{2n_{ef}-1.5} \exp(-0.06 n_{ef} E_z/E) \tag{2.14}$$

where $\nu_z = 1.6\omega_{au}Z^2/n_{ef}^{4.5}$, $E_z = 10.9E_{au}Z^3/n_{ef}^4$ is the atomic electric field, $\omega_{au} = eE_{au}a_B/\hbar$ is the atomic frequency, $a_B = \hbar^2/m_e e^2$ is the Bohr radius, $n_{ef} = Z\sqrt{J_H/J_i}$ is the main effective quantum number, J_z is the ionisation potential of the ion with the charge Z and $J_H = 0.5eE_{au}a_B$ is the ionisation potential of hydrogen.

The ion acceleration sets in during the time of the order of a few inverse ion plasma frequencies:

$$\omega_{pi} = \left(\frac{4\pi Z^2 e^2 n_i}{Am_p}\right)^{1/2} \tag{2.15}$$

This time, ω_{pi}^{-1}, defines also the duration of the maximum electric field at the rear target surface. Correspondingly, the ion charge and density are defined by the ionisation during this time, $n_i \approx n_0 \nu_{fi}(Z)/\omega_{pi}$. Here, n_0 are the atomic densities of species before the ionisation. Moreover, the free electron density $n_e = Zn_i$ and their thermal velocity, υ_{Te} define the electron plasma frequency,

$$\omega_{pe} = \left(\frac{4\pi e^2 n_e}{m_e}\right)^{1/2} \tag{2.16}$$

the penetration depth of the electric field inside the target,

$$l_s \approx \upsilon_{Te}/\omega_{pe} \tag{2.17}$$

and the total number the ions which are accelerated at the first stage

$$N_{ia} \approx S n_e l_s \tag{2.18}$$

To conclude, the physical mechanism underlying the radiation interaction with an opaque target can be described as follows (see Fig. 2.3) [3]. Laser radiation affects the target surface and begins to evaporate its material at time τ_v, later, at the moment of time τ_b, the vapor breakdown occurs, producing plasma. Until that moment, the vapor was nearly transparent to the laser radiation, whereas the plasma absorbs it intensively. Laser radiation is nearly totally absorbed by low density plasma, which is thus heated. The plasma is dissipated and becomes less dense. A new portion of radiation is incident on the target, again leading to material evaporation from the surface. However, the vapor produced is no longer transparent but represents plasma. At a high laser intensity, the times τ_v and τ_b become extremely small. Evaporation and vapor breakdown occur only at the pulse front (pre-pulse), while most of the radiation interacts with the plasma. By *hot overdense laser plasma*, we understand plasma produced by high power laser radiation when the transient processes of evaporation and vapor breakdown do not take much of the pulse front. We will consider in the following sections only such plasma supposing that it has maximal solid density with sharp gradient and its ion charge is determined by laser intensity.

Because any kind of interaction of a main pulse starts from energy deposition into an object we will begin our report from the analysis of laser pulse absorption in overdense plasma.

2.3 Absorption of High Intensity Laser Pulse in Overdense Plasma

Intense laser pulse can be absorbed in overdense plasma by different mechanisms. We shall begin by reviewing the mechanisms of absorption of laser radiation in dense, hot plasma, as the transfer of laser energy to different channels of interaction (see Fig. 2.2) depends on this.

2.3.1 Linear Absorption in Inhomogeneous Plasma

The equations for EM wave propagation in inhomogeneous plasma are well known (see [7, 8]):

$$\nabla^2 \mathbf{E} - \frac{1}{c^2}\frac{\partial^2 \mathbf{E}}{\partial t^2} = \frac{4\pi}{c^2}\frac{\partial \mathbf{j}}{\partial t} + \nabla(\nabla \cdot \mathbf{E}), \quad \nabla^2 \mathbf{B} - \frac{1}{c^2}\frac{\partial^2 \mathbf{B}}{\partial t^2} = -\frac{4\pi}{c}\nabla \times \mathbf{j} \tag{2.19}$$

where electron current density $\mathbf{j} = en\upsilon_e$ is determined by the equation of motion

$$\frac{m_e \partial \vec{\upsilon}_e}{\partial t} = -e\left(\mathbf{E} + \frac{\vec{\upsilon}}{c} \times \mathbf{B}\right) - m_e \nu_{ei} \vec{\upsilon}_e \tag{2.20}$$

Here ν_{ei} is the electron-ion collision frequency. Next we linearise the equations and assume that all fields and hydro-parameters have harmonic time dependence with laser frequency ω as:

$$f(\mathbf{r},t) = f_0(\mathbf{r}) + f_1(\mathbf{r})\exp(i\omega t) + f_2(\mathbf{r})\exp(2i\omega t) + .., \tag{2.21}$$

Inserting this approximation into above equations allows us to write induced current: $\mathbf{j}_1 = \sigma_e \mathbf{E}_1$, where electrical conductivity is given by:

$$\sigma_e = \frac{i\omega_p^2}{4\pi\omega(1+i\nu_{ei}/\omega)} \tag{2.22}$$

Substituting this current into wave equation gives:

$$\nabla^2\mathbf{E}_1 - \frac{\omega^2}{c^2}\mathbf{E}_1 = \frac{\omega_p^2}{c^2}\frac{\mathbf{E}_1}{1+i\nu} + \nabla(\nabla \cdot \mathbf{E}), \tag{2.23}$$

where $\nu = \nu_{ei}/\omega$. For s-polarised light the absorption coefficient in the limit $kL > 1$ is $\eta_c = 1 - \exp(-\frac{8\nu_{ei}L}{3c}\cos^3\theta)$, where L is the scale of plasma inhomogeneity and θ is the angle of the wave incidence on the plasma. As is well known [9, 10], for laser intensities I greater than 10^{15} W/m^2, the electron temperature rises rather quickly $T_e \sim I^{1/2}$, so collisional absorption becomes ineffective $\nu_{ei} \sim I^{-3/4}t^{-3/2}$. Besides, the oscillation velocity of electrons becomes comparable to their thermal velocity, which also reduces the effective collision frequency:

$$\nu_{ef} \approx \nu_{ei}\frac{\upsilon_T^2}{(\upsilon_E^2 + \upsilon_T^2)^{3/2}} \tag{2.24}$$

Thus, at intensities higher than 10^{16} W/m^2, the collision-less mechanisms of absorption begin to be significant. The movement of ions must be taken into account at times: $t > \frac{\upsilon_T}{\omega c_s}$, thus we should note, that even for subpicosecond laser pulses the plasma does not have a sharp boundary, because even laser pulses with high contrast ratio have enough energy in the pre-pulse to create a plasma density gradient. Although the scale of plasma inhomogeneity is in this case $L < \lambda$, the absorption can be determined by L. For example with laser intensity less than 10^{17} W/cm^2, resonant absorption plays a basic role, modified by the mechanism of plasma wave breaking. In this case, electrons – accelerated by a plasma wave field in vacuum – are reflected from an ambipolar barrier, and returned to the target, heating it to a depth of their free path. The absorption coefficient in this case is approximately [3]: $\eta_r \approx \frac{\sin^2\theta}{\cos\theta}kL$, where θ is the angle of incidence to the target, $k = \omega/c$, L is the scale of plasma density inhomogeneity in vicinity of critical concentration and $n_c = \omega^2/4\pi e^2$.

High intensity laser pulses absorb in over dense plasma by different nonlinear mechanisms. In the following the non-linear relativistic equations for laser plasma interactions are considered to obtain analytical solutions for such cases.

2.3.2 *Basic Set of Equations*

Let the plane linearly polarised electromagnetic wave propagate along the X axis normal to the semi-limited plasma. The plasma temperature is T_e, the density grows from zero to n_e over a distance L and the plasma is over-dense (max $n_e \gg n_{cr}$).

The wave with amplitude E_0 and frequency ω is chosen in such a way that the quiver velocity $\upsilon_E = eE_0/m\omega$, is larger than the thermal velocity $(T_e/m)^{1/2} = \upsilon_T$. We suppose that during the laser pulse (< 100 fs) the movement of ions is negligible and the plasma edge preserves its sharpness. Movement of the plasma electron component is described by the self consistent set of equations, consisting of the collision-less kinetic Boltzmann equation ($0 < x < l_s$) and Maxwell equations for electromagnetic fields (in covariant form) [2,9]:

$$\begin{aligned} &\frac{p^\mu}{m_e}\frac{\partial f}{\partial x^\mu} + \frac{eF^{\mu\nu}p_\nu}{c}\frac{\partial f}{\partial p^\mu} = 0 \\ &\frac{\partial F^{\mu\nu}}{\partial x_\mu} = J^\nu \\ &J_\nu = e\int d^4 p c p_\nu \theta(p_0)\delta(p_\mu p^\mu - mc^2) f \end{aligned} \tag{2.25}$$

here $F^{\mu\nu}$ is the Maxwell tensor; and of the kinetic Fokker-Planck equation for $x > l_s$:

$$\frac{\partial f}{\partial t} + \upsilon_x \frac{\partial f}{\partial x} + eE_a(x,t)\frac{\partial f}{\partial p_x} = \nu_{ei}\frac{\partial^2 f}{\partial \phi^2} \tag{2.26}$$

Ambipolar field E_a is determined from the zero current along the axis X condition or Poisson equation.

2.3.3 Oblique Incidence of Laser Radiation on Plasma Boundary

Oblique incidence is 'boosted' into normal incidence [10]. Denoting the boost (S) frame quantities by primes, the inverse Lorentz transformations for the wave frequency and k vector are since $k_y = ksin\theta = \omega c \sin\theta$ *and* $\upsilon_0 = c\sin\theta$ we have $k_y{}' = 0$, $\omega' = \omega/\gamma_0$, $k' = k/\gamma_0$ where $\gamma_0 = 1/\cos\theta$. For the space and time coordinates, we have $t = \gamma_0\left(t' + \frac{\upsilon_0}{c^2}y'\right)$, $x = x'y = \gamma_0(y' + \upsilon_0 t')$, $z = z'$. Initially, $j_{ey} = j_{iy} = 0$, so in the boost frame, $(\rho_0)'_{e,i} = \gamma(\rho_0)_{ei}$ and $(j_{yo})'{}_{e,i} = -(\rho_0)'{}_{e,i} c\sin\theta$. Thus, the normalised, time interval $\omega' t' = \omega t$ is invariant, as is the wave phase $\omega t - k \cdot r$. Note that the critical density in the simulation frame transforms as $\frac{n_c'}{n_c} = \frac{\omega'^2}{\omega^2} = \frac{1}{\gamma_0^2}$. Hence the initial normalised unperturbed electron density is $\frac{n_e(t=0)'}{n_c'} = \gamma_0^3 \tilde{n}_e$. Finally, to launch the EM wave, we must specify its amplitude $a_0 = \upsilon_E/c$ at the left-hand simulation boundary. Since $\upsilon_E/c = eE_0/m\omega c$, we have for a p-polarised wave,

$$a_0' = \frac{eE'_y}{m\omega' c} = \frac{eE_0}{m\omega c}\gamma_0\cos\theta = a_0 \tag{2.27}$$

We let $\tilde{E}'_y(x'=0) = \tilde{B}'(x'=0) = a_0$, so since $E'_x = 0$, we obtain $\tilde{E}_x = -\upsilon_0 a_0 = -a_0 \sin\theta$, $\tilde{E}_y = a_0/\gamma_0 = a_0 \cos\theta$, and $\tilde{B}_z = a_0$. It should be stressed that the boost technique cannot always be used to model oblique incidence interaction. In the general 2-D geometry, all physical quantities and the distribution function depend separately on t, x, y, p_x and p_y. The transformation to the system corresponding to normal incidence with one spatial coordinate is only possible if the distribution function and other physical values depend on this set of variables in the following way:$f\left(t - \frac{\sin\theta}{c} y; x; p_x; p_y\right)$.

2.3.4 Analytical Modelling

Let's consider now the analytical model for absorption coefficient. Solution of Eq. (2.26) can be written in the general form as [1]:

$$f(x;t;p_z;p_y) = \int \delta(\bar{p} - \bar{p}(x_0;\bar{p}_0;t))\delta(x - x(x_0;\bar{p}_0;t)) f(\bar{p}_0;x_0) d\bar{p}_0 dx_0 \quad (2.28)$$

where $\bar{p}(x_0;\bar{p}_0;t);x(x_0;\bar{p}_0;t)$ is the phase trajectory of a separate electron with the initial impulse $\bar{p}_0$ and the coordinate x_0; $\bar{p}(x_0;\bar{p}_0;t)$; $x(x_0;\bar{p}_0;t)$: - result of solution of equations of electron motion

$$\bar{p}_\perp = \bar{p}_{0\perp} - e\bar{A}(x,t)/c$$

$$\frac{d}{dt}\dot{x}\sqrt{\frac{m^2 + (\bar{p}_{0\perp}/c - e\bar{A}(x,t)/c^2)^2}{1 - \dot{x}^2/c^2}} = -e\frac{\partial \varphi(x,t)}{\partial x} - \frac{\partial}{\partial x}\sqrt{\frac{m^2c^4 + (\bar{p}_{0\perp}c - e\bar{A}(x,t))^2}{1 - \dot{x}^2/c^2}} \quad (2.29)$$

These equations define the phase trajectory of electrons with initial momentum $\bar{p}_0$ and coordinate x_0 in self-consistent electromagnetic fields. Here $\boldsymbol{A}(x,t)$ is the vector potential of the transverse ($\mathrm{div}A = 0$) electromagnetic fields, $\varphi(x,t)$ is the scalar potential of the longitudinal fields satisfying the Maxwell equations:

$$\left(\frac{\partial^2}{\partial x^2} - \frac{\partial^2}{c^2 \partial t^2}\right)\mathbf{A}(x,t) = -\frac{4\pi e}{c}\int \mathbf{v} f(x;t,p_z;p_y)\, d\mathbf{p}$$
$$\frac{\partial^2 \varphi(x,t)}{\partial x^2} = -4\pi e\left(Zn_i - \int f(x;t,p_z;p_y)\, d\mathbf{p}\right) \quad (2.30)$$

We assume spatial one-dimensionality of our task and it allows us to use the law of conservation of transverse canonical momentum to find $p_\perp(p_{0\perp}, x_0, t)$. The initial distribution function $f(\bar{p}_0;x_0)$ in Eq. (2.28) is the Maxwell-Boltzmann one:

$$f(\bar{p}_0;x_0) = \frac{1}{2\pi m T_e}\exp\left(-\frac{\bar{p}_0^2}{2mT_e}\right) Zn_{i0}\exp\left(-\frac{e\varphi(x_0)}{T_e}\right) \quad (2.31)$$

where the potential $\varphi(x_0)$ satisfies the Poisson equation:

$$\frac{\partial^2\varphi}{\partial x_0^2} = -4\pi eZn_{i0}\left[\exp\left(-\frac{e\varphi(x_0)}{T_e}\right) - \frac{n_i(x_0)}{n_{i0}}\right] \tag{2.32}$$

The profile of ion concentration $n_i(x)$ is assumed specified function with the typical scale L growing from $n_i = 0$ at $x = 0$ to the constant value $n_{i0} = n_i|_{x\to\infty}$.

2.3.5 Analytical Model for the Anomalous Skin Effect

Now we consider the analytical model for the anomalous skin effect regime as collisionless mechanism of absorption at the normal incidence of a laser wave. According to the numerical calculation, the electromagnetic field in plasma can be approximated by the following equations [1]:

$$\begin{aligned} E_y(x;t) &= \frac{E_0\sin(\omega t)}{1+x^2/l_s^2} \\ B_z(x;t) &= -\frac{2cE_0x\cos(\omega t)}{\omega l_s^2(1+x^2/l_s^2)^2} \\ E_x(x) &= E_0\left[\Theta(x)\exp(-x/l) - \Theta(-x)\exp(x/l)\right], \end{aligned} \tag{2.33}$$

where $\Theta(x)$ is the Heavyside step-like function. The fields are symmetrical with respect to replacement $(x \to -x)$; this symmetry is caused by the requirement of mirror-like reflection of the electrons by the plasma edge. Note, that the primary role in the laser radiation absorption is played by the electrons, whose velocity is parallel to the plasma edge. The normal component of their velocity is small, excluding the possibility to overcome the potential barrier (field E_x) on the edge. The lengths l_s and l (i.e., the lengths of the skin layer for the transverse E_y and longitudinal E_x fields) are determined by the comparison with numerical simulation results. Approximate values of these lengths are $l_s = (c^2V_T/\omega\cdot\omega_p^2)^{1/3}$ and $l \approx L$. The non-exponential character of the transverse wave feeding with plasma depth is the result of the anomalous skin effect. The requirement of equilibrium of electrostatic force and this ponderomotive pressure makes it possible to evaluate E_x and characteristic scale l:

$$\frac{E_0^2}{4\pi}(1+R) = e\int_0^\infty E_x(x)n_i(x)dx \tag{2.34}$$

resulting in characteristic evaluation:

$$E_{x0} \approx E_0\left(\frac{cL}{\omega l^2}\right)\left(\frac{\omega^2}{\omega_p^2}\right)\left(\frac{eE_0}{m\omega c}\right)(1+R) \tag{2.35}$$

Distribution function can be modelled by evaluation of the electron movement in the fields and taking Maxwell distribution as the starting one:

$$f(x;p_x;p_y;t) = \frac{n_e(x)}{2\pi mT}\exp\left[\frac{mc^2-(m^2c^4+p_x^2(0)c^2+p_y^2(0)c^2)^{1/2}}{T}\right] \tag{2.36}$$

Solution of the relativistic motion equations looks like [1]:

$$\begin{aligned}
P_y(0) &= p_y+e/c\,[A_y(x-V_xt;0)-A_y(x,t)]\\
P_x(0) &= p_x+e\int_0^t E(x(\tau);\tau)d\tau+\frac{e(p_y-e/cA_y(x;t)}{mc\gamma}\int_0^t\frac{\partial A_y(x(\tau);\tau)}{\partial x}\partial\tau\\
&+\frac{e^2}{2m\gamma c^2}\int_0^t\frac{\partial A_y^2(x(\tau);\tau)}{\partial x}d\tau
\end{aligned} \tag{2.37}$$

$$A(x;t) = \frac{cE_0\cos(\omega t)}{\omega(1+x^2/l_s^2)},\quad \gamma=\left(1+\frac{p_x^2+p_y^2}{m^2c^2}\right)^{1/2}$$

The lengths l_s and l are much shorter than the length of free path of electron in plasma; hence the trajectory $x(\tau)$ can be approximated by the straight line : $x(\tau)=x+V_x\tau=x-V_x(t-\tau)$. Hence, the Eqs. (2.34), (2.36) and (2.37) make it possible to determine the analytical expression for the distribution function.

In the case of oblique laser pulse incidence on plasma, as we already mentioned it is convenient to calculate the distribution function in the coordinate frame, moving with the speed $V=c\sin(\theta)$ (θ is the incidence angle) along the plasma edge. As was explained in this system the incidence is normal. Longitudinal field E_x recalculated to this set of coordinates results in the additional constant magnetic field and, hence, in the following field configuration:

$$\begin{aligned}
E_x(x) &= E_0\,[\Theta(x)\exp(-x/l)-\Theta(-x)\exp(x/l)]\,/\cos(\theta)\\
A_y(x;t) &= \frac{cE_0\cos(\omega t)}{\omega(1+x^2/l_s^2)}\cos(\theta)-tg(\theta)E_0\,[\Theta(x)\exp(-x/l)-\Theta(-x)\exp(x/l)]
\end{aligned} \tag{2.38}$$

The solution for oblique incidence can be drawn out of the solution of Eq. (2.37) by replacement of γ in Eq. (2.37) by:

$$\gamma=\left[1+\frac{p_x^2}{m^2c^2}+\left(\frac{p_y}{mc}+\sin(\theta)\right)^2\right]^{1/2} \tag{2.39}$$

Noteworthy, that while for low intensities the absorption depends on the incidence angle with the characteristic maximum for $\theta\approx\pi/2$, for higher intensities such dependence is smoother. In this case the absorption coefficient tends its value for the smaller angle of incidence.

In the case of high temperatures and short laser pulses for the anomalous skin effect conditions the absorption coefficient is as follows [1]: $\eta_a \approx 2.8kl_s/\cos\theta$.

For normal skin effect, an electron oscillates in a laser field and absorbs the energy of its electromagnetic field when it collides with an ion, thus: $\eta_n \sim kl_s$. This is possible, if $\frac{\upsilon_T}{\nu_{ei}} = l_{ei} < l_s$. When T_e increases, we obtain $l_{ei} > l_s : \upsilon_T/\omega > l_S$. Given these requirements, the laser field penetrates a length into plasma l_{sa}. Substituting $\nu_{ef} = \upsilon_T/l_{sa}$ into the equation for the length of the skin-layer (for $\nu_{ef} \geq \omega$) $l_s = (c/\omega_p)(\nu_{ef}/\omega)^{1/2}$, we deduce, that the field penetrates to the thickness of the anomalous skin-layer:

$$l_{sa} = \frac{c}{\omega_p}\left(\frac{\upsilon_T\omega_p}{c\omega}\right)^{1/3} \tag{2.40}$$

The absorption coefficient again: $\eta_a \approx k_0 l_{sa} \approx \alpha_a^{2/3}(\nu_T/c)$, $\alpha_a \equiv c\omega/\nu_T\omega_p$. For $\alpha > 1$; $l_s > \nu_T/\omega$ we have the regime of anomalous skin effect at high frequency, or Sheath Inverse Bremsstrahlung (SIB) regime [1].

2.3.6 Analytical Solution for SIB

In the zero approximation in the parameter ν_T/υ_E we obtain a well known set of hydrodynamic equations for cold plasma from the system Eqs. (2.25) and (2.26). Consider now the situation when $\omega c/\omega_p\nu_T > 1$, i.e., the case of SIB. The conservation law for the transverse canonical impulse permits to reduce the system to two equations for the vector potential of normally incident electromagnetic wave $\bar{A}(x;t)$ and the longitudinal electric field $E_x(x;t)$ in plasma:

$$\left(\frac{\partial^2}{\partial\xi^2} - \frac{\omega^2}{\omega_p^2}\frac{\partial^2}{\partial\tau^2}\right)\vec{a} = \left(\eta_i(\xi) + \frac{\partial E}{\partial\xi}\right)\frac{\vec{a}}{\sqrt{1+a^2}}\sqrt{1-\nu^2}$$

$$\frac{\omega}{\omega_p}\frac{\partial}{\partial\tau}\nu\sqrt{\frac{1+a^2}{1-\nu^2}} = E - \frac{\partial}{\partial\xi}\sqrt{\frac{1+a^2}{1-\nu^2}};\nu = -\frac{\omega}{\omega_p}\frac{\partial E/\partial\tau}{\eta_i(\xi)+\partial E/\partial\xi} \tag{2.41}$$

The system Eq. (2.41) is written in the following dimensionless variables: $\xi = \frac{\omega_p}{c}x$; $\tau = \omega t$; $\vec{a} = e\vec{A}/mc^2$; $E = eE_x/mc\omega_p$; $\eta_i(\xi) = n_i(\xi)/n_i(\xi = \infty)$. Consider the case for a wave with circular polarisation state:

$$\vec{a}(\xi;\tau) = a(\xi)(\bar{e}_z\cos\tau + \bar{e}_y\sin\tau) \tag{2.42}$$

In this case, Eq. (2.41) is reduced to one equation and the plasma profile can be chosen as $\eta_i(\xi) \equiv \Theta(\xi)$. This equation can be easily integrated, and its decreasing solution has the following form:

$$a(\xi)=\frac{2Nch[N(\xi+\xi_0)]}{ch^2[N(\xi+\xi_0)]-N^2},\quad \xi>0,\quad \xi_0=\frac{1}{N}arcch\frac{a_0N}{\sqrt{1+a_0^2}-1} \tag{2.43}$$

where $N=\sqrt{1-\omega^2/\omega_p^2}$ and $a_0=a(\xi=0)$ is the value of the field on the plasma surface.

In vacuum, the following incident and reflected waves correspond to the solution of Eq. (2.41):

$$\vec{a}(\xi;\tau)=\left(a_0\cos\frac{\omega}{\omega_p}\xi-\frac{\omega}{\omega_p}\sqrt{1+a_0^2}\sqrt{2(\sqrt{1+a_0^2}-1)-\frac{\omega^2}{\omega_p^2}a_0^2}\sin\frac{\omega}{\omega_p}\xi\right)$$
$$\times(\bar{e}_x\cos\tau+\bar{e}_y\sin\tau) \tag{2.44}$$

At $a_0\ll 1$, solution Eq. (2.43) becomes a simple shielding law $a(\xi)=a_0\exp(-\xi)$. The electron has energy $\varepsilon_e=mc^2\sqrt{1+a^2}$, its velocity is $\upsilon_\perp^2=\frac{c^2a^2}{1+a^2}$ and its longitudinal velocity is $v=0$. The stationary solution exists not for all values of a_0. When a_0 reaches the value $\frac{3}{2}\frac{\omega^2}{\omega_p^2}$ the electron concentration becomes zero at the point $\xi=0$. At greater a_0, solution Eq. (2.43) loses its physical meaning, because the concentration becomes negative. Stationary solutions in this case are impossible: plasma does not hold the incident wave, and the field penetrates in it in the form of separate filaments – solitons [1].

The equation set Eq. (2.41) permits to find fields in plasma in the zero approximation in $\upsilon_T^2/\upsilon_E^2$. By means of these fields, one can find the equation for the electron phase trajectory $\bar{p}(x_0;\bar{p}_0;t)$; $x(x_0;\bar{p}_0;t)$ and to construct the distribution function. The obtained result is valid in the first order of $\upsilon_T^2/\upsilon_E^2$. Now consider absorption connected with Landau damping on separate particles. Using distribution function Eq. (2.28), the dissipated power can be presented as [1]:

$$Q=e\int\bar{v}(x_0;\bar{p}_0;t)\bar{E}(t;x(x_0;\bar{p}_0;t))f(\bar{p}_0;x_0)d\bar{p}_0dx_0$$
$$=2\int v_x\frac{\partial}{\partial x}\sqrt{e^2A^2+m^2c^4}\,|_{x(x_0;\bar{p}_0;t)}\,f(\bar{p}_0;x_0)d\bar{p}_0dx_0 \tag{2.45}$$

In dimensionless units, the absorption coefficient can be written as:

$$\eta_a=2\frac{v_T}{c}\frac{\omega_p^2}{\omega^2}\frac{1}{A_0^2}\int_{-\infty}^{+\infty}\int_0^{\infty}v\frac{\partial}{\partial\xi}\sqrt{1+a^2}|_{\xi(\xi_0;v_0;\tau)}(\theta(\xi_0)$$
$$+\frac{\partial^2}{\partial\xi_0^2}\sqrt{1+a^2(\xi_0)})d\xi_0\frac{\exp(-v_0^2)}{\sqrt{\pi}}dv_0 \tag{2.46}$$

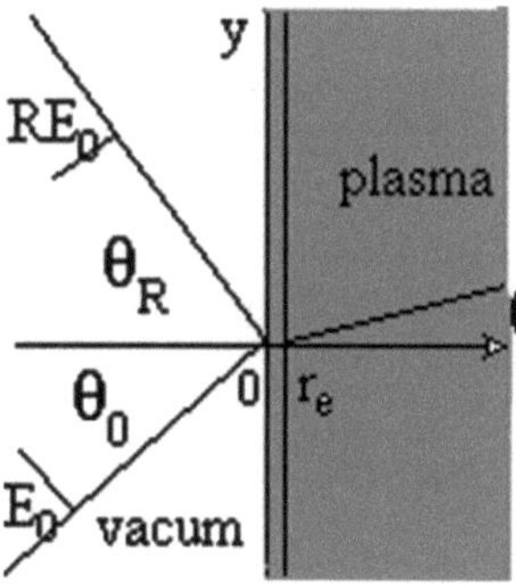

Fig. 2.4 Schematic picture of laser pulse interaction with overdense plasma at oblique incidence

where $\xi(\xi_0; v_0; t)$ and $v(\xi_0; v_0; t)$ determine the law of the longitudinal motion of an electron from the equation:

$$\frac{d\dot{\xi}}{d\tau\sqrt{1-\dot{\xi}^2}} = E(\xi(\tau);\tau) - \frac{\partial}{\partial \xi}\sqrt{\frac{1+a^2}{1-\dot{\xi}^2}}|_{\xi=\xi(\tau)} \tag{2.47}$$

For small $\dot{\xi} \sim \upsilon_T/c$, solution is $\xi(\xi_0; \upsilon_0) = \xi_0 + \frac{\upsilon_T}{c}\frac{\omega_p}{\omega}\upsilon_0\tau$. In the next orders, small oscillations are superimposed on the uniform motion. In weak fields ($a \ll 1$), the absorption coefficient determined from Eq. (2.44) has the form: $\eta_{SIB}^{(0)} = \frac{4}{\sqrt{\pi}}\left(\frac{\upsilon_{T\parallel}}{c}\right)^3 \frac{\omega_p^2}{\omega^2}$ and at $a > 1$, the estimations show that $\eta_{SIB} \sim \eta_{SIB}^{(0)}/a$ [1].

2.3.7 *Brunel Absorption*

In the case of oblique incidence when the amplitude of electrons in the laser field $\upsilon_E/\omega > L$, we have Brunel absorption [10]. To analyse this case in detail we suppose that laser field pulls out electrons from an ionised plasma layer, after which they are accelerated in the self-consistent field over a distance on the order of a laser wavelength across the plasma surface. Absorption occurs because of transfer of laser pulse energy to fast electron flux; the ions are assumed to be stationary on this timescale. The exchange of energy and momentum between fields and particles can be completely described in differential 4-form as follows: $\partial T^{\alpha\beta}/\partial x^\beta = 0$, where $T^{\alpha\beta}$ is the symmetric energy-momentum tensor of the entire field-particle system, and the Einstein summation convention applies. This tensor contains both the energy density flux vector as well as with the 3-dimensional momentum flux tensor. Since the interaction area is a thin layer, the calculation of absorption can be reduced to determination of the boundary conditions for the momentum and energy fluxes of field and particles respectively. We consider the two regions in Fig. 2.4: one in vacuum on the left side ($x = 0$) where there are no particles and a second at the beginning of the skin-layer $0 < x < r_e$, where some fraction of the plasma electrons undergoes acceleration, thus generating a relativistic electron flux. We suppose that all functions are uniform along the y-axis, thus integrating in yx plane and using Gauss's theorem, one obtains balance relations for the y-components of fluxes. For the x-component of the energy flux, the momentum fluxes normal (xx) and parallel (yx) to the surface respectively, these relations give:

$$I(\cos\theta_0 - R^2\cos\theta_R) = v\cos\theta_e\left[m_e n_e c^2(\gamma-1)+\gamma^2(\varepsilon+P)\right]$$
$$\frac{I}{c}(\cos^2\theta_0 + R^2\cos^2\theta_R) = \frac{v^2}{c^2}\cos^2\theta_e\left[m_e n_e c^2\gamma+\gamma^2(\varepsilon+P)\right]+P+\frac{E_y^2-E_x^2}{8\pi}$$
$$\frac{I}{c}(\cos\theta_0\sin\theta_0 - R^2\cos\theta_R\sin\theta_R)$$
$$=\frac{v^2}{c^2}\cos\theta_e\times\sin\theta_e\left[m_e n_e c^2\gamma+\gamma^2(\varepsilon+P)\right]-E_yE_x \tag{2.48}$$

Here $I = cE_0^2/4\pi$, E_0 the laser field, R the amplitude reflection coefficient, $\gamma = (1 - \sum_\alpha v_\alpha^2/c^2)^{-1/2}$ the electron relativistic factor; angles $\theta_{0,R,e}$ are defined as in Fig. 2.4, v is the modulus of the electron velocity.

Formally we suppose that absorption takes place within the layer $[0,\ r_e]$ and the energy acquired here by the electrons is retained up to $z = \infty$. This approximation is valid for an overdense plasma with a step-like profile. It is worth noting that Eqs. (2.48) are generally valid, and implicitly include all absorption mechanisms satisfying the geometrical constraints of Fig. 2.4. On the other hand, they only include part of the integrals of motion of the full system.

We now consider the solution of the system of conservation laws for oblique incidence of laser radiation on overdense plasma with sharp boundary at $x = 0$. We neglect pressure and internal energy of electron gas in skin layer, restricting the analysis to collisionless plasma and strong laser fields. Numerical calculations confirm that $\theta_R \approx \theta_0$ up to around $I \approx 10^{22}$ W/cm^2, beyond which ion motion may become significant even over a few laser cycles, thus we will henceforth take $\theta_R \approx \theta_0$. The balance equations for momentum flux are equivalent to the electrons equation of motion and in the region $0 < x < r_e$ reduce to the familiar forms, where the field components can be written as $B_x \approx (R+1)E_0 = fE_0, E_y \approx \cos\ \theta_0 fE_0$. The x-component of the electron current density can be found with the help of Poisson's equation in the region $x > r_e$, giving $E_{am} \approx 4\pi e n_e x$ so that $\dot{E}_{am} \approx C_3\omega E_{am}/2\pi$, where $C_3 = $ const. This field is balanced by the laser field, yielding a net hot electron flux:

$$vn_e\cos\theta_e = \frac{C_3\omega\left(-v_yB_x/c+\sin\theta_0 fE_0\right)}{8\pi^2 e} \tag{2.49}$$

Translational invariance along the y-axis results in the following integral of motion:

$$p_y - \frac{eA_y}{c} - m_e c\gamma \sin\theta_0 = -m_e c \sin\theta_0 \tag{2.50}$$

Assuming harmonic behaviour allows us to write: $d/dt \approx \omega C_1$, where $C_1 =$ const. Then choosing dimensionless variables $p \rightarrow p/m_e c,\ v \rightarrow v/c$ the equations reduce to:

$$C_1 p_z = f a_0 (\sin\,\theta_0 - v_y),\ p_y = \sin\,\theta_0 (\gamma - 1) - f a_0 \cos\,\theta_0 \tag{2.51}$$

Here $a_0 = C_2 eE_0/m\omega c$, where $C_2 =$ const. A simple Fresnel approximation gives $C_2 \approx \omega/\omega_{pe\_cold}$: corrections of order unity are to be expected through plasma compression, relativistic transparency and so on. At ultra-high laser intensities this coefficient increases owing to the relativistic reduction in the effective plasma frequency, but still remains less than unity. Rearranging we can find the electron velocity components and obtain an algebraic equation for the hot electron energy $\gamma(a_0, \theta_0)$, yielding the following set of equations for the hot electron flux:

$$\gamma^2 - 1 - \left(\sin\,\theta_0 \frac{f a_0}{C_1 \gamma} + \frac{f^2 a_0^2}{C_1 \gamma} \cos\,\theta_0\right)^2 - (\sin\,\theta_0(\gamma - 1) - f a_0 \cos\,\theta_0)^2 = 0$$

$$v_x = \sin\,\theta_0 \frac{f a_0}{C_1 \gamma^2} + \frac{f^2 a_0^2}{C_1 \gamma^2} \cos\,\theta_0,\ v_y = \sin\,\theta_0 (1 - 1/\gamma) - f a_0 \cos\,\theta_0/\gamma \tag{2.52}$$

This can be reduced further to a fourth order algebraic equation for γ, which has the following solutions in the low and high intensity limits respectively:

$$\gamma - 1 \approx f^2 a_0^2 \left(\frac{\sin^2\theta_0}{C_1^2} + \cos^2\theta_0\right)/2,\ a_0 \ll 1$$

$$\gamma \approx f a_0\, g(\theta_0),\ a_0 \gg 1 \tag{2.53}$$

where the function $g(\theta_0)$ is determined from:

$$g^4(\theta_0) - g^2(\theta_0)(g(\theta_0)\sin\theta_0 - \cos\theta_0)^2 - \frac{\cos^2\theta_0}{C_1^2} = 0 \tag{2.54}$$

From above one can also get the dependence between θ_0 and the electron entry angle θ_e:

$$\sin\theta_e = \frac{p_y}{p} = \sqrt{\frac{\gamma - 1}{\gamma + 1}} \sin\theta_0 - \frac{f a_0 \cos\theta_0}{\sqrt{\gamma^2 - 1}} \tag{2.55}$$

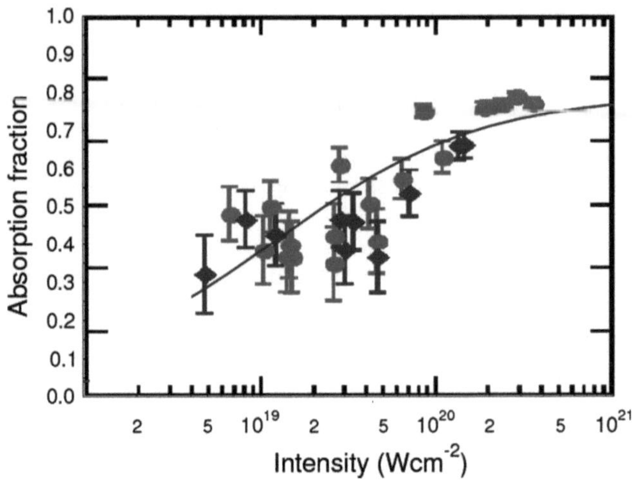

Fig. 2.5 Theoretical dependence of absorption coefficient (for $C_2 = 0.1$ and $\kappa_{\max} = 0.8$) on laser intensity compared with experimental data (*circle* and *diamond* symbols) at angle of incidence 45°

From Eq. (2.48), one can derive an absorption coefficient:

$$\kappa = \frac{v\cos\theta_e m_e n_e c^2}{I\cos\theta_0}(\gamma - 1) = \frac{C_3(1-1/\gamma)f}{2\pi a_0}(\tan\,\theta_0 + f a_0) \tag{2.56}$$

where $\gamma(a_0,\theta_0)$ is determined from Eq. (2.52). This coefficient depends on the numerical constants $C_{1,3} \sim O(1)$. To determine these we make use of the low intensity limit. Thus in Eq. (2.56) for $a_0 < 1$ and choosing $C_3/C_1^2 = 2$, one obtains: $\kappa \approx a_0\left(\frac{4\sin^3\,\theta_0}{\pi\cos\theta_0} + \frac{2C_1^2}{\pi}\sin\,2\theta_0\right)$, which reduces to Brunel's result. For $a_0 \gg 1$, Eq. (2.56) reduces to:

$$\kappa \approx \left(1+\sqrt{1-\kappa}\right)^2 F(\theta_0) \tag{2.57}$$

where $F(\theta_0) = \frac{C_1^2}{\pi}\left(\frac{\tan\theta_0}{a_0}+1\right)$. Introducing $\kappa_{\max} = \kappa(\theta_0 = 0; a_0 \to \infty)$, then $C_1 = \sqrt{\pi\kappa_{\max}}/(1+\sqrt{1-\kappa_{\max}})$. Thus the arbitrary constants $C_{1,3}$ can be fixed by the asymptotic behaviour of the absorption coefficient at large and small laser intensities.

Comparison of our model calculation [11] with experimental and simulation results of Ref. [12] is displayed in Fig. 2.5, which as we see, shows good agreement in the high-intensity regime.

2.3.8 Ponderomotive Absorption in a Non-uniform Plasma

Strong absorption is attributed to the mechanism called J×B heating [1, 2]. This mechanism becomes important at high intensities, particularly when electrons become relativistic. It should be pointed out that an increase in the laser radiation intensity reduces the difference between the absorption of the s- and p- polarised light, as found recently on several occasions in inhomogeneous plasma.

We consider now laser intensity up to 10^{18} W/cm^2. Result of decomposition Eq. (2.41) at $a < 1$ looks like:

$$\frac{\partial^2 a^{(0)}}{\partial \xi^2} - \frac{\omega^2}{\omega_p^2}\frac{\partial^2 a^{(0)}}{\partial \tau^2} = \eta_i(\xi)a^{(0)}; \quad E^{(0)} = 0$$

$$\frac{\partial^2 E^{(1)}}{\partial \tau^2} + \frac{\omega_p^2}{\omega^2}\eta_i(\xi)E^{(1)} = \frac{\omega_p^2}{\omega^2}\eta_i(\xi)\frac{\partial}{\partial \xi}\frac{a^{(0)2}}{2} \tag{2.58}$$

For linear polarisation of a incident wave a ponderomotive force depends on time. Then the solution of zero approximation of system has the standard form. The function $a(\xi)$ depends on a specific structure of concentration. The solution of the equation of the first approximation for a longitudinal field E at a given right part in view of a resonance looks as follows:

$$E^{(1)}(\xi,\tau) = \frac{1}{4}\frac{\partial a^2(\xi)}{\partial \xi}\left[1 + \frac{\omega_p^2\eta_i(\xi)/\omega^2}{\omega_p^2\eta(\xi)/\omega^2 - 4}\cos\left(2\tau - \cos\left(\omega_p\sqrt{\eta_i(\xi)}\tau/\omega\right)\right)\right] \tag{2.59}$$

In the point $\eta_i(\xi) = 4\omega^2/\omega_p^2$ there is the resonance. Energy of a longitudinal field per unit of the area of plasma:

$$W = \frac{m^2c^2\omega_p}{8\pi e^2}\int E^{(1)^2} d\xi \approx \frac{m^2c^3\omega\tau}{2^7 e^2}\left[\frac{\partial a^2}{\partial \xi^*}\right]^2\left[\frac{\partial\sqrt{\eta_i(\xi)}}{\partial \xi^*}\right]^{-1} \tag{2.60}$$

By dividing absorbed power on Pointing vector of laser wave:

$$<\gamma> = \frac{\omega^2}{32\pi c}\left(a^2(0) + \omega_p^2 a'^2(0)/\omega^2\right) \tag{2.61}$$

we receive required absorption coefficient:

$$\eta_a = \pi a^2(\xi^*)\left[\frac{\partial a^2(\xi^*)}{\partial \xi^*}\right]^2\left[\left(a^2(0) + a'^2(0)\right)\frac{\partial\sqrt{\eta_i(\xi^*)}}{\partial \xi^*}\right]^{-1} \tag{2.62}$$

at condition: $1 > a^2 \gg \upsilon_T\omega_p/c\omega$.

Consider $\eta_i(\xi) = \exp(\alpha\xi)$, where $\alpha = c/L\omega_p$ - plasma density scalelength. From above we receive the following result for absorption coefficient:

$$\eta_a = \frac{64}{\pi} a_0^2 x_l^3 sh^2 (2\pi x_l) K'^2_{2i\Omega/\alpha}(4x_l); \quad x_l = \omega L/c \tag{2.63}$$

The stroke designates derivative of the McDonald function [13]. The maximum $\eta_a = 0.38$ absorption coefficient reaches at $x_{lmax} = 0.26$, as well as in the case of oblique incidence and at large L it exponentially decreases on L. Thus, at sufficient laser intensity in spatially non-uniform plasma ($L > c/\omega_p$) this mechanism of absorption is dominant. The important effects breaking the one-dimensional pattern, are 'hole boring' and rifling of a surface. As a curved surface is formed, absorption and temperature of fast electrons increase, since the density gradient formed in parallel with the laser electric field. We should also note the generation of magnetic fields in such plasma. A strong magnetic field can change requirements of absorption, connected with direct acceleration of electrons, excitation of plasma waves and vortexes.

2.4 Reflection of a Short Laser Pulse from an Overdense Plasma Layer

Beside absorption, reflection of laser light is also important parameter for laser plasma interaction analysis. In the present section we study the interaction of short intense laser pulse with thin plasma foil because its second boundary give us a more interesting picture compared to a reflection from bulk target having only one boundary.

2.4.1 Analytical Model for a Laser-Driven Thin Foil

We consider a laser pulse of relativistic intensity propagating along the X-axis, which interacts with a foil of submicron thickness located at the $Y-Z$ plane. Based on our PIC-code calculations we found that quasi-equilibrium distributions of fields and particles are established along the X-axis during the first laser periods. In an ultra-thin target, only a short time interval is sufficient to change the plasma conditions. These distributions are sustained during the laser pulse. In particular, an unchangeable ion density profile during this time can be considered even though the target is accelerated by the laser pulse pressure. During a few laser pulse cycles a quasi-stationary state is established, and the plasma cloud expands with its centre of mass moving at almost constant velocity along the X-axis. As a result of these assumptions, the following equations for the electron density and the laser field can be obtained [14]:

$$2\theta_{eh}\frac{\partial^2 \eta_e}{\partial \xi^2} - \eta_e = -\eta - \frac{\partial^2 \sqrt{1+a^2}}{\partial \xi^2}$$

$$\left(\frac{\partial^2}{\partial \xi^2} + \Omega^2 - \left(\frac{\partial \psi}{\partial \xi}\right)^2\right) a = \eta_e \frac{a}{\sqrt{1+a^2}}, \qquad \frac{\partial}{\partial \xi}\left(a^2 \frac{\partial \psi}{\partial \xi}\right) = 0 \qquad (2.64)$$

The adiabatic equation of state for an electron gas with an initial temperature θ_{eh} and an adiabatic constant $\gamma = 2$ is used. The normalised vector potential of a circularly polarised wave is:

$$\mathbf{a}(\xi;\tau) = a(\xi,\tau)\{\mathbf{e}_z \cos[\tau - \psi(\xi,\tau)] + \mathbf{e}_y \sin[\tau - \psi(\xi,\tau)]\}. \qquad (2.65)$$

The ion dynamic is described by the system of hydro-equations. The following dimensionless variables and functions are used: $\Omega = \omega/\omega_p$, $\omega_p = (4\pi Z n_{i0} e^2/m)^{1/2}$ is the electron plasma frequency for the initial density Zn_{i0}, $\eta_e = n_e/Zn_{i0}$ is the normalised electron density, $\eta = n_i/n_{i0}$ is the normalised ion density and $\delta_e = m_e/m_i$, $\upsilon_i = v_i/c$. The normalised (in units $m_e c^2$) electron temperature T_{eh} is estimated as $\theta_{eh} \approx \kappa^* \Omega I_{18} \tau_{eff}/l_f$, where l_f is the foil thickness in units of c/ω_p, and I_{18} - laser intensity in units of 10^{18} W/cm^2, $\tau_{eff} \approx 0.3\, t_L$ - is the effective time of temperature formation ($t_L = \tau_L/\omega$ is the laser pulse duration), κ^* - is the absorption coefficient of the laser energy transformed into the electron energy.

The pressure of the laser pulse generates an electron shock wave, which in turn initiates an ion shock. This shock wave reaches the rear surface of an ultra-thin foil in a time comparable to a laser period, and forms a compressed layer (Fig. 2.6). Our simulations reveal that the initial ion profile of thickness l_f with an ion density $\eta = 1$ transforms into a compressed ion layer of thickness σ with a density η_0 and $l_g = l_f - \sigma$ with a density η_1 after only two periods. All electrons and an unknown part of the ions are now located in the compressed layer. The laser pulse interacts with such a target until its termination. We observed in the simulation that the target starts to move within a time of a few laser cycles. Thus, we incorporate a target movement $s(\tau)$ besides the simultaneously changing ion density profile. Therefore, the density profiles of Fig. 2.6 are transformed into a moving reference frame $\tilde{\xi} = \xi - s(\tau)$. The equations of our analytical model describe the target dynamics at a time when quasi-stationary electron and ion density profiles are set up. Therefore we can suppose an ion density profile $\eta(\xi,\tau)$, with a shape as depicted in Fig. 2.6.

2.4.2 Nonlinear Reflection of a Laser Wave from a Thin Layer of Cold Plasma

The target electrons are pushed away from the ions by the laser radiation pressure. There are two points $\xi_{1,2}$ which mark the points where the electron density vanishes

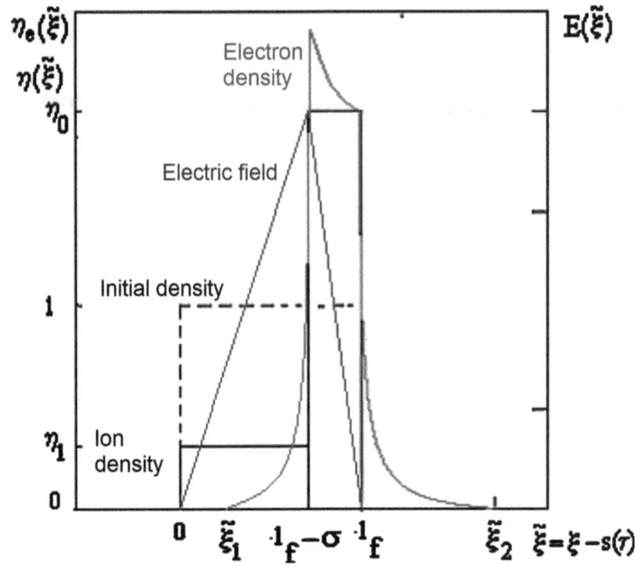

Fig. 2.6 Model of density profiles of a thin foil, which correlates with our PIC simulations: *dashed line* – initial density profile. The electron density profile, ion density profile and the electric field distribution for the case of a sharp electron density profile are also plotted (as *solid lines*)

in the reference frame of the target (see Fig. 2.6). Considering a cold plasma, we write the solution of the system (2.64) in the interval of constant ion density η_0:

$$\psi(\tilde{\xi}) = C\int_{l_g}^{\tilde{\xi}} \frac{d\tilde{\xi}}{a^2(\tilde{\xi})} + \psi(l_g)$$

$$\tilde{\xi}(a) - l_g = \int_{a(l_g)}^{a} \frac{da}{\sqrt{1+a^2}\sqrt{C_1 + 2\eta_0\left(\sqrt{1+a^2}-1\right) - \Omega^2 a^2 + C^2 a^{-2}}} \tag{2.66}$$

where C, C_1, $a(l_g)$ are unknown constants. The solution has the analogous form in the intervals $\tilde{\xi} \in [\tilde{\xi}_1; l_g]$ and $\tilde{\xi} \in [l_f; \tilde{\xi}_2]$, providing another set of constants.

For $\xi < \tilde{\xi}_1$ we can express the vector potential with incident and reflected vacuum waves:

$$\mathbf{a}(\tilde{\xi};\tau) = a_v(\bar{e}_x\cos(\tau - \Omega\tilde{\xi}) + \bar{e}_y\sin(\tau - \Omega\tilde{\xi})) + Ra_v(\bar{e}_x\cos(\tau + \Omega\tilde{\xi} - \alpha) + \bar{e}_y\sin(\tau + \Omega\tilde{\xi} - \alpha)) \tag{2.67}$$

where a_v is the amplitude of incident wave. Then, R, α are amplitude and phase of the reflection coefficient, respectively. For $\xi > \xi_2$ the transmitted light wave propagates into the vacuum:

$$\mathbf{a}(\xi;\tau) = a(\xi_2)(\bar{e}_x \cos(\tau - \Omega\xi - \psi_t) + \bar{e}_y \sin(\tau - \Omega\xi - \psi_t)) \tag{2.68}$$

with $\psi_t = \psi(\xi_2)$.

In the next step we have to equalise the values of the functions and their derivatives at the points l_g, l_f to find the unknown constants. Additionally, the condition of quasi-neutrality should be fulfilled because the charge of all electrons is equal to the ion charge. As a result, the constants C_1, C and σ can be expressed as functions of the reflection coefficient. Then, R and α are determined by a system of algebraic equations [14]:

$$\begin{aligned}
&\frac{2a_v R\Omega \sin\alpha}{\mu} \\
&= -\sqrt{1+a_v^2\mu^2}\sqrt{2\eta_0\sqrt{1+a_v^2\mu^2} - 2\eta_0\sqrt{1+a_v^2T^2} - \Omega^2 a_v^2\mu^2 + \frac{\Omega^2 a_v^2 T^4}{\mu^2}} \\
&\eta_0 \int_{a_vT}^{a_v\mu} \frac{da}{\sqrt{1+a^2}\sqrt{2\eta_0\sqrt{1+a^2} - 2\eta_0\sqrt{1+a_v^2T^2} - \Omega^2a^2 + a_v^4\Omega^2T^4a^{-2}}} \\
&= l_f + \frac{2a_v^2 R\Omega \sin\alpha}{\sqrt{1+a_v^2\mu^2}} \\
&C = a_v^2\Omega(1-R^2), \quad C_1 = 2\eta_0[1 - \sqrt{1+a_v^2(1-R^2)}] \\
&\qquad T = \sqrt{1-R^2},\ \mu^2 = 1 + R^2 + 2R\cos\alpha.
\end{aligned} \tag{2.69}$$

In the limit $\eta_0 \to \infty$, $\sigma \to 0(\sigma\eta_0 = const \neq 0)$, a rectangular density profile at $[l_g, l_f]$ can be changed into $\eta(\xi) = \eta_0\delta(\xi/\sigma) = \eta_0\sigma\delta(\xi)$. In this case, the system described by Eq. (2.69) can be simplified and an expression for the reflection coefficient is obtained:

$$\begin{aligned}
&R = \sqrt{\frac{l_f^2 + 4\Omega^2(1+a_v^2) - \sqrt{\left(l_f^2 + 4\Omega^2(1+a_v^2)\right)^2 - 16a_v^2\Omega^2 l_f^2}}{8a_v^2\Omega^2}}, \\
&\eta_0\sigma = l_f - \frac{2R\Omega a_v^2\sqrt{1-R^2}}{\sqrt{1+a_v^2(1-R^2)}}, \quad \eta_1(l_g) = \frac{2R\Omega a_v^2\sqrt{1-R^2}}{\sqrt{1+a_v^2(1-R^2)}}
\end{aligned} \tag{2.70}$$

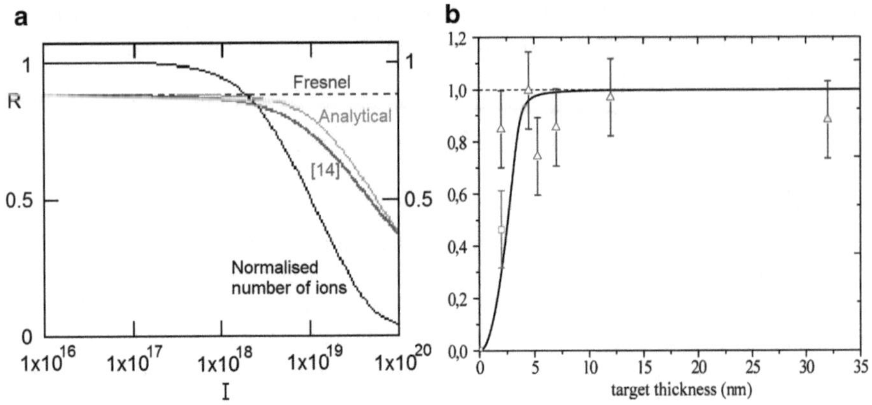

Fig. 2.7 (**a**) The dependence of the reflection coefficient of a thin foil on the laser intensity. A hydrogen foil with a thickness of 10 nm and a density 6×10^{22} cm^{-3} was used as a target. The *dashed line* is the Fresnel limit. Lines corresponding to the analytical result and the calculation in [14] are shown. The dependence of the normalised number of ions in the compression layer on laser intensity is also plotted as a *solid line*; (**b**) The dependence of the reflection coefficient of the thin foil on the foil thickness: *solid line* – analytical model; symbols – experiment. The laser intensity is 5×10^{19} W/cm^2

$\cos\alpha = -R$, $\sin\alpha = -\sqrt{1-R^2}$, $\psi(l_f) = -\arcsin(R)$. In the limit of a weak field $a_v \ll 1$, these formulae merge into the Fresnel formula: $R \to l_f/\sqrt{l_f^2 + 4\Omega^2}$. In the limit of a strong field: $a_v \gg 1$ [15]: $R \approx l_f/2\Omega a_v$, these formulae also describe the relativistic transparency and the onset of electron layer compression. The dependencies of the reflection coefficient of a thin foil on the laser intensity and the foil thickness are shown in Fig. 2.7a, b.

References

1. A.A. Andreev, *Generation and Application of Ultra-High Laser Fields* (NOVA Science Publishers Inc., New York, 2001)
2. G. Mourou, T. Tajima, S. Bulanov, Rev. Mod. Phys. **78**(2), 309 (2006)
3. A.A. Andreev, A.A. Mak, N.A. Solovyev, *An Introduction to Hot Laser Plasma Physics* (NOVA Science Publishers Inc., New York, 2000)
4. Ya.B. Zeldovich, Yu.P. Raiser, *Physics of Shock Waves and High-Temperature Hydrodynamic Phenomena* (Nauka, Moscow, 1966)
5. Y. Raizer, *Gas Discharge* (Nauka, Moscow, 1974)
6. I.I. Sobelman, L.A. Vainstein, E.A. Yukov, *Excitation of Atoms and Broadening of Spectral Lines* (Springer, Heidelberg, 1995)
7. V.L. Ginzburg, *The Propagation of Electromagnetic Waves in Plasmas* (Pergamon, New York, 1964)
8. W.L. Kruer, *The Physics of Laser Plasma Interaction* (Adison-Wesley, New York, 1998)

9. I.P. Shkarofsky, T.W. Johnston, M.P. Bachynski, *The Particle Kinetics of Plasmas* (Addison-Wesley, New York, 1966)
10. P. Gibbon, *Short Pulse Laser Interaction with Matter* (Imperial College Press, London, 2005)
11. P. Gibbon, A.A. Andreev, K.Yu. Platonov, Plasma Phys. Contr. F. **54**, 045001 (2012)
12. Y. Ping et al., Phys. Rev. Lett. **100**, 085004 (2008).
13. Z.X. Wang, D.R. Guo, *Special Functions* (World Scientific Publishing Co., Pte., Ltd., Singapore, 1989)
14. A. Andreev, S. Steinke, M. Schnuerer, et al., Phys. Plasmas **17**, 123111 (2010)
15. V. Vshivkov et al., Phys. Plasmas **5**, 2727 (1998)

Part II
High Energy Density Physics

Chapter 3
Shock Waves and Equations of State Related to Laser Plasma Interaction

Shalom Eliezer

Abstract Equations of state (EOS) of are fundamental to numerous fields of science, such as astrophysics, geophysics, plasma physics, inertial confinement physics and more. Laser induced shock waves techniques enable the study of equations of states and related properties, expanding the thermodynamic range reached by conventional gas gun shock waves and static loading experiments. Two basic techniques are used in laser-induced shock wave research, direct drive and indirect drive. In direct drive, one or more beams irradiate the target. In the indirect drive, thermal x-rays generated in laser heated cavities create the shock wave. Most of the laser induced shock waves experiments in the last decade used the impedance matching. Both direct and indirect drive can be used to accelerate a small foil-flyer and collide it with the studied sample, creating a shock in the sample, similar to gas-gun accelerated plates experiments. These lectures describe the physics of laser induced shock waves and rarefaction waves. The different formulae of the ideal gas EOS are used in connection with shock waves and rarefaction waves. The critical problems in laser induced shock waves are pointed out and the shock wave stability is explained. A general description of the various thermodynamic EOS is given. In particular the Gruneisen EOS is derived from Einstein and Debye models of the solid state of matter. Furthermore, the very useful phenomenological EOS, namely the linear relation between the shock wave velocity and the particle flow velocity, is analysed. This EOS is used to explain the $\approx$1 Gbar pressures in laser plasma induced shock waves. Last but not least, the shock wave jump conditions are derived in the presence of a magnetic field.

S. Eliezer (✉)
Plasma Physics Department, Soreq NRC, Yavne 81800, Israel
Institute of Nuclear Fusion, Polytechnic University of Madrid, Madrid, Spain
e-mail: shalom.eliezer@gmail.com

P. McKenna et al. (eds.), *Laser-Plasma Interactions and Applications*,
Scottish Graduate Series, DOI 10.1007/978-3-319-00038-1_3,

3.1 Introduction

The equation of state (EOS) describes a physical system by the relation between its thermodynamic quantities, such as pressure, energy, density, entropy, specific heat, and is related to both fundamental physics and the applied sciences [1–6]. The knowledge of EOS is necessary to understand the science of extreme concentration of energy and matter, behaviour of systems at high pressure, phase transitions, strongly coupled plasma systems, etc. The knowledge of EOS is required for many applications such as inertial fusion energy, astrophysics, geophysics and planetary science, new materials including nano-particles. The EOS describes Nature over all possible values of pressure, density and temperatures where local thermodynamic equilibrium can be sustained. Since it is not yet known from basic principles how to describe quantitatively material at every available thermodynamic state, including all phases of matter, it is necessary to introduce simplified methods whose range of applicability is limited.

The science of high pressure is studied experimentally in the laboratory by using static and dynamic techniques. In static experiments the sample is squeezed between pistons or anvils. The conditions in these static experiments are limited by the strength of the construction materials. In the dynamic experiments shock waves are created. Since the passage time of the shock is short in comparison with the disassembly time of the shocked sample, one can do shock wave research for any pressure that can be supplied by a driver, assuming that a proper diagnostic is available. In the scientific literature, the following shock wave generators are discussed: a variety of guns (such as rail guns and two stage light-gas guns) that accelerate a foil to collide with a target, magnetic compression, chemical explosives, nuclear explosions and high power lasers [7]. The dimension of pressure is given by the scale defined by the pressure of one atmosphere at standard conditions $\approx 1\,\text{bar} = 10^6\,\text{dyne/cm}^2$ (in c.g.s. units) = 10^5 Pascal (in M.K.S. units, Pascal = N/m^2).

In 1974 the first direct observation of a laser-driven shock wave was reported [8]. A planar solid hydrogen target was irradiated with a 10 J, 5 ns, Nd laser ($1.06\,\mu$m wavelength) and the spatial development of the laser driven shock wave was measured using high-speed photography. The estimated pressure in this pioneer experiment was 2 Mbar. Twenty years after the first published experiment, The NOVA laser from Livermore [9] laboratories in USA created a pressure of $750 \pm 200\,$Mbar. This was achieved in a collision between two gold foils, where the flyer (Au foil) was accelerated by a high intensity x-ray flux created by the laser plasma interaction. The highest laser induced pressures, $\approx 10^9$ atmospheres have been obtained during the collision of a target with an accelerating foil. This acceleration was achieved by laser-produced plasma, or by x-rays from a cavity produced by laser plasma interactions.

A shock wave is created in a medium that suffers a sudden impact (for example, a collision between an accelerated foil and a target) or in a medium that releases large amounts of energy in a short period of time (for example, high explosives). When a pulsed high power laser interacts with matter very hot plasma is created. This plasma exerts a high pressure on the surrounding material, leading to the formation of an intense shock wave, moving into the interior of the target. The momentum of the out-flowing plasma balances the momentum imparted to the compressed medium behind the shock front. For very high laser intensities ($I > 10^{15}\,\mathrm{W/cm^2}$) also the laser momentum I/c (where c is the speed of light) has to be taken into account [7]. The thermal pressure together with the laser momentum and the momentum of the ablated material drives the shock wave.

Shock waves in laser-plasma interactions are derived in (a) direct drive, (b) indirect drive by x-rays or ion beams, and (c) by the impact of a flyer plate accelerated by the laser beam (directly or indirectly). The main requirements for the EOS measurements are the creation of a one dimensional uniform, steady state shock wave where the initial target state is known and well defined, namely preheating by fast electrons for example is not permitted. Furthermore the diagnostics is crucial for accurate EOS measurements. For example, in order to achieve accuracy of the order of 1 %, a 1 mm target size during a 1 ns measurement requires a 10 μm spatial and 10^{-11} seconds temporal resolutions.

3.2 Sound Waves and Rarefaction Waves

The starting points in analysing the one-dimensional flow in a fluid is the equations describing the conservation laws of mass, momentum and energy:

$$\begin{aligned}
&\text{mass conservation: } \frac{\partial \rho}{\partial t} = -\frac{\partial}{\partial x}(\rho u) \\
&\text{momentum conservation: } \frac{\partial}{\partial t}(\rho u) = -\frac{\partial}{\partial x}\left(P + \rho u^2\right) \\
&\text{energy conservation: } \frac{\partial}{\partial t}\left(\rho E + \frac{1}{2}\rho u^2\right) = -\frac{\partial}{\partial x}\left(\rho E u + P u + \frac{1}{2}\rho u^3\right)
\end{aligned} \tag{3.1}$$

The motion of the fluid and the changes of density of the medium caused by a small pressure change ΔP describe the physics of sound waves [10]. For equilibrium pressure P_0 and density ρ_0 the changes in pressure ΔP and density $\Delta \rho$ due to the existence of a sound wave are extremely small. The motion in a sound wave is isentropic, $S(x) = \text{const.}$, therefore the change in the pressure is given by:

$$\Delta P = \left(\frac{\partial P}{\partial \rho}\right)_S \Delta \rho \equiv c_s^2 \Delta \rho \tag{3.2}$$

where c_s is the speed of sound. The mass and momentum conservations of Eq. 3.1 for small changes together with Eq. 3.2 yield the wave equation for pressure ΔP, density $\Delta\rho$ and flow velocity $\Delta\mu$:

$$\frac{\partial^2(F)}{\partial t^2} - c_S{}^2\frac{\partial^2(F)}{\partial x^2} = 0; F = \Delta\rho \text{ or } \Delta P \text{ or } \Delta u \tag{3.3}$$

The changes ΔP, $\Delta\rho$ and $\Delta\mu$ have two families of solutions f and g. The disturbances $f(x - c_s t)$ are moving in the positive x-direction while $g(x + c_s t)$ propagates in the negative x-direction.

If the undisturbed gas is not stationary, then the flow stream carries the waves. A transformation from the coordinates moving with the flow (velocity u in $+x$ direction) to the laboratory coordinates means that the sound wave is travelling with a velocity $u + c_s$ in the $+x$ direction and $u - c_s$ in the $-x$ direction. The curves dx/dt in the $x - t$ plane are called characteristic curves. We consider two characteristics: $C_+ : dx/dt = u + c_s$ and $C_- : dx/dt = u - c_s$.

Using the mass conservation and the momentum conservation given in Eq. 3.1 for an isotropic process ($S =$ const.) we get Riemann invariants J_+ and J_- given in Eq. 3.4. These invariants are occasionally used to solve numerically the flow equations for an isentropic process since J_+ and J_- are constants along the characteristics C_+ and C_- accordingly.

$$J_+ = u + \int \frac{dP}{\rho c_s} = u + \int \frac{c_s d\rho}{\rho};\ J_- = u - \int \frac{dP}{\rho c_s} = u - \int \frac{c_s d\rho}{\rho} \tag{3.4}$$

We now analyse the rarefaction wave where the pressure is suddenly dropped in an isentropic process. For example, after the high power laser is switched off and the ablation pressure drops. Another interesting case is after the laser induced high-pressure wave has reached the backside of a target and near the interface with the vacuum there is a sudden drop in pressure (note that pressure always vanishes at the vacuum-target boundary). In these cases, if one follows the variation in time for a given fluid element one gets $D\rho/Dt < 0$ and $DP/Dt < 0$ where $D/Dt = \partial/\partial t + u\partial/\partial x$.

We consider the behaviour of a gas, confined in a cylinder, caused by a receding piston, in order to visualise the phenomenon of a rarefaction wave. The piston is moving in the $-x$ direction so that the gas is continually rarefied as it flows (in the $-x$ direction). The disturbance, called a rarefaction wave, is moving forward, in the $+x$ direction. One can consider the rarefaction wave to be represented by a sequence of jumps $d\rho$, dP, du, so that we can use the Riemann invariant in order to solve the problem. The forward rarefaction wave moves into an undisturbed material defined by pressure P_0, density ρ_0, flow u_0 and the speed of sound c_{s0}. Using the Riemann invariants defined in Eq. 3.4 one gets:

$$u - u_0 = \int_{P_0}^{P} \frac{dP}{\rho c_s} = \int_{\rho_0}^{\rho} \frac{c_s d\rho}{\rho} \quad \text{rarefaction moving in} + \text{x direction}$$

$$u - u_0 = -\int_{P_0}^{P} \frac{dP}{\rho c_s} = -\int_{\rho_0}^{\rho} \frac{c_s d\rho}{\rho} \quad \text{rarefaction moving in} - \text{x direction} \tag{3.5}$$

As an example we calculate some physical quantities for a rarefaction wave in an ideal gas. Since in a rarefaction wave the entropy is constant, one can use the Riemann invariant with the EOS between the pressure, the density and the speed of sound to get:

$$\frac{P}{P_0} = \left(\frac{\rho}{\rho_0}\right)^{\gamma}; \frac{c_s}{c_{s0}} = \left(\frac{\rho}{\rho_0}\right)^{\frac{\gamma-1}{2}}$$

$$\mathrm{u}-\mathrm{u}_0 = \int_{\rho_0}^{\rho} \frac{c_s d\rho}{\rho} = \int_{c_{s0}}^{c_s} \frac{2dc_s}{(\gamma-1)} = \frac{2}{(\gamma-1)}(c_s - c_{s0}) \tag{3.6}$$

$$\Rightarrow c_s = c_{s0} + \frac{1}{2}(\gamma-1)(u-u_0)$$

where γ is defined as the ratio of the specific heat at constant pressure to the specific heat at constant volume C_p/C_V. From the last of equations it is evident that the speed of sound is decreased since u is negative. This implies that the density and the pressure are decreasing as expressed mathematically by $D\rho/Dt < 0$ and $DP/Dt < 0$ in a rarefaction wave.

3.3 Shock Waves

The development of singularities, in the form of shock waves, in a wave profile due to the nonlinear nature of the conservation equations have been already discussed by B. Riemann, W.J.M. Rankine and H. Hugoniot in the second half of the nineteenth century (1860–1890).

It is convenient to analyse a shock wave by inspecting a gas in a tube compressed by a piston moving into it with a constant velocity u. The medium has initially (the undisturbed medium) a density ρ_0, a pressure P_0 and it is at rest, $u_0 = 0$. A shock wave starts moving into the material with a velocity denoted by u_s. Behind the shock front the medium is compressed to a density ρ_1 and a pressure P_1. The gas flow velocity in the compressed region is equal to the piston velocity, $u = u_1$ (denoted also by $u_p = u_1$ and usually called the particle velocity). The initial mass before it is compressed, $\rho_0 A u_s t$ (A is the cross sectional area of the tube), equals the mass after compression, $\rho_1 A(u_s - u_1)t$, implying the mass conservation law:

$$\rho_0 u_s = \rho_1 (u_s - u_1) \tag{3.7}$$

The momentum of the gas put into motion, $(\rho_0 A u_s t)u_1$ equals the impulse due to the pressure forces, $(P_1 - P_0)At$, yielding the momentum conservation law (equivalent to the Newton's second law):

$$\rho_0 u_s u_1 = P_1 - P_0 \tag{3.8}$$

The increase of internal energy [energy/mass] and of kinetic energy per unit mass due to the piston-induced motion is $(\rho_0 A u_s t)(E_1 - E_0 + u_1^2/2)$. This increase in energy is supplied by the piston work, thus the energy conservation implies:

$$\rho_0 u_s \left(E_1 - E_0 + \frac{1}{2} {u_1}^2 \right) = P_1 u_1 \tag{3.9}$$

In the shock wave frame of reference, the undisturbed gas flows into the shock discontinuity with a velocity $v_0 = -u_s$ and leaves this discontinuity with a velocity $v_1 = -(u_s - u_1)$.

The jump conditions, usually called the Hugoniot equations, in the laboratory frame of reference are given in Eqs. 3.7, 3.8, and 3.9 and for a fluid initially at rest. In the more general case, the material is set into motion before the arrival of the shock wave (for example, by another shock wave). If the initial flow velocity is $u_0 \neq 0$, then the conservation laws (mass, momentum and energy) in the laboratory frame of reference can be written as:

$$\begin{aligned}
&\rho_0 (u_s - u_0) = \rho_1 (u_s - u_1) \\
&\rho_0 (u_s - u_0)(u_1 - u_0) = P_1 - P_0 \\
&\rho_0 (u_s - u_0) \left(E_1 - E_0 + \frac{1}{2} {u_1}^2 - \frac{1}{2} {u_0}^2 \right) = P_1 u_1 - P_0 u_0
\end{aligned} \tag{3.10}$$

These relations are used to determine the state of the compressed solid behind the shock front. Assuming that the initial state is well defined and the quantities E_0, u_0, P_0, and $\rho_0 = 1/V_0$ are known, one has five unknowns E_1, u_1, P_1, $\rho_1 = 1/V_1$ and u_s with three equations (occasionally the specific volume V is used instead of the density ρ). Usually the shock wave velocity is measured experimentally, and if the equation of state is known (in this case one has four equations) $\mathrm{E} = \mathrm{E}(P, \rho)$ then the quantities of the compressed state can be calculated. If the equation of state is not known, then one has to measure experimentally two quantities of the shocked material, for example u_s and u_1 in order to solve the problem.

If $E(P, \rho)$ is known then from Eq. 3.10 one can write (the notation of P_1 is changed to P_H and ρ_1 is ρ)

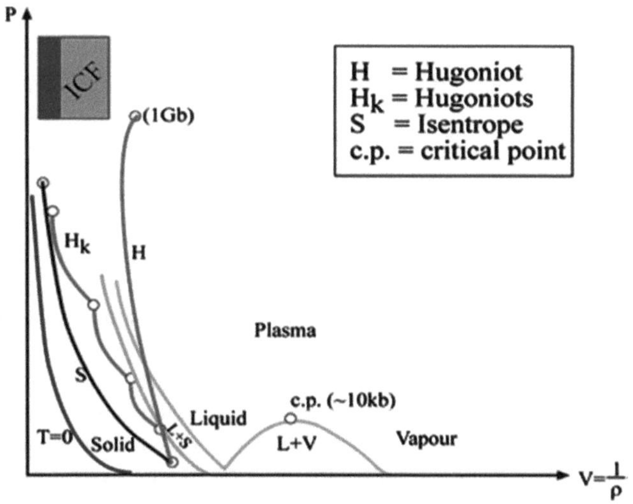

Fig. 3.1 Pressure–specific volume (=1/density ρ) diagram of Hugoniot and some thermodynamic curves on the background of the four phases of state: solid, liquid, vapor (gas) and plasma. The mixture domain liquid-solid (L + S) and liquid-vapor (L + V) are also shown. The schematic domain of inertial confinement fusion (ICF) ignition domain is denoted in this figure

$$P_H = P_H(\rho;\rho_0,P_0) \tag{3.11}$$

This curve is known in the literature as the Hugoniot curve. The Hugoniot curve is a two parameter (ρ_0,P_0) family of curves, so that for each initial condition (ρ_0,P_0) there is a different curve. The Hugoniot curve is not a thermodynamic function, it does not show the pressure-volume (or density) trajectory of a shock wave development, but it is a plot of all possible final shocked states for a given initial state (ρ_0,P_0). For example, the Hugoniot curve is different than the isentropic curves of the pressure $P_S(\rho)$, which describes the thermodynamic trajectory of pressure- density for any given entropy S. It is interesting to note that for a given final pressure the compression ($\rho/\rho_0 = V_0/V$) is higher for an isentrope relative to the Hugoniot and the isothermal compression is the highest.

In Fig. 3.1 we can see schematically the pressure – specific volume (=1/density ρ) diagram of Hugoniot and some thermodynamic curves on the background of the 4 phases of state: solid, liquid, vapor (gas) and plasma. The mixture domain liquid-solid ($L+S$) and liquid-vapor ($L+V$) are also shown. The schematic domain of inertial confinement fusion (ICF) ignition domain is denoted in this figure. A series of Hugoniot curves with different initial conditions is denoted by H_k. It is interesting to realise that a series of shock waves (Hugoniots H_k) are approaching the isentrope curve S.

It is useful to consider the shock wave relations for an ideal gas with constant specific heats. In this case the equations of state are:

$$E = C_V T = \frac{PV}{\gamma - 1}; \; S = C_V \ln\left(PV^{\gamma}\right) \tag{3.12}$$

where γ is defined as the ratio of the specific heat at constant pressure to the specific heat at constant volume C_p/C_V. Using EOS from Eq. 3.12 and the Hugoniot relations, and the Hugoniot curve for an ideal gas equation of state is obtained

$$\frac{P_1}{P_0} = \frac{(\gamma+1)V_0 - (\gamma-1)V_1}{(\gamma+1)V_1 - (\gamma-1)V_0}$$
$$\frac{P_1}{P_0} \to \infty \Rightarrow \left(\frac{\rho_1}{\rho_0}\right)_{\max} = \left(\frac{V_0}{V_1}\right)_{\max} = \frac{\gamma+1}{\gamma-1} \tag{3.13}$$

For example, the maximum density caused by a planar shock wave in a medium with $\gamma = 5/3$ is $4\rho_0$.

Using the EOS for a constant entropy S (the second equation of 3.12), the definition of the speed of sound defined in Eq. 3.2 and the Hugoniot relations , and one gets the ratio M of the shock velocity to the sound velocity (or equivalently, the flow velocity (v_0 and v_1) to the sound velocity in the shock wave frame of reference) which is known as the Mach number:

$$\mathrm{M}_0{}^2 \equiv \left(\frac{u_s}{c_{s0}}\right)^2 = \left(\frac{v_0}{c_{s0}}\right)^2 = \frac{1}{2\gamma}\left[(\gamma-1)+(\gamma+1)\frac{P_1}{P_0}\right] > 1$$
$$\mathrm{M}_1{}^2 \equiv \left(\frac{v_1}{c_{s1}}\right)^2 = \frac{1}{2\gamma}\left[(\gamma-1)+(\gamma+1)\frac{P_0}{P_1}\right] < 1 \tag{3.14}$$

The meaning of these relations is that in the shock frame of reference, the fluid flows into the shock front at a supersonic velocity ($M_0 > 1$) and flows out at a subsonic velocity ($M_1 < 1$). In the laboratory frame of reference, one has the well-known result that the shock wave propagates at a supersonic speed (with respect to the undisturbed medium), and at a subsonic speed with respect to the compressed material behind the shock. Although this phenomenon has been proven here for an ideal gas equation of state, this result is true for any medium, independent of the equation of state [11].

In a shock wave the entropy always increases. For example, in an ideal EOS with the Hugoniot relation the increase in entropy during a shock wave process is given by:

$$S_1 - S_0 = C_V \ln\left(\frac{P_1 V_1{}^\gamma}{P_0 V_0{}^\gamma}\right)$$
$$= \left[\frac{P_0 V_0}{(\gamma-1)T_0}\right] \ln\left\{\left(\frac{P_1}{P_0}\right)\left[\frac{(\gamma-1)\frac{P_1}{P_0}+(\gamma+1)}{(\gamma+1)\frac{P_1}{P_0}+(\gamma-1)}\right]^\gamma\right\} > 0 \tag{3.15}$$

The increase in entropy indicates that a shock wave is not a reversible process, but a dissipative phenomenon. The entropy jump of a medium (compressed by shock wave) increases with the strength of the shock wave (defined by the ratio

P_1/P_0). The larger P_1/P_0 the larger is $S_1 - S_0 = \Delta S$. The value of ΔS is determined by the conservation laws (mass, momentum and energy) and by the equation of state, however, the mechanism of this change is described by viscosity and thermal conductivity [1].

Figure 3.2 describes the space-time ($x-t$) diagram for a shock wave followed by rarefaction waves moving into a given medium. At the origin of $x-t$ the piston starts moving with a constant velocity creating a shock wave described by the straight line between the pressures P_0 and P_1 domains. At a time t_1 the piston stops moving, the line describing the piston becomes vertical (x =constant) and therefore a set of rarefaction waves are generated. Three rarefaction waves are described by the lines a, b and c. The rarefaction waves move faster than the shock wave and they decrease the domain of the shocked material. In Fig. 3.2a the space (x)-profile of three pressures are described at times t_1, t_2 and t_3. In Fig. 3.2b the time profile of three pressures are given at positions x_1, x_2 and x_3. The influence of the rarefaction waves (lines a, b and c) on the profiles are shown explicitly by the points a, b and c on the last profiles in Fig. 3.2a, b. In these figures the pressure profiles decrease linearly in time or space, however this is not generally true and in fact the profiles depend on the time duration of the shock waves (t_1 in this case) and on the equations of state.

We end this section with a discussion on shock wave stability. One can see from isentropic speed of sound in Eq. 3.6 that different disturbances of density travel with different velocities, so that the larger the density ρ the faster the wave travels. Therefore, an initial profile $\rho(x,0)$ becomes distorted with time. This is true not only for the density but also for the pressure $P(x,0)$, for the flow velocity $u(x,0)$, etc. In this way a smooth function of these parameters will steepen in time due to the nonlinear effect of the wave propagation (higher amplitudes move faster). Therefore, a compression wave is steepened into a shock wave because in most solids the sound velocity is an increasing function of the pressure. In the laboratory frame of reference, the speed of a disturbance is the sum of the flow velocity and the sound velocity ($c_s + u$). Therefore, a higher-pressure disturbance will catch up with the lower pressure disturbance causing a sharpening profile of the wave. In reality there are also dissipative mechanisms such as viscosity and thermal transport. Therefore the sharpening profile mechanism can only increase until the dissipative forces become significant, and they begin to cancel out the effect of increasing sound speed with pressure. When the sum of these opposing mechanisms cancels out the wave profile does not change in time anymore and it becomes a steady shock wave.

As already stated above, a disturbance moves at the speed $c_s + u$ in a compression wave. Therefore, a disturbance behind the shock front cannot be slower than the shock velocity; because in this case it will not be able to catch the wave front, and the shock would decay (namely the shock is unstable to small disturbances behind it). Similarly, a small compressive disturbance ahead of the shock must move slower than the shock front in order not to create another shock wave. Thus the conditions for a stable shock wave can be summarised in the following way:

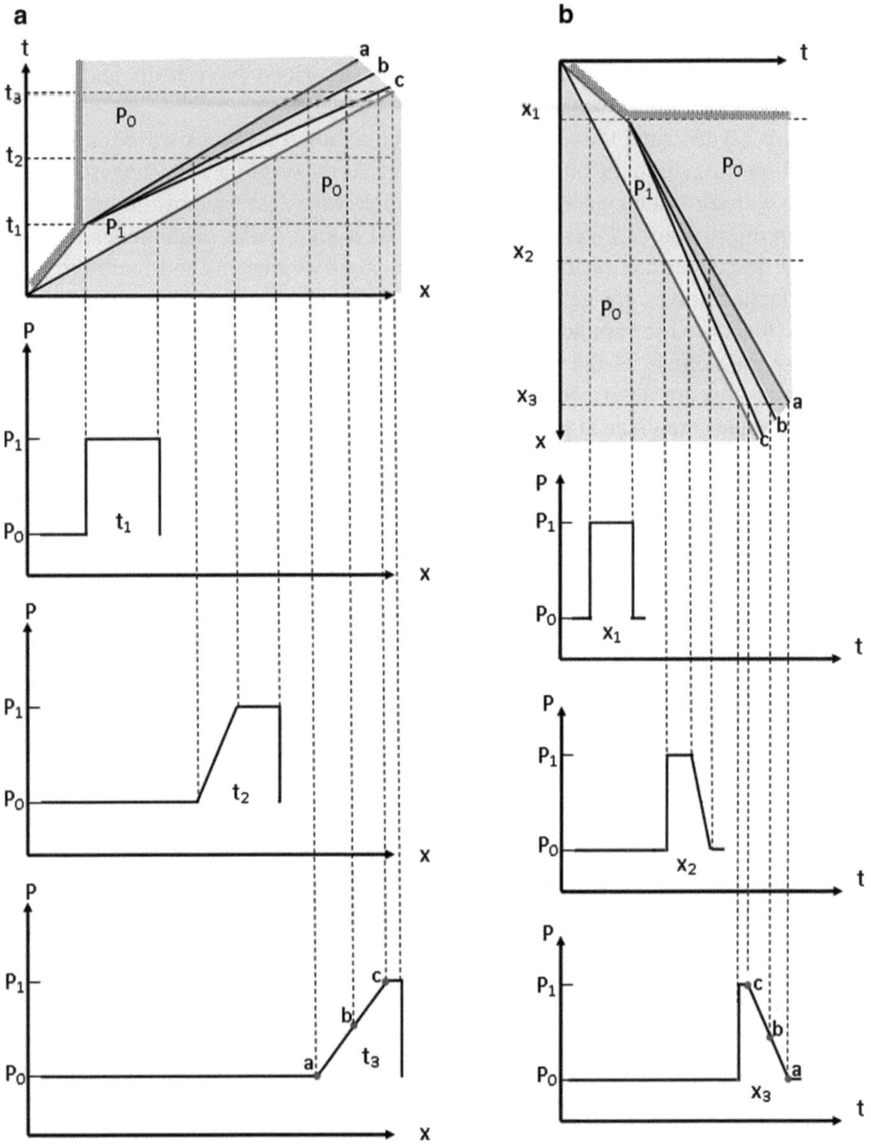

Fig. 3.2 The space-time $(x - t)$ diagram for a shock wave followed by rarefaction waves moving into a given medium. At the origin of $x - t$ the piston starts moving with a constant velocity creating a shock wave. At a time t_1 the piston stops moving and therefore a set of rarefaction waves are generated. In figure (**a**) the space (x)-profile of three pressures are described at times t_1, t_2 and t_3. In figure (**b**) the time (t) profile of three pressures are given at positions x_1, x_2 and x_3

$$\frac{dc_s}{dP} > 0; \; c_s + u_p \geq u_s; \; u_s > c_{s0} \tag{3.16}$$

The first of these equations states that the speed of sound increases with increasing pressure. The second equation describes the fact that the shock wave is subsonic (Mach number smaller than one) with respect to the shocked medium. The last equation of is the well known phenomenon that a shock wave is supersonic (Mach number larger than one) with respect to the unshocked medium.

In the domain of phase transitions (solid-solid due to change in symmetry or solid-liquid) the shock wave can split into two or more shock waves. However, in these cases the stability criteria can be satisfied for each individual shock wave.

3.4 Critical Problems

When a high power laser interacts with matter very hot plasma is created. This plasma exerts a high pressure on the surrounding material, leading to the formation of an intense shock wave, moving into the interior of the target [7].

The problems with the high pressure laser induced shock waves are the small size of the targets ($\approx 100\,\mu$m), the short laser pulse duration (≈ 1 ns), the poor spatial uniformity of a coherent electromagnetic pulse (the laser), and therefore the non-uniformity of the created pressure. The main critical problems can be summarised as: (a) the planarity (1D) of the shock wave regardless of the laser irradiance non-uniformity. (b) Steady shock wave during the diagnostic measurements in spite of the laser short pulse duration. (c) Well-known initial conditions of the shocked medium. This requires to control (namely, to avoid) the fast electron and x-ray preheating. (d) Good accuracy (≈ 1 %) of the measurements.

The planarity of the shock wave is achieved by using optical smoothing techniques [12–16]. With these devices the laser is deposited into the target uniformly, within ≈ 2 % of energy deposition. For example [12], one technique denoted as 'induced spatial incoherence' (ISI), consists of breaking each laser beam into a large number of beam lets by reflecting the beam off a large number of echelons. The size of each beamlet is chosen in such a way that its diffraction limited spot size is about the target diameter. All of the beam lets are independently focused and overlapped on the target. Another technique [13] divides the beam into many elements that have a random phase shift. This is achieved by passing the laser beam through a phase plate with a randomly phase shifted mask.

The focal spot of the laser beam on target has to be much larger than the target thickness in order to achieve a 1D steady state shock wave. For any planar target with thickness d, irradiated by a laser with a focal spot area = $\pi R_L^2 a$ lateral rarefaction wave enters the shocked area and reduces the pressure and density of the shocked area. This effect distorts the one-dimensional character of the wave, since the shock front is bent in such a way [17] that for very large distances ($\gg d$) the shock wave front becomes spherical. The rarefaction wave propagates toward the symmetry axis with the speed of sound c_s (in the shock-compressed area), which is larger than the shock wave velocity u_s. Therefore, the undisturbed (by the rarefaction wave) one-dimensional shocked area on the back surface of the target equals $\pi(R_L-(d/u_s)c_s)^2$. Therefore, in order to have a one-dimensional shock wave one requires that $R_L \approx 2d$ at least, so that the laser focal spot area $A \approx 10d^2$. This constraint implies very large laser focal spots for thick targets.

The second constraint requires a steady shock wave, namely the shock velocity has to be constant as it traverses the target. A rarefaction wave (RW), initiated at a time $\Delta\tau \approx \tau_L$ (the laser pulse duration) after the end of the laser pulse, follows a shock wave (SW) into the target. It is necessary that the rarefaction wave does not overtake the shock wave at position $x = d$ (the back surface) during the measurement of the shock wave velocity, implying $\tau_L > d/u_s$. For strong shocks, the shock velocity is of the order of the square root of the pressure therefore $\tau_L > d/u_s \approx d/(P)^{1/2}$. Hot electrons can appear during the laser-plasma interaction causing preheating of the target. This preheats the target before the shock wave arrives, therefore 'spoiling' the initial conditions for the high-pressure experiment. Since it is not easy to measure accurately the temperature of the target due to this preheating, it is necessary to avoid preheating. By using shorter wavelengths (0.5 μm or less) the fast electron preheat is significantly reduced. It is therefore required that the target thickness d is larger than the hot electron mean free path λ_e. Using the scaling law for the hot electron temperature T_h one has $d \gg \lambda_e \approx T_h^2 \approx (I_L\lambda_L)^{0.6}$. Taking into account these constraints and using the experimental scaling law $P \approx I_L^{0.8}$, one gets the scaling of the laser energy $W_L = I_L A \tau_L \approx I_L^{2.4} \approx P^3$. Therefore, in order to increase the one-dimensional shock wave pressure by a factor two it is necessary to increase the laser energy by an order of magnitude.

A more elegant and efficient way to overcome these problems is to accelerate a thin foil. The foil absorbs the laser, plasma is created (ablation) and, the foil is accelerated like in a rocket [7]. In this way, the flyer stores kinetic energy from the laser during the laser pulse duration (the acceleration time) and delivers it, in a shorter time during the collision with a target, in the form of thermal energy. The flyer is effectively shielding the target so that the target initial conditions are not changed by fast electrons or by laser-produced x-rays. For these reasons the laser driven flyer can achieve much higher pressures on impact than the directly laser induced shock wave [9, 18, 19].

The accuracy of measurements in the study of laser induced high pressure physics require diagnostics with a time resolution better than 100 ps, and occasionally better than 10 ps, and a spatial resolution of the order of few microns. The accurate measurements of shock wave speed and particle flow velocity are usually obtained with optical devices, including streak camera [20–22] and velocity interferometers [23, 24].

3.5 EOS and the Thermodynamic Equations

We assume that X describes the state of a system defined by a potential $F(X)$. The conjugate variable of X is $P = dF/dX$. If X is replaced by P as independent variable by the Legendre transformation, $\Psi(P) = F - PX$ then $\Psi(P)$ is also a potential. The Legendre transformation for several variables is defined by:

$$\Psi(P_1, P_2, \ldots P_n) = F(X_1, X_2, \ldots X_n) - \sum_{i=1}^{n} P_i X_i;\; P_i = \left(\frac{\partial F}{\partial X_i}\right)_{j \neq i_i}$$

$$d\Psi(P_1, P_2, \ldots P_n) = dF - \sum_{i=1}^{n} (P_i dX_i + dP_i X_i) \tag{3.17}$$

For example, the conjugate variables of entropy and specific volume (S, V) are the temperature and pressure (T, P) accordingly. Assuming a system with a constant number of particles, $N =$ const., the Gibbs potential G is derived from the internal energy E by the following Legendre transformation

$$G(T,P) = E(S,V) - \left[\left(\frac{\partial E}{\partial S}\right)_V S + \left(\frac{\partial E}{\partial V}\right)_S V\right];\Rightarrow G = E - TS + PV \tag{3.18}$$

A summary of the thermodynamic potentials, derived from each other by a Legendre transformation is given in Table 3.1. The thermodynamic potentials are: internal energy E, enthalpy H, Helmholtz free energy F, Gibb's free energy G and the grand potential Φ. The appropriate variables of the potentials are denoted by the specific volume V (= 1ρ where ρ is the density), the temperature T, the pressure P, the entropy S, the chemical potential μ and the number of particles N.

The various equations of state derived from these potentials are summarised in Table 3.2. As one can see from this table there are many possible presentations of EOS. Some specific examples will be used in this chapter. In particular, for ideal gas EOS the following relations are given [2, 3]: the Helmholtz free energy F, the pressure P, the internal energy E, the heat capacity at constant volume C_V, the entropy S and the Gibb's free energy G and the chemical potential μ:

Table 3.1 Thermodynamic potentials

Quantity	Variables	Relations
E [internal energy]	S,V,N	$E = TS - PV + \mu N$
H [Enthalpy]	S,P,N	$H = E + PV$
F [Helmholtz free energy]	T,V,N	$F = E - TS$ $F = PV + \mu N$
G [Gibb's free energy]	T,P,N	$G = \mu N$
Φ [Grand potential]	T,V,μ	$\Phi = -\mathrm{PV}$ $\Phi = F - \mu N$

Table 3.2 The EOS derived from the different thermodynamic potentials

	Potential	EOS
E	$dE = TdS - PdV + \mu dN$	$\mu = \left(\frac{\partial E}{\partial N}\right)_{S,V}$; $T = \left(\frac{\partial E}{\partial S}\right)_{V,N}$; $P = -\left(\frac{\partial E}{\partial V}\right)_{S,N}$
H	$dH = TdS + VdP + \mu dN$	$\mu = \left(\frac{\partial H}{\partial N}\right)_{S,P}$; $T = \left(\frac{\partial H}{\partial S}\right)_{P,N}$; $V = \left(\frac{\partial H}{\partial P}\right)_{S,N}$
F	$dF = -SdT - PdV + \mu dN$	$\mu = \left(\frac{\partial F}{\partial N}\right)_{T,V}$; $S = -\left(\frac{\partial F}{\partial T}\right)_{V,N}$; $P = -\left(\frac{\partial F}{\partial V}\right)_{T,N}$
G	$dG = -SdT + VdP + \mu dN$	$\mu = \left(\frac{\partial G}{\partial N}\right)_{T,P}$; $S = -\left(\frac{\partial G}{\partial T}\right)_{P,N}$; $V = \left(\frac{\partial G}{\partial P}\right)_{T,N}$
Φ	$d\Phi = -SdT - PdV - Nd\mu$	$N = -\left(\frac{\partial \Phi}{\partial \mu}\right)_{T,V}$; $S = -\left(\frac{\partial \Phi}{\partial T}\right)_{V,\mu}$; $P = -\left(\frac{\partial \Phi}{\partial V}\right)_{T,\mu}$

$$
\begin{aligned}
F(T,V,N) &= -Nk_BT\ln\left[\left(\frac{mk_BT}{2\pi\hbar^2}\right)^{3/2} V\right] \\
P &= -\left(\frac{\partial F}{\partial V}\right)_T = \left(\frac{N}{V}\right)k_BT; \\
E &= -T^2\left[\frac{\partial(F/T)}{\partial T}\right]_V = \frac{3}{2}Nk_BT; \\
C_V &= \left(\frac{\partial E}{\partial T}\right)_V = \frac{3}{2}Nk_B \\
S &= -\left(\frac{\partial F}{\partial T}\right)_V = Nk_B\ln\left[\left(\frac{mk_BT}{2\pi\hbar^2}\right)^{3/2} V\right] + \frac{3}{2}Nk_B \\
G(T,P,N) &= F + PV = -Nk_BT\ln\left[\left(\frac{mk_BT}{2\pi\hbar^2}\right)^{3/2}\left(\frac{k_BT}{P}\right)\right] \\
\mu &= \frac{G}{N}
\end{aligned}
\tag{3.19}
$$

where k_B and $2\pi\hbar = h$ are Boltzmann and Planck constants accordingly, m is the mass of the ideal gas particles, and all other variables are defined above.

3.6 Gruneisen EOS for the Solid

3.6.1 Einstein Model of Solids

In 1907 [25] Albert Einstein suggested a model for the solid in order to explain the experimental observations that the heat capacity of the solid decreases at low temperatures below the Dulong-Petit value of $3R$ per mole ($R = 8.31\ \text{JK}^{-1}\ \text{mole}^{-1}$).

Einstein assumed that a solid can be described by a lattice of N atoms vibrating as a set of $3N$ independent harmonic oscillators in one dimension. The vibrations were quantised by Einstein!

In order to calculate the heat capacity one needs the equation of state for the solid. The EOS is calculated from basic principles if the energy eigenvalues are known and the partition function Q, related to the free energy F ($F = -\beta ln\ Q$ where $\beta = 1/k_B T$), can be calculated. In Einstein's model the energy eigenvalues, the partition function Q and the Helmholtz free energy F are:

$$\varepsilon_{j,n} = \left(n_j + \frac{1}{2}\right) h\nu_j \begin{cases} j = 1,2,\ldots 3N \\ n_j = 0,1,2,\ldots \end{cases} \Rightarrow E_n = \sum_{j=1}^{3N} n_j h\nu_j + E_c;$$

$$Q = \sum_n e^{-\beta E_n} = e^{-\beta E_c} \sum_{n_1=0}^{\infty} e^{-\beta n_1 h\nu_1} \sum_{n_2=0}^{\infty} e^{-\beta n_2 h\nu_2} \ldots \sum_{n_{3N}=0}^{\infty} e^{-\beta n_{3N} h\nu_{3N}}$$

$$= e^{-\beta E_c} \prod_{j=1}^{3N} [1 - \exp(-\beta h\nu_j)]^{-1}$$

$$F = -\frac{1}{\beta} \ln Q = E_c + \frac{1}{\beta} \sum_{j=1}^{3N} \ln[1 - \exp(-\beta h\nu_j)] \tag{3.20}$$

where h is the Planck constant and E_C is the cold energy. From the free energy F all thermodynamic variables can be calculated. In particular the energy of the system E and the heat capacity C_V are:

$$E = F - T\left(\frac{\partial F}{\partial T}\right)_V \Rightarrow E = E_c + \sum_{j=1}^{3N} \frac{h\nu_j}{e^{\beta h\nu_j} - 1} = E_c + \frac{3Nh\nu}{e^{h\nu/k_B T} - 1}$$

$$C_V = \left(\frac{\partial E}{\partial T}\right)_V = 3Nk_B \left(\frac{h\nu}{k_B T}\right)^2 \frac{e^{h\nu/k_B T}}{\left(e^{h\nu/k_B T} - 1\right)^2} \tag{3.21}$$

In deriving the energy E, Einstein assumed that all ν_j are equal. According to the Bose-Einstein statistics these solid oscillations are described by scalar particles with spin 0 with energy $h\nu$ and distribution $f(\nu) = 1/[\exp(\beta h\nu) - 1]$. These oscillations were later recognised as the famous phonons in the solid.

As experiments suggested, Einstein's model predicts $C_v \to 0$ for $T \to 0$, however this model gives only qualitative agreement with experiments. Einstein suggested in 1911 that a large number of frequencies will improve his model as was done in 1912 by Debye.

3.6.2 Debye Model of Solids

In Debye's model the Einstein single frequency is replaced by a spectrum of frequencies. In order to do that the number of oscillating modes $g(p)dp$ is taken as the phase space $g(p)dp = V4\pi p^2 dp = g(\nu)d\nu$ where $p = h\nu/c$ is the momentum of a zero mass scalar particle (the phonon) moving with the sound velocity c. In this case the density of state in the frequency space is given by:

$$g(\nu)d\nu = V(1/c^3)4\pi\nu^2 d\nu \to V(1/{c_L}^3 + 2/{c_t}^3)4\pi\nu^2 d\nu \tag{3.22}$$

where c_L and c_t are the longitudinal and transverse sound velocity in the solid. Debye assumed a maximum possible frequency, denoted by ν_D, determined by the requirement that in the solid are only $3N$ modes, namely:

$$\begin{aligned} 3\text{N} &= 4\pi\left(\frac{1}{{c_L}^3} + \frac{2}{{c_t}^3}\right) V \int_0^{\nu_D} \nu^2 d\nu \\ &= \frac{4\pi V {\nu_D}^3}{3}\left(\frac{1}{{c_L}^3} + \frac{2}{{c_t}^3}\right) \\ &\Rightarrow g(\nu)d\nu = \frac{9N\nu^2 d\nu}{{\nu_D}^3} \end{aligned} \tag{3.23}$$

It is convenient to define also a Debye temperature T_D equal to $h\nu_D = k_B T_D$. For example, $T_D = 390\,\text{K}$ for aluminium and 150 K for Na. Changing the sum in Eq. 3.21 to an integral with a density of states (3.23) one obtains the following energy and the heat capacity in the Debye model:

$$\begin{aligned} E_T &= \int_0^{\nu_D} \frac{h\nu g(\nu)d\nu}{e^{h\nu/k_B T} - 1} = \frac{9Nh}{{\nu_D}^3}\int_0^{\nu_D} \frac{\nu^3 d\nu}{e^{h\nu/k_B T} - 1} = 9Nk_B T_D \left(\frac{T}{T_D}\right)^4 \int_0^{T_D/T} \frac{\xi^3 d\xi}{e^{\xi} - 1} \\ C_V &= \left(\frac{\partial E_T}{\partial T}\right)_V \end{aligned} \tag{3.24}$$

The energy and the heat capacity in the high temperature limit, $T \to \infty$, are the ideal EOS $E_T = 3Nk_B T$ and $C_V = 3Nk_B$, (the Dulong-Petit value). For $T \to 0$ (i.e. $T_D/T \to \infty$) one gets:

$$T \to 0:\ E_T = \frac{3\pi^4 N k_B T_D}{5}\left(\frac{T}{T_D}\right)^4; C_V = \frac{12\pi^4 N k_B}{5}\left(\frac{T}{T_D}\right)^3 \tag{3.25}$$

This is the famous $C_V \propto T^3$ law derived by Debye in order to explain satisfactorily the experiments.

3.6.3 Gruneisen Model of Solids

Using the Einstein free energy thermodynamic function F given in Eq. 3.20 Gruneisen derived the pressure P:

$$\gamma_j \equiv -\frac{V}{\nu_j}\left(\frac{\partial \nu_j}{\partial V}\right)_T = -\left(\frac{\partial \ln \nu_j}{\partial \ln V}\right)_T$$

$$\Rightarrow P = -\left(\frac{\partial F}{\partial V}\right)_T = -\frac{dE_c}{dV} + \frac{1}{V}\sum_{j=1}^{3N}\frac{\gamma_j h\nu_j}{e^{\beta h\nu_j}-1} = P_c + P_T \tag{3.26}$$

where P_C and P_T are the cold and thermal pressures in the solid accordingly. γ_j does not vanish therefore the frequency depends on density ρ (= specific volume V). The compression of a solid makes it harder and thus the restoring force becomes grater, which in turn implies an increase in the vibration frequencies. Therefore, one expects ν_j to increase with decreasing volume so that γ_j has positive values. Using Debye model and assuming $\gamma_j = \gamma$ for $j = 1, 2, \ldots, 3N$, Gruneisen derived from Eq. 3.25 the following relation between the thermal pressure and the thermal energy E_T in the solid:

$$P_T = \frac{\gamma}{V}E_T;\; E_T = \sum_{j=1}^{3N}\frac{h\nu_j}{e^{\beta h\nu_j}-1} \tag{3.27}$$

The first of equations is known in the literature as **Gruneisen EOS** and γ is the **Gruneisen coefficient**. This coefficient can be related to the following measurable quantities: α the linear expansion coefficient, the compressibility κ_T and C_V. Taking the derivative of the Gruneisen EOS with respect to T for constant V, and using a thermodynamic identity one gets:

$$\alpha = (1/3V)(\partial V/\partial T)_P$$

$$\kappa_T = -(1/V)(\partial V/\partial P)_T$$

$$(\partial P/\partial T)_V = \gamma C_V/V$$

$$(\partial V/\partial T)_P = -(\partial V/\partial P)_T(\partial P/\partial T)_V$$

$$\Rightarrow \gamma = \frac{3\alpha V}{\kappa_T C_V} \tag{3.28}$$

The quantities on the right hand side can be measured experimentally. For example one has $\gamma(\rho_0) = 2.17$ for Al, $\gamma(\rho_0) = 1.60$ for Fe, $\gamma(\rho_0) = 1.96$ for Cu.

3.6.4 Slater-Landau Calculation of γ

Using the theory of elasticity the sound velocities are function of density:

$$c_L = \left[\frac{3(1-\sigma)}{\kappa_T \rho (1+\sigma)}\right]^{1/2} ; c_t = \left[\frac{3(1-2\sigma)}{2\kappa_T \rho (1+\sigma)}\right]^{1/2}$$

$$\sigma = \text{Poisson ratio} = -(\delta y/y_0)/(\delta x/x_0) \tag{3.29}$$

where x_0 and y_0 are the initial length and thickness of the sample accordingly. Assuming that the Poisson ratio is independent of the specific volume V and using the equations (3.23) and one gets $v_D = \text{const.}V^{1/6}\kappa_T^{-1/2}$, implying:

$$\gamma = \frac{d\ln v_D}{d\ln V} \Rightarrow \gamma = -\frac{2}{3} - \frac{1}{2}V\frac{(\partial^2 P/\partial V^2)_T}{(\partial P/\partial V)_T} \tag{3.30}$$

Since this relation is valid for every temperature T, it is convenient to take $T = 0$, where $P = P_c$. If P_c is known then $\gamma(V)$ can be calculated. Furthermore, Eq. 3.30 for $P = P_c$ is known as the Landau Slater differential equation for γ.

3.7 $(u_s - u_p)$EOS

It was found experimentally [6] that for many solid materials, initially at rest, the following linear relation between the shock velocity u_s and the particle velocity u_p is valid to a very good approximation:

$$u_s = c_0 + s u_p \tag{3.31}$$

The values of c_0 and s for some elements are given in Table 3.3. This equation is considered an EOS on the Hugoniot since one has the following mass and momentum conservation:

$$\left.\begin{aligned} \rho_0 u_s &= \rho(u_s - u_p) \\ P_H &= \rho_0 u_s u_p \end{aligned}\right\} \Rightarrow \left\{\begin{aligned} u_p &= (P_H/\rho_0)^{1/2}\left(\frac{1}{\rho_0} - \frac{1}{\rho}\right)^{1/2} \\ u_s &= (P_H/\rho_0)^{1/2}\left(1 - \frac{\rho_0}{\rho}\right)^{-1/2} \end{aligned}\right. \tag{3.32}$$

Substituting Eq. 3.32 into Eq. 3.31 one gets the following EOS on the Hugoniot i.e. the pressure P and density ρ are on the Hugoniot curve:

Table 3.3 The experimental fit to $u_s = c_0 + s u_p$ on the Hugoniot curve

	Element	Z	ρ_0 [g/cm^3]	c_0 [cm/μs]	s
Li	Lithium	3	0.534	0.477	1.066
Be	Beryllium (S200)	4	1.85	0.800	1.124
Mg	Magnesium	12	1.78	0.452	1.242
Al	Aluminium (6061-T6)	13	2.703	0.524	1.40
Ni	Nickel	28	8.90	0.465	1.445
Cu	Copper	29	8.93	0.394	1.489
Zn	Zinc	30	7.139	0.303	1.55
Nb	Niobium	41	8.59	0.444	1.207
Mo	Molybdenum	42	10.2	0.5143	1.255
Ag	Silver	47	10.49	0.327	1.55
Sn	Tin	50	7.287	0.259	1.49
Ta	Tantalum	73	16.69	0.341	1.2
W	Tungsten	74	19.3	0.403	1.237
Pt	Platinum	78	21.44	0.364	1.54
Au	Gold	79	19.3	0.308	1.56
Th	Thorium	90	11.7	0.213	1.278
U	Uranium	92	19.05	0.248	1.53

ρ_0 is the initial density of the element with an atomic number Z

$$P = \left(\rho c_0^2\right)\left[\frac{(\rho/\rho_0 - 1)}{(s-1)^2\left(\frac{s}{s-1} - \frac{\rho}{\rho_0}\right)^2}\right]$$

$$P \equiv P_H \to \infty \Rightarrow (\rho/\rho_0)_{\max} = \frac{s}{s-1} \tag{3.33}$$

For example $s = 1.4$ for Al implying a maximum shock wave compression of $(\rho\rho_0)_{max} = 3.5$.

It is convenient to describe the Hugoniot curve in the pressure-particle speed space, $P - u_p$. In particular, for the equation of state the Hugoniot is a parabola. When the shock wave reaches the back surface of the solid target, the free surface starts moving (into the vacuum or the surrounding atmosphere) with a velocity u_{FS} and a release wave, in the form of a rarefaction wave, is back-scattered into the medium. Note that if the target is positioned in vacuum, then the pressure of the back surface (denoted in the literature as the free surface) is zero, a boundary value fixed by the vacuum. If an atmosphere surrounds the target, then a shock wave will run into this atmosphere. In our analysis we do not consider this effect and take $P = 0$ at the free surface. This approximation is justified for analysing the high pressure shocked target that is considered here. If the target A is bounded by another solid target B, see Fig. 3.3a, then a shock wave passes from A into B and a wave is back-scattered (into A). The impedances $Z = \rho_0 u_s$ of A and B are responsible for the

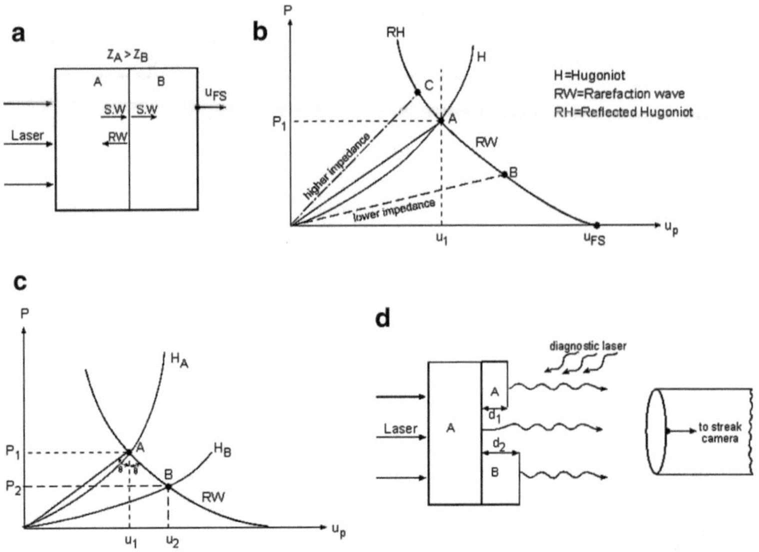

Fig. 3.3 Laser induced shock waves (**a**) into target *A* with impedance smaller than target *B*. (**b**) The Hugoniot (*H*) P-up curve, the rarefaction wave (*RW*) and the reflected Hugoniot (*RH*). (**c**) The Hugoniot curves HA and HB for shock waves in targets A and B. (**d**) A schematic setup for an impedance matching experiment (Adapted from Eliezer [7])

character of this reflected wave. If $Z_A > Z_B$ then a rarefaction wave is back-scattered (into A) while in the $Z_A < Z_B$ case a shock wave is back-scattered at the interface between A and B. Note that in both cases a shock wave goes through (into medium B). These possibilities are shown schematically in Fig. 3.3b. The main laser beam creates a shock wave. The Hugoniot of A is denoted by H, and point A describes the pressure and particle flow velocity of the shock wave (just) before reaching the interface between the targets. If $Z_A > Z_B$ then the lower impedance line (the line $P = Zu_p$) meets the rarefaction wave (RW) curve at point B, while point C describes the case $Z_A < Z_B$ (a higher impedance) where at the interface a shock wave is back-scattered. The final pressure and final flow velocity (just) after the wave passes the interface is determined by point C for the higher impedance ($Z_B > Z_A$) and by point B for the lower impedance ($Z_B < Z_A$). This later case is shown in detail in Fig. 3.3c.

If the impedances of A and B are not very different, impedance matching, then to a very good approximation the RW curve in Fig. 3.3c and RH-RW curve in Fig. 3.3b are the mirror reflection (with respect to the vertical line at u_1 = constant) of the Hugoniot H_A and H curves accordingly. For Fig. 3.3c one has:

$$Z \equiv \rho_0 u_s$$

$$P_1 = Z_A u_1 = \rho_{0A} u_{sA} \mathrm{u}_1$$

$$u_{sA} = c_{0A} + s_A u_1$$

$$P_2 = Z_B u_2$$

$$\tan\theta = \frac{u_2 - u_1}{P_2 - P_1} = \frac{u_1}{P_1} \Rightarrow \frac{P_2}{P_1} = \frac{2Z_B}{Z_A + Z_B} \approx \frac{2\rho_{0B}c_{0B}}{\rho_{0A}c + \rho_{0B}c_{0B}} \tag{3.34}$$

where the last approximate equality is for weak shock waves. A similar result is obtained in the case with higher impedance ($Z_B > Z_A$).

In Fig. 3.3d, a schematic setup of an impedance matching experiment is given. When a shock wave reaches the interface with the vacuum it irradiates according to the temperature of the shock wave heated medium. If the shock wave temperature is high ($\approx$ few thousands degrees K) then the self-illumination may be large enough to be detected by a streak camera (or other appropriate optical collecting device with a fast information recording). If the detecting devices are not sensitive to the self-illumination then the measurement of a reflected (diagnostic) laser may be more useful, since the reflection changes significantly with the arrival of the shock wave. The shock wave velocities in A and B are directly measured in this way by recording the signal of shock breakthrough from the base of A, and from the external surfaces of the stepped targets. The time t_1 that the shock wave travels through a distance d_1 in A and the time t_2 that the shock wave travels through a distance d_2 in B yields the appropriate shock velocities in both targets, namely $u_{sA} = d_1/t_1$ and $u_{sB} = d_2/t_2$. Since the initial densities are known, the impedances of A and B are directly measured: $Z_A = \rho_{0A}u_{sA}$ and $Z_B = \rho_{0B}u_{sB}$. Using Eq. 3.34, P_1 is known (from the measurement of u_{sA} and using the $u_s - u_p$ EOS, where ρ_{0A} and s_A are known, to calculate u_1), and P_2 is directly calculated from the measurements of both impedances. In this way, the difficult task of measuring two parameters in the unknown (equation of state) material B is avoided.

As it has been shown it is quite straightforward to measure the shock wave velocity (assuming that the shock wave is steady, one-dimensional and the measurement device is very accurate). It is also possible to measure indirectly the particle flow velocity by measuring the free surface velocity. Accurate optical devices, called VISAR = Velocity Interferometer System for Any Reflector and ORVIS = Optically Recording Velocity Interferometer System (practically, very fast recording 'radar' devices in the optical spectrum), have been developed to measure accurately the free surface velocity. After the shock wave reaches the back surface of the target a release wave with the characteristics of a rarefaction wave is back scattered into the target. Since this isentrope is almost the mirror image of the Hugoniot (the 'mirror' is at u_1 =constant) one gets $u_1 = u_{FS}/2$. Therefore the measurement of u_s and u_{FS} determines all the parameters in the compressed medium (assuming the initial state is accurately known).

The highest experimental pressure $P \approx 10^9$ bars in the laboratory has been achieved during the collision of a target with an accelerating foil. In 1994 in an indirect drive experiment at Livermore a pressure of about 1 Gbar was created by accelerating a foil with soft x rays from the indirect drive [9], while in 2005 at Osaka in Japan [19] the 1 Gbar pressure was derived by the impact of a foil accelerated directly by the laser drive.

The flyer has a known (i.e. measured experimentally) initial velocity before impact, u_f. The initial state before collision for the target B is $u_p = 0$ and $P = 0$, while for the flyer A it is $u_p = u_f$ and $P = 0$. Upon impact, a shock wave moves forward into B, and another shock wave goes into the flyer in the opposite direction. The pressure and the particle velocity are continuous at the interface of target-flyer. Therefore, the particle velocity of the target changes from zero to u, while the particle velocity in the flyer changes from u_f to u. Moreover, the pressure in the flyer plate A equals the pressure in the target plate B, and if the equations of states are known and given by Eq. 3.30 and the second equation is the pressure from the Hugoniot relations $P_H = \rho_0 u_s u_p$, one gets:

$$P_H = \rho_{OB} u \left(c_{OB} + s_B u\right) = \rho_{OA} (u_f - u) \left[c_{OA} + s_A \left(u_f - u\right)\right] \tag{3.35}$$

This is a quadratic equation in u, with the following solution:

$$\mathrm{u} = \frac{-\mathrm{b} - \sqrt{\mathrm{b}^2 - 4ac}}{2a} \left\} \begin{array}{l} a \equiv \rho_{OA} s_A - \rho_{OB} s_B \\ b \equiv -\left(\rho_{OA} c_{OA} + \rho_{OB} c_{OB} + 2\rho_{OA} s_A u_f\right) \\ c \equiv \left(\rho_{OA} c_{OA} + \rho_{OA} s_A u_f\right) u_f \end{array} \right. \tag{3.36}$$

From the knowledge of u we derive the pressure using Eq. 3.34. Note that if the targets are identical, namely A=B then $u = u_f/2$. If the equation of state of the target B is not known, then it is necessary to measure the shock wave velocity u_{sB} as explained above. In this case, the pressure equality in the flyer and the target yields (Note that in this equation u_{sB} is known):

$$\begin{aligned} P_H &= \rho_{OB} u_{sB} u = \rho_{OA} \left[c_{OA} + s_A \left(u_f - u\right)\right] \left(u_f - u\right) \\ u &= u_f + w - \left[w^2 + \left(\frac{\rho_{OB}}{\rho_{OA} s_A}\right) u_{sB} u_f\right]^{1/2} ; w \equiv \frac{1}{2 s_A} \left(c_{OA} + \frac{\rho_{OB} u_{sB}}{\rho_{OA}}\right) \end{aligned} \tag{3.37}$$

In these types of experiments it is occasionally convenient to measure the free surface velocity of the target and to study also the dynamic strength of materials including spall.

Spall is a dynamic fracture of materials, extensively studied in ballistic research. The term spall, as used in shock wave research, is defined as planar separation of material parallel to the wave front as a result of dynamic tension perpendicular to this plane. The reflection of a shock wave pulse from the rear surface (the free surface) of a target causes the appearance of a rarefaction wave into the target. Tension (i.e. negative pressure) is induced within the target by the crossing of two opposite rarefaction waves, one coming from the front surface due to the fall of the input pressure and the second due to reflection of the shock wave from the back surface. If the magnitude and duration of this tension are sufficient then internal rupture, called spall, occurs [26–32].

In a cross section with a spall of an aluminium (6061) target, 100 μm thick is shown in Fig. 3.4. A laser created shock wave in the aluminium target induced the

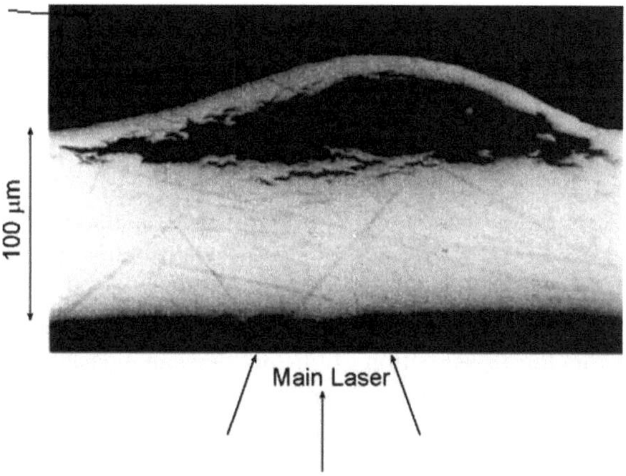

Fig. 3.4 Cross section in an Al 6061 target after the spall creation (Adapted from Eliezer [7])

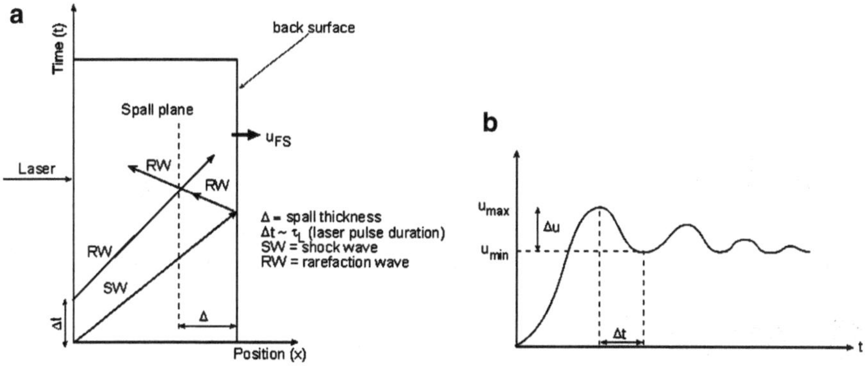

Fig. 3.5 (**a**) Space-time $(x-t)$ diagram of laser induced shock wave (*SW*) and two rarefaction waves (*RW*), one from the back (free) surface and the second from the front surface after the laser pulse ends. At the interface of the two RWs a spall may be created. (**b**) Typical free surface velocity as measured by a VISAR when a spall is created (Adapted from Eliezer [7])

spall. This typical metallurgical cross section was taken after the experiment was finished. In Fig. 3.5a one can see the space-time $(x-t)$ diagram of laser induced shock wave (SW) and two rarefaction waves (RW), one from the back (free) surface and the second from the front surface after the laser pulse ends. At the interface of the two RWs a spall may be created. Figure 3.5b describes a typical free surface velocity as measured by a VISAR when a spall is created. Using the Riemann invariance:

$$\left.\begin{array}{l} u(P=0,u=u_{\min}) \equiv u_{\min} = u'_0 - \int_{\Sigma}^{0} \frac{dP}{\rho c_s} \\ u(P=0,u=u_{\max}) \equiv u_{\max} = u'_0 + \int_{\Sigma}^{0} \frac{dP}{\rho c_s} \end{array}\right\} \Rightarrow \Delta u \equiv u_{\max} - u_{\min} = 2\int_{\Sigma}^{0} \frac{dP}{\rho c_s} \tag{3.38}$$

Assuming that the negative pressure is not too large, then to a good approximation $\rho = \rho_0$ and $c_S = c_0$, implying that the spall strength σ_{spall} is:

$$\Sigma = -\frac{\rho_0 c_0 (u_{\max} - u_{\min})}{2} = -\frac{1}{2}\rho_0 c_0 \Delta u \equiv -\sigma_{spall} \tag{3.39}$$

The strain ε that has been formed at the spall area is defined by $\varepsilon(1D) = \Delta l/l$; $\varepsilon(3D) = \Delta V/V = -\Delta\rho/\rho$, where Δl is the difference between the final and original lengths of the target in one dimension (1D) and l is the original length, while in three dimensions (3D) the strain is defined by the relative change in the volume. From the cross section of Fig. 3.4 one can measure directly the dynamic strain. One of the important parameters, for the different models describing the spall creation, is the strain rate $\dot{\varepsilon} = d\varepsilon/dt$. High power short pulse lasers have been used to create strain rates [30] as high as $5 \times 10^8\,\mathrm{s}^{-1}$. The strain ε and the strain rate $d\varepsilon/dt$ can be approximated by

$$\dot{\varepsilon} = \frac{u_p}{c_0};\; \varepsilon = \frac{d\varepsilon}{dt} \approx \frac{1}{2c_0}\frac{du_{FS}}{dt} \approx \frac{1}{2c_0}\frac{\Delta u}{\Delta t} \tag{3.40}$$

When the shock wave reaches the back surface of the solid target bounded by the vacuum (or the atmosphere) the free surface develops a velocity $u_{FS}(t)$. This velocity is given by the sum of the particle flow velocity u_p and the rarefaction wave velocity U_r. The material velocity increase U_r is given by the Riemann integral along an isentrope from some point on the Hugoniot (pressure P_H) to zero pressure, namely:

$$u_{FS} = u_p + U_r$$
$$U_r = \int_{\rho_0}^{\rho} \frac{c_s d\rho}{\rho} = \int_{V(P_H)}^{V(P=0)} \left(-\frac{dP_S}{dV}\right)^{1/2} dV \tag{3.41}$$

Layers of the target adjacent to the free surface go into motion under the influence of the shock wave transition from V_0, $P_0 = 0$ to V, P_H, and subsequent isentropic expansion in the reflected rarefaction wave from V,P_H to V_2, $P_0 = 0$ where $V_2 > V_0$. Although these two processes are not the same, it turns out that for $u_p \ll u_s$ one has to a very good approximation, $u_p \approx U_r \Rightarrow u_{FS} \approx 2u_p$. It was found experimentally, for many materials, that this relation is very good (within 1 %) up to shock wave pressures of about one mega bar. Therefore, from the free surface velocity measurements, one can calculate the particle flow velocity of the shock wave compressed material. This free surface velocity together with the experimental measurement of the shock wave velocity might serve as the two necessary quantities, out of five (P_H, $V = 1/\rho$, E_H, u_s, u_p), to fix a point on the Hugoniot.

A typical free surface velocity measurement, in the case of the creation of a spall, is given in Fig. 3.5b. u_{max} (related to u_p in the above discussion) in this figure is the maximum free surface velocity. At later times the free surface velocity decreases to u_{min} until a second shock arrives from the spall 'the new free surface'. When a rarefaction wave reaches the internal rupture of the target (the spall) a shock wave is reflected towards the free surface, causing an increase in the free surface velocity. These reverberation phenomena are repeated until the free surface reaches an asymptotic constant velocity.

3.8 Shock Waves in Magnetic Fields

Mega-gauss magnetic fields are easily achieved in laser plasma interactions [7]. These large magnetic fields that are created in the corona have a large pressure, $P_B(\text{Mbar}) \approx 0.04[B(\text{Mgauss})]^2$. The magnetic fields do not penetrate into the solid on the time scale of the shock wave transient. However, if shock waves are created in the corona, or between the critical surface and the ablation surface, then the magnetic pressure, the thermal pressure and the shock wave pressure might be comparable:

$$\beta \equiv \frac{P_T}{P_B} \approx 4\left(\frac{n_e}{10^{20}cm^{-3}}\right)\left(\frac{T_e}{keV}\right)\left(\frac{MGauss}{B}\right)^2 \tag{3.42}$$

For small magnetic fields, $\beta \gg 1$ and the magnetic fields are not important. However, for $\beta \approx 1$ or smaller, a state that is possible to achieve with a few mega-gauss magnetic field, the creation of a shock wave requires the analysis of shock waves in the presence of a magnetic field. Since the magnetic field has in general a direction not parallel to the shock wave velocity, it is necessary to consider the directions normal and parallel to the shock wave front (the discontinuity). In the following the shock wave (front) frame of reference is used. The normal components to the shock front are denoted by a subscript 'n', while the tangential components (i.e. the components parallel to the shock wave surface) have the subscript 't'. The variables before the shock (upstream) and after the shock (downstream) get the subscript 0 and 1 respectively. In this case one has the following jump equations (across the shock wave discontinuity): mass conservation, momentum conservation normal and parallel to the shock front, energy conservation, and continuity (over the shock front) of the normal magnetic induction field B_n and of the parallel electric field E_t. We use the magneto-hydrodynamic equations (in Gaussian units) to derive conservation laws in the form:

$$\begin{gathered}
\frac{\partial}{\partial t}[\mathrm{X}] + \nabla\cdot\Gamma = 0 \\
\lim_{V\to 0}\iiint_V \frac{\partial}{\partial t}[X] = 0 \\
\Rightarrow \iiint_V \nabla\cdot\Gamma dV = \oint\!\!\!\oint_A \Gamma\cdot n dA = 0 \Rightarrow (\Gamma_0 - \Gamma_1)\cdot n = 0
\end{gathered} \tag{3.43}$$

Where V is a volume containing the shock wave singularity, A is the area enclosing the volume V (Gauss divergence theorem) and n is a unit vector perpendicular to the area under consideration. The area A is taken as a small box surrounding the shock-front, where the thickness of the box tends to zero (therefore V goes to zero). In this case only the two faces with directions n and $-n$, on both sides of the shock front, contribute to the integral. Γ_0 and Γ_1 are the values of Γ on both sides of the shock surface (the discontinuity). The end result of these classes of equations is the jump conditions across the shock wave front.

The Maxwell equation $\nabla \cdot \mathbf{B} = 0$, that describes the fact that there are no magnetic poles, gives immediately a jump condition, $B_{0n} = B_{1n}$. Another Maxwell equation $\partial \mathbf{B}/\partial t = -c\nabla \times \mathbf{E}$, zero Lorentz force $(E + v \times B/c = 0)$ and no turbulence yields the equation of the continuity of the tangential component of the electric field in the following form $B_{0t}v_{0n} - v_{0t}B_{0n} = B_{1t}v_{1n} - v_{1t}B_{1n}$.

The mass momentum conservation gives $\rho_0 v_{0n} = \rho_1 v_{1n}$. The momentum jump conditions (two equations) are obtained from the following combined momentum and Maxwell equations. The last equation is the energy conservation equation:

$$\rho d\mathbf{v}/dt = -\nabla P + \mathbf{J} \times \mathbf{B}/c; \ \nabla \times \mathbf{B} = 4\pi \mathbf{J}/c \tag{3.44}$$

$$\partial/\partial t\left[\rho\left(\varepsilon + \mathrm{v}^2/2\right)\right] + \nabla \cdot \left[\rho \mathbf{v}\left(\varepsilon + P/\rho + \mathrm{v}^2/2\right)\right] - \mathbf{E} \cdot \mathbf{J} = 0 \tag{3.45}$$

After some vector and tensor manipulations the two momentum conservation laws and the energy conservation are obtained. Summarising the six jump conditions for a shock wave in a magnetic field yields:

$$\begin{aligned}
&\text{(i)}\ \ \rho_0 \mathrm{v}_{0n} = \rho_1 \mathrm{v}_{1n} \\
&\text{(ii)}\ \ P_0 + \rho_0 \mathrm{v}_{0n}{}^2 + \frac{B_{0t}{}^2}{8\pi} = P_1 + \rho_1 \mathrm{v}_{1n}{}^2 + \frac{B_{1t}{}^2}{8\pi} \\
&\text{(iii)}\ \ \rho_0 \mathrm{v}_{0n}\mathrm{v}_{0t} - \frac{B_{0n}B_{0t}}{4\pi} = \rho_1 \mathrm{v}_{1n}\mathrm{v}_{1t} - \frac{B_{1n}B_{1t}}{4\pi} \\
&\text{(iv)}\ \ \frac{1}{2}\left(\mathrm{v}_{0n}{}^2 + \mathrm{v}_{0t}{}^2\right) + \varepsilon_0 + \frac{P_0}{\rho_0} + \frac{\mathrm{v}_{0n}B_{0t}{}^2 - \mathrm{v}_{0t}B_{0n}B_{0t}}{4\pi\rho_0 \mathrm{v}_{0n}} = \\
&\qquad \frac{1}{2}\left(\mathrm{v}_{1n}{}^2 + \mathrm{v}_{1t}{}^2\right) + \varepsilon_1 + \frac{P_1}{\rho_1} + \frac{\mathrm{v}_{1n}B_{1t}{}^2 - \mathrm{v}_{1t}B_{1n}B_{1t}}{4\pi\rho_1 \mathrm{v}_{1n}} \\
&\text{(v)}\ \ B_{0n} = B_{1n} \\
&\text{(vi)}\ \ \mathrm{v}_{0n}B_{0t} - \mathrm{v}_{0t}B_{0n} = \mathrm{v}_{1n}B_{1t} - \mathrm{v}_{1t}B_{1n}
\end{aligned} \tag{3.46}$$

The first equation is the mass conservation, the second and third equations are the momentum conservation and the fourth equation is the energy conservation. The fifth equation is the continuity of the normal component of B, while the last equation is the continuity of the tangential component of the electric field (note that $E = B \times v/c$).

The shock wave surface is at an angle θ relative to the magnetic induction B, namely, the shock front propagates with an angle θ relative to the magnetic induction B. The variables before the shock and after the shock in the shock wave frame of reference are:

$$\begin{aligned}&\textbf{upstream}: \rho_0, \mathrm{P}_0, E_0; \mathrm{v}_{0n}, \mathrm{v}_{0t}; \mathrm{B}_{0n} = \mathrm{B}\cos\theta, \mathrm{B}_{0t} = B\sin\theta \\ &\textbf{downstream}: \rho_1, \mathrm{P}_1, E_1; \mathrm{v}_{1n}, \mathrm{v}_{1t}; \mathrm{B}_{1n}, \mathrm{B}_{1t}\end{aligned} \tag{3.47}$$

Assuming that the initial conditions are known, one has six equations with seven unknowns. Therefore it is necessary to measure one parameter. For example, if the plasma satisfies the ideal gas equation of state, then a measurement of the temperature behind the shock wave gives the pressure.

3.9 Experiments in Laser Induced Shock Waves

We summarise a few of the major achievements in laser plasma shock wave experiments.

EOS points on the principal Hugoniot of copper up to 20 Mbar and gold and lead up to 10 Mbar have been made with accuracy of 1 % in shock velocity, using the HELEN laser [33]. The experiments were performed in the indirect drive configuration and used the impedance match method. Shock breakout from base and steps was detected by monitoring light emission from the rear surface of the target with optical streak cameras and shock velocities were derived from the transit times across known-height steps.

Absolute measurements of the equation of state of iron at pressures in the range 1–8 Mbar, relevant to planetary physics, were performed with step targets at Luli Laser [34]. The shock velocity and the free surface velocity have been simultaneously measured by self-emission and VISAR diagnostics.

The Hugoniot of tantalum up to pressures of 40 Mbar was measured with the Gekko/Hyper laser [35]. Tantalum is a material typically used in dynamic high pressure studies to study the reflected-shock for a material or projectile. EOS measurements of tantalum are limited up to 10 Mbar by conventional techniques such as gas-gun. The laser induced shock wave measurements were based on the impedance match method, and the shock breakout was detected from the self-emission and the reflection of a probe laser from the rear surface. A radiation pyrometer based on a colour temperature measurement was used as well.

Plastics and dielectric materials play very important roles as shell materials in ICF and their EOS data are needed for target design and analysing the experimental data. Plastics are important in laser induced shock waves experiments, since they are constituents of diverse targets. Unlike metals they are largely transparent to high energy x-rays, i.e. x-rays can be used to backlit relatively thick samples of plastic and provide information on the sample as a function of time.

EOS of dielectric materials, sapphire (Al_2O_3) and lithium fluoride (LiF) up to 20 Mbar was measured using the Omega laser [36] and two line-imaging VISAR. The measured Hugoniot data indicated that the SESAME EOS provides a good description of the EOS of both sapphire and lithium fluoride.

Foams, low-density porous materials have many applications in the physics of high pressures, in particular related to ICF and astrophysics. In laser irradiated foam buffered targets an efficient thermal smoothing of laser energy is achieved. In indirect drive, low density foam placed inside the hohlraum, prevents cavity closure due to the inward motion of the high Z plasma from the wall. In EOS experiments the use of foams enable to reach states of matter with higher temperatures at lower than solid densities. Moreover foams may be used to increase pressures due to impedance mismatch on foam-solid interface. Temperature and shock velocities in $800\,\mathrm{mg/cm^3}$ foams shocked to pressures of few Mbar were performed with the LULI laser [37]. The experiments were based on the impedance matching method with a VISAR. The pyrometry diagnostics for temperature measurements was also used.

In experiments performed with the PALS laser [38] the EOS of lower density foams in the range 60–130 mg/cm^3 up to pressures of 3.6 Mbar was measured. The EOS data was obtained using aluminium as reference material and the shock breakout from double layer Al/foam targets. Samples with different values of initial density were used, enabling the study of a wide region of the phase diagram. Shock acceleration when the shock crossed the Al/foam interface was measured as well. The experimental results showed that Hugoniot of low density foams at high pressures is close to that of a perfect gas with the same density.

To conclude, the EOS research with lasers has become a very important tool at very high pressures, densities and temperatures. Although many new diagnostics have been developed the laser-EOS research it is still lacking the accuracy achieved with gas gun induced shock waves.

References

1. Ya.B. Zel'dovich, Yu.P. Razier, *Physics of Shock Waves and High-Temperature Hydrodynamic Phenomena* (Academic, New York, 1966 and Dover Publications, New York, 2002)
2. S. Eliezer, A. Ghatak, H. Hora, *Fundamentals of Equations of State* (World Scientific, Singapore, 1986)
3. S. Eliezer, A. Ghatak, H. Hora, *An Introduction to Equations of State: Theory and Applications* (Cambridge University Press, Cambridge, 2002)
4. S. Eliezer, R.A. Ricci (eds.), *High Pressure Equations of State: Theory and Applications* (North Holland Pub, Amsterdam, 1991)
5. A.V. Bushman, G.I. Kanel, A.L. Ni, V.E. Fortov, *Intense Dynamic Loading of Condensed Matter* (Taylor and Francis, Washington, DC, 1993)
6. R.G. McQueen, *High Pressure Equations of State: Theory and Applications*, pp 101–216, S. Eliezer, R.A. Ricci (eds.), (North-Holland Pub, Amsterdam, 1991)
7. S. Eliezer, *The Interaction of High-Power Lasers with Plasmas* (Institute of Physics Publishing, Bristol, 2002)

8. C.G.M. van Kessel, R. Sigel, Phys. Rev. Lett. **33**(17), 1020 (1974)
9. R. Cauble, D.W. Phillion, T.J. Hoover et al., Phys. Rev. Lett. **70**(14), 2102 (1993)
10. R. Courant, K.O. Friedrichs, *Supersonic Flow and Shock Waves* (Interscience Pub, New York, 1948)
11. L.D. Landau, E.M. Lifshitz, *Fluid Mechanics* (Pergamon Press, Oxford, 1987)
12. R. Lehmberg, S. Obenschain, Opt. Comm. **46**(1983), 27 (1983)
13. Y. Kato, K. Mima, N. Miyanaga et al., Phys. Rev. Lett. **53**(11), 1057 (1984)
14. S. Skupsky, R.W. Short, T. Kessler et al., J. Appl. Phys. **66**(8), 3456 (1989)
15. M. Koenig, B. Faral, J.M. Boudenne et al., Phys. Rev. E **50**(5), R3314 (1994)
16. D. Batani, S. Bossi, A. Benuzzi et al., Laser Part. Beams **14**(2), 211 (1996)
17. M. Werdiger, B. Arad, Z. Henis et al., Laser Part. Beams **14**(02), 133 (1996)
18. R. Fabbro, B. Faral, J. Virmont et al., Laser Part. Beams **4**(3–4), 413 (1986)
19. H. Azechi, T. Sakaiya, T. Watari et al., Phys. Rev. Lett. **102**(23), 235002 (2009)
20. L. Veeser, J. Solem, A. Lieber, Appl. Phys. Lett. **35**(10), 761 (1979)
21. A. Ng, P. Celliers, D. Parfeniuk, Phys. Rev. Lett. **58**(3), 214 (1987)
22. A.M. Evans, N.J. Freeman, P. Graham et al., Laser Part. Beams **14**(113), 113 (1996)
23. E. Moshe, E. Dekel, Z. Henis, S. Eliezer, Appl. Phys. Lett. **69**, 1379 (1996)
24. P. Celliers, G. Collins, L. Da Silva, D. Gold, R. Cauble, Appl. Phys. Lett. **73**, 1320 (1998)
25. A. Einstein, Ann. Phys. **22**, 180 (1907)
26. I. Gilath, S. Eliezer, M. Dariel, L. Kornblit, Appl. Phys. Lett. **52**(15), 1207 (1988)
27. S. Eliezer, I. Gilath, T. Bar-Noy, J. Appl. Phys. **67**(2), 715 (1990)
28. M. Boustie, F. Cottet, J. Appl. Phys. **69**(11), 7533 (1991)
29. V. Fortov, V. Kostin, S. Eliezer, J. Appl. Phys. **70**(8), 4524 (1991)
30. E. Moshe, S. Eliezer, Z. Henis et al., Appl. Phys. Lett. **76**, 1555 (2000)
31. D. Grady, J Mech. Phys. Solids **36**(3), 353 (1988)
32. E. Dekel, S. Eliezer, Z. Henis et al., J. Appl. Phys. **84**, 4851 (1998)
33. S. Rothman, A. Evans, C. Horsfield, P. Graham, B. Thomas, Phys. Plasmas **9**, 1721 (2002)
34. A. Benuzzi-Mounaix, M. Koenig, G. Huser et al., Phys. Plasmas **9**(6), 2466 (2002)
35. N. Ozaki, K.A. Tanaka, T. Ono et al., Phys. Plasmas **11**(4), 1600 (2004)
36. D.G. Hicks, P.M. Celliers, G.W. Collins, J.H. Eggert, S.J. Moon, Phys. Rev. Lett. **91**(3), 035502 (2003)
37. M. Koenig, A. Benuzzi-Mounaix, D. Batani, T. Hall, W. Nazarov, Phys. Plasmas **12**, 012706 (2005)
38. R. Dezulian, F. Canova, S. Barbanotti et al., Phys. Rev. E **73**(4), 047401 (2006)

Chapter 4
The Effect of a Radiation Field on Excitation and Ionisation in Non-LTE High Energy Density Plasmas

Steven J. Rose

Abstract We look at the direct effect of an ambient radiation field on excitation and ionisation in a non-LTE high energy density plasma. The equations that determine the excitation and ionisation are presented together with a comparison between theory and experiment for a number of cases. In particular we look at so-called photo-ionised plasmas which are also of interest in astrophysics.

4.1 Introduction

In this chapter we consider the effect of an ambient radiation field on the state of bound-electron excitation and ionisation in a non-LTE high energy density plasma. We look at both broad and narrowband radiation fields and at both experiments and theory but restrict our consideration to incoherent radiation thereby excluding any consideration of the interaction between high-power optical, XUV or x-ray laser light and bound electrons which is a separate field that has been covered elsewhere. We shall concentrate our attention on the direct influence that the radiation field has on the population of the different states through processes in which a photon in the radiation field is absorbed or emitted by these states. In experimental situations to provide an ambient (imposed) radiation field, the field can be incident from outside the plasma or can be generated within the plasma. An example of the former case would be the plasma inside a high-power laser-heated hohlraum. Such radiation will heat the plasma, raising, for example, the temperature characterising the free electron distribution. However this heating process is not of direct concern to us here; rather we shall take the temperatures characterising the free electrons and ions (T_e and T_i) from experiment, or from calculations and we shall concentrate

S.J. Rose (✉)
Blackett Laboratory, Imperial College, London SW7 2BZ, UK
e-mail: s.rose@imperial.ac.uk

P. McKenna et al. (eds.), *Laser-Plasma Interactions and Applications*,
Scottish Graduate Series, DOI 10.1007/978-3-319-00038-1_4,

on the direct effect that radiation has on the excitation and ionisation. An example of the latter case would be a laser-heated plasma of a low-Z material where the lines are typically optically thick so the radiation generated by the material also directly influences the excitation and ionisation within the plasma. As before we assume knowledge of the temperatures characterising the free electrons and ions (T_e and T_i) and concentrate on the direct effect that radiation has on the excitation and ionisation. In the first case, because the radiation is imposed there is no need to consider the spatial extent of the material, whereas in the latter case some knowledge of the spatial extent is needed to understand whether the radiation within the plasma interacts with the bound electrons or not.

4.2 Population Kinetics Including a Radiation Field

The rate equation that defines the time evolution of the population of each level α [1] is given by:

$$\frac{dn_\alpha(t)}{dt} = -n_\alpha(t)\sum_\beta\left(R^c_{\alpha\to\beta}+R^r_{\alpha\to\beta}\right)+\sum_\beta n_\beta(t)\left(R^c_{\beta\to\alpha}+R^r_{\beta\to\alpha}\right) \quad (4.1)$$

Here we use the usual notation for these equations in which the number density of ions of a particular level α in the plasma is denoted $n_\alpha(t)$. $R^c_{\alpha\to\beta}$ is the collisional rate for a transition from level α to β (the sum over β involves all states in all ionisation stages that are considered). $R^c_{\alpha\to\beta}$ includes electron collisional excitation and de-excitation, electron collisional ionisation and three-body recombination and auto-ionisation and dielectronic recombination. $R^r_{\alpha\to\beta}$ is the radiative rate from α to β and includes photo-excitation and de-excitation (both spontaneous and stimulated) and photo-ionisation and photo-recombination (also both spontaneous and stimulated). In addition to the rate equations

$$\sum_\alpha Z_\alpha(t)n_\alpha(t) = n_e \quad (4.2)$$

gives the total electron number density where Z_α is the degree of ionisation (the number of ionised electrons arising from the ion in level α) and n_e is the electron number density. In some cases where there is reason to believe that the populations have reached a steady-state then the problem can be simplified by solving Eqs. (4.1) and (4.2) together with

$$\frac{dn_\alpha(t)}{dt} = 0 \quad (4.3)$$

The radiative rates require a knowledge of the radiation intensity $I(\nu)$, which we assume to be independent of position (together with T_e and T_i) throughout the plasma where ν is the photon frequency involved. For photo-excitation, the cross-section for excitation from level α to β is $\sigma_{\alpha\to\beta}$, where in the equations below α and β are levels of the same ionisation stage and level α lies below level β in energy. The photo-excitation rate is given by Mihalas [1] as

$$R^r_{\alpha\to\beta} = 4\pi \int_0^\infty \frac{\sigma_{\alpha\to\beta}(\nu) I(\nu)}{h\nu} d\nu \tag{4.4}$$

For the downward rate [1]

$$R^r_{\beta\to\alpha} = 4\pi \left(\frac{\Omega_\alpha}{\Omega_\beta} e^{U_{\alpha\to\beta}}\right) \int_0^\infty \frac{\sigma_{\alpha\to\beta}(\nu)}{h\nu} \left(\frac{2h\nu^3}{c^2} + I(\nu)\right) e^{-h\nu/kT_e} d\nu \tag{4.5}$$

where Ω_α is the degeneracy of level α,

$$U_{\alpha\to\beta} = \frac{E(\beta) - E(\alpha)}{kT_e} = \frac{\varepsilon_0}{kT_e} \tag{4.6}$$

and $E(\alpha)$ is the energy of level α. For the case of a Planckian radiation field given by

$$I(\nu) = \frac{2h\nu^3}{c^2} \frac{1}{e^{h\nu/kT_r} - 1} \tag{4.7}$$

with a radiation temperature equal to the electron temperature ($T_r = T_e$), then the rates are in detailed balance. The cross section for photo-excitation is related to the oscillator strength $f_{\alpha\to\beta}$ and line shape $\phi_{\alpha\to\beta}(\nu)$ (normalised to unity) by

$$\sigma_{\alpha\to\beta}(\nu) = \frac{\pi e^2}{m_e c} f_{\alpha\to\beta} \phi_{\alpha\to\beta}(\nu) \tag{4.8}$$

For photo-ionisation from level α to β'

$$R^r_{\alpha\to\beta'} = 4\pi \int_0^\infty \frac{\sigma_{\alpha\to\beta'}(\nu) I(\nu)}{h\nu} d\nu \tag{4.9}$$

where $\sigma_{\alpha\to\beta'}(\nu)$ is the photo-ionisation cross-section and level β' denotes a level in the next higher ionisation stage than that of level α. For radiative recombination

$$R^r_{\beta'\to\alpha} = 4\pi \left(\frac{n_e}{2} \left(\frac{h^2}{2\pi m_e k T_e}\right)^{3/2} \frac{\Omega_\alpha}{\Omega_{\beta'}} e^{U_{\alpha\to\beta'}}\right) \times \int_0^\infty \frac{\sigma_{\alpha\to\beta'}(\nu)}{h\nu} \left(\frac{2h\nu^3}{c^2} + I(\nu)\right) e^{-h\nu/kT_e} d\nu \tag{4.10}$$

As with the case of photo-excitation and de-excitation, for a Planckian radiation field given by Eq. (4.7), with $T_r = T_e$, detailed balance between the radiative recombination and photo-ionisation rates is found. The integrals in Eqs. (4.9) and (4.10) need in general to be evaluated numerically, although expansion in special functions can be found for the specific case $\sigma(\nu) \propto \nu^{-3}$, which is an approximation that is often used [1]. However for the photo-excitation / de-excitation integrals in Eqs. (4.4) and (4.5), if we assume that the radiation field does not alter appreciably over the line width (which is usually a good approximation for broadband radiation, but not always for narrowband radiation) then we can take the line shape to be a delta function

$$\phi_{\alpha \to \beta}(\nu) = \delta(\nu - \nu_0) \tag{4.11}$$

and this allows the integrals to be easily evaluated analytically giving:

$$R^r_{\alpha \to \beta} = A_{\alpha \to \beta} n_{ph} \tag{4.12}$$

$$R^r_{\beta \to \alpha} = A_{\beta \to \alpha}(1 + n_{ph}) \tag{4.13}$$

where n_{ph} is the modal photon density covering the transition $\alpha \to \beta$. It is given by, for the case of an external source of photons (as is considered, for example, in the calculations of [2])

$$n_{ph} = \frac{1}{e^{h\nu_0/kT_b} - 1} \tag{4.14}$$

where the brightness temperature of the radiation field at the line centre is kT_b.

For the case of no radiation field, the radiative rates between levels α and β is just the spontaneous radiative decay rate:

$$R^r_{\beta \to \alpha} = A_{\beta \to \alpha} \tag{4.15}$$

and the Einstein A-value given by

$$A_{\beta \to \alpha} = \frac{8\pi^2 e^2 \varepsilon_0^2}{h^2 m c^3} \frac{\Omega_\alpha}{\Omega_\beta} f_{\alpha \to \beta} \tag{4.16}$$

where $f_{\alpha \to \beta}$ is the absorption oscillator strength. The radiative recombination rate is given by (4.10) with $I(\nu)$ set to zero:

$$R^r_{\beta' \to \alpha} = 4\pi \left(\frac{n_e}{2} \left(\frac{h^2}{2\pi m_e k T_e} \right)^{3/2} \frac{\Omega_\alpha}{\Omega_{\beta'}} e^{U_{\alpha \to \beta'}} \right) \int_0^\infty \frac{\sigma_{\alpha \to \beta'}(\nu)}{h\nu} \left(\frac{2h\nu^3}{c^2} \right) e^{-h\nu/kT_e} d\nu \tag{4.17}$$

The terms involving radiation processes all reduce to zero for the case of no radiation field except for the terms involving spontaneous radiative de-excitation or spontaneous radiative recombination given by Eqs. (4.15), (4.16) and (4.17)

The equations above apply whether the radiation field is imposed externally or generated internally; as long as one knows the radiation intensity $I(\nu)$ the equations above determine the excitation and ionisation. However in the latter case the situation is complicated by the fact that the radiation field that determines the excitation and ionisation is generated by the same plasma and so is itself determined by that excitation and ionisation. This presents difficulties that are only properly dealt with theoretically and computationally by fully considering the non-linearity of the system and the reader is directed to books by Mihalas [1], Mihalas and Mihalas [3] and Castor [4] to consider this. However there are two limiting cases which can be dealt with more easily. The first limiting case arises when the radiation field generated by the plasma is to a good approximation not produced by bound-bound or bound-free transitions but rather by free-free transitions from the ionised electrons, in which case the equations above can be used for the excitation and ionisation of the bound electrons. The second limiting case assumes that only the radiation field generated by bound-bound (line) transitions has any effect on the excitation and ionisation. In this case a photon generated by a particular bound-bound transition can excite only that transition in another ion and this simplification allows the use the escape factor approximation to modify the spontaneous decay rate for the bound-bound transition. In terms of the equations we effectively set the radiation field to zero in Eqs. (4.4), (4.5), (4.9) and (4.10) and then modify Eqs. (4.12) and (4.13) to account for the photon trapping in the lines:

$$R^r_{\alpha\to\beta} = 0 \tag{4.18}$$

$$R^r_{\beta\to\alpha} = A_{\beta\to\alpha} g(\tau_0) \tag{4.19}$$

where the escape factor $g(\tau_0) \to 0$ for the line centre optical depth $\tau_0 \to \infty$ and $g(\tau_0) \to 1$ for $\tau_0 \to 0$. The line centre optical depth is given by

$$\tau_0 = \frac{\pi e^2}{mc} f_{\alpha\to\beta}\phi(\nu_0)\left(n_\alpha - \frac{\Omega_\alpha}{\Omega_\beta} n_\beta\right) y \tag{4.20}$$

where y is a characteristic distance in the plasma. The connection of this characteristic distance with the geometry of the plasma and the dependence of g on the plasma geometry is discussed by [5] and [6]. In Eq. (4.20) ϕ is the line profile and the functional form of g is dependent on this profile (discussed in detail by, for example, [1]) If the number density is sufficiently low that the Doppler profile is a good approximation then:

$$\phi(\nu) = \frac{1}{\sqrt{\pi}\Delta\nu_D} e^{-x^2} \tag{4.21}$$

$$x = \frac{\nu - \nu_0}{\Delta\nu_D} \tag{4.22}$$

$$\Delta\nu_D = \sqrt{\frac{2kT_i}{m_i}} \frac{\nu_0}{c} \tag{4.23}$$

It should be noted that the escape factor approximation assumes a uniform plasma.

4.2.1 No or Small Radiation Field

There have been several experiments in which the ionisation distribution has been measured for a non-LTE plasma of high energy density with no (or relatively small) imposed radiation field and where the sample is so small that the radiation field generated within the plasma can be neglected. These include the work of Foord et al. [7, 8], Glenzer et al. [9], Chenais-Popovics et al. [10], Wong et al. [11] and Heeter et al. [12]. Each of these experimental measurements has been analysed by several groups using different codes and generally there is fairly good agreement between theory and experiment. There have been very few experiments in which the ambient radiation field has been imposed and is sufficiently high to significantly alter the distribution of excitation and ionisation. These experiments fall into two categories.

4.2.2 Narrow-Band Photo-Pumping

For the case of narrow-band photo-pumping, the majority of these experiments have had as their aim XUV and x-ray laser action where the mechanism often considered is line coincidence photo-pumping, in which a strong, narrow-band source of x-rays produced by line radiation from one plasma is used to pump directly a resonant (or nearly resonant) transition in a physically separate plasma. The shortest wavelength at which significant gain has been observed is in Be-like C at a wavelength of 2,163 Å [13]. All attempts to produce such gain at shorter wavelength have so far been unsuccessful. However there has been much theoretical work on both the degree to which the two lines coincide (for example [2, 14]) and on the calculation of the operation of the scheme (for example [2, 15, 16]) Possible reasons for this lack of experimental success include the differential Doppler shift between the pumping and pumped plasma and an overestimation of the brightness in the pumping line (possibly due to velocity gradients in the pumping plasma producing much larger effective line widths than have been assumed) Wark et al. [17], Patel et al. [18–20], Beer et al. [21], Almiev et al. [22, 23] discuss these issues. Whilst not resulting in gain, experiments have demonstrated direct photo-pumping of one plasma by another. Direct line coincidence photo-pumping of the He-α resonance line in one aluminium laser-generated plasma by the same line in a separate aluminium plasma has been diagnosed by measurements of enhanced fluorescence [24]. There have also been three experiments in which there is some evidence that line-coincidence photo-pumping involving two different species has resulted in enhanced fluorescence. The first involved the Ly-β line in H-like Al pumping the $2p^6 - 2p^5 3d\,^3D_1$ in neon-like Se [25]. The second is that of a $3d - 2p$ transition in Be-like Mn line pumping the Ly-β line in H-like F [26] and the third involved the Ly-α line of Al pumping the $1s^2 2s - 1s^2 5p$ line in Li-like Fe [27]. In addition to experiments in the laboratory, there is considerable evidence that line-coincidence photo-pumping produces lasing action in the vicinity of the star η-Car [28–30]. This laser operates in the visible region of the spectrum; however there is no evidence for astrophysical lasing action at shorter wavelengths.

4.2.3 Broad-Band Photo-Pumping

The effect of a broad-band radiation field on the distribution of excitation and ionisation has received attention both theoretically and experimentally. The theoretical work was initiated in astrophysical calculations starting with the work of Tarter et al. [31] and Tarter and Salpeter [32] who performed calculations for photo-ionised astrophysical plasmas. There has been considerable work in the astrophysical community since that time culminating in the development of several codes that calculate the distribution of excitation and ionisation for photo-ionised plasmas, such as the code CLOUDY [33]. There have also been a number of calculations relating to experiments undertaken with high-power lasers and pulsed-power machines. For example the effect of the radiation field within a hohlraum on the excitation and ionisation of open L- and M-shell ions in the pusher of an ICF capsule was calculated by Rose [34] using the code GALAXY [35]. Experimentally, broad-band photo-pumping in which radiation from a high-Z plasma pumped He-like Al was observed by Renaudin et al. [36]. Continuum radiation from one filtered high-Z plasma pumping transitions in a separate open L-shell aluminium plasma was observed by Smith et al. [37]. Experiments in which the broadband photopump arises from a collapsed Z-pinch have been described by Bailey et al. [38] and Foord et al. [39]. Foord et al. show comparison between the experimentally measured ionisation distribution and the predictions of a number of models including CLOUDY and GALAXY. In general the agreement is good. In a later paper by Rose et al. [40] the comparison was extended to a simple average-atom model NIMP [41] and good agreement was also found suggesting that it is not the accuracy of the atomic data that was critical to the agreement but rather the completeness of the levels included in the model (Fig. 4.1).

This set of experiments is of particular importance in that, for the first time, a photo-ionised plasma was produced with a value of photo-ionisation parameter ξ [31, 32] which was similar to that believed to occur in an astrophysical photo-ionised plasma. The photo-ionisation parameter characterises the degree to which a broad-band radiation field directly influences the ionisation and comes about from a simplified analysis of the rate equations. Assuming sufficiently low density and sufficiently high radiation field then, for a steady state, the ionisation is dominated by photo-ionisation and the recombination by two-body processes (dielectronic recombination and radiative recombination) rather than three-body recombination. So

$$n_\alpha R^r_{\alpha\to\beta'} = n_{\beta'} n_e \gamma_{\beta'\to\alpha}(T_e) \tag{4.24}$$

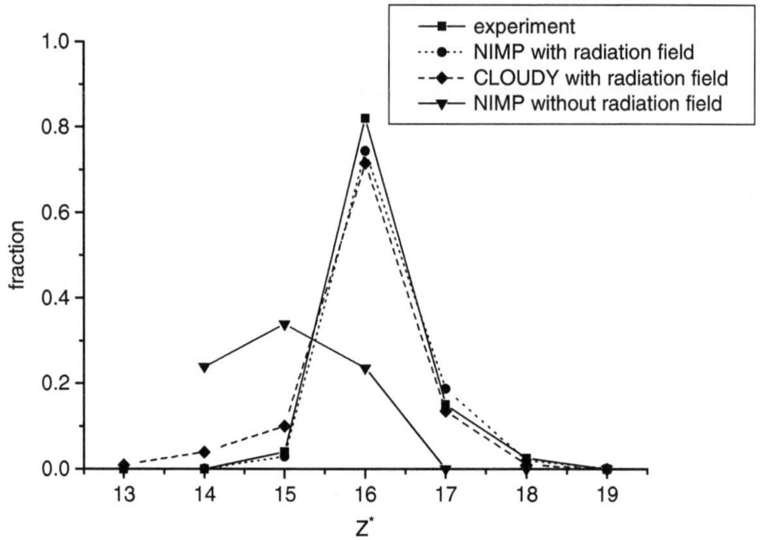

Fig. 4.1 Comparison between the distribution of ionisation fraction for Fe measured in the Z-machine experiment [39] with the model CLOUDY [33] and the average-atom model NIMP [40]. The calculation for NIMP without a radiation field is included to show its effect on the distribution of ionisation. The plasma has an electron number density $n_e = 2 \times 10^{19}\,\mathrm{cm}^{-3}$, an electron temperature $T_e = 150\,\mathrm{eV}$, a radiation temperature $T_r = 165\,\mathrm{eV}$ (although radiation is incident on the plasma from a fraction $\alpha = 0.01$ of 4π steradians). This results in a photo-ionisation parameter $\xi \approx 20$ erg cm s^{-1}

where $R^r_{\alpha\to\beta'}$ is given by Eq. (4.9), $\gamma_{\beta'\to\alpha}(T_e)$ is the sum of the dielectronic and radiative recombination rates and α and β' are the ground states of adjacent ionisation states. We note that the stimulated radiative recombination has been neglected in comparison with the spontaneous radiative recombination and hence $\gamma_{\beta'\to\alpha}$ only has a dependence on T_e and not $I(\nu)$. Consequently the ratio of ion number densities in β' and α is given by

$$\frac{n_{\beta'}}{n_\alpha} = \frac{4\pi \int_0^\infty \sigma_{\alpha\to\beta'}(\nu) I(\nu) d\nu / h\nu}{n_e \gamma_{\beta'\to\alpha}(T_e)} \tag{4.25}$$

Although dependent on the spectral shape of $I(\nu)$, the integral in Eq. (4.25) is proportional to the flux F (erg cm^{-2}s^{-1}) arriving at the plasma from the source, so for fixed spectral shape and fixed T_e

$$\frac{n_{\beta'}}{n_\alpha} \propto \frac{F}{n_e} \tag{4.26}$$

The photo-ionisation parameter ξ (erg cm s^{-1}) is defined [31, 32] as

$$\xi = \frac{4\pi F}{n_e} \tag{4.27}$$

and determines the ionisation in the plasma (for given spectral shape function and electron temperature). Although preliminary experiments had been undertaken by Morita et al. [42] and by Bailey et al. [38], the experiment of Foord et al. [39] was the first experiment to report an ionisation parameter $\xi \sim 20\,\mathrm{erg\,cm\,s^{-1}}$ which is similar to that believed to be found around low-mass x-ray binary stars. Higher values of ξ are believed to exist near massive black holes ($\xi \gtrsim 300\,\mathrm{erg\,cm\,s^{-1}}$) To obtain these values in laboratory experiments requires a higher radiation field and / or lower density than in the experiments of Foord et al. High-power lasers can generally provide higher radiation fields than Z-pinches although the plasma densities involved need to be higher. This is because the plasmas are smaller and higher densities are needed to achieve a steady-state in the disassembly time of the target. A nitrogen plasma with a photo-ionisation parameter $\xi \sim 10\,\mathrm{erg\,cm\,s^{-1}}$ has recently been produced in a laser-driven hohlraum target (H. Takabe and F. Wang, 2007, "private communication"). Fujioka et al. [43] have used a laser-driven implosion to generate a radiation field resulting in a photo-ionisation parameter $\xi \sim 6\,\mathrm{erg\,cm\,s^{-1}}$. The importance of the completeness of the atomic data used in analysis of photo-ionised plasma experiments has been discussed by Hill and Rose [44] and the value of the photo-ionisation parameter alone as a means of determining the relevance of laboratory photo-ionised plasma experiments to astrophysical situations, has been questioned by Hill and Rose [45].

To obtain the most extreme values of the photo-ionisation parameter in the laboratory, target designs are being developed that will use the largest high-power lasers available (such as the National Ignition Facility) However Hill and Rose [46] have proposed that narrow-band radiation can also be used as the ambient radiation field in photo-ionised experiments. These will provide a high value of the radiation field in a target that is large enough to allow a low density plasma to reach a steady-state. It should also be noted that the radiation field within a burning capsule at NIF will provide a very high radiation field and the densities are so high that the excitation and ionisation of a dopant ion at low concentration is predicted to come to a steady-state within the time for disassembly. However although the radiation field is high, the density is also very high and the requirement that the two-body recombination significantly exceeds the three-body recombination cannot be met. Consequently the plasma cannot be characterised as a photo-ionised plasma in the sense originally introduced by Tarter et al. [31] and Tarter and Salpeter [32].

4.3 Conclusions

A series of non-LTE code comparison workshops [47–49] have highlighted the differences between different approximations to solving the Eqs. (4.1), (4.2) and (4.3) These differences come about because of differences in the values of energy levels, in the radiative and collisional rates used and also in the number of levels and ionisation stages considered. In general the differences between models are more pronounced when radiation is included than when it is not, although the number of cases studied is limited. It is apparent that there is a need for more experimental data against which the complex and diverse models used for both laboratory and astrophysical plasmas in the presence of intense radiation can be tested. Over the next few years new high-power lasers, such as the National Ignition Facility will allow experiments to investigate plasmas in which the influence of the radiation field is much more pronounced than in previous experiments and in that way provide more stringent tests of theory.

References

1. D. Mihalas, *Stellar Atmospheres* (W H Freeman and Company, San Francisco, 1978)
2. W.E. Alley et al., J. Quant. Spectrosc. Radiat. Transf. **27**, 257 (1982)
3. D. Mihalas, B. Weibel-Mihalas, *Foundations of Radiation Hydrodynamics* (Dover, Mineola, 1999)
4. J. Castor, *Radiation Hydrodynamics* (Cambridge University Press, New York, 2004)
5. F.M. Kerr et al., Astrophys. J. **629**, 1091 (2005)
6. G. J. Phillips et al. High Energy Density Phys. **4**, 18 (2008)
7. M.E. Foord et al., Phys. Rev. Lett. **85**, 992 (2000)
8. M.E. Foord et al. J. Quant. Spectrosc. Radiat. Transf. **65**, 231 (2000)
9. S.H. Glenzer et al., Phys. Rev. Lett. **87**, 045002 (2001)
10. C. Chenais-Popovics et al., Phys. Rev. E. **65**, 046418 (2002)
11. K.L. Wong et al., Phys. Rev. Lett. **90**, 235001 (2003)
12. R.F. Heeter et al., Phys. Rev. Lett. **99**, 0195001 (2007)
13. N. Qi, M. Krishnan, Phys. Rev. Lett. **59**, 2051 (1987)
14. S.J. Rose, J. Quant. Spectrosc. Radiat. Transf. **33**, 603 (1985)
15. J. Nilsen, J.H. Scofield, E.A. Chandler, Appl. Opt. **31**, 4950 (1992)
16. J. Nilsen, App. Opt. **31**, 4957 (1992)
17. J.S. Wark et al., Phys. Rev. Lett. **72**, 1826 (1994)
18. P.K. Patel et al., J. Quant. Spectrosc. Radiat. Transf. **57**, 683 (1997)
19. P.K. Patel et al., J. Quant. Spectrosc. Radiat. Transf. **58**, 835 (1997)
20. P.K. Patel et al., J. Quant. Spectrosc. Radiat. Transf. **65**, 429 (2000)
21. M.E. Beer et al., J. Quant. Spectrosc. Radiat. Transf. **65**, 71 (2000)
22. A.R. Almiev, S.J. Rose, J.S. Wark, J. Quant. Spectrosc. Radiat. Transf. **70**, 11 (2001)
23. A.R. Almiev, S.J. Rose, J.S. Wark, J. Quant. Spectrosc. Radiat. Transf. **71**, 129 (2001)
24. C.A. Back, C. Chenais-Popovics, R.W. Lee, Phys. Rev. A. **44**, 6730 (1991)
25. P. Monier, C. Chenais-Popovics, J.P. Geindre, J.C. Gauthier, Phys. Rev. **38**, 2508 (1988)
26. D.M. ONeill et al., Inst. Phys. Conf. Ser. No. 116, International Colloquium on x-ray Lasers (1990)
27. A. Gouveia et al., J. Quant. Spectrosc. Radiat. Transf. **81**, 199 (2003)
28. S. Johansson, V. Letokhov, Phys. Rev. Lett. **90**, 011101-1 (2003)
29. S. Johansson, V. Letokhov, MNRAS **364**, 731 (2005)
30. S. Johansson, V. Letokhov, *Astrophysical Lasers* (Oxford University Press, Oxford/New York, 2009)
31. C.B. Tarter, W.H. Tucker, E.E. Salpeter, Astrophys. J. **156**, 943 (1969)
32. C.B. Tarter, E.E. Salpeter, Astrophys. J. **156**, 953 (1969)
33. G.J. Ferland et al., Publ. Astron. Soc. Pac. **110**, 761 (1998)
34. S.J. Rose, J. Quant. Spectrosc. Radiat. Transf. **54**, 333 (1995)
35. S.J. Rose, J. Phys. B: Atom. Molec. Opt. Phys. **31**, 2129 (1998)
36. P. Renaudin et al., Phys. Rev. E **50**, 2186 (1994)

37. C.C. Smith et al., J. Quant. Spectrosc. Radiat. Transf. **54**, 387 (1995)
38. J.E. Bailey et al., J. Quant. Spectrosc. Radiat. Transf. **71**, 151 (2001)
39. M.E. Foord et al., Phys. Rev. Lett. **93**, 055002 (2004)
40. S.J. Rose et al., J. Phys. B: Atom. Molec. Opt. Phys. **37**, L337 (2004)
41. S.J. Rose, Rutherford Appleton Laboratory Report, RAL-TR-97-020 (1997)
42. Y. Morita et al., J. Quant. Spectrosc. Radiat. Transf. **71**, 519 (2001)
43. S. Fujioka et al., Nat. Phys. **5**, 821 (2009)
44. E. Hill, S.J. Rose, High Energy Density Phys. **5**, 302 (2009)
45. E. Hill, S.J. Rose, Phys. Plasmas **17**, 103301 (2010)
46. E. Hill, S.J. Rose, High Energy Density Phys. **7**, 377 (2011)
47. C. Bowen, R.W. Lee, Yu Ralchenko, J. Quant. Spectrosc. Radiat. Transf. **99**, 102 (2006)
48. J.G. Rubiano et al., High Energy Density Phys. **3**, 225 (2007)
49. C.J. Fontes et al., High Energy Density Phys. **5**, 15 (2009)

Chapter 5
Energetic Electron Generation and Transport in Intense Laser-Solid Interactions

Paul McKenna and Mark N. Quinn

Abstract In the interaction of a power laser pulse with a dense target a significant fraction of the laser pulse energy is absorbed to produce an intense beam of energetic (MeV) electrons. The physics of the generation and transport of this large current (multi-mega-Ampere) of fast electrons within the target is of fundamental importance to many topics in high intensity laser-solid interactions, including ion and radiation source development, warm dense matter physics and advanced schemes for inertial fusion energy. A review of the underlying physics governing energetic electron generation and transport in solids is given, together with recent examples of progress in this field of research. Prospects for controlling the transport of energetic electrons are also discussed.

5.1 Introduction

The generation and transport of large (mega-ampere) currents of laser-generated relativistic (MeV) electrons in dense plasma is of fundamental importance to many topics in high power laser-solid interactions. It plays a defining role in controlling the properties of sheath-accelerated ion beams [1], in the production of high energy X-ray sources [2] and in energy deposition to produce states of warm dense matter [3, 4]. It is also central to the success of the fast ignition approach to inertial fusion energy (IFE) [5]. In that scheme, fast electrons are the primary transporter of energy from an ignition laser pulse to the compressed deuterium-tritium fuel. The electrons must be transported over a few hundred microns of dense plasma (with density increasing over four orders of magnitude) without significant divergence or energy loss. Thus the quest to understand the properties of the generation and

P. McKenna (✉) • M.N. Quinn
Department of Physics, SUPA, University of Strathclyde, Glasgow G4 0NG, UK
e-mail: paul.mckenna@strath.ac.uk

P. McKenna et al. (eds.), *Laser-Plasma Interactions and Applications*,
Scottish Graduate Series, DOI 10.1007/978-3-319-00038-1_5,

transport of beams of energetic electrons has motivated a large international effort in this research field.

Understanding the transport of large currents of relativistic (fast) electrons in dense plasma requires knowledge not only of beam properties, but also the physics of collective effects such as self-generated electric and magnetic fields and the influence these have on the beam transport. A population of fast electrons is produced at the focus of the laser pulse and injected into the target. The transport of the electrons is diagnosed using a variety of approaches, typically involving measurement of the escaping electrons and secondary optical, X-ray and ion emission from the target.

In this chapter, the physics of high current generation and transport in solids irradiated by intense laser pulses is outlined, including discussion of collective effects and the influence these have on electron beam propagation. Examples of recent experimental progress are given and new schemes for potentially controlling the transport of high currents of energetic electrons are given. The chapter is not intended as a comprehensive review of the subject, but provides an overview of important physical concepts and a summary of selected recent results to which the authors have contributed.

5.2 Fast Electron Generation

5.2.1 Laser Energy Absorption to Fast Electrons

The physics of the absorption of laser pulse energy by electrons is discussed in detail in Chap. 2 in this text. Here we summarise key features, as shown schematically in Fig. 5.1. Consider a solid target irradiated by a plane polarised intense laser pulse. Evaporation and ionisation of the target surface creates a plasma extending out into vacuum, with a density profile that is typically exponential for a one-dimensional isothermal expansion. At a distance x from the surface of the target with initial density n_{solid}, the plasma density is:

$$n_e(x) = n_{solid} \exp\left(-\frac{x}{L_s}\right) \tag{5.1}$$

The steepness of the profile is described by the density scale length L_s. This is the distance at which the density drops by a factor of $1/e$, where e is Euler's number.

The laser pulse is able to propagate through the increasingly dense plasma up to a critical density, n_c, at which the frequency of the plasma oscillation is equal to the laser drive frequency. At this region the pulse energy is partially absorbed with some fraction reflected. Here plasma electrons oscillating in the electromagnetic field can be accelerated into the overdense target away from the restoring force of the laser. The accelerating force is described by the Lorentz equation:

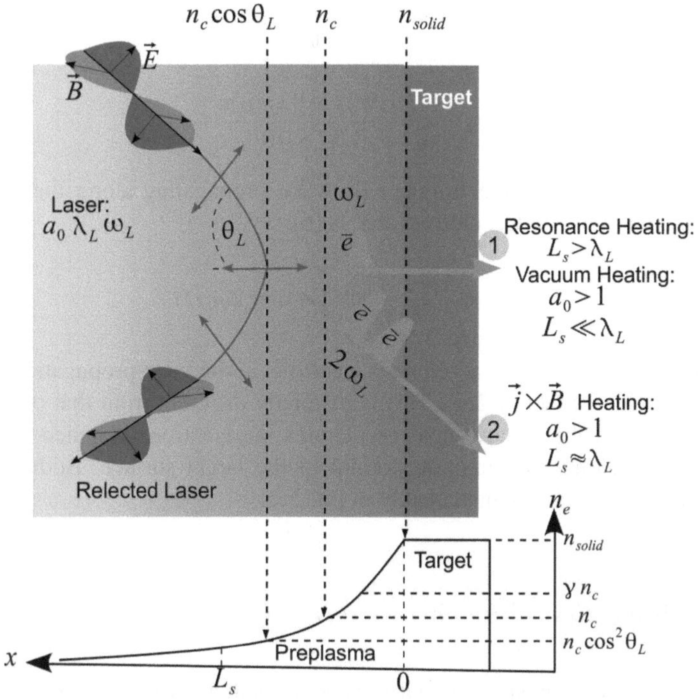

Fig. 5.1 Schematic illustrating resonance and $\mathbf{J} \times \mathbf{B}$ absorption mechanisms in the interaction of a linearly polarised laser pulse with a solid target with preformed plasma density gradient

$$\mathbf{F_L} = \frac{d\mathbf{p}}{dt} = m_e \frac{d(\mathbf{v}\gamma)}{dt} = -e \cdot (\mathbf{E} + \mathbf{v} \times \mathbf{B}) \tag{5.2}$$

where $\mathbf{E}$ and $\mathbf{B}$ are the laser electric and magnetic fields, and $\mathbf{p}$ and $\mathbf{v}$ are the electron momentum and velocity, respectively. Both terms on the right hand side can result in the injection of fast electrons at distinct frequencies and directions into the target. The electric field component, with amplitude E_0, leads to oscillations of the electrons in the electron field direction transverse to the laser propagation with velocity:

$$v_\perp = \frac{eE_0}{m_e \omega_L} \tag{5.3}$$

At high laser intensities the force on the electrons due to the $e(\mathbf{v} \times \mathbf{B})$ component, which accelerates electrons in the direction of laser propagation, becomes comparable to the force resulting from the electric field.

The spatial gradient in intensity across the focused laser pulse results in a relativistic ponderomotive force. Expressed relative to a volume of charge with density n_e, this force component is usually written as $\mathbf{j} \times \mathbf{B}$, where current density $\mathbf{j} = en_e\mathbf{v}$. The regime of relativistic electron motion is reached when the ratio of the classical electron momentum and the relativistic electron momentum is equal to unity. This ratio is termed the normalised laser amplitude, a_0, and can be quantified as:

$$a_0 = \sqrt{\frac{I_L[\mathrm{W/cm^2}]\lambda_L^2[\mu\mathrm{m}^2]}{1.37\times 10^{18}}} \tag{5.4}$$

For a linearly polarised electromagnetic wave propagating along the z-axis, the z-component of the ponderomotive force is [6]:

$$F_p = -\frac{m_e}{4}\frac{d}{dx}v_\perp^2(1-\cos(2\omega_L t)) \tag{5.5}$$

This oscillating $\mathbf{j}\times\mathbf{B}$ force accelerates electrons in the laser propagation direction in bunches with frequency $2\omega_L$. This is distinctly different from that of resonance heating in which the electric field accelerates the electrons bunched at the laser frequency ω_L along the direction normal to the target surface. Both absorption processes are illustrated schematically in Fig. 5.1.

5.2.2 Properties of the Fast Electron Source

The absorption mechanisms summarised above result in the coupling of a significant fraction of the laser energy to the super-heating of a population of plasma electrons to relativistic velocities. The random or stochastic nature of the electron acceleration in the laser field results in fluctuations in their trajectories and hence their acquired energies. The averaging of these single particle distributions over time results in a Maxwellian distribution, with a characteristic fast electron temperature k_BT_f (where k_B is the Boltzmann constant). A single-temperature Maxwellian distribution can be expressed as a function of fast electron energy, E_f, as:

$$f(E_f) = N_f\sqrt{\frac{4E_f}{\pi(k_BT_f)^3}}\exp\left(-\frac{E_f}{k_BT_f}\right) \tag{5.6}$$

where N_f is the total number of fast electrons. Collective effects that influence the overall absorption can result in a departure from the pure single-temperature Maxwellian distribution, producing for example a dual temperature distribution [7,8]. At laser intensities for which k_BT_f approaches or exceeds m_ec^2, the electron energy spectrum is given by the Maxwell-Juttner distribution [9]:

$$f(\gamma) = N_f\frac{\gamma^2\beta}{\frac{k_BT_f}{m_ec^2}K_2(m_ec^2/k_BT_f)}\exp\left(-\gamma/\frac{k_BT_f}{m_ec^2}\right) \tag{5.7}$$

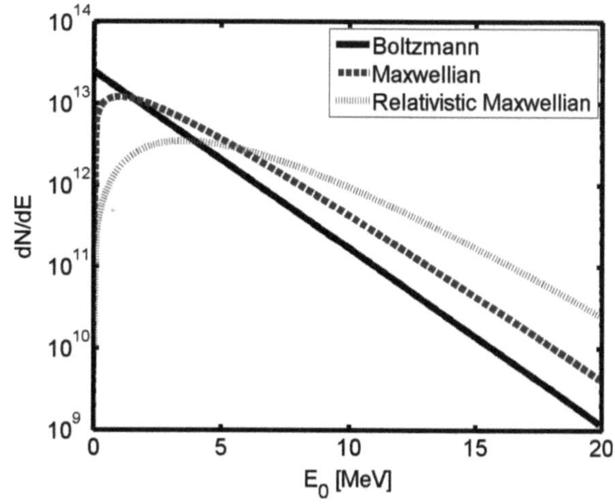

Fig. 5.2 Example fast electron energy distributions (at source), all with $k_B T_f = 2\,\mathrm{MeV}$ and $N_f = 5 \times 10^{13}$

where $\beta = v/c$, $\gamma = \frac{1}{\sqrt{1-\beta^2}}$ and K_2 is a second order Bessel function. The relative shapes of both distributions are shown in Fig. 5.2, and are compared with a standard Boltzmann exponential function for the same N_f (equal to 5×10^{13} in this example). N_f is typically estimated as:

$$N_f = \frac{\eta_{L\to e} E_L}{k_B T_f} \tag{5.8}$$

where $\eta_{L\to e}$ is the laser-to-fast electron energy conversion efficiency and E_L is the laser pulse energy.

$\eta_{L\to e}$ is a function of the absorption processes and therefore laser pulse parameters and target properties such as density scale length at the front surface. Recent examples of investigations aimed at quantifying $\eta_{L\to e}$ include Davies et al. [10], Myatt et al. [11], Nilson et al. [12] and Chen et al. [13]. While the collective measurements exhibit relatively large variations due to the differences in laser pulse parameters, $\eta_{L\to e}$ is typically stated to vary between 0.1 and 0.4 for I_L in the range $10^{18} - 10^{20}\,\mathrm{W/cm^2}$.

The fast electron temperature, $k_B T_f$, is a measure of the mean energy of the initial fast electron population and scales with laser irradiance. Various power law scalings are reported in the literature, depending on the absorption regime accessed. In the relativistic regime the ponderomotive $\mathbf{j} \times \mathbf{B}$ scaling, as determined by Wilks et al. [14], can be applied, which for a linear polarised laser field is:

$$k_B T_f = m_e c^2 \left(\sqrt{1 + a_0^2/2} - 1 \right) \tag{5.9}$$

As it is not possible to directly measure the fast electron energy distribution within solid density targets, the energy spectrum is typically inferred from measurements of secondary particle and radiation emission, in conjunction with modelling.

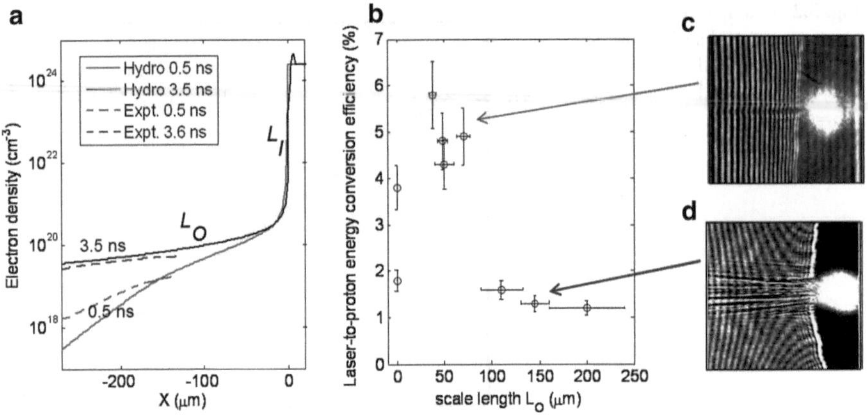

Fig. 5.3 (**a**) Example preplasma density profiles at the front surface of a Cu target for different expansion times (0.5 and 3.5 ns). The experimental measurements (denoted by Expt.) made with an transverse optical probe are limited by diffraction to probing densities of the order of $> 10^{19}$ cm^{-3} and below, and the density scale length in this 'outer' (underdense) region is labelled L_O. Numerical simulations (denoted by Hydro) enable the full density profile to be calculated which shows the steepened scale length L_I in the 'inner' region near the critical density surface. (**b**) Laser-to-proton energy conversion efficiency as a function of L_O, demonstrating an optimum scale length for absorption to electrons. (**c**) Example interferometric probe image showing self-focusing / channeling of the CPA laser beam in relatively short scale-length preplasma corresponding to the case of optimum absorption. (**d**) Example interferometric probe image showing filamentation of the CPA laser beam in longer scale-length preplasma, for which the laser absorption to electrons has decreased. The laser pulses are incident from the *left* in (c, d), and self emission at the critical surface is observed as a bright spot (Adapted from McKenna et al. [15])

5.2.3 *Effects of Front Surface Density Gradient on Absorption*

The sensitivity of laser energy absorption to the plasma density scale length has been investigated experimentally by producing controlled and well-characterised preformed plasma on the front (irradiated) surface of targets and measuring the change in energy coupling to ions and X-ray emission. In the experiment reported in reference [15], controlled 1-D plasma expansion was achieved by irradiating the target with a low intensity laser pulse focused to a relatively large spot size of 500 μm. Phase plates were used to ensure a smooth focal spot distribution. Fast electron generation was driven by a separate high intensity laser pulse, with high intensity-contrast (low background amplified spontaneous emission pedestal), focused to a small spot size, to an intensity of $> 10^{20}$ Wcm^{-2}. Example plasma expansion profiles are shown in Fig. 5.3a (obtained via hydrodynamic simulations and experimental measurements for two different temporal separations (0.5 and 3.5 ns) of the plasma generation and fast electron drive laser pulse).

In this example, measurements of proton acceleration are used to diagnose the laser energy absorption to fast electrons – the protons are accelerated by the sheath field established at the rear surface of the target by the fast electrons (the target normal sheath acceleration, TNSA, mechanism). The total laser-to-proton energy conversion efficiency as a function of the density scale length, L_O, is shown in Fig. 5.3b. An optimum density gradient is observed, at which the laser pulse is found to 'channel' through the underdense plasma, as shown in Fig. 5.3c. At larger scale lengths the laser pulse is observed to filament, Fig. 5.3d, resulting in lower energy coupling to fast electrons and hence protons. Thus the absorption to fast electrons is sensitive not only to the physics occurring in the region of the critical density but also to laser pulse propagation physics in the expanding underdense plasma region.

5.3 Fast Electron Beam Propagation and Stopping

5.3.1 Current Neutrality

An important consideration in the transportation of a large current of relativistic electrons in a solid is the charge neutrality requirement to enable the beam to propagate. The electric field generated in the laser absorption region, i.e. the source of the electrons, is a function of the fast electron current density, $\mathbf{j}_f$, and can be approximated as:

$$\frac{\partial \mathbf{E}}{\partial t} = -\frac{\mathbf{j}_f}{\varepsilon_0} \tag{5.10}$$

If we consider, for example, that a 100 J laser pulse with duration equal to 1 ps produces a fast electron current of the order of 10 MA, the resulting magnitude of the induced electric field is of the order of 10^{15} V/m. Similarly, self-generated magnetic fields, arising from charge separation, act to reverse the electron motion with respect to the initial propagation direction, confining the fast electron current near the laser absorption region and preventing propagation into the target. The maximum current which can propagate is given by the Alfvén limit [16]:

$$I_A \simeq \frac{\beta\gamma m_e c^2}{e} = \beta\gamma 1.7 \times 10^4\,\mathrm{A} \tag{5.11}$$

which equates to $I_A = 65$ kA for these example parameters. Therefore, in order for a multi-mega-Ampere current of relativistic electrons to propagate within the target, local charge neutrality must exist. This is achieved when a much larger return current of slowly moving electrons is drawn from the background to neutralise the fast current density and the inhibiting fields, such that [17]:

$$\mathbf{j}_f + \mathbf{j}_r = 0 \tag{5.12}$$

where $\mathbf{j}_r$ is the return current density. This is shown schematically in Fig. 5.4. Both the fast electron current and the colder return current each have a self-generated magnetic field, although oppositely directed. Although the currents must be spatially

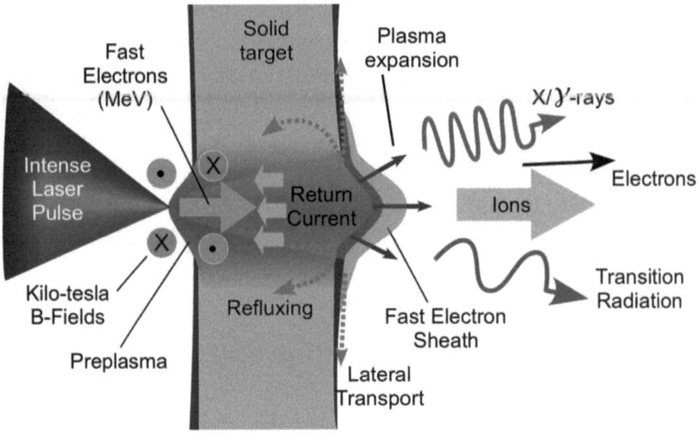

Fig. 5.4 Schematic illustrating fast electron generation at the front side of a solid target, propagation of the fast electrons and the counter-streaming colder 'return' current within the target and the generation of a sheath field giving rise to ion acceleration at the rear side. Self-generated magnetic fields within the target and the lateral transport of fast electrons along the target rear surface are also shown

localised to balance, the B-fields are generally not co-located in space. Charge neutrality occurs over a time given by:

$$\mathrm{t} = \frac{2\pi}{\omega_{pe}} \tag{5.13}$$

where ω_{pe} is the plasma electron frequency, given by:

$$\omega_{pe} = \sqrt{\frac{e^2 n_e}{\varepsilon_o \bar{\gamma} m_e}} \tag{5.14}$$

The charge neutrality time at a density of $\approx 10^{22}\,\mathrm{cm}^{-3}$ in a metallic target is of the order of 0.1 fs. In insulators the charge neutralisation can take longer due to the need to ionise the material to provide the return current. In targets for which there is a restriction in the number of background electrons (e.g. insulators and lower density plasmas) the fast electron propagation into the target can be inhibited by the stronger fields generated. The fast electron transport properties of conductors and insulators are therefore different. Thus it can be seen that the local charge neutrality condition plays an important role in defining the fast electron transport physics in solids.

5.3.2 *Collisions and Heating*

The fast electron beam transport is affected by collisions with the electrons and ions within the target. The fast electrons can directly undergo scattering, either elastic

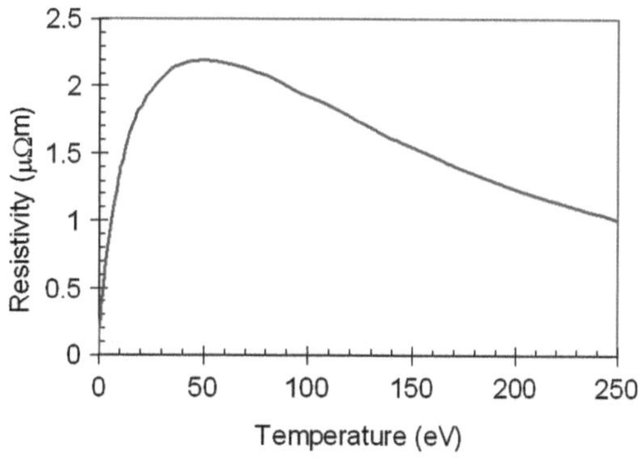

Fig. 5.5 Resistivity of solid aluminium as a function of temperature

or inelastic, and lose energy via excitation, ionisation and radiation emission. The energy loss is quantified by the stopping power, dE/ds, which is a function of the incident electron energy and the material characteristics. The total stopping power is the sum of the individual effects arising from collisions and radiative losses:

$$\left(\frac{dE}{ds}\right)_{total} = \left(\frac{dE}{ds}\right)_{ion} + \left(\frac{dE}{ds}\right)_{rad} \tag{5.15}$$

However, in the case of solid density plasmas for which the number of return-current electrons is much larger than the fast electrons, the vast majority of collisions resulting in target heating occurs indirectly due to the return current. These electrons are moving much more slowly than the fast electrons and thus lose energy much more effectively through scattering. It is these electrons which produce the vast majority of the target heating. The heating changes the resistivity of the target material, as shown in Fig. 5.5. The increase in resistivity of metals at low temperatures (up to 10 eV) is driven by electron collisions. At high temperature the resistivity drops again as the material enters into the Spitzer regime, corresponding to a high degree of ionisation. In cold insulators there are no conduction electrons and thus the resistivity is high. As processes such as field ionisation and fast electron impact ionisation occur the resistivity decreases and the material undergoes a transition to a conductor.

The return current is not only largely responsible for heating the target, but also plays a key role in stopping the fast electrons, even though the fast electrons may undergo few direct collisions. Given that the cold return current not only induces resistivity changes in the target but is also directly affected by the resistivity and considering the condition for charge neutrality, as given by Eq. 5.12, there exists an Ohmic potential and therefore an electric field within the material which acts to transfer energy from the fast electrons to drive the return current. This is often referred to as Ohmic stopping. The precise nature of this stopping mechanism is complex and is a function of the number and mean energy of the fast electrons and the density and resistivity of the background material. The complexity of this self-consistent problem requires sophisticated high performance computational modelling.

5.4 Beam Divergence

The divergence of the beam of fast electrons is of particular interest in terms of optimising beam transport. It defines a number of parameters for the fast ignition approach to IFE, including the maximum size of the electron source, and hence the ignition laser spot size, and the distance between the source and the fuel assembly. It is also a crucial parameter for enhancing the energy of sheath-accelerated ions, which is a function of the beam density of fast electrons at the target rear surface. Accurate knowledge and control of the fast electron beam divergence is therefore important to some of the principal applications of intense laser-solid interactions.

5.4.1 Self-Generated Magnetic Fields

As a current of fast electrons propagates into a solid density plasma a magnetic field, **B**, is generated according to:

$$\frac{\partial \mathbf{B}}{\partial t} = -\nabla \times \mathbf{E} = -\nabla \times (\eta \mathbf{j}_f) \tag{5.16}$$

The magnetic field is a function of the plasma resistivity due to its dependency on the electric field, **E**, required to draw the return current. The **B**-field is azimuthal and grows if perfect current neutralisation does not occur, i.e. $\mathbf{j}_f + \mathbf{j}_r \neq 0$. The growth rate is given by (combining Ohm's law, $\mathbf{E} = -\eta \mathbf{j}_f$, with Faraday's law):

$$\frac{\partial \mathbf{B}}{\partial t} = \eta(\nabla \times \mathbf{j}_f) + \nabla\eta \times \mathbf{j}_f \tag{5.17}$$

Thus the magnetic field develops in regions with spatial variations in current density or resistivity, and in the case of a cylindrically symmetric electron beam the gradients are in the radial direction, giving rise to azimuthal magnetic fields. The first term on the right hand side gives rise to a **B**-field which acts to push fast electrons to regions of higher current density. As the beam density is highest on-axis, the resulting radial force has a collimating or pinching effect on the fast electron beam, as shown schematically in Fig. 5.6. The second term arises from resistivity gradients in the background plasma. Such resistivity gradients arise due to gradients in heating induced by the density gradient in the fast electron beam and therefore spatially localised return current. The direction of the **B**-field resulting from this second term depends on the shape of the material-specific resistivity curve. If the material is less resistive at the higher temperatures existing along the beam propagation axis then the net result is a magnetic field which acts to push electrons away from the axis, in effect giving rise to beam hollowing [18].

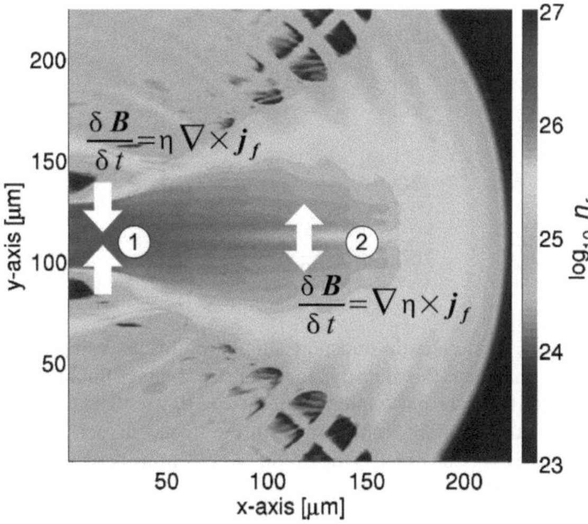

Fig. 5.6 Two-dimensional map of fast electron density at a fixed time in a plastic target, derived from simulations. The fast electrons propagate from *left* to *right*. The effects of both magnetic field components in Eq. 5.17 on the fast electron density is shown: (*1*) pinching at the injection region ($\nabla\eta = 0$) and (*2*) beam hollowing deep within the target where the temperature is relatively cool ($\nabla\eta = \max$)

If the resistivity change is very small over the temperature gradient then this effect is small. Experimental evidence of beam hollowing in plastic targets has been reported [19].

5.4.2 Beam Divergence in Homogeneous Targets

The effects of self-generated **B**-fields and their potential to lead to fast electron beam collimation has been studied in some detail by Bell and Kingham [20]. The strength of the field can be estimated as:

$$\partial\mathbf{B} \approx \frac{\eta \mathbf{j}_f \partial t}{r_f} \tag{5.18}$$

where r_f is the fast electron beam radius. The field strength exceeds 1 kT at the electron beam source, where the beam diameter is typically $\approx 5\,\mu$m and $\partial t = \tau_L \approx 1$ ps (typical laser pulse duration). As the beam spreads within the target the field strength quickly decreases, as shown in the hybrid simulation result in Fig. 5.7. According to the formalism by Bell and Kingham [20], collimation is considered to occur if the **B**-field deflects the fast electrons through an angle $\theta_{1/2}$ over the distance $r_f/\theta_{1/2}$ in which the beam radius doubles. For small $\theta_{1/2}$ the ratio of the fast electron beam radius to the gyroradius, r_g, should satisfy the condition:

$$\frac{r_f}{r_g} > \theta_{1/2}^2 \tag{5.19}$$

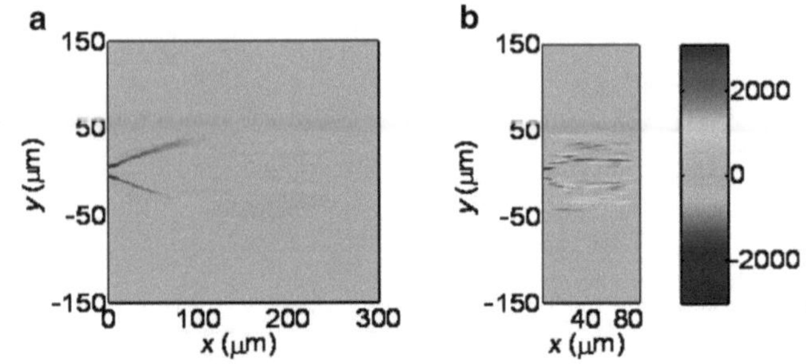

Fig. 5.7 Fast-electron transport simulation results using the LEDA code: two-dimensional map of the magnetic flux density (in Tesla) for a target thickness of (**a**) 300 μm and (**b**) 80 μm, at 1.4 ps after the laser pulse interaction (at $x = 0$, $y = 0$). An azimuthal B-field is observed to a depth of 100 μm from the front surface in the thicker target. The field is perturbed in the case of the thinner target due to refluxing of the fast electrons between the target surfaces (Adapted from Yuan et al. [21])

where

$$r_g = \frac{m_e \mathbf{v}_f}{e\mathbf{B}} \tag{5.20}$$

Thus:

$$\frac{r_f}{r_g} = \frac{r_f e\mathbf{B}}{\gamma m_e \mathbf{v}_f} > \theta_{1/2}^2 \tag{5.21}$$

Collimation is therefore considered to occur if $\Gamma > 1$, where:

$$\Gamma = \frac{r_f e\mathbf{B}}{\gamma m_e \mathbf{v}_f \theta_{1/2}^2} \tag{5.22}$$

Substitution of typical electron beam parameters shows that a **B**-field strength of the order of a kilo-Tesla is required for collimation, which can only realistically occur at the electron source where the beam density is highest.

There is no direct experimental evidence of fast electron beam collimation in homogeneous solid density targets. However, there is some experimental evidence that self-generated magnetic fields are likely to pinch the electron beam, acting to restrict transverse spreading within the target. By investigating proton acceleration from the rear surface of solid targets ranging in thickness from 50 μm to 1.5 mm, Yuan et al. [21] conclude that higher than expected maximum proton energies measured with thick targets are likely to result from an increased fast electron beam density at the target rear surface, arising from a **B**-field pinching effect on fast electron transport within the target. The fact that this is only observed in relatively thick targets is attributed to the refluxing of fast electrons within the target. This effect, which is described later in this chapter, acts to curb the growth of the self-generated **B**-field, by destroying the field distribution, as shown in Fig. 5.7.

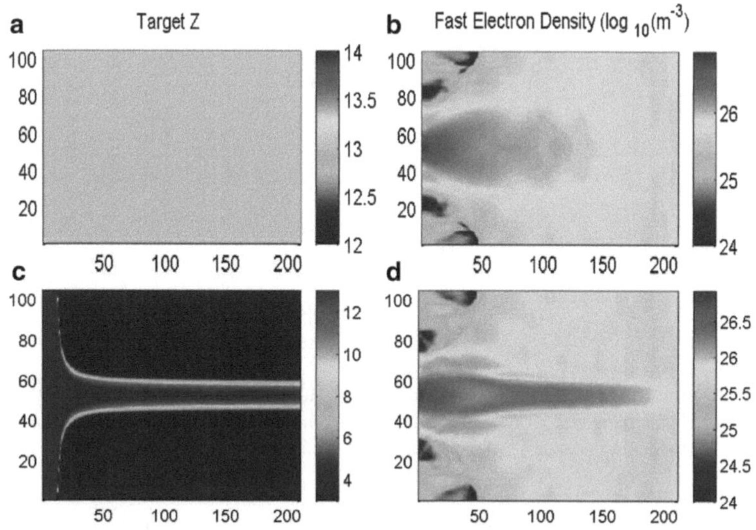

Fig. 5.8 Hybrid-PIC simulation results of fast electron transport. Comparison of fast electron divergence in the case of a homogeneous foil target (**a**, **b**) and a 'structured collimator' target (**c**, **d**). In both cases the target Z-number (a, c) and fast electron density at 750 fs (b, d) are shown (Courtesy of Dr. A.P.L. Robinson, Rutherford Appleton Laboratory; Adapted from Robinson and Sherlock [22])

5.4.3 Controlling Beam Divergence

Recently a scheme for collimating fast electron transport in solids using magnetic fields has been introduced and experimentally demonstrated. The scheme, called the 'structured collimator', was proposed by Robinson and Sherlock [22] and involves engineering the target to generate azimuthal magnetic fields at the boundary between two materials of differing resistivity. By producing a controlled resistivity gradient in a target at the interface of two different materials, the second term in Eq. 5.17 is exploited. If the target is engineered to consist of a fiber of high resistivity material, surrounded by lower resistivity material, then a magnetic field is generated that acts to collimate the fast electron transport along the fiber. Example hybrid-PIC simulation results illustrating the magnetic field generation within the target and the resulting effect of the fast electron beam are shown in Fig. 5.8.

Experimental demonstrations of this scheme have been achieved, involving 1-D collimation in a thin high-resistive tin layer sandwiched between two low resistivity aluminium slabs [23] and 2-D collimation in a high-resistivity-core low-resistivity-cladding structure analogous to optical waveguides [24]. The fast electron beam was shown to be confined to a diameter of 50 μm, over a transport distance of 250 μm. Figure 5.9 shows an example measurement of the lateral extent of the fast electron beam at the target rear surface, which is almost a factor of 2 smaller in the case of the structured collimator compared a homogeneous target.

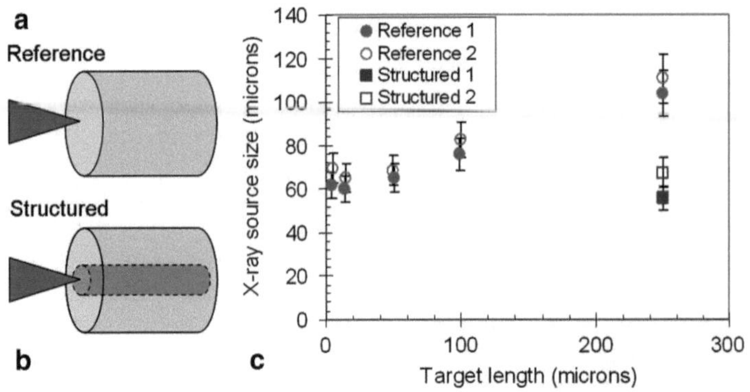

Fig. 5.9 A comparison of fast electron transport in (**a**) plain 'reference' and (**b**) 'structured' targets. (**c**) Measurements of x-ray source size are shown using pinhole camera data (*circles*) and spectral line broadening (*squares*) (Adapted from Ramakrishna et al. [24])

The 'structured collimator' idea has recently been extended in the form of a complex array of resistivity gradients, using alternating layers of different Z materials, to focus fast electrons in cases in which the source is highly divergent. This scheme has been investigated numerically and is described in reference [25].

Other new schemes involving engineering the temporal profile of the laser pulse and the use of dual laser pulses to control fast electron beam divergence have also recently been demonstrated [26].

5.5 Transport Instabilities and Filamentation

As discussed above, the availability of a neutralising return current plays an important role in defining the physics of fast electron beam transport. As the beam propagates it is subject to a number of different types of instability, including both longitudinal and transverse modes, which can cause it to filament, resulting in increased energy losses and changes to the electron beam angular divergence. In the context of fast ignition, these instabilities could critically affect the efficient delivery of energy from the ignitor laser pulse to the fuel. In this section, some of the main types of instabilities are outlined, together with some recent experimental results on the effect of target parameters on beam filamentation.

5.5.1 Beam Filamentation Mechanisms

Two main types of transport instabilities occur where beams of electrons undergo counter-streaming in plasma. These are the electromagnetic Weibel-like filamentation and the electrostatic two-stream instability. The latter is characterized by a growth rate lower than the plasma collision frequency for high density plasma and hence will not be considered in detail here.

The collisionless Weibel instability is a transverse instability mode, produced by magnetic repulsion between counter-propagating currents. Any modulation in the fast electron beam density profile gives rise to localised magnetic field generation which acts to enforce pinching of the beam into filaments, resulting in beam breakup. This type of instability can be seeded by the counter-propagation of the return current to the fast electron beam. Any small fluctuation in the magnetic field acts on the electrons via the Lorentz force, leading to inhomogeneities in the current density distribution. This in turn positively reinforces the modulations in the magnetic field, and thus the current density modulations evolve into transverse filaments.

The timescale over which these filaments form is on the order of the beam plasma frequency ω_f and the spatial scale of the order of the plasma skin depth $c\omega_f$. A critical factor governing the growth rate of Weibel instabilities is the ratio of the beam density n_f to the background plasma density n_e [27]:

$$\Gamma_w = \omega_f \left(\frac{n_f}{\gamma n_e} \right)^{\frac{1}{2}} \times \frac{v_f}{c} \, [s]^{-1} \tag{5.23}$$

Since the fast electron beam density is $\approx 0.01 n_e$, the growth of the Weibel instability occurs on a timescale of Γ_w^{-1}, which equates to a propagation length of the order of 10–100 μm. In the context of fast ignition physics, this transverse instability mechanism could be a problem in the region of the low density corona plasma.

Gremillet et al. [28] has shown that there is a transverse beam temperature dependence of the Weibel instability growth rate. The instability will grow as long as the local pinching force created by the modulations in the magnetic field is greater than the outward transverse forces created by the transverse beam temperature. Thus it follows that if the transverse beam temperature is sufficiently high then the growth of this instability can be sufficiently suppressed, stabilising the fast electron beam propagation.

Resistive filamentation instabilities occur when the return current is collisional. This type of transverse instability is essentially an extension to the Weibel instability, in that it is driven by magnetic field generation arising from counter-propagating electron currents. The magnetic fields are a function of the resistivity of the plasma, as shown in Eq. 5.17. As the plasma resistivity increases the magnetic fields grow stronger, increasing the growth rate of this instability. The typical timescale of the **B**-field generation associated a hot filament of radius r_F scales with the magnetic diffusion time:

$$\tau_r = \frac{\mu_0 r_F^2}{\eta} \tag{5.24}$$

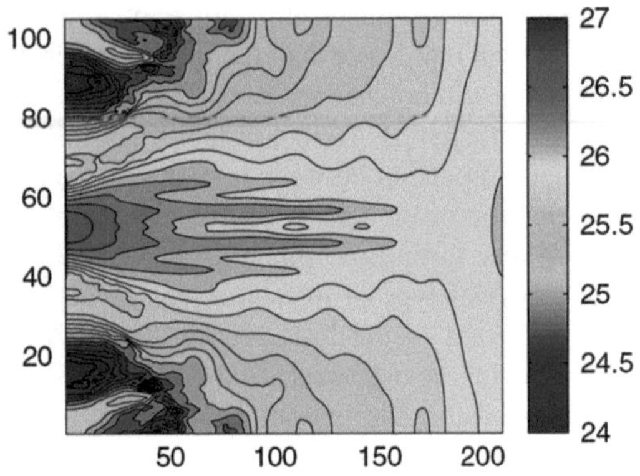

Fig. 5.10 Hybrid simulation of fast electron density in a solid target exhibiting beam filamentation. The beam is propagating from *left* to *right* and filaments are clearly observed deep within the target (Courtesy of Dr. A.P.L. Robinson, Rutherford Appleton Laboratory)

Thus the magnetic field grows faster for smaller filaments. Diffusion times in the range $0.25 - 6\,\text{ps}$ are obtained for filament radii in the range 10–100 μm for a background resistivity equal to $10^{-6}\,\Omega\text{m}$.

Target resistivity also plays an important role in the ionisation instability. As discussed previously, whereas in metals there is a population of free electrons readily available to form a return current, in insulators ionisation must occur to achieve this. As described by Krasheninnikov et al. [29] and Debayle and Tikhonchuk [30], perturbations in the ionisation rate along the fast electron beam front can lead to filamentation of the beam. The strong electrostatic field at the leading edge of the fast electron beam, resulting from charge separation, results in ionisation. The free electrons generated form the return current and produce further ionisation via collisions. Modulations in the local density of the fast electron beam lead to local variations in the ionisation rate and therefore return current formation, which act to corrugate the ionisation front giving rise to instability growth.

In addition, there are macroscopic instabilities, such as the hosing instability. These arise due to the resistive phase difference between the magnetic axis and the beam displacement and can effect the overall fast electron beam divergence.

5.5.2 Recent Experimental Investigations of Transport Instabilities

A number of indirect approaches have been applied to diagnose fast electron beam filamentation in dense plasma. These generally rely on making spatially resolved measurements of coherent transition radiation (CTR) and proton emission from the target rear surface and K_α emission from a buried fluorescent layer. Examples include a comparison of the extent of fast electron beam filamentation in thin foils of different materials [31] and the onset of filamentation in lower density foam targets [27].

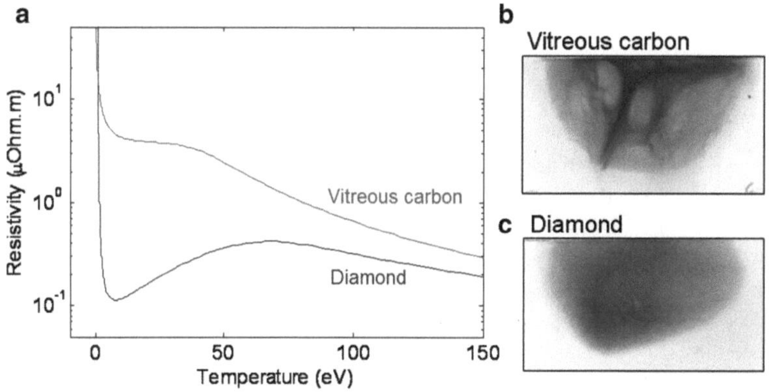

Fig. 5.11 (**a**) Resistivity as a function of temperature for vitreous (disordered) carbon and diamond (highly ordered carbon lattice) calculated using quantum molecular dynamics simulations, as described in [33]. Example measurements of proton beam uniformity from (**b**) vitreous carbon, showing structure induced by fast electron beam filamentation, and (**c**) diamond, showing a smooth proton beam and hence no filamentation

We have recently shown that lattice structure can have a strong influence on fast electron transport instabilities in solids [32, 33]. This experiment involved comparing electron transport patterns in three different forms (allotropes) of the same element, carbon: single-crystal diamond, vitreous carbon, and pyrolytic carbon, and used measurements of the uniformity of proton emission to diagnose electron beam filamentation. Smooth electron transport was observed with diamond, whereas beam filamentation, arising from resistive instabilities, is observed with the less ordered forms of carbon, as shown in Fig. 5.11. The highly ordered lattice structure of diamond is shown to result in a transient state of warm dense carbon with metallic-like conductivity at temperatures of the order of 1–100 eV, leading to suppression of electron beam filamentation. Using quantum molecular dynamics simulations, lattice structure is shown to be important in defining the electric conductivity under these highly non-equilibrium conditions. 3D fast electron transport simulations with the ZEPHYROS particle-based hybrid code, using the calculated conductivities, reveals that this has a defining role on the fast electron beam transport pattern. The comparatively low resistivity of diamond in this transient warm dense matter state results in a lower resistive instability growth rate compared to other forms of carbon.

5.6 Refluxing and Lateral Spreading in Thin Foils

So far, factors influencing the longitudinal transport of the beam of fast electrons from the front to the rear sides of a solid target have been considered. Large currents of fast electrons have also been shown to reflux between the two surfaces and to spread transversely on the sides of thin foil targets.

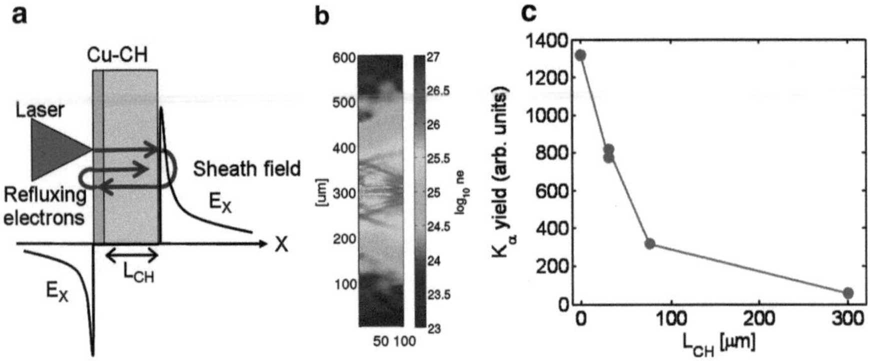

Fig. 5.12 (**a**) Schematic showing fast electron refluxing in a layered Cu-CH target. Electrons are reflected by the sheath fields formed on both surfaces. (**b**) A hybrid simulation result showing a 2-D map of fast electron density within a thin foil target. The electron reflection and spreading within the target is clearly observed. (**c**) Measurements of Cu K_α photon yield as a function of the thickness of a CH transport layer (the layered target geometry is shown in **a**). The significant enhancement in K_α emission as L_{CH} is decreased is attributed to fast electron refluxing

5.6.1 *Refluxing*

After transversing the target to the rear surface, a small fraction of the most energetic electrons escape, establishing a quasi-electrostatic space-charge sheath field which reflects the majority of the fast electrons back into the target. This field is responsible for ionisation and ion acceleration via the TNSA mechanism, as described in Chap. 12. A similar sheath potential forms at the front surface of the target, reflecting fast electrons returning to the front surface. The resulting effect is that the fast electrons reflux (or recirculate) within the target, as described by Sentoku et al. [34]. This generally occurs when the target thickness $L < t_L c/2$, where t_L is the laser pulse duration and the relativistic fast electrons are assumed to have a velocity close to c, the speed of light. The basic idea is summarised in Fig. 5.12. Refluxing has been inferred in a number of experimental investigations of fast electron transport [11, 35–37] and to explain observed enhancements in TNSA-ion beam properties [34, 38].

In a recent experimental study by Quinn et al. [39] significant refluxing of fast electrons is demonstrated to occur in thin foil targets irradiated by intense picosecond laser pulses. By variation of the thickness of the target, it is shown that refluxing electrons contribute significantly to K_α emission from a fluorescence layer at the front surface, as shown in Fig. 5.12. The measured enhancements are consistent with an analytical model of electron refluxing and demonstrates the extent to which refluxing occurs and the importance of decoupling this effect in fast electron transport studies. Refluxing electrons are likely to play an important role in seeding fast electron transport instabilities involving counter-streaming currents

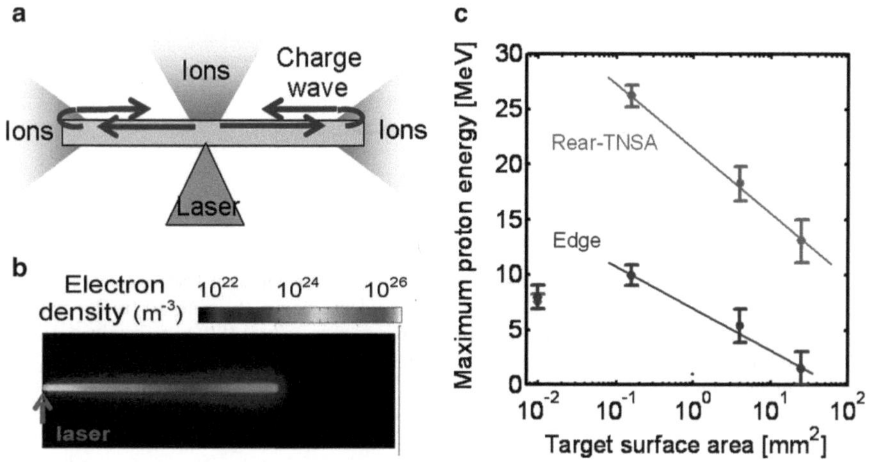

Fig. 5.13 (**a**) Schematic illustrating transverse refluxing and ion acceleration from the rear and edges of a foil target. (**b**) Simulation result showing fast electron density in a thin foil target, 2.5 ps after the start of the laser pulse. Fast electrons are observed to 'pile-up' at the target edge before being reflected back towards the laser focal spot. (**c**) Maximum energy of protons accelerated from the rear (TNSA) and one edge of a foil target as a function of target surface area. As the target size is decreased, increased electron transverse refluxing increases the ion maximum energy

and, as shown in Fig. 5.7, numerical simulation results suggest that the refluxing electrons also effect the growth of the self-collimating magnetic fields and hence the electron beam divergence.

5.6.2 *Lateral Transport*

One of the effects of fast electron refluxing in thin foils is that the direction of the electrons can be altered as illustrated in the simulation result in Fig. 5.12b. Upon reflection in the sheath fields the axial component of the electron velocity is reduced, while the transverse velocity is largely unaltered. As the electrons reflux many times the charge cloud can expand out laterally forming a charge disk [40]. The radial expansion requires a return current to flow. The expansion therefore continues until the charge wave reaches the target edge. At the edge the return current is no longer supported and the expansion cannot continue. It is stopped by the radial component of the electrostatic field, arising due to the build-up of net negative charge. Simulations of this effect, using the code LSP (an implicit particle-in-cell model with fluid background), are shown in Fig. 5.13 [40]. The electrons effectively 'pile up' at the target edge and are reflected after a time of the order of the plasma period.

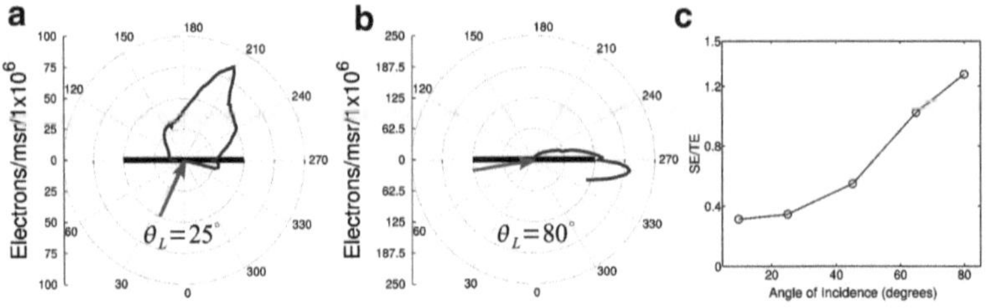

Fig. 5.14 Angular measurements of escaping fast electrons for different laser incidence angles. Example angular distributions are shown for laser incidence angle, θ_L, equal to (**a**) 25° and (**b**) 80°, relative to target normal. (**c**) Ratio of fast electron surface current (*SE*) to transmitted current (*TE*) as a function of laser incidence angle. The percentage of fast electrons transported along the surface increases with θ_L

It has been experimentally shown that the strength of the transient field at the target edges is of the order of 0.2 TV/m. This field builds up long after the laser pulse has passed (due to the millimeter distances over which the electrons travel to reach the target edge) and results in ionisation and acceleration. The generation of multi-MeV protons and carbon ions at the edge of 4 mm-wide metallic foils irradiated by ultra-intense ($\approx 6 \times 10^{20}\,\mathrm{W/cm^2}$), picosecond laser pulses has been demonstrated [40].

In addition to driving ion acceleration at the target edges, the radially expanding fast electron current which is reflected from the edges has been shown to enhance the energies of TNSA-accelerated ions formed on the rear surface (opposite to the laser focal spot) [41,42]. Figure 5.13 shows an example measurement illustrating that the maximum energies of protons accelerated from both the rear and edges of thin foil targets increase as a function of decreasing target surface area. It has been shown in numerical simulations that this occurs because the duration of the electrostatic fields are enhanced by the radially refluxing current. The effect can be used not only to enhance ion acceleration, but potentially also to shape the spatial-intensity distribution of the resulting proton beam, as demonstrated by Tresca et al. [42]. Furthermore, the laterally-transported fast electrons can be used to induce secondary fields in specially engineered targets to induce further acceleration of the protons, for example via the use of hollow hemisphere targets as proposed by Burza et al. [43], and to induce focusing of the protons, as demonstrated by Kar et al. [44].

The lateral transport of large currents of fast electrons can proceed not only via refluxing within the target, but also along the target surface due to the growth of quasi-static magnetic and electric fields [45–47]. These surface currents can play an important role in the cone-guided approach to fast ignition. Fast electron transport along the inner walls of the cone is predicted to enhance energy coupling into the compressed fuel. In a recent experiment using intense picosecond laser pulses the sensitivity of the surface transport to laser pulse parameters, including intensity, polarisation and angle of incidence was investigated. As shown in Fig. 5.14, large angles of incidence with respect to target normal drive a significant fast electron current along the target surface.

5.7 Future Directions

Considerable progress has been made over the past decade in terms of increasing our understanding of the physics of the transport of huge currents of energetic electrons in dense plasma. As discussed above, this new knowledge of the collective processes involved has in the past few years led to new ideas involving target ('structured collimator') and laser pulse ('multi-pulse') engineering to control the divergence and other properties of the fast electron beam. These novel approaches will potentially have important impact on the fast ignition approach to IFE. Further experimental and theoretical work is required to develop and apply these new ideas.

The recent new insight on the role of lattice structure effects in defining the fast electron transport pattern could have important implications for understanding the physics of transient states of warm dense matter and in potentially controlling fast electron beam transport instabilities. The work reveals that the choice of allotrope of a given element is important when considering the design of advance fusion target schemes and novel targets for ion acceleration. Developing a deeper understanding of the role of lattice structure effects, including the temporal evolution of the electrical conductivity of transient states of warm dense matter is the subject of ongoing work.

Finally the new insight into the role of lateral transport of fast electrons in thin foil targets is already yielding dividends in terms of enhancing the maximum energy of ions accelerated by the TNSA mechanism. Factors of greater than two have already been demonstrated by controlling the extent of lateral refluxing occurring. Techniques based on target shaping to 'shape' the ion beam and to produce the staged acceleration of ions have already been proposed and will be the subject of future work.

Acknowledgements Paul McKenna gratefully acknowledges the principal contributions made by past and present members of his research group to many of the example experimental results presented in this chapter, together with collaborators from the Central Laser Facility, Queen's University Belfast, Imperial College London, the University of Lund, GSI-Darmstadt, Sandia National Laboratories and the Chinese Academy of Sciences in Beijing. The 'structured collimator' work was led by collaborators at the Central Laser Facility (theory) and Queen's University Belfast (experiment).

References

1. S.C. Wilks, A.B. Langdon, T.E. Cowan et al., Phys. Plasmas **8**, 542 (2001)
2. C. Reich, P. Gibbon, I. Uschmann, E. Förster, Phys. Rev. Lett. **84**, 4846 (2000)
3. P. Patel, A.J. Mackinnon, M. Key et al., Phys. Rev. Lett. **91**, 125004 (2003)
4. M. Koenig, A. Benuzzi-Mounaix, A. Ravasio et al., Plasma Phys. Contr. F. **47**, B441 (2005)
5. M. Tabak, J. Hammer, M.E. Glinsky et al., Phys. Plasmas **1**, 1626 (1994)
6. W.L. Kruer, K. Estabrook, Phys. Fluids **28**, 430 (1985)
7. P. Gibbon, Phys. Rev. Lett. **73**, 664 (1994)
8. L.M. Chen, J. Zhang, Q.L. Dong et al., Phys. Plasmas **8**, 2925 (2001)

9. F. Jüttner, Annalen der Physik **339**, 856 (1911)
10. J.R. Davies, Plasma Phys. Contr. F. **51**, 014006 (2009)
11. J. Myatt, W. Theobald, J.A. Delettrez et al., Phys. Plasmas **14**, 056301 (2007)
12. P.M. Nilson, W. Theobald, J. Myatt et al., Phys. Rev. E **79**, 016406 (2009)
13. C.D. Chen, P.K. Patel, D.S. Hey et al., Phys. Plasmas **16**, 082705 (2009)
14. S.C. Wilks, W. Kruer, M. Tabak, A. Langdon, Phys. Rev. Lett. **69**, 1383 (1992)
15. P. McKenna, D.C. Carroll, O. Lundh et al., Laser Part. Beams **26**, 591 (2008)
16. H. Alfvén, Phys. Rev. **55**, 425 (1939)
17. A.R. Bell, J.R. Davies, S. Guerin, H. Ruhl, Plasma Phys. Contr. F. **39**, 653 (1997)
18. J.R. Davies, J.S. Green, P.A. Norreys, Plasma Phys. Contr. F. **48**, 1181 (2006)
19. P.A. Norreys, J.S. Green, J.R. Davies et al., Plasma Phys. Contr. F. **48**, L11 (2006)
20. A.R. Bell, R.J. Kingham, Phys. Rev. Lett. **91**, 035003 (2003)
21. X.H. Yuan, A.P.L. Robinson, M.N. Quinn et al., New J Phys. **12**, 063018 (2010)
22. A.P.L. Robinson, M. Sherlock, Phys. Plasmas **14**, 083105 (2007)
23. S. Kar, A.P.L. Robinson, D.C. Carroll et al., Phys. Rev. Lett. **102**, 055001 (2009)
24. B. Ramakrishna, S. Kar, A.P.L. Robinson et al., Phys. Rev. Lett. **105**, 135001 (2010)
25. A.P.L. Robinson, M.H. Key, M. Tabak, Phys. Rev. Lett. **108**, 125004 (2012)
26. R.H.H. Scott, C. Beaucourt, H.-P. Schlenvoigt et al., Phys. Rev. Lett. **109**, 015001 (2012); arXiv:1012.2029v2
27. R. Jung, J. Osterholz, K. Löwenbrück et al., Phys. Rev. Lett. **94**, 195001 (2005)
28. L. Gremillet, G. Bonnaud, F. Amiranoff, Phys. Plasmas **9**, 941 (2002)
29. S.I. Krasheninnikov, A.V. Kim, B.K. Frolov, R. Stephens, Phys. Plasmas **12**, 073105 (2005)
30. A. Debayle, V.T. Tikhonchuk, Phys. Rev. E **78**, 066404 (2008)
31. M. Storm, A. Solodov, J. Myatt et al., Phys. Rev. Lett. **102**, 23500 (2009)
32. M.N. Quinn, D.C. Carroll, X.H. Yuan et al., Plasma Phys. Contr. F. **53**, 124012 (2011)
33. P. McKenna, A.P.L. Robinson, D. Neely et al., Phys. Rev. Lett. **106**, 185004 (2011)
34. Y. Sentoku, T.E. Cowan, A.J. Kemp, H. Ruhl, Phys. Plasmas **10**, 2009 (2003)
35. W. Theobald, K. Akli, R. Clarke et al., Phys. Plasmas **13**, 043102 (2006)
36. H.S. Park, D.M. Chambers, H.K. Chung et al., Phys. Plasmas **13**, 056309 (2006)
37. P.M. Nilson, W. Theobald, J. Myatt et al., Phys. Plasmas **15**, 056308 (2008)
38. A. Mackinnon, Y. Sentoku, P.K. Patel et al., Phys. Rev. Lett. **88**, 215006 (2002)
39. M.N. Quinn, X.H. Yuan, X.X. Lin et al., Plasma Phys. Contr. F. **53**, 025007 (2011)
40. P. McKenna, D.C. Carroll, R.J. Clarke et al., Phys. Rev. Lett. **98**, 145001 (2007)
41. S. Buffechoux, J. Psikal, M. Nakatsutsumi et al., Phys. Rev. Lett. **105**, 15005 (2010)
42. O. Tresca, D.C. Carroll, X.H. Yuan et al., Plasma Phys. Contr. F. **53**, 105008 (2011)
43. M. Burza, A. Gonoskov, G. Genould et al., New J. Phys. **13**, 013030 (2011)
44. S. Kar, K. Markey, M. Borghesi et al., Phys. Rev. Lett. **106**, 225003 (2011)
45. T. Nakamura, S. Kato, H. Nagatomo, K. Mima, Phys. Rev. Lett. **93**, 265002 (2004)
46. Y.T. Li, X.H. Yuan, M. Xu et al., Phys. Rev. Lett. **96**, 2 (2006)
47. J. Psikal, V.T. Tikhonchuk, J. Limpouch, O. Klimo, Phys. Plasmas **17**, 013102 (2010)

Part III
Inertial Confinement Fusion

Chapter 6
The Physics of Implosion, Ignition and Propagating Burn

John Pasley

Abstract The subject of inertial confinement fusion centres around the achievement of ignition and propagating burn in a fuel mass that has been imploded by some form of driver. Whether this driver is a laser, a hohlraum radiating soft x-rays, or a charged particle beam, this theme of implosion followed by ignition and propagating burn is a common one. In this chapter we shall consider the process by which the fuel is compressed, as well as looking at the process of ignition. We shall consider situations in which burning proceeds from a region that is at the same density as the surrounding fuel and also the more typical situation of central hotspot ignition, in which the ignition region is at a lower density than the surrounding material. Burn-up of fuels other than 50:50 deuterium-tritium will be briefly considered, and some discussion of burn wave propagation will be presented. Finally we shall consider some of the requirements placed upon the implosion, and show that in order to achieve the required high densities in the imploded fuel it is necessary to employ a series of shock waves to accelerate the shell.

6.1 Introduction

In comparing inertial confinement fusion (ICF) [1] to other approaches such as tokamak based magnetic confinement fusion (MCF) [2], the principle difference that will always be mentioned is that of the very much higher fuel density that is needed in ICF. In ICF, the density of the fuel ions is on the order of $10^{26}\,\mathrm{cm}^{-3}$, where as in MCF it is more typically $10^{14}\,\mathrm{cm}^{-3}$. The first question that any student is likely to raise, therefore, is that of why such a high density is necessary in the inertial confinement approach to fusion?

J. Pasley (✉)
Department of Physics, York Plasma Institute, University of York, York, YO10 5DQ UK
e-mail: john.pasley@york.ac.uk

P. McKenna et al. (eds.), *Laser-Plasma Interactions and Applications*, Scottish Graduate Series, DOI 10.1007/978-3-319-00038-1_6,

To get at an answer to this question, let us consider that, at the moment of ignition, the final fuel configuration in an ICF scheme is a sphere with some uniform density (which could be high or low). Furthermore, let us suggest that the fuel is burning at some uniform temperature, T.

Neglecting other reactions beside the primary D-T fusion reaction, the thermonuclear reaction rate is given by:

$$\frac{dn}{dt} = N_D N_T \langle\sigma v\rangle \tag{6.1}$$

where $\langle\sigma v\rangle$ is the DT reaction cross-section averaged over a Maxwellian at temperature T, and n is the number of thermonuclear reactions occurring per second. If we again assume that the only reaction taking place is the DT reaction (a reasonable approximation) and the fuel starts off with equal populations of deuterium and tritium nuclei, then at some point during the burn we can write that:

$$N_D = N_T = \frac{N_0}{2} - n \tag{6.2}$$

where N_0 is the initial ion number density. The change of the burn fraction, $\phi = 2n/N_0$, is then given by

$$\frac{d\phi}{dt} = \frac{N_0}{2}(1-\phi)^2\langle\sigma v\rangle \tag{6.3}$$

If we further assume that $\langle\sigma v\rangle$ is constant throughout the burn, then integration yields:

$$\frac{\phi}{1-\phi} = \frac{N_0\tau}{2}\langle\sigma v\rangle \tag{6.4}$$

where τ is the confinement time.

In ICF, the burn is halted by the propagation of a rarefaction wave inward from the surface of the compressed fuel. Therefore we can approximate the confinement time by dividing the radius of the dense imploded fuel, r, by the sound speed, c_s. However in reality the overwhelming bulk of the fuel mass in an imploded ICF capsule is located near the surface of the dense region [1,3]. This is in part due to the natural distribution of mass within a sphere, and also the fact that spherical implosion of a hollow shell inevitably results in a lower density region in the centre of the imploded mass. Therefore, we shall suggest that it might be more sensible to take:

$$\tau \cong \frac{r}{3c_s} \tag{6.5}$$

This gives us an expression for the burn efficiency of the form:

$$\frac{\phi}{1-\phi} = \frac{N_0 r}{6c_s}\langle\sigma v\rangle \tag{6.6}$$

Now, it turns out that at typical burn temperatures the ratio of $\langle\sigma v\rangle$ to c_s is approximately constant [3]. Inserting numerical values for $\langle\sigma v\rangle$ and c_s gives us the approximate relationship between burn fraction and fuel ρr:

$$\phi = \frac{\rho r}{\rho r + 6(\mathrm{gcm}^{-2})} \tag{6.7}$$

So, if we suggest that a reasonable criterion for success would be a burn fraction of greater than $1/3$, then the product of fuel density and fuel radius should be approximately 3 gcm^{-2} or more. In fact of course the fuel in an ICF capsule is neither sitting at a uniform density initially, nor does it burn in the volumetric manner suggested by this simple picture. However, it turns out that sophisticated numerical simulations, that more accurately reflect reality, approximately reproduce this result [3].

Taking then this value of $\rho r \geq 3\mathrm{gcm}^{-2}$ then, let us consider what this number actually suggests from a practical standpoint. The density of uncompressed deuterium-tritium ice is approximately 0.22 gcm^{-3}. The $\rho r \geq 3$ gcm^{-2} criterion therefore suggests a fuel radius of around 13.6 cm if the fuel is at this density. This equates to a fuel mass of 2.3 kg, and, in case the reader is not already discouraged, a thermonuclear energy yield (for a burn-up of 33 %) of approximately 2.6×10^{14} J, or somewhat more than the energy released by the detonation of 60,000 tonnes of the high explosive TNT [4]. It is impractically expensive to contain an explosive energy release on this scale [5], and therefore inertial fusion energy schemes based upon fuel at normal solid or liquid densities are not a realistic proposition. However, before moving on, we should also notice that there is a further difficulty. As will be discussed in more detail later in this chapter, the success of ignition in an ICF scheme relies upon a portion of the fuel, with a $\rho r \geq 0.3$ gcm^{-2}, being raised to a high temperature of around 10 keV (or in excess of 100 million kelvin.) If we again insert the density of uncompressed DT into this relationship, known as the ignition criterion, we find that the radius of this hot region should be rather in excess of a centimetre, and have a mass of approximately 10 g. Given the heat capacity of DT is around 100 MJ/keV/g this implies that many gigajoules must be given to the fuel in order that it ignite. This must be done on an extremely short time scale since the hot region will otherwise lose much of this energy to its surroundings, before it has reached the necessary temperature. With the exception of nuclear explosives, there are no energy sources capable of achieving such feats of violent heating of gram quantities of materials. Therefore, again, we are directed away from the use of DT at its normal density.

It may not be obvious to all how compression can help. If we take a flat, slab-like sample of DT, of thickness x and density ρ, then compression along the x-axis results in an increase in density, but it also results in a reduction in thickness such that the quantity ρx remains constant. However this is not true for cylindrical, or spherical, compression. In these cases the ρr scales respectively with r^{-1} or r^{-2}, such that these forms of compression can give raise to a dramatic increase in ρr. Or, to put it another way, the ρr of the fuel in a spherically imploding ICF capsule climbs from a very small value initially, to roughly 3 gcm^{-2}, at the moment of peak compression.

How much then should the fuel be compressed? The answer here is 'as much as possible'. The reason being that, as previously mentioned, ignition relies upon heating a region of $\rho r \geq 0.3$ gcm^{-2} to a very high temperature. In a spherical geometry, the mass corresponding to this ρr scales with ρ^{-2}, and hence so does the energy required to heat the ignition region. In addition to this point, it is worth considering that while it might be possible to have a very dense ignition region surrounded by less dense fuel, a power plant is likely to be aiming for an output power on the order of 1 GWe, so we might wish to aim for a capsule that produced somewhere in the range of 100 MJ to 10 GJ. Assuming 33 % burn-up this leads us to a fuel mass on the order of a milligram, and compressions in excess of a factor of 10^3.

The compression required then is an extremely large one. Before we go on to consider how we can achieve such a compression, let us look first at the other key criteria that must be satisfied: the ignition criteria.

6.2 Ignition

In the context of fusion research in general (i.e. including other forms of fusion technology such as tokamaks), ignition refers to the attainment of a situation in which thermonuclear burning is self-perpetuating. That is to say that no external energy source is required in order to maintain the burn. In a magnetic confinement fusion system it is thought to be possible to have a reactor that operates usefully without achieving ignition [6]. Such a reactor can have an energy output that far exceeds its energy input, while still relying upon external heating to perpetuate the burn. This is not the case for inertial fusion. Clearly, however, the mechanics of an inertial fusion reactor are rather different to those of a tokamak, due to the fact that the burn in ICF occurs in discrete pulses, as each capsule is injected, compressed, heated and burnt [7].

What then do we mean by 'self-perpetuating' burn, then, in a system in which the burn, by definition, rapidly extinguishes itself?

In ICF, the term ignition is used to refer to one of the necessary stages of the evolution of the fuel. In ICF, ignition is the process of the fuel self-heating from the comparatively modest temperatures that it is left in by the action of the driver (typically 5–10 keV) to the very much higher temperatures associated with widespread burning (60–110 keV). It demarcates the time period in which heating is overwhelmingly performed by the action of the driver from the time period in which all significant heating is performed by the deposition of thermonuclear alpha particle, and neutron, energy [8]. A few tens of picoseconds later, the burn will rapidly extinguish, as the burn wave runs outward toward the low density plasma that surrounds the dense core. The burn ceases because the burn wave no longer has dense fuel to propagate into. However the burn wave is entirely self-sufficient provided dense fuel remains for it to consume. Therefore we can see that the burning ICF , capsule satisfies our earlier, more general definition of an ignited plasma.

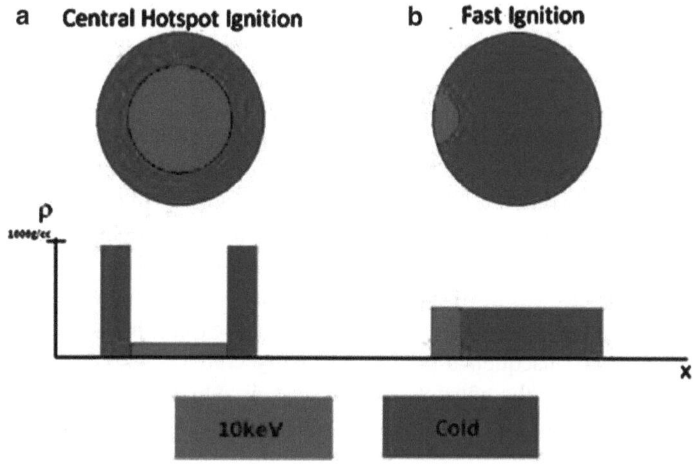

Fig. 6.1 A schematic diagram illustrating the fuel configuration (**a**) at the end of the implosion phase in a conventional central hotspot ignited ICF capsule and (**b**) after the application of the ignitor pulse in fast ignition

It is common, in studies of ICF, to discount the role of neutron heating in burn propagation. However, according to the relationship presented in [9] a capsule with a ρr of 3 gcm^{-2} will recapture around 13 % of the neutron energy released, making the assumption of uniform burn. However since four-fifths of the energy released by the DT reaction is released as neutrons, this is far from being a negligible effect. The ignition process, however, is overwhelmingly a result of alpha-heating, since it typically occurs in a hot region of fuel whose ρr is on the order of 0.3–0.6 gcm^{-2}. Reference [8] give the neutron mean range as approximately 4.7 gcm^{-2} in DT, so little neutron energy will be redeposited within this hot region, and the effect on ignition will be marginal.

6.2.1 *The Ignition Environment*

The ignition process is similar irrespective of whether we are considering conventional central hotspot, fast ignition or shock ignition. However the environment in which the ignition occurs will differ, according to the approach taken.

6.2.1.1 Conventional Central Hotspot Ignition

At the end of the implosion phase, the fuel is roughly divided into two regions, as depicted in Fig. 6.1a [1, 3].

In grasping the energetics of ICF it is crucial to appreciate that the region of fuel heated by implosion, known as the hotspot, contains only around 1 % of the total fuel mass. The bulk of the fuel is contained in the dense region, which, at the end of the implosion, is comparatively cool. The significance of this is that the dense fuel is not heated by the driver, but by the propagating burn from the hot spot. If the driver were required to heat the whole of the fuel to ignition temperatures then the achievable gain would be insufficient for us to contemplate power generation.

In conventional, central hotspot ignition, the density of the hotspot is typically around 100 gcm^{-3} whilst that of the surrounding cold fuel is around 1,000 gcm^{-3}. The implosion process results in a configuration that is approximately isobaric. That is to say that the pressure of the cold fuel is comparable to the pressure of the central hotspot. This is a consequence of the imploding cold fuel being decelerated by piston-like action upon the hotspot. This process, in which the cold fuel is brought to a halt by the mounting pressure in the central region, is known as stagnation.

6.2.1.2 Fast Ignition

In fast ignition [10, 11], the picture is rather different. Here every effort is made to assemble the fuel to a uniform high density i.e. there is no central hotspot. In reality a central hotspot will inevitably form in the violent collapse of a hollow shell. However it is possible to greatly limit the heating of this region as compared to the conventional central ignition case. In fast ignition, the hotspot is rather formed near the surface of the dense fuel blob, by the action of an 'ignitor'. A variety of different forms have been suggested for this heating device [10–12]; most commonly an ultra-high-intensity laser generated beam of relativistic electrons is proposed [10, 11]. For our purposes however the nature of this heating mechanism is immaterial. The consequence is the formation of an intensely heated region of DT near the surface of the dense fuel blob, as depicted in Fig. 6.1b.

The formation of the hotspot in the dense fuel results in a heated region that is at a much higher pressure than its surroundings. This results in far more rapid hydrodynamic disassembly of the hotspot than in the case of conventional central hotspot ignition. This has a direct impact upon the requirements for ignition, as will be discussed further below.

Fuel in this state of approximately uniform density is sometimes labelled as being 'isochoric', though this literally refers to a system held at constant volume.

6.2.2 Hotspot Power Balance

The success of ignition, which is the rapid self-heating of the hotspot by alpha particle deposition, is dependent upon the balance of heating and cooling terms in the hotspot. The only significant heating process after stagnation is that of alpha particle deposition, assuming that neutron heating may be neglected, as previously discussed in Sect. 6.2 above. The cooling terms are those of radiative loss, electronic heat conduction out of the hotspot, and the hydrodynamic work done on the fluid. This latter term is enhanced in fast ignition, where the hotspot sits adjacent to cold fuel of comparable density.

Equation 6.8 relates the necessary condition for hotspot self-heating:

$$W_\alpha > W_{\mathrm{brem}} + W_{\mathrm{cond}} + W_{\mathrm{hydro}} \tag{6.8}$$

Here W_α unexpected represents the power deposition by alpha particles, W_{brem} represents the (predominantly bremsstrahlung) radiative losses, W_{cond} the conduction losses and finally W_{hydro} the hydrodynamic losses.

It turns out that this inequality can only be satisfied if the density-radius product of the hotspot is somewhat greater than the alpha-particle range in the hotspot [1, 3, 8]. If the hotspot meets this criterion, then a significant fraction of the alpha particle energy released by fusion reactions will be redeposited within the hotspot, causing the temperature to increase rapidly. At temperatures in the low tens of keV range this increase in temperature leads to a rapid increase in the rate of thermonuclear burning, and further heating. In this way the hotspot temperature rises rapidly to a temperature on the order of 70 keV, at which point a combination of factors cause the temperature increase to plateau. A key point is that the range of alpha particles in dense fuel is quite a strong function of temperature [3, 8, 13]. In higher temperature fuel, alpha particles have an increased range. Therefore alpha particles that would have stopped in the hotspot at 10 keV, now proceed into the surrounding dense fuel. This limits the self-heating of the hotspot; however the effect is actually a useful one, since it acts to concentrate the heating power of the now vigorously burning hotspot upon the innermost regions of the dense fuel [8].

6.2.2.1 Alpha-Particle Heating

Alpha particle stopping in dense plasma is a non-trivial process, and can only be treated approximately by consideration of binary collisions with individual ions and electrons in the background fluid. Many-body plasma effects play a role in decelerating the alpha particles, and this is still a topic of active research. However during the early stages of ignition, when the plasma temperature is still quite low, the effects of ion-electron scattering dominate [8]. The range of an alpha particle in dense DT is shown in Fig. 6.2.

6.2.3 Ignition Criteria

Detailed numerical simulations incorporating fusion reactions, alpha particle stopping, radiation transport, electron conduction and hydrodynamics allow the

Fig. 6.2 Alpha particle range in DT under a range of conditions relevant to ICF, taken from Ref. [3], and based upon the calculations presented in reference [13]

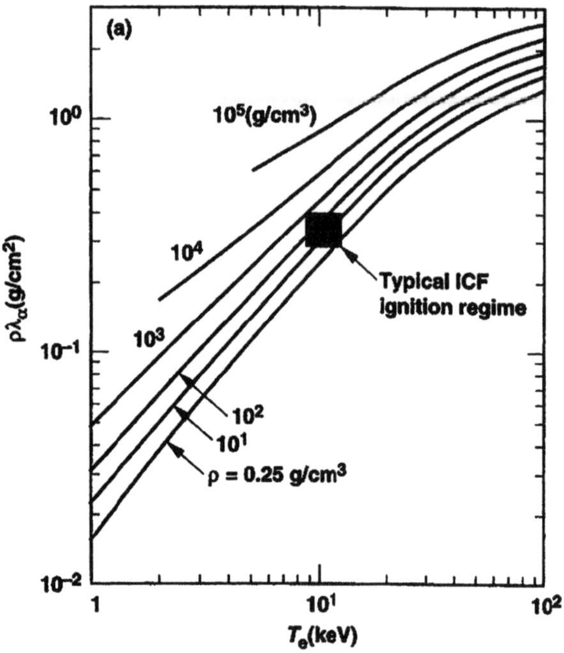

ignition requirements for a given fuel assembly and hotspot temperature to be found by a process of iteration [8]. The results differ somewhat depending upon whether the fuel is assembled in the isobaric or the isochoric state. The requirements for isochoric ignition are somewhat more demanding due to the far greater amount of mechanical work done by the hotspot in this case. Typical ignition ρr for isobaric fuels are around 0.3 gcm^{-2}, while for isochoric fuels a figure of 0.5 gcm^{-2} would be more typical at temperatures of around 10 keV [8]. It should be pointed out however, that the heating of the central hotspot by piston-like compression in conventional ICF is woefully inefficient ($\approx 1\ \%$). The potential of fast ignition lies principally in the possibility of heating the hotspot far more efficiently by action of the ignitor.

A second criterion for ignition, that is less commonly mentioned, is one concerning the overall fuel ρr. The process of self-heating of the hotspot takes a finite amount of time typically on the order of 10–20 ps. Since the fuel disassembly by rarefaction from the outer surface proceeds simultaneously with the ignition and burn, this implies that sufficient ρr of dense fuel must be present around the hotspot to prevent free-surface expansion from rarefying the hotspot before it has had time to ignite. This places a lower bound upon the total assembled fuel ρr that is required for ignition to take place in a hot, centrally located, region of that fuel blob. This criterion dictates a minimum ρr of approximately 1 gcm^{-2}, including contributions from both the hotspot and the surrounding cold fuel.

6.2.3.1 Establishing Ignition Criteria

The ρr required for a particular configuration of fuel to barely ignite is often calculated in order to indicate what set of parameters must be achieved at the end of an implosion, or fast ignition heating pulse, in order to achieve ignition. These simulations are often performed using a radiation hydrodynamics code with 1D spherical symmetry [8]. The simulations commence from some idealised configuration at a high density, some central spherical portion of which is initialised at a high temperature. If ignition results, then the radius of the hot central region is iterated until the minimum hotspot radius for successful ignition is arrived at. This value of ρr at which ignition just occurs is sometimes referred to as the critical ρr for ignition [14]. The outer radius of the fuel, in such calculations, is usually set at so large a radius that it has no bearing upon the result.

As an aside, it is worth noting that this critical value of ρr amounts to a critical mass at a given density. One can then make an interesting comparison with criticality in nuclear fission, by noting that both fusion and fission based schemes require a critical mass of fuel (which is density dependent). However fusion also requires that this critical mass be raised to some critical temperature. Given that this temperature is on the order of 100 million kelvin, this renders fusion a substantially more challenging proposition than fission!

It is possible to emulate either isochoric or isobaric fuel configurations in this way, depending on whether the hot region is set at the same, or a substantially lower, initial density. Such calculations may also be employed to investigate other less standard situations, such as ignition in fuels that are other than 50:50 DT, or ignition in DT that has been contaminated by some other impurity which cannot contribute to the burning, such as may occur in fast ignition when material from the cone becomes mixed with the fuel.

Some care must be taken in interpreting the results of such calculations. Close to the boundary of ignition failure it may take a substantial period of time for ignition to establish itself, as is discussed briefly in the following section. In ICF, as has been mentioned several times, ignition must establish itself promptly or else failure will be inevitable due to the disassembly of the compressed fuel by rarefaction. Establishing ignition criteria for a range of fuel assemblies and conditions as discussed here is therefore more useful for determining which fuel configurations can be ignited most readily, rather than for ascertaining the precise value of hotspot ρr required. Further simulations, that afford a more realistic representation of an ignition experiment, must be performed for this purpose.

6.2.3.2 Modes of Ignition

In simple 1-D spherical ignition calculations, such as are described in the previous section, ignition can manifest in several different ways, depending on how far above the critical ρr for ignition the fuel is sitting [8]. Far above the boundary for ignition,

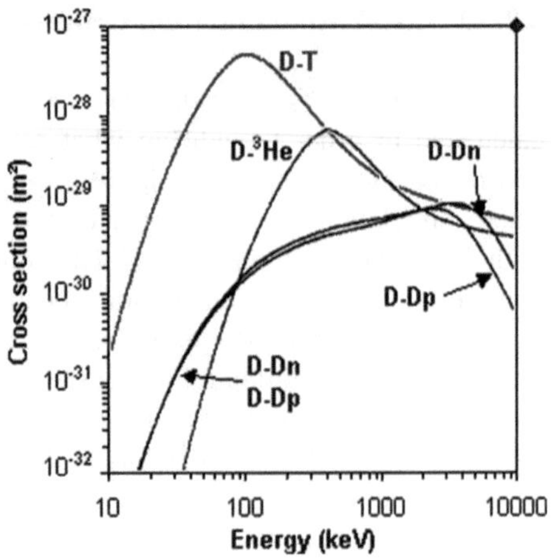

Fig. 6.3 Fusion reaction cross sections for some key reactions

self-heating is exceptionally rapid, and the fuel ignites with little hydrodynamic motion occurring. This may be referred to as volumetric ignition, although the reader should note that this term is used in several different contexts in the field of ICF. However, close to the ignition boundary, heating can be sluggish and ignition may only manifest after a period of $\approx$ 100 ps. In this case ignition is typically preceded by the formation of a strong shock wave at the surface of the hotspot. This shock wave forms a shell of compressed hot fuel around the hotspot, that tends to insulate the central region while at the same time increasing the rate fusion reactions in the compressed layer (as burn rate is proportional to the square of density). While the physics of this delayed, shock-led, ignition is interesting it is not a useful mechanism in the context of ICF since fuel disassembly will prevent such a drawn-out ignition process from ever reaching completion.

6.2.4 Ignition of Fuels Other Than 50:50 DT

There are several reasons why it is interesting to explore ignition of fuels other than 50:50 DT. The first of these is that tritium is not a naturally occurring isotope, and must be created by nuclear processes. In a fusion reactor the plan is to breed tritium from neutron reactions with lithium; however, this is a challenging task. Furthermore, tritium is not ideal as a fuel material since it is both radioactive and has applications in nuclear weapons. This latter point is relevant since it means that the possession of tritium by some states would raise proliferation concerns. Ideally any new power generation technology should be capable of being sited anywhere on earth, to reduce the potential of conflict driven by inequalities in the availability of energy supplies.

It would be helpful therefore if some less problematic material could be employed as a fuel for fusion reactors. However, as shown in Fig. 6.3, the fusion

Table 6.1 A list of some of the most important fusion reactions from the standpoint of power generation, and their Q values

Reaction	Q [MeV]
$D + T \rightarrow \alpha + n$	17.6
$D + D \rightarrow T + p$ (50 %)	4.04
$D + D \rightarrow {}^3He + n$ (50 %)	3.27
$T + T \rightarrow \alpha + 2n$	11.3
$D + {}^3He \rightarrow \alpha + p$	18.35

reaction cross section for DT is approximately two orders of magnitude higher than any other fusion reaction at temperatures below 50 keV. Given the difficulty of achieving ignition in 50:50 DT it therefore seems unrealistic, at present, to contemplate ignition in any other material. Exceptions might be cases in which the ignition of some alternative fuel is seeded by a burn wave propagating out of 50:50 DT or schemes in which DT is employed throughout but with a slightly larger proportion of deuterium than tritium. Such capsules would inevitably have a lower yield than capsules which only relied upon burn in a similar mass of 50:50 DT. However the relaxation of the demands placed on the tritium breeding cycle might make such designs worthwhile. For reference purposes, some of the relevant reactions are shown in Table 6.1.

6.2.4.1 Primary and Secondary Fusion Reactions

When considering burn in fuel materials other than 50:50 DT it is worth noting that several different types of fusion reaction may be important. Reactions may be of two different types:

Thermonuclear, or primary, reactions: these are fusion reactions between thermal ions in the plasma; for example between the D and T ions in an ICF hotspot.

Beam-like, or secondary, reactions: these are fusion reactions between a daughter ion from an earlier fusion reaction and thermal ion in the plasma. For instance, the production of an energetic tritium ion from a DD fusion reaction in deuterium fuel can be followed by this high energy tritium ion fusing with one of thermal deuterium ions in the background plasma.

Both types of reactions occur even in the burning of 50:50 DT fuel. The reason being that DD burning will occur in such fuel, and this will result in some secondary reactions between tritium or helium-3 and the thermal deuterons. However the contribution of such reactions to the yield will be insignificant. In the burning of other fuel materials, the contribution of such reactions to the yield can dominate the energy release, as the secondary reactions may be significantly more exothermic than the primary ones.

In some circumstances, the following neutron induced reaction may also play a role in determining the yield of a burning assembly [14]:

$$^3He + n \rightarrow T + p \quad Q = 0.76\,\mathrm{MeV}$$

This reaction can convert He-3 into the more reactive tritium, when the burn takes place in a neutron rich environment.

6.3 Burn Propagation

Once ignition has occurred in the hotspot, burn will readily progress into the surrounding dense fuel. As mentioned in Sect. 6.2.2, the range of alpha particles increase as the fuel is heated. This results in the alpha particles being increasingly able to escape from the hot spot as the self-heating progresses [3, 8]. Consequently an increasing fraction of the thermonuclear energy being liberated in the hotspot will act to heat the surrounding dense fuel.

This picture, of the burn region expanding in radius as successive layers are heated and thus become less prone to alpha self-heating, resulting in heating of the next layer of cooler dense fuel, continues until the burn front encounters the rarefaction wave propagating inward from the outer surface of the fuel. At this point the burn rapidly falters since the speed of the rarefaction wave scales with $T^{1/2}$ and the burn rate scales with n^2. As the burn front is invariably propagating either up or down a density gradient, the propagation velocity of the burn front will vary with time. The duration of the burn in a typical ICF capsule is on the order of a few tens of picoseconds.

In some circumstances, strong shock waves will form at the leading edge of the burning region [8]. This can only occur if the heat front driven by alpha-heating propagates subsonically relative to the hot central region.

6.4 Implosion

The goals of the implosion in ICF include, at least, compression of the bulk of the fuel to a high density. In the conventional central hotspot ignited approach to ICF another aim of the implosion is to form a central hotspot, which satisfies the ignition ρr criterion. The implosion is a complex process, which we must simplify substantially in order to bring some degree of sense to it.

During the implosion, the outermost regions of the capsule are ablated by the driver. The rest of the material is accelerated inward. This is often termed rocket-like acceleration, and indeed this is a fair analogy. However the student should be careful not to extend the analogy of a space rocket too far! For instance, there is a significant delay between the commencement of the ablation of the exterior of the capsule, and the time at which the inner surface of the capsule begins to implode, due to the time taken for the shock waves that are accelerating the material to pass through the thickness of the shell.

The bulk of the compression in an ICF implosion actually occurs at late times, as the fuel stagnates against itself in the centre. While the shock waves which are driven through the capsule do indeed result in significant compression, the key criteria for an implosion are the final implosion velocity and the adiabat of the fuel [3, 8]. The first parameter determines how much energy there is available during stagnation for driving the compression. The second parameter determines the degree to which the fuel will resist compression; essentially it is a measure of how much the fuel has been heated during the implosion. Hot fuel is harder to compress. Therefore all efforts must be made to limit the heating of the fuel during the implosion, be this through excessive shock heating (which we shall discuss shortly) or other forms of preheat (radiative/hot particle). So one could paraphrase and suggest that the ICF capsule designer must go to great lengths to ensure that the fuel remains cold. The question that we must address first then is that of 'how cold is "cold"?'

6.4.1 Cold Fuel and Fermi-Degeneracy

It turns out that the answer to our question of 'how cold is "cold"?' lies in an area of physics that one might not necessarily expect to have great relevance to ICF. Electrons are fermions, and fermions cannot occupy the same state as one another. They must remain degenerate, or in different states. Fermi statistics dictate that if a large number of electrons occupy a given volume, then many of them must be in higher energy states in order to remain degenerate [15]. If the volume of the gas is reduced, and the number of electrons remains the same, then the electrons will be forced to ever higher energies. What this implies is that compressing an electron gas inevitably requires at least the energy required to force the electrons into the higher energy states dictated by Fermi statistics. There is no way around this. The mean energy of electrons in such a system is given by:

$$E_{Favg} = \frac{3}{5} E_{Fermi} \tag{6.9}$$

where,

$$E_{Fermi} = 3.65 \times 10^{-15} n^{2/3} \text{ eV} \tag{6.10}$$

One can also associate a pressure and a temperature with the degenerate electron gas:

$$\begin{aligned} P_{Fermi} &= \frac{2}{5} n E_{Fermi} \\ &= 2.34 \times 10^{-39} n^{5/3} \text{ Mbar} \end{aligned} \tag{6.11}$$

and,

$$T_{Fermi} = \frac{E_{Fermi}}{k} \tag{6.12}$$

And this presents us with an answer to our question: if the temperature of the dense fuel in an ICF capsule is kept well below this Fermi temperature, then it can be compressed without significant extra work being required than that which is unavoidable. Putting numbers into Eq. 6.13 we can also see that, for example, the Fermi temperature of deuterium at 1,000 gcm^{-3} is around 18.5 million kelvin. The answer to our question 'how cold is "cold"?' is therefore 'very hot!'

6.4.2 Implosion Velocity

It is actually quite straightforward to obtain a simple estimate of the required implosion velocity for an ICF capsule, by equating the kinetic energy of the imploding shell to the required total Fermi energy (the average Fermi energy of an electron in the imploded state, multiplied by the total number of electrons in the dense fuel core). Doing this for fuel at around 1,000 gcm^{-3} leads to an estimate of the implosion velocity of around $2.7 \times 10^7\,\mathrm{cm/s}$. As would be expected, this is an underestimate. Typical implosion velocities for central hotspot ignition are close to $4 \times 10^7\,\mathrm{cm/s}$, however much of the energy in that scheme is required to heat the hotspot.

While not the topic of this chapter, the reader is probably aware that hydrodynamic instabilities, in particular the Rayleigh-Taylor instability [16], make it unfeasible to implode a capsule whose initial thickness is an extremely small fraction of its initial radius. This limits the distance over which the fuel can be accelerated, and therefore means that the acceleration must be quite violent in order that the fuel reaches the required final implosion velocity. This in turn suggests that the pressure required to accelerate the fuel must be very large. An alternative way to look at this would be to say that if we set the $P\Delta V$ work done on the capsule equal to the required total Fermi energy of the fuel, then the pressure required to accelerate the fuel will be large: around 20 Mbar. However, as will be shown in the following sections, applying such a large pressure suddenly will result in the fuel being heated dramatically, preventing efficient compression to high density.

6.4.3 Shock Waves

A shock wave [17] is a discontinuity in the gas dynamic variables which propagates through the fluid at a velocity that is greater than the speed of sound ahead of the wave, and less than the speed of sound behind the wave. Such waves form inevitably from pressure disturbances in fluids, except in the case where the disturbances are very weak (such weak disturbances are known as sound waves). If we imagine the situation of a sinusoidal pressure disturbance propagating in a fluid, the cause of the formation of such waves can be seen by examining how the sound speed in the wave varies with position. Regions of the wave that are at a higher pressure have associated with them a higher local sound speed than regions that are at a lower pressure. This results in progressive steepening of the wave front, and finally the formation of a sudden jump in the pressure from peak to trough. This sudden discontinuous jump in pressure is known as a shock wave. It can be shown that the density and temperature follow a similar trend, passing from the low to the high pressure portions of the wave. The degree to which the shock wave front is actually discontinuous is dependent upon the dominant mode of energy transport across the wave front. At low temperatures, viscosity and thermal conduction dominate the energy transport, and therefore the shock front thickness is governed by relevant electron mean free paths. At low temperatures, in high density fluids, these mean free paths are on the order of nanometers, and therefore the jump does, from a macroscopic standpoint, indeed appear to be discontinuous. At very high temperatures, however, radiation can play a dominant role, and therefore the shock front thickness can be better approximated by the much longer mean free paths of the relevant photon populations that are conveying the energy from the hot to the cold region of the fluid. In this latter case the discontinuous nature of the shock front transition tends to disappear [17].

6.4.3.1 Hugoniot Curves

Hugoniot curves [17] represent the locus of all possible final states (pressure P_1, and specific volume V_1) that may be produced by the passage of a single shock wave, from some initial state (P_0, V_0) :

$$V_1 = \left[\frac{P_0V_0(\gamma-1)+P_1V_0(\gamma-1)}{P_0(\gamma-1)+P_1(\gamma+1)}\right] \tag{6.13}$$

where γ is the adiabatic exponent of the fluid, such that $P_1V_1^\gamma = P_0V_0^\gamma$ If P is plotted against V then, from some initial state, the Hugoniot curve lies above the isentropic compression curve, but diverges significantly from it only if the ratio of P_1 to P_0 is large. It is also simple to show from Eq. 6.13 that the limiting compression produced by a single shock wave is given by the ratio $(\gamma+1)/(\gamma-1)$ by suggesting that the initial pressure is negligible.

Some insight can be gained by considering the relationships between initial and final temperatures and entropies generated by the passage of a single shock wave:

$$\frac{T_1}{T_0} = \frac{P_1V_1}{P_0V_0} \tag{6.14}$$

and,

$$S_1 - S_0 = C_V \ln\frac{P_1V_1^\gamma}{P_0V_0^\gamma} \tag{6.15}$$

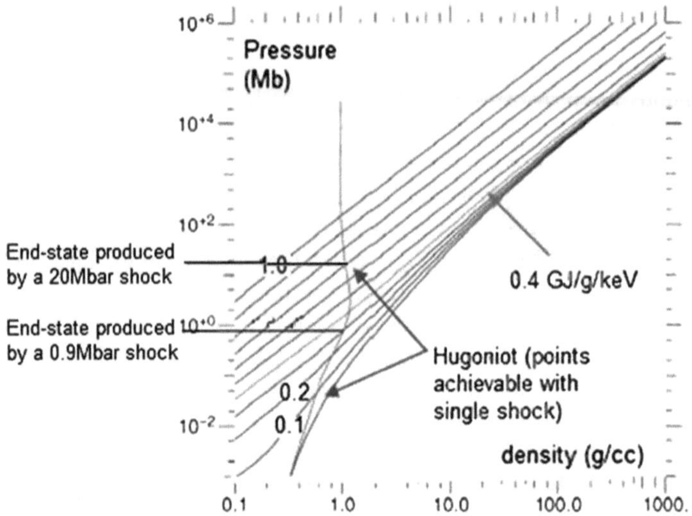

Fig. 6.4 A set of isentropes for DT are plotted (*black lines*), with the Hugoniot curve for a single shock from DT-ice at normal solid densities (*purple line*) overlaid. The extent to which it is acceptable to raise the fuel pressure by the passage of a single shock going into DT-ice is limited. At high densities the cold, quasi-Fermi-degenerate curve lies very close to isentropes with a specific entropy of less than 0.4 GJ/g/keV. Therefore, since entropy cannot be lost from the fuel during the implosion, we must aim to maintain the dense imploding fuel below this isentrope (*red line*) throughout the implosion (Color figure online)

These two relationships suggest that the parameter that is critical in determining the degree of any heating which takes place due to the passage of a shock wave is the ratio of post-to pre-shock pressure, P_1/P_0 and not the absolute magnitude of the final pressure. This is critical since it suggests that we might reduce the amount of heating required to achieve a given final pressure by using a series of shock waves; each shock wave limiting the ratio of P_x to P_{x-1}, however the final pressure after many shocks being allowed to reach a high value. Since, in ICF, we need to apply very large pressures to our fuel, without heating it excessively, this suggests that we must resort to the use of multiple shock waves.

6.4.3.2 Target Design

Figure 6.4 illustrates the difficulty in accelerating the fuel by application of a sudden pressure jump from zero to tens of megabars, as might be suggested by our simple calculation in Sect. 6.4.2. Looking at the isentropes for DT fuel, it can be seen that at very high densities isentropes with a specific entropy of less than 0.4 GJ/g/keV lie very close to the cold quasi-Fermi-degenerate curve (represented by a specific entropy of 0.0 in Fig. 6.4). Therefore, since it is not possible to lose entropy from the fuel during the course of the implosion, it is necessary to

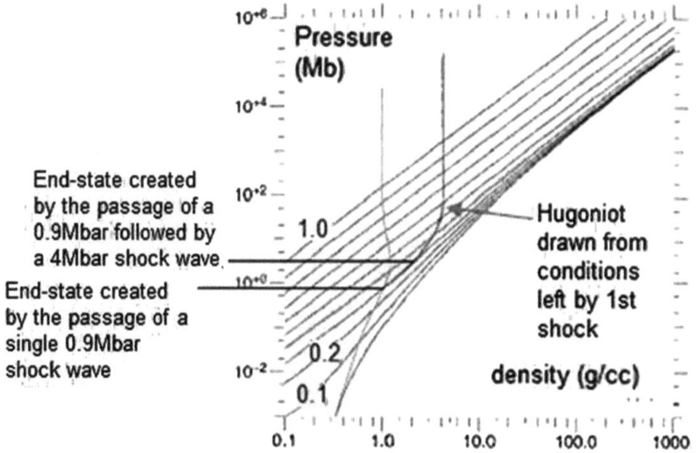

Fig. 6.5 This figure builds upon the information contained in Fig. 6.4, showing the Hugoniot curve for a second shock wave starting from the conditions created by the passage of a 0.9 Mbar shock wave into cold DT-ice. It can be seen that this second shock can have a strength of approximately 4 Mbar without raising the entropy of the fuel above the critical 0.4 GJ/g/keV level

maintain the dense fuel below this value of specific entropy during the course of the implosion. By considering the Hugoniot curve from cold-DT ice (plotted as a purple line in Fig. 6.4), it can be seen that a single shock of strength $\approx$ 20 Mbar, such as is suggested by our naïve calculation in Sect. 6.4.2, leaves the fuel with a very much higher specific entropy than this (approximately 0.7 GJ/g/keV). This implies that substantially more work than necessary will need to be performed upon the fuel in order to compress it to the desired density (thereby invalidating the simple calculation performed in Sect. 6.4.2 and encouraging us to go to even higher pressures- which would exacerbate the problem further).

By consideration of Fig. 6.4, it can be seen that an acceptable strength for the first shock wave going into the fuel might be around 0.9 Mbar. This is approximately the strength of the first shock employed at the National Ignition Facility [3]. Figure 6.5 illustrates how this principle is extended to the second shock wave. A new Hugoniot curve is plotted from the fuel conditions created by the passage of the first 0.9 Mbar shock wave.

It is found that the shock pressure may be increased by approximately a factor of four by each successive shock wave, without raising the fuel entropy significantly above the level 'set' by the first shock wave. NIF uses a series of four shock waves [3], to raise the pressure to the levels required to achieve a satisfactory implosion velocity. In reality, for central hotspot ignition, this implosion velocity is on the order of $4 \times 10^7\,\mathrm{cm/s}$. In central hotspot ignition, the velocity is required to be significantly higher than in the case of fast ignition, since the stagnation of the shell is required to heat the central hotspot, in addition to compressing the dense fuel. Figure 6.6, taken from reference [3] shows the complete trajectory of the dense fuel and hotspot in a NIF-like central hotspot ignited target.

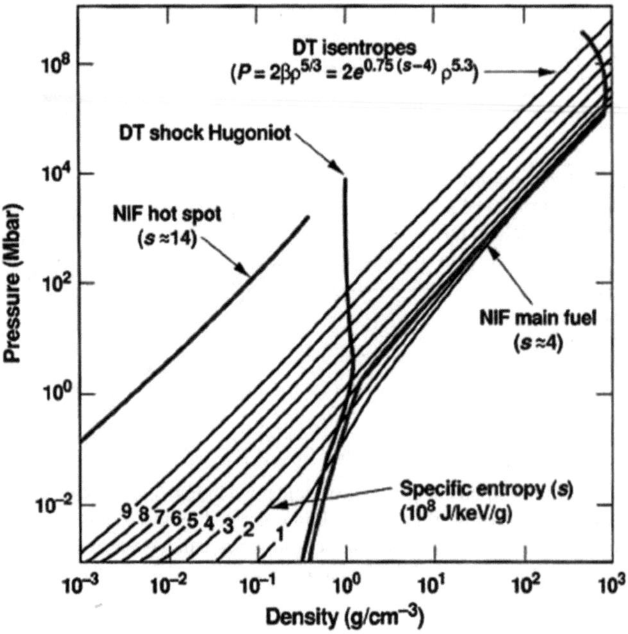

Fig. 6.6 Shows the trajectories of the dense fuel and central hot-spot in a NIF-like implosion, taken from reference [3]. Notice that the entropy units are in 10^8 J/keV/g unlike in the previous two figures

6.5 Conclusions

In this chapter, we have attempted to give the reader an understanding of the fundamental requirements for ignition and propagating burn in an ICF fuel pellet. We have seen that in order for ignition to take place in a centrally ignited target, the ρr of the hot-spot must exceed approximately 0.3 gcm^{-2}. For high gain burn, the ρr of the fuel as a whole should exceed approximately 3 gcm^{-2}. In Sect. 6.4 we considered the requirements placed upon the implosion in order to render the fuel in the desired state. We showed that it is important to limit the entropy of the fuel by accelerating it with multiple shock waves, such that the fuel pressure during stagnation does not significant exceed the unavoidable Fermi back-pressure, thereby limiting the compression.

References

1. J.H. Nuckolls et al., Nature **239**, 139 (1972)
2. J. Wesson, *Tokamaks* (Oxford University Press, New York, 2004)

3. J.D. Lindl, Phys. Plasmas **2** (11), 3933–4024 (1995) or J.D. Lindl, *Inertial Confinement Fusion – the Quest for Ignition and Wnergy Gain Using Indirect Drive* (AIP Press, New York, 1998)
4. P.W. Cooper, S.R. Kurowski, *Introduction to the Technology of Explosives* (Wiley-VCH, New York 1996)
5. F. Reines, B. Atom. Sci. **15**, 118–122 (1959)
6. B.J. Green et al., Plasma Phys. Contr. F. **45**, 687–706 (2003)
7. S. Nakai, K. Mima, Rep. Prog. Phys. **67**, 321–349 (2004)
8. S. Atzeni, J. Meyer-ter-Vehn, *The Physics of Inertial Fusion*, (Oxford University Press, New York, 2004)
9. E.L. Avrorin, Sov. J. Plasma Phys. **6**, 527–531 (1980)
10. M. Tabak et al., Phys. Plasmas **1**, 1626 (1994)
11. S. Hatchett et al., Bull. Am. Phys. Soc. **46**, 47 (2001)
12. M. Roth et al., Phys. Rev. Lett. **86**, 436–439 (2001)
13. G.S. Fraley et al., Phys. Fluids **17**, 474 (1974)
14. A.M. Frolov, Plasma Phys. Contr. F. **40**, 1417–1428 (1998)
15. H. Kroemer, C. Kittel, *Thermal Physics* (Freeman, San Francisco, 1980)
16. R. Betti et al., Phys. Plasmas **5**, 1446, 1998
17. Y.B. Zel'dovich, Yu.P. Razier, *Physics of Shock Waves and High-Temperature Hydrodynamic Phenomena* (Academic, New York, 1966 and Dover Publications, New York, 2002)

Chapter 7
Cryogenic Deuterium and Deuterium-Tritium Direct–Drive Implosions on Omega

Valeri N. Goncharov

Abstract The success of ignition target designs in inertial confinement fusion (ICF) experiments critically depends on the ability to maintain the main fuel entropy at a low level while accelerating the shell to ignition-relevant velocities of $V_{\rm imp} > 3 \times 10^7$ cm/s. The University of Rochester's Laboratory for Laser Energetics has been imploding cryogenic deuterium and deuterium–tritium targets on the Omega Laser System for over a decade. Fuel entropy is inferred in these experiment by measuring fuel areal density near peak compression. Measured areal densities up to $\langle \rho R \rangle_{\rm n} \sim 300\,{\rm mg/cm^2}$ (larger than 85 % of predicted values) are demonstrated in the cryogenic implosion with $V_{\rm imp}$ approaching 3×10^7 cm/s and peak laser intensities of $8 \times 10^{14}\,{\rm W/cm^2}$. Scaled to the laser energies available at the National Ignition Facility, implosions, hydrodynamically equivalent to these Omega designs, are predicted to achieve $\langle \rho R \rangle_{\rm n} > 1.2\,{\rm g/cm^2}$, sufficient for ignition demonstration in direct-drive ICF experiments.

7.1 Introduction

To ignite the deuterium–tritium (DT) fuel in a conventional, hot-spot ignition scheme in inertial confinement fusion (ICF), ion temperature and areal density of the central, lower-density region (hot spot) of the final fuel assembly must be sufficient to create fuel self-heating by alpha particles produced as a result of fusing D and T [1, 2]. In addition, the areal density (ρR) of the main fuel must be large enough to provide confinement time sufficient to burn a significant portion of that fuel. A typical target consists of a higher-density shell filled with a lower-density fuel vapor. The shell has an outer layer of ablator material and an inner layer of

V.N. Goncharov (✉)
Laboratory for Laser Energetics, Department of Mechanical Engineering, University of Rochester, Rochester, NY 14623, USA
e-mail: vgon@lle.rochester.edu

P. McKenna et al. (eds.), *Laser-Plasma Interactions and Applications*, Scottish Graduate Series, DOI 10.1007/978-3-319-00038-1_7,

frozen fuel. To compress the main fuel layer and initiate a burn wave propagating from the vapor through the main fuel, the shell is accelerated inward by a shaped pressure drive created by laser energy that is delivered either directly to the target (direct drive) or indirectly by converting its energy to x-rays inside the hohlraum (indirect drive) [1, 2]. As pressure in the vapor builds up due to convergence, the shell begins to decelerate when the vapor pressure exceeds shell pressure and an outgoing shock wave is launched into the incoming shell. During deceleration, hot-spot areal density and temperature increase as the shell's kinetic energy is converted into internal energy of the hot spot and main fuel. Achieving ignition conditions requires the areal density of the hot spot to exceed stopping range of the alpha particles produced by fusing D and T. This leads to $(\rho R)_{\rm hs} \geq 0.3\,{\rm g/cm^2}$ [1, 2]. In addition, the hot-spot ion temperature $T_{\rm hs}$ must be larger than $\sim$5 keV so that the alpha heating exceeds bremsstrahlung losses [1,2]. Since both hot-spot areal density and temperature depend on in-flight shell kinetic energy, there is a threshold value of this energy below which target fails to ignite.

Target designing starts by calculating how much energy the drive pressure must provide to the shell so ignition requirements are met at stagnation. Numerical simulations give the following expression for the minimum shell kinetic energy required for ignition [3, 4]:

$$E_{\rm k,min}({\rm kJ}) = 51\alpha^{1.9}\left(\frac{V_{\rm imp}}{3\times 10^7}\right)^{-5.9}\left(\frac{p_{\rm a}}{100\ {\rm Mbar}}\right)^{-0.8} \tag{7.1}$$

This expression depends on the following in-flight hydrodynamic parameters, crucial for achieving ignition: (1) the peak in mass-averaged main fuel velocity (implosion velocity) $V_{\rm imp}$; (2) the in-flight fuel adiabat α [defined as the ratio of the shell pressure p to the Fermi-degenerate pressure at shell density ρ; for DT fuel, $p \simeq \mu\alpha\rho^{5/3}$ and $\mu = 2.2\,{\rm Mbar/(g/cm^3)^{5/3}}$], and (3) the drive (ablation) pressure $p_{\rm a}$. Even though Eq. 7.1 provides a very useful scaling law, it gives very little insight into the physics that determines this scaling. To provide such an insight, a simplified model of hot-spot formation is developed and presented next.

7.1.1 A Simple Ignition Model

To calculate minimum shell kinetic energy of an igniting target, nearly all this energy is assumed to be converted into the internal hot-spot and fuel energy at stagnation,

$$E_k \sim p_{\rm max}R^3 \sim (\rho_{\rm hs}T_{\rm hs}R)^3/p_{\rm max}^2, \tag{7.2}$$

where $p_{\rm max} \sim \rho_{\rm hs}T_{\rm hs}/m_{\rm i}$ is the peak hot-spot pressure and $m_{\rm i}$ is ion mass. Since the minimum value of product $(\rho R)_{hs}T_{hs}$ is 0.3 $\,{\rm g/cm^2}\times 5$ keV, as described earlier, then [2]:

$$E_{\rm k,min} \sim 1/p_{\rm max}^2 \tag{7.3}$$

and calculation of $E_{\rm k,min}$ reduces to determining the peak hot-spot pressure.

The maximum pressure is calculated by assuming that the hot-spot radius at peak convergence is R, and a fraction $f_{\rm shl}$ of shell kinetic energy $E_k = MV_{\rm imp}^2/2$ has been transferred at that time to the hot-spot internal energy $2\pi p_{\rm max}R^3$, where M is the unablated shell mass. Then, the maximum hot-spot pressure is

$$p_{\rm max} \sim f_{\rm shl}E_k/R^3. \tag{7.4}$$

With the goal of expressing $E_{\rm k,min}$ and $p_{\rm max}$ in terms of in-flight shell parameters, stagnation variables must be related to these at the beginning of shell deceleration. Using the fact that the hot spot is adiabatic during deceleration [4, 5], $p_{\rm max}$ can be written in terms of vapor pressure p_d and radius of vapor region R_d at the beginning of shell deceleration:

$$p_{\rm max} = p_d(R_d/R)^5. \tag{7.5}$$

Equating right-hand sides of Eqs. 7.4 and 7.5 gives a hot-spot convergence ratio during deceleration,

$$\frac{R_d}{R} \sim \sqrt{\frac{f_{\rm shl}E_k}{p_dR_d^3}}. \tag{7.6}$$

Then, using Eqs. 7.5 and 7.6 defines the maximum hot-spot pressure as a ratio of the shell kinetic energy to the internal energy of the vapor at the beginning of deceleration [5]:

$$p_{\rm max} \sim p_d\left(\frac{f_{\rm shl}E_k}{p_dR_d^3}\right)^{5/2} \sim p_d\left(\frac{f_{\rm shl}M}{p_dR_d^3}\right)^{5/2} V_{\rm imp}^5. \tag{7.7}$$

For $f_{\rm shl} = 1$, Eqs. 7.3 and 7.7 give $p_{\rm max} \sim V_{\rm imp}^5$ and $E_{\rm k,min} \sim V_{\rm imp}^{-10}$, similar to the result of the isobaric model [6]. The fraction $f_{\rm shl}$, however, is smaller than unity and depends on in-flight shell parameters. Keeping in mind that the shell is decelerated by the outgoing shock wave, $f_{\rm shl}$ can be defined as a fraction of the shell mass (an effective mass $M_{\rm eff}$) overtaken by this shock while the hot spot converges inward. In the strong shock limit, the Hugoniot conditions across the shock give:

$$M_{\rm eff} \equiv f_{\rm shl}M \sim \sqrt{\rho_{\rm shl}p_{\rm max}}R^2\Delta t, \tag{7.8}$$

where $\rho_{\rm shl}$ is the shell density ahead of the shock front. The hot-spot time of confinement by the shell inertia is determined by Newton's law, $M_{\rm eff}R/(\Delta t)^2 \sim p_{\rm max}R^2$, which yields [7]:

$$\Delta t \sim \sqrt{M_{\rm eff}/p_{\rm max}R}. \tag{7.9}$$

Then, Eqs. 7.8 and 7.9 lead to:

$$M_{\rm eff} \sim \rho_{\rm shl}R^3. \tag{7.10}$$

With the help of the latter equation, Eq. 7.4 yields intuitively simple scaling

$$p_{\max} \sim \rho_{\text{shl}} V_{\text{imp}}^2. \tag{7.11}$$

The maximum pressure, however, does not scale as V_{imp}^2, as Eq. 7.11 would suggest, since ρ_{shl} is different from the in-flight shell density. As the unshocked part of the incoming shell keeps converging during deceleration, its density ρ_{shl} increases inversely proportional to surface area:

$$\rho_{\text{shl}} \simeq \rho_d \left(\frac{R_d}{R}\right)^2. \tag{7.12}$$

Combining Eqs. 7.5, 7.11, and 7.12 defines the hot-spot convergence ratio in terms of in-flight shell quantities:

$$\frac{R_d}{R} \sim \left(\frac{V_{\text{imp}}^2 \rho_d}{p_d}\right)^{1/3}. \tag{7.13}$$

Substituting Eq. 7.13 into Eqs. 7.10 and 7.12 gives the effective shell mass and ρ_{shl}:

$$M_{\text{eff}} \sim \rho_d R_d^3 \frac{R}{R_d} \sim \rho_d R_d^3 \left(\frac{p_d}{\rho_d V_{\text{imp}}^2}\right)^{1/3}, \tag{7.14}$$

$$\rho_{\text{shl}} \sim \rho_d \left(\frac{V_{\text{imp}}^2 \rho_d}{p_d}\right)^{2/3}. \tag{7.15}$$

Finally, the scaling for the maximum pressure is obtained by combining Eqs. 7.7 and 7.14:

$$p_{\max} \sim \rho_d V_{\text{imp}}^2 \left(\frac{V_{\text{imp}}^2 \rho_d}{p_d}\right)^{2/3} = p_d \left(\frac{V_{\text{imp}}^2 \rho_d}{p_d}\right)^{5/3}. \tag{7.16}$$

Pressure at the beginning of the deceleration phase is proportional to the drive ablation pressure, $p_d \sim p_{\text{a}}$, and shell density is related to the drive pressure through the in-flight shell adiabat α, $p_d(\text{Mbar}) \sim 2.2\alpha\rho_d^{5/3}$. This gives:

$$p_{\max} \sim \frac{p_{\text{a}}^{1/3} V_{\text{imp}}^{10/3}}{\alpha}. \tag{7.17}$$

This scaling of $p_{\max}$ with V_{imp} is similar to that derived using self-similar analysis [8] which leads to $p_{\max}^{\text{self-similar}} \sim V_{\text{imp}}^3$. Substituting Eq. 7.17 back into Eq. 7.3 gives a scaling law similar to that obtained using simulation results [see Eq. 7.1]:

$$E_{\text{k,min}} \sim 1/p_{\max}^2 \sim V_{\text{imp}}^{-20/3} p_{\text{a}}^{-2/3} \alpha^2. \tag{7.18}$$

Equation 7.17 shows that the maximum pressure has a weaker implosion velocity dependence than $V_{\rm imp}^5$ obtained assuming that all kinetic energy of the shell is transferred to the internal energy of the fuel at stagnation. The weaker dependence is due to the fact that the kinetic energy fraction contributing to the fuel internal energy is proportional to the fraction of the shell mass overtaken by the outgoing shock wave during the hot-spot confinement time. Several competing effects define this fraction: First, the mass flux per unit areal across the shock increases with hot-spot convergence since both shell density $\rho_{\rm shl}$ and maximum pressure $p_{\rm max}$ increase with R_d/R [see Eqs. 7.5 and 7.12], so $\sqrt{\rho_{\rm shl} p_{\rm max}} \sim \sqrt{\rho_d p_d}(R_d/R)^{7/2}$. Multiplied by the surface area of the shock front, the mass flux across the shock is $\sqrt{\rho_{\rm shl} p_{\rm max}} R^2 \sim \sqrt{\rho_d p_d} R_d^2 (R_d/R)^{3/2}$. The convergence ratio increases with the implosion velocity, as shown in Eq. 7.13, giving:

$$\text{mass flux} \sim \sqrt{\rho_{\rm shl} p_{\rm max}} R^2 \sim V_{\rm imp}. \tag{7.19}$$

The confinement time, on the other hand, decreases with convergence ratio and implosion velocity. Indeed, writing $\Delta t \sim R/V_{\rm imp}$ [this can be obtained by substituting Eqs. 7.10 and 7.11 into Eq. 7.9] and using Eq. 7.13 gives:

$$\text{confinement time} \sim \frac{R}{V_{\rm imp}} \sim \left(\frac{R_d}{R}\right)^{-5/2} \sim V_{\rm imp}^{-5/3}. \tag{7.20}$$

Then, the product of mass flux and confinement time gives the effective mass and fraction of kinetic energy that contributes to the stagnation pressure $M_{\rm eff} \sim f_{\rm shl} \sim V_{\rm imp}^{-2/3}$, in agreement with Eq. 7.14. Negative power in velocity dependence of the effective mass changes pressure scaling from $V_{\rm imp}^5$ to $V_{\rm imp}^{10/3}$.

The maximum pressure, on other hand, has a stronger dependence on $V_{\rm imp}$ than that given by the dynamic pressure argument $p_{\rm max} \sim \rho_{\rm shl} V_{\rm imp}^2$. This is due to convergence effects and an increase in the unshocked shell density during deceleration. Since $\rho_{\rm shl}$ rises with convergence ratio and implosion velocity, as shown in Eq. 7.12, the maximum pressure scales as $p_{\rm max} \sim (\rho_{\rm inflight} V_{\rm imp}^{2\times 2/3}) V_{\rm imp}^2 \sim V_{\rm imp}^{10/3}$, in agreement with Eq. 7.17.

Since $E_{\rm k,min}$ has strong dependence on implosion velocity, as shown in Eqs. 7.1 and 7.18, it is crucial that a shell reaches the designed value of $V_{\rm imp}$ to achieve ignition in an experiment. The minimum $V_{\rm imp}$ can be estimated by the following argument: Balancing a fraction of the kinetic energy of the shell and the internal energy of the fuel yields:

$$M V_{\rm imp}^2/2 > 2\pi p_{\rm max} R^3. \tag{7.21}$$

For fully ionized gas with ion charge Z and ion mass $m_{\rm i}$, $p_{\rm max} = (1+Z)\rho_{\rm hs} T_{\rm hs}/m_{\rm i}$. For DT fuel this gives $p_{\rm max} \simeq 4\rho_{\rm hs} T_{\rm hs}/5m_P$, where m_p is proton mass. Finally, writing shell mass at stagnation as $M \sim 4\pi R^2 \rho_{\rm fuel} \Delta$ leads to:

$$V_{\rm imp} > \sqrt{\frac{4}{5}\frac{(\rho R)_{\rm hs}}{(\rho\Delta)_{\rm fuel}}\frac{T_{\rm hs}}{m_p}}, \tag{7.22}$$

where ρ_{fuel} and Δ are the density and thickness of compressed fuel, respectively. To create a hot spot and trigger burn propagation into the cold fuel, the hot-spot areal density and temperature must exceed, as discussed earlier, $(\rho R)_{hs} \times T_{hs} > 0.3\,\text{g/cm}^2 \times 5\,\text{keV}$. To burn enough cold fuel and achieve gain=fusion energy/laser energy > 1 requires, on the other hand, $(\rho\Delta)_{\text{fuel}} > 1\ \text{g/cm}^2$ [1,2]. Substituting these three conditions back into Eq. 7.22 gives:

$$V_{\text{imp}} > 3 \times 10^7\ \text{cm/s}. \tag{7.23}$$

This leads to a requirement on stagnation pressure $p_{\max}$. Indeed, the ablation pressure in an ICF implosion is $p_a \sim 100$ Mbar, and the effective dynamic pressure of the accelerated shell at $V_{\text{imp}} = 3 \times 10^7$ cm/s and $\alpha = 1$ is $\rho V^2 \simeq (100/2.2)^{3/5}$ $[3 \times 10^7]^2 \simeq 9$ Gbar. In general,

$$\text{dynamic pressure}_{\text{inflight}} \simeq 9\left(\frac{p_a}{100\ \text{Mbar}}\right)^{3/5} \alpha^{-3/5} \left(\frac{V_{\text{imp}}}{3 \times 10^7}\right)^2 \text{Gbar}. \tag{7.24}$$

An additional amplification in dynamic pressure is due to shell convergence during deceleration. As described earlier, unshocked-shell density amplification is proportional to the hot-spot convergence ratio to the second power [see Eq. 7.12]. According to Eq. 7.13, the hot spot converges by a factor of ~4.4 during deceleration for $\alpha \sim 1$ and $V_{\text{imp}} \sim 3 \times 10^7$ cm/s. This gives an additional increase by a factor of $4.4^2 = 20$ in the dynamic pressure, leading to a maximum hot-spot pressure in an igniting target of $p_{\max} > 200$ Gbar, or for a given implosion velocity and drive pressure,

$$p_{\max} \simeq 180\left(\frac{p_a}{100\ \text{Mbar}}\right)^{1/3} \alpha^{-1} \left(\frac{V_{\text{imp}}}{3 \times 10^7}\right)^{10/3} \text{Gbar}. \tag{7.25}$$

Using the numerical factor obtained in Eq. 7.25, one can recover a numerical factor in Eq. 7.18 as well,

$$E_{\text{k,min}} \simeq 30\,\alpha^2 \left(\frac{V_{\text{imp}}}{3 \times 10^7}\right)^{-20/3} \left(\frac{p_a}{100\ \text{Mbar}}\right)^{-2/3} \text{kJ}. \tag{7.26}$$

The numerical coefficient in Eq. 7.26 is 40 % smaller than that in the fitting formula shown in Eq. 7.1. This is a consequence of the fact that only a fraction f_{shl} of the total shell kinetic energy is transferred to the fuel at stagnation. Typically, $f_{\text{shl}} \sim 0.5$–0.6, which brings the numerical coefficient in Eq. 7.26 in closer agreement with the numerical result.

7.1.2 Sensitivity of Ignition Condition on Implosion Parameters

The minimum shell kinetic energy required for ignition has a strong dependence on shell's velocity and adiabat [see Eq. 7.1]. When a particular target design is considered for an ignition experiment, one of the important design parameters is margin [this is also referred to as an ignition threshold factor (ITF)] [9] defined as the ratio of the shell kinetic energy E_k to its minimum value required for ignition, $E_{\mathrm{k,min}}$,

$$\mathrm{ITF} = \frac{E_k}{E_{\mathrm{k,min}}}. \tag{7.27}$$

In using Eq. 7.1 to determine $E_{\mathrm{k,min}}$, one needs to keep in mind that Eq. 7.1 does not account for asymmetry effects (such as shell and hot-spot nonuniformity growth, mix of ablator material and fuel, etc.). A more-complete analysis using two- and three-dimensional hydrodynamic simulations results in correction factors related to these effects (for details, see [9]). Since the main purpose of this paper is to address accuracy in the modelling of average one-dimensional (1-D) hydrodynamic parameters, the terms proportional to multidimensional effects will be neglected.

Robustness of a particular design is determined by how much uncertainty in velocity, adiabat, and the drive pressure it can tolerate before the probability of achieving ignition becomes very small. Such maximum uncertainty values depend on ITF.

The target fails to ignite if the shell kinetic energy E_k in an experiment is lower than the ignition energy threshold $E_{\mathrm{k,min}}$ or the actual energy threshold $E_{\mathrm{k,min}}$ is higher than calculated $E_{\mathrm{k,min}}$ due to inaccuracies in modelling of hydrodynamic quantities. If E_k^{design} and $E_{\mathrm{k,min}}^{\mathrm{design}}$ are design values of shell kinetic energy and energy threshold, respectively, and ITF$= E_k^{\mathrm{design}}/E_{\mathrm{k,min}}^{\mathrm{design}}$, then the maximum deviations in V_{imp}, α, and p_{a} (which are denoted as δV_{imp}, $\delta\alpha$, and δp_{a}, respectively) from predictions are determined from condition $E_k^{\mathrm{limit}}/E_{\mathrm{k,min}}^{\mathrm{limit}} = 1$, where $E_k^{\mathrm{limit}} = M(V_{\mathrm{imp}} - \delta V_{\mathrm{imp}})^2/2$, $E_{\mathrm{k,min}}^{\mathrm{limit}} = E_{\mathrm{k,min}}(\alpha + \delta\alpha, V_{\mathrm{imp}} - \delta V_{\mathrm{imp}}, p_{\mathrm{a}} - \delta p_{\mathrm{a}})$, $E_k^{\mathrm{design}} = MV_{\mathrm{imp}}^2/2$, and $E_{\mathrm{k,min}}^{\mathrm{design}} = E_{\mathrm{k,min}}(\alpha, V_{\mathrm{imp}}, p_{\mathrm{a}})$. This reads as

$$1 = \mathrm{ITF}\left(1 - \frac{\delta V_{\mathrm{imp}}}{V_{\mathrm{imp}}}\right)^{7.9}\left(1 + \frac{\delta\alpha}{\alpha}\right)^{-1.9}\left(1 - \frac{\delta p_{\mathrm{a}}}{p_{\mathrm{a}}}\right)^{0.8}. \tag{7.28}$$

Since it is very difficult to assess the fuel adiabat by a direct measurement, the adiabat increase $\delta\alpha$ is replaced in this analysis with an equivalent amount of energy deposited in the fuel, ΔE, expressed in terms of a fraction ε_E of the shell kinetic energy $\Delta E = \varepsilon_E E_{k,0}$. To relate $\delta\alpha$ and ΔE, we write internal energy as a product of pressure and volume, $E = 3/2pV$. Replacing pressure by the drive ablation pressure p_{a} and the fuel volume by fuel mass over shell density, $V = M/\rho$, gives $E = 3p_{\mathrm{a}}M/2\rho$. Shell density is related to the ablation pressure as $\rho \sim (p_{\mathrm{a}}/\alpha)^{3/5}$. Then, collecting all appropriate numerical coefficients leads to

$$E(\mathrm{kJ}) = 1.5\left(\frac{p_{\mathrm{a}}}{100\ \mathrm{Mbar}}\right)^{2/5}\alpha^{3/5}M(\mathrm{mg}). \tag{7.29}$$

Fixing shell mass and drive pressure gives $1+\delta\alpha/\alpha=(1+\Delta E/E_0)^{5/3}$. Then, Eq. 7.28) takes the form

$$\left(1-\frac{\delta V_{\mathrm{imp}}}{V_{\mathrm{imp}}}\right)^{-7.9}\left[1+\frac{30\varepsilon_E(V_{\mathrm{imp}}/3\times10^7)^2}{(p_{\mathrm{a}}/100\,\mathrm{Mbar})^{2/5}\alpha^{3/5}}\right]^{3.1}\left(1-\frac{\delta p_{\mathrm{a}}}{p_{\mathrm{a}}}\right)^{-0.8}=\mathrm{ITF}. \quad (7.30)$$

Figure 7.1 shows plots of maximum-allowable reduction in shell velocity (a), shell preheat as a percentage fraction of the shell kinetic energy (b), and reduction in drive pressure (c) as functions of ITF.

Figure 7.1 shows that for NIF-scale ignition designs with ITF ~3.5–5, ignition fails if reduction in velocity is greater than ~15 %, the shell is preheated by more than ~1 % of the shell kinetic energy. The drive pressure, according to Fig. 7.1c, can be reduced as much as 80 % before ignition will fail, but this number does not account for a reduction in the implosion velocity associated with reduced drive. Therefore, (c) must be used in combination with (a). In addition to ignition failure due to a significant deviation from predicted 1-D hydrodynamic parameters (velocity, adiabat, drive pressure), other failure mechanisms are due to asymmetries in an implosion. Nonuniformity sources are target imperfections, such as ice roughness and ablator roughness, and asymmetry in laser illumination. Multiplied by the Rayleigh–Taylor (RT) and Richtmyer–Meshkov (RM) instabilities [1, 2] during an implosion, such nonuniformities could either disrupt the shell or lead to significant hot-spot distortions. The distortion region width inside the hot spot exceeding 20–40 % of the 1-D hot-spot radius is typically sufficient to reduce alpha-particle production and ion temperature and quench the burn [7].

Even though control of the multidimensional effect is one of the main challenges for any ignition design, validation of code ability to adequately model target drive efficiency and the amount of the fuel preheat is a primary goal of the ICF experiments. This paper will describe how these global hydrodynamic parameters predicted by hydro-simulations were experimentally validated using direct-drive implosions on Omega.

7.2 Early Direct-Drive Target Designs and Target Stability Properties

7.2.1 All-DT, Direct-Drive, NIF-Scale Ignition Target Design

The original direct-drive target design [10, 11] for the National Ignition Facility (NIF) Laser System [12] is a 350-μm-thick, solid-DT layer inside a very thin (<3-μm) plastic shell (shown in Fig. 7.2).

Because the plastic shell ablates early in the pulse and the DT layer acts as both the main fuel and ablator, this design is referred to as an 'all-DT' design. The fact that the ablator and the main fuel are the same material (DT) has several

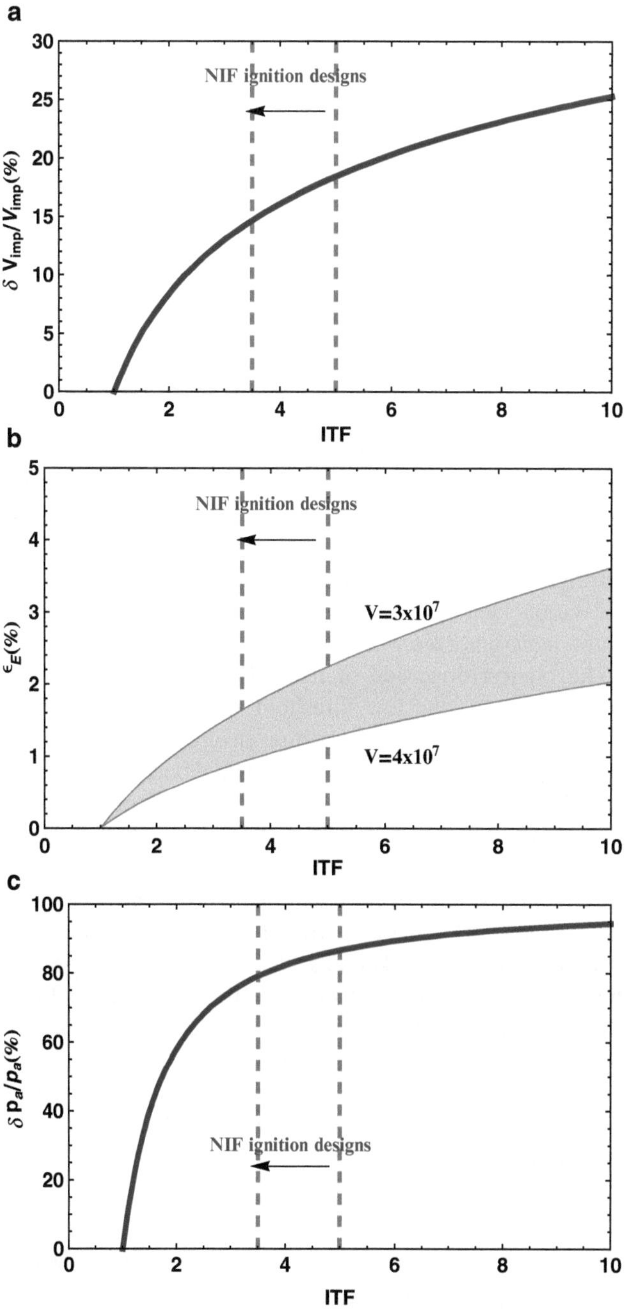

Fig. 7.1 Maximum velocity reduction (**a**), maximum preheat energy as fractions of the shell kinetic energy (**b**), and maximum pressure reduction (**c**) versus ITF

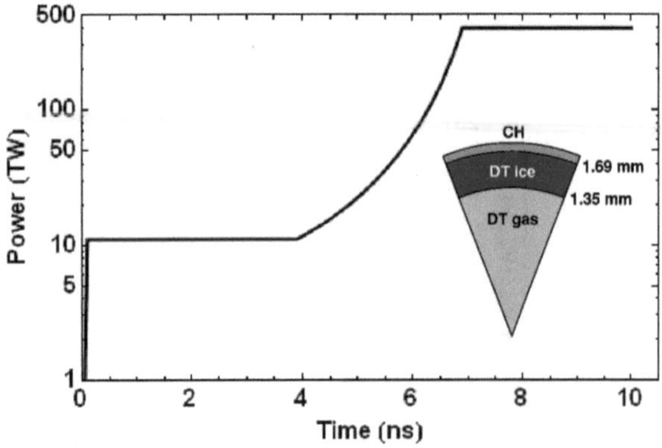

Fig. 7.2 An $\alpha = 3$, 'all-DT', 1.5-MJ, direct-drive–ignition target design with a 1-D gain of 45

advantages: (1) This eliminates the interface between the fuel and ablator. Any mismatch in density or opacity between two neighbouring materials in the shell usually leads to an enhancement in the early-time perturbation growth or the RT instability growth factor [13]. (2) Because of its initial low density, DT gives both the lowest in-flight aspect ratio (IFAR) for the same shell mass and the largest ablative stabilisation factor in the RT instability growth rate formula compared to other ablator materials (see Sect. 7.2.2.1 for more details on design stability properties). The biggest downside of using DT as an ablator, as demonstrated in Omega experiments, is the low threshold for the two-plasmon-decay (TPD) instability [14], which generates suprathermal electrons that preheat the fuel. At the time of writing this article, there is no experimental demonstration of low-adiabat, high fuel compression in direct-drive designs with DT or D_2 ablators driven at ignition-relevant intensities above 3×10^{14} W/cm^2 (this will be discussed further in Sect. 7.5). In the design presented in Fig. 7.2, the fuel is accelerated by 1.5 MJ of laser energy to a peak velocity of $V_{\text{imp}} = 4.3 \times 10^7$ cm/s at adiabat $\alpha = 3$. The target ignites and gives 1-D gain of 45 with ITF = 5. This design uses a continuous pulse shape (as opposed to the picket pulse described in the next section), launching the initial shock that sets the in-flight shell adiabat. Later, at $t = 4$ ns, the intensity gradually rises, launching a compression wave. The head of this wave catches up with the first shock in the vapor region, soon after it breaks out of the shell. Timing the first shock and compression wave breaking out of the fuel and preventing the compression wave from turning into a shock inside the fuel are crucial for achieving ignition in this design.

7.2.2 Target Stability Properties: Rayleigh–Taylor Instability Growth and Target IFAR

A shell kinetic energy required to ignite DT fuel in an ICF implosion is strongly dependent on the maximum shell velocity. According to Eq. 7.1, increasing the shell velocity to well above the minimum value of $V_{\text{imp}} \sim 3 \times 10^7$ cm/s seems to be very beneficial for reducing the laser-energy requirement. Increasing implosion velocity, however, must be achieved without compromising shell stability. To understand how V_{imp} scales with target parameters, we start by writing

$$V_{imp} \sim g t_{accel}, \tag{7.31}$$

where g is shell acceleration and t_{accel} is the acceleration time. The acceleration is determined from Newton's law,

$$M_{shell} g \sim 4\pi R^2 p_a \rightarrow g \sim 4\pi \frac{p_a R^2}{M_{shell}}, \tag{7.32}$$

where M_{shell} is the initial shell mass, R is shell radius, and p_a is ablation pressure. For a given laser energy E_{laser} and drive intensity I, the acceleration time is

$$t_{accel} \sim \frac{E_{laser}}{4\pi R^2 I}. \tag{7.33}$$

Substituting Eqs. 7.32 and 7.33 into Eq. 7.31 gives $V_{imp} \sim p_a E_{laser}/M_{shell} I$. Results of simulations lead to a numerical factor of 0.8 in the latter equation. Therefore,

$$V_{imp} \simeq 0.8 \frac{p_a E_{laser}}{M_{shell} I}. \tag{7.34}$$

Since $p_a \sim I^{0.8}$ to $I^{0.7}$ [1, 2], implosion velocity increases, for a given shell mass and laser energy, by reducing drive intensity. This intuitively contradictory result can be explained by noting that a lower laser drive is overcompensated by the duration of the shell acceleration, as shown in Eq. 7.33. The acceleration distance is longer for lower intensity drives: $R \sim V_{imp} t_{accel} \sim p_a E_{laser}^2/M_{shell} I^2 R^2$, so $R^3 \sim p_a E_{laser}^2/M_{shell} I^2 \sim I^{-1.2}$. The implosion velocity can also be increased, according to Eq. 7.34, by reducing shell mass. An increase in V_{imp}, however, is beneficial for reducing $E_{k,min}$ only up to the point where multidimensional effects (asymmetry growth) start to affect target performance. Hydrodynamic instabilities put severe constraints on target designs, limiting the values of the shell mass and adiabat used in a robust target design. To determine such constraints, we next identify target parameters that affect the target stability.

7.2.2.1 Rayleigh–Taylor Instability

The dominant hydrodynamic instability in an ICF implosion is the Rayleigh–Taylor (RT) instability [1, 2]. The RT instability develops in systems where the heavier fluid is accelerated by the lighter fluid [15]. In an ICF implosion, the heavier shell material is accelerated by the lighter blowoff plasma, creating the condition for the RT instability. This instability amplifies shell distortions, seeded by both the ablator

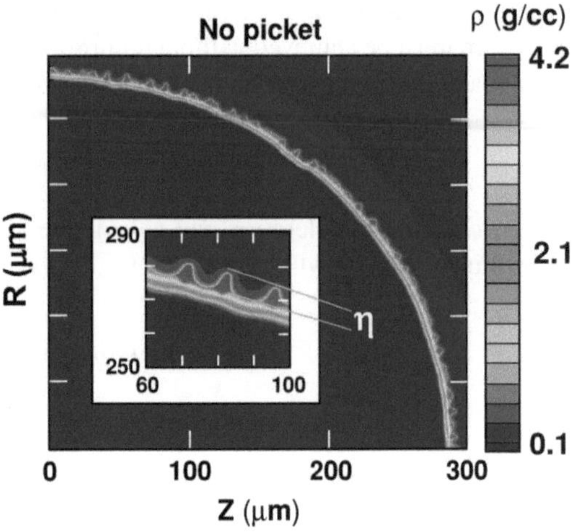

Fig. 7.3 Two-dimensional simulation of a direct-drive implosion using hydrocode DRACO. Shell distortions with amplitude η grow in time due to the Rayleigh–Taylor instability

and ice roughness, and laser illumination nonuniformities (laser 'imprint' [13]). Excessive growth of these perturbations leads to shell breakup during acceleration, limiting the final compression and hot-spot temperature. An example of a direct-drive implosion simulation is shown in Fig. 7.3.

Shell distortions developed due to the RT instability during acceleration are clearly visible in this simulation. The small initial perturbation amplitude η_0 grows in time as

$$\eta \sim \eta_0 e^{\gamma_{\mathrm{RT}} t}, \tag{7.35}$$

where the γ_{RT} is the growth rate. In the classical RT configuration where a heavier fluid with density ρ_2 is supported by a lighter fluid of density ρ_1 in a gravitational field g directed from heavier to lighter fluids, the RT growth rate is

$$\gamma_{\mathrm{RT,classical}} = \sqrt{A_T kg}, \quad A_T = \frac{\rho_2 - \rho_1}{\rho_2 + \rho_1}, \tag{7.36}$$

where A_T is Atwood number, $k = 2\pi/\lambda$ is the perturbation wave number, and λ is the perturbation wavelength. In an ICF implosion, the thermal conduction (electron dominant in direct-drive implosions and x-ray-radiation–dominant in indirect-drive implosions) that drives the ablation process significantly reduces the growth rate from its classical value [16]. The full expression for the growth rate in this case is rather complicated and can be found in [17]. Here, we show the growth rate in the limit $kL_0 < 1$, where L_0 is the effective thickness of the ablation front,

$$\gamma_{\mathrm{RT,ICF}} \simeq \sqrt{kg - \Omega_{\mathrm{bl}}^2 + \Omega_a^2} - \Omega_a, \quad \Omega_{\mathrm{bl}} = k\sqrt{V_a V_{\mathrm{bl}}}, \quad \Omega_a = 2kV_a, \tag{7.37}$$

where V_a and V_{bl} are the ablation and blowoff velocities, respectively (for definition of V_{bl} see [17]). Because mass density in the plasma blowoff region is much smaller than shell density, $A_T \simeq 1$ for modes with $kL_0 < 1$. There are two stabilising terms

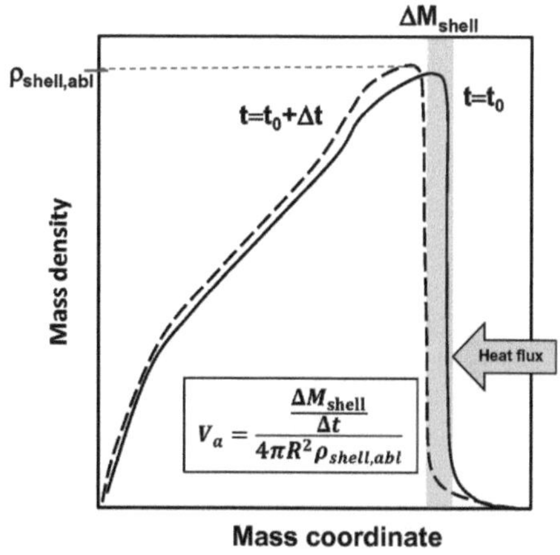

Fig. 7.4 Ablation velocity is defined as mass ablation rate, $\Delta M/\Delta t \simeq \mathrm{d}M/\mathrm{d}t$, divided by the ablation region surface area $4\pi R^2$ and the shell density at the ablation front $\rho_{\mathrm{shell,abl}}$

in $\gamma_{\mathrm{RT,ICF}}$: the first is proportional to Ω_{bl} and the other to Ω_a. Both of them are due to the mass ablation driven by thermal conduction; physical mechanisms of the two, however, are different.

The ablation process is characterised by the ablation velocity V_a, defined as the ratio of the mass ablation rate per unit area of target surface, $(\mathrm{d}M/\mathrm{d}t)/(4\pi R^2)$, and the shell density at the ablation front $\rho_{\mathrm{shell,abl}}$ (see Fig. 7.4),

$$V_a = \frac{\mathrm{d}M}{\mathrm{d}t} \Big/ \left(4\pi R^2 \rho_{\mathrm{shell,abl}}\right), \tag{7.38}$$

where R is the ablation-front radius.

When mass ablation is included, several physical mechanisms reduce the ablation-front perturbation growth and, in some cases, totally suppress it. These are illustrated in Fig. 7.5.

First, different plasma blowoff velocities at different parts of the corrugated ablation region create modulation in the dynamic pressure or 'rocket effect' that leads to a stabilising restoring force [13, 18, 19]. Indeed, as a result of the perturbation growth, the peaks [point A in Fig. 7.5a] of the ablation-front ripple protrude into the hotter plasma corona, and the valleys [point B in Fig. 7.5a] recede toward the colder shell material. Since the temperature is uniform along the ablation front [16], the temperature gradients and the heat fluxes are slightly enhanced at the peaks and reduced at the valleys, as shown in Fig. 7.5a. An excess/deficiency in the heat flux speeds up/slows down the ablation front. This is illustrated in Fig. 7.5b, where the solid and dashed lines indicate the positions of the ablation front at two instances in time separated by Δt. The ablation front at the peaks (point A) propagates further into the shell than at the valleys (point B). This increases velocity of the blowoff material ('exhaust' velocity, if an analogy of the ablatively driven shell with a rocket is used) at point A and reduces it at point B. A modulation in the blow-off velocity leads to a modulation in the dynamic pressure, creating a restoring

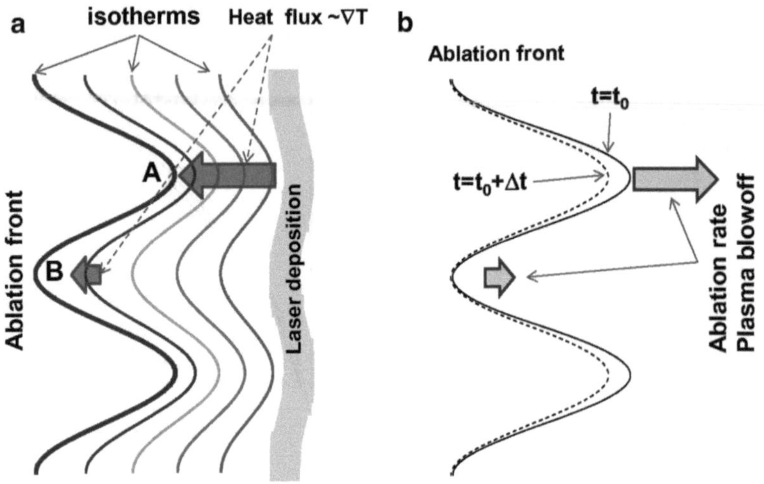

Fig. 7.5 (**a**) Ablation-front modulation creates stronger temperature gradients at perturbation peaks (*A*) and weaker gradients at valleys (*B*). Since heat flux is proportional to such gradients, this leads to a slightly enhanced heat flux at A and a reduced heat flux at B. (**b**) Modulation in heat flux results in modulation in the mass ablation rate. The mass removed by ablation at point A is larger than that at point B, leading to both a fire-polishing effect and a restoring force caused by dynamic overpressure

force and reducing perturbation growth [see terms with $\Omega_{\rm bl}^2$ in Eq. 7.37]. The second stabilising mechanism caused by ablation is an increased mass ablation rate at the perturbation peaks in comparison with the valley. This leads to faster mass removal at point A and slower removal at point B (so-called 'fire-polishing' effect). This, together with convection of vorticity from the unstable ablation front toward the blowoff region, gives the stabilising terms proportional to Ω_a in Eq. 7.37.

Since the ablation and blow-off velocities are inversely proportional to the shell density at the ablation front, and density and ablation pressure are related as $\rho_{\rm shell,abl} \sim (p_{\rm a}/\alpha_{\rm abl})^{3/5}$, the velocities scale with the adiabat near the ablation front $\alpha_{\rm abl}$ as

$$V_a \sim V_{\rm bl} \sim \alpha_{\rm abl}^{3/5}. \tag{7.39}$$

Equation 7.39 shows that reducing shell density or increasing shell adiabat at the ablation front enhances shell stability.

7.2.2.2 Target in-Flight Aspect Ratio (IFAR)

The other important parameter characterising shell stability is the shell in-flight aspect ratio (IFAR) defined as the ratio of the shell radius R to the in-flight shell thickness $\Delta_{\rm inflight}$ (see Fig. 7.6).

Designs with thicker shells are less sensitive to the instability growth because they break up at a larger distortion amplitude and have smaller seeding of the

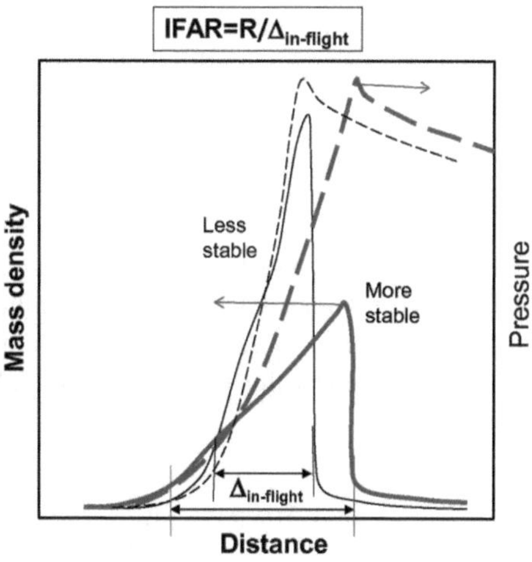

Fig. 7.6 In-flight aspect ratio is defined as the ratio of the shell radius to the in-flight shell thickness. Designs with smaller IFAR are less sensitive to the shell distortion growth since they break up at larger distortion amplitudes

deceleration RT instability. Such an instability develops as lower-density vapor pushes against the higher-density shell. During the shell acceleration, perturbations feed through from the unstable ablation front to the inner shell surface, $\eta_{\text{inner}} \sim \eta_{\text{ablation}} e^{-k\Delta_{\text{inflight}}}$. As the shell decelerates, the inner surface distortions start to grow from η_{inner}, leading to hot-spot deformation at peak compression. Thus, the thicker the shell, the smaller the feed through factor, and the smaller the finite hot-spot deformation.

Next, we find a scaling of IFAR with implosion parameters. As defined, IFAR$=R/\Delta_{\text{inflight}}$. The in-flight shell thickness is the initial shell thickness Δ_0 reduced by shell compression during acceleration (effect of mass ablation is neglected in this analysis),

$$\Delta_{\text{inflight}} \simeq \Delta_0 \frac{\rho_0}{\langle\rho\rangle_{\text{inflight}}} \frac{R_0^2}{R^2}, \tag{7.40}$$

where ρ_0 and $\langle\rho\rangle$ are initial and average in-flight shell densities, respectively, and R_0 is the initial shell radius. For the all-DT design where the shell consists mainly of DT, $\langle\rho\rangle_{\text{inflight}} = [p_a(\text{Mbar})/2.2\langle\alpha\rangle]^{3/5}$, where $\langle\alpha\rangle$ is the mass-averaged shell adiabat. This gives

$$\text{IFAR} = \frac{R}{\Delta_{\text{inflight}}} \simeq 10 \frac{R_0}{\rho_0 \Delta_0} \left(\frac{R}{R_0}\right)^3 \left(\frac{p_a}{100\,\text{Mbar}}\right)^{3/5} \langle\alpha\rangle^{-3/5}. \tag{7.41}$$

Initial shell radius in an optimised design is proportional to shell velocity times acceleration time, $R_0 \sim V_{\text{imp}} t_{\text{accel}}$. Together with Eq. 7.33 this gives $R_0^3 \sim E_{\text{laser}} V_{\text{imp}}/4\pi I$. Fitting the latter formula with the simulation results gives the numerical factor of 0.7. Thus,

$$R_0 \simeq \left(0.7 \frac{E_{\text{laser}} V_{\text{imp}}}{4\pi I}\right)^{1/3}. \tag{7.42}$$

Multiplying numerator and denominator of Eq. 7.41 by $4\pi R_0^2$ and replacing $4\pi R_0^2 \Delta_0 \rho_0 \simeq M_{\rm shell}$ with the shell mass expressed using Eq. 7.34 yields

$$\mathrm{IFAR} = 80\frac{R^3}{R_0^3}\left(\frac{V_{\rm imp}}{3\times 10^7}\right)^2\left(\frac{p_{\rm a}}{100\ \mathrm{Mbar}}\right)^{-2/5}\langle\alpha\rangle^{-3/5}. \tag{7.43}$$

Equation 7.43 shows that IFAR's value decreases as the shell implodes (the ratio R/R_0 gets smaller), reaching its peak value at the beginning of the shell acceleration, when drive intensity reaches its peak value. Then, the stability property of a design is characterised by this peak IFAR value. Fit to the results of numerical simulations gives [20]

$$\max(\mathrm{IFAR}) \simeq 60\left(\frac{V_{\rm imp}}{3\times 10^7}\right)^2\left(\frac{p_{\rm a}}{100\ \mathrm{Mbar}}\right)^{-2/5}\langle\alpha\rangle^{-3/5}, \tag{7.44}$$

which can be recovered from Eq. 7.43 by using $R \simeq 0.9R_0$. Numerical simulations of directly driven cryogenic implosions (both on Omega and the NIF) show that to keep the shell from breaking up because of the short-scale perturbation growth during the acceleration, IFAR should not exceed

$$\mathrm{IFAR}_{\max} \simeq 40. \tag{7.45}$$

Using Eq. 7.44, we conclude that increasing implosion velocity by reducing the drive intensity alone, as Eq. 7.34 suggests, is not the best strategy from a stability point of view since two factors cause IFAR to increases in this case: (1) an increase in $V_{\rm imp}$ and (2) a reduction in $p_{\rm a}$. The fact that reduction in drive pressure increases IFAR is a consequence of the larger travelled distances required to accelerate a shell to a given $V_{\rm imp}$ if the drive pressure is lower. Larger acceleration distances mean larger initial shell radius and higher IFAR. This is illustrated in Fig. 7.7 where initial shell dimensions are schematically shown for different drive intensities.

The smallest drive intensity requires the largest initial and in-flight aspect ratios.

Increasing the implosion velocity by reducing shell mass has a lesser effect on IFAR since the latter increases only as a result of larger $V_{\rm imp}$, [see Eq. 7.43]. This approach, however, has limited beneficial effects: As the IFAR exceeds the maximum value set by the stability considerations, the target performance begins to degrade. Improving shell stability while reducing shell mass can be accomplished, according to Eq. 7.44, by increasing the average shell adiabat $\langle\alpha\rangle$. This must be done, however, without raising adiabat of the unablated fuel since that is set by the condition on maximum fuel pressure at stagnation, as shown in Eq. 7.25. An adiabat-shaping technique [21] was proposed and implemented in the direct drive designs to raise the adiabat only at the outer part of the shell, without degrading the adiabat at the inner part of the fuel. The designs with adiabat shaping will be discussed in Sect. 7.5.

Fig. 7.7 Initial shell dimensions for all-DT designs driven at indicated intensities using $E_{\text{laser}} = 1.5$ MJ

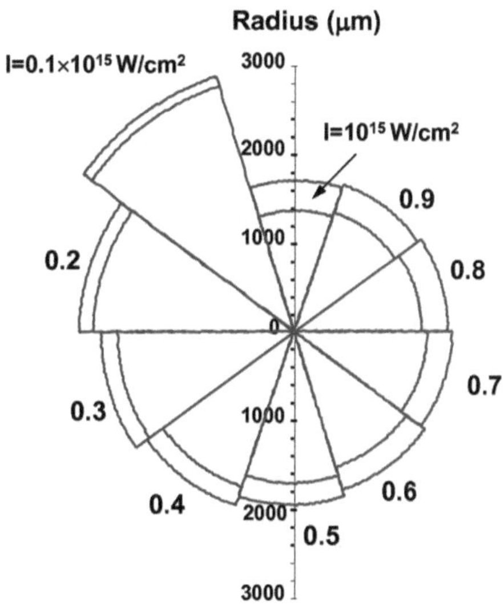

7.3 Experimental Cryogenic Program on Omega

The experimental cryogenic program on Omega is designed to study fundamental physics of direct-drive ICF implosions. In particular, the following key questions were considered:

1. Is a low-adiabat compression of cryogenic fuel possible in a spherical implosion driven by direct laser illumination?
2. Can cryogenic fuel be accelerated to velocities in excess of 3×10^7 cm/s in such implosions?
3. At what drive intensities does the laser drive become inefficient in accelerating low-adiabat fuel, creating an excessive amount of fuel preheat due to suprathermal electrons, and scattering a significant fraction of the incident laser light as a result of laser-plasma interaction?
4. Can asymmetry growth be controlled during an implosion, so

   (a) the short-scale perturbations with wavelength $\lambda \sim \Delta_{\text{inflight}}$ do not break up the shell, and
   (b) hot-spot deformation is not severe enough to significantly reduce hot-spot ion temperature and quench the yield?

To address these questions, various experimental techniques were developed and used to diagnose Omega implosions. Selecting a specific technique was based on measurement accuracy, which must be high enough to be able to tune the physics models and to meet the predictive accuracy goals discussed in Sect. 7.1.2. Next, we list the experimental techniques that were used to address these key questions.

7.3.1 Adiabat

The shell adiabat during an implosion can be inferred from shell density and temperature measurements. Two techniques have been developed and used on Omega implosions to measure these quantities: spectrally resolved x-ray scattering [22, 23] and time-resolved x-ray absorption spectroscopy [24]. X-ray scattering requires large scattering volumes to keep signal-to-noise ratio at acceptable levels. This significantly limits the accuracy of measuring the adiabat at inner parts of the shells in designs with spatial adiabat gradients. The x-ray absorption technique, on the other hand, is designed to be much more local since the temperature and density are inferred by analysing the spectral shapes of a backlighter source attenuated by a buried mid-Z tracer layer inside the shell. Hydrodynamic instabilities developed during shell implosion, however, redistribute the signature layer material throughout the shell, making temperature and density measurements dependent on the accuracy of mix models.

A significant progress in understanding how to infer the fuel adiabat in a spherical implosion was made after Ref. [20] demonstrated that the peak in areal density in an optimised implosion depends mainly on laser energy and the average adiabat of the unablated mass,

$$\max(\rho R)_{\text{optimized}} \simeq 2.6\frac{[E_{\text{laser}}(\text{MJ})]^{1/3}}{\alpha^{0.54}}. \tag{7.46}$$

This scaling can be understood based on the following consideration: The unablated mass at the beginning of shell deceleration can be written as

$$M \sim \rho_d \Delta_d R_d^2, \tag{7.47}$$

where $\Delta_d = R_d/A_d$ is the shell thickness and A_d is the shell aspect ratio at the start of shell deceleration, respectively. The mass is related to drive pressure (or shell pressure at the beginning of deceleration, p_d) using Newtons law,

$$M\frac{R_d}{t_{\text{accel}}^2} \sim p_d R_d^2, \quad \to M \sim p_d R_d t_{\text{accel}}^2, \tag{7.48}$$

where t_{accel} is defined in Eq. 7.33. Equating the right-hand sides of Eqs. 7.47 and 7.48 yields

$$R_d \sim t_{\text{accel}}\sqrt{\frac{p_d}{\rho_d}A_d} \sim \left(\frac{E_{\text{laser}}}{V_{\text{imp}}^2 I}\right)^{1/3}\sqrt{\frac{p_d}{\rho_d}A_d}. \tag{7.49}$$

At peak compression, the main contribution to areal density is given by the shock-compressed region. Thus, rewriting Eq. 7.14 as

$$M_{\text{eff}} \sim (\rho\Delta)_{\text{shocked}} R^2 \sim \rho_d R_d^2 \frac{R}{R_d} \tag{7.50}$$

leads to

$$\max(\rho R) \sim (\rho\Delta)_{\text{shocked}} \sim \rho_d R_d \left(\frac{R_d}{R}\right). \tag{7.51}$$

Substituting Eqs. 7.49 and 7.13 into Eq. 7.51 results in

$$\max(\rho R) \sim p_d \left(\frac{\rho_d}{p_d}\right)^{5/6} \frac{E_{\text{laser}}^{1/3}}{I^{1/3}} \sqrt{A_d}. \tag{7.52}$$

Finally, replacing ρ_d with $\sim (p_d/\alpha)^{3/5}$ gives

$$\max(\rho R) \sim \frac{p_d^{2/3}}{I^{1/3}} \sqrt{A_d} \frac{E_{\text{laser}}^{1/3}}{\sqrt{\alpha}}. \tag{7.53}$$

Shell aspect ratio at the start of deceleration phase has a weak dependence on implosion parameters: For an implosion with a higher shell adiabat, the shell is thicker but the deceleration phase starts while the shell is at larger radius, so the ratio R_d/Δ_d is constant $A_d \simeq 2$ for all implosion conditions. For a well-tuned implosion when the drive pressure keeps pushing the shell up to the beginning of shell deceleration (shell coasting is minimised), $p_d \sim p_a$. Since $p_a \sim I^{2/3}$, Eq. 7.53 becomes

$$\max(\rho R)_{\text{optimized}} \sim I^{1/9} \sqrt{A_d} \frac{E_{\text{laser}}^{1/3}}{\sqrt{\alpha}}, \tag{7.54}$$

which agrees with the numerical fit shown in Eq. 7.46 taking into account the weak dependence of $\sqrt{A_d} I^{1/9}$ on implosion parameters. When ablation drive is terminated early and the shell starts to decompress during the coasting phase, then p_d drops, reducing the maximum areal density [see Eq. 7.53].

Equation 7.53 shows that the adiabat of an unablated mass in an implosion without a significant coasting phase can be inferred by measuring the areal density close to the shell's peak convergence. The areal density in an ICF implosion is measured using either x-ray backlighting [25], Compton radiography [26], or charged-particle spectrometry [27,28]. While the first two techniques are still under development, the areal density in current cryogenic experiments is inferred by measuring the spectral shapes of fusion-reaction products. Areal density in D_2 fuel is determined from energy downshift in secondary protons [27] created in D^3He reactions [primary reaction creates a neutron and ^3He ion, D+D→n(2.45 MeV)+^{3}He (0.82 MeV), and a secondary reaction creates an α particle and a proton, ^{3}He+D→ α(6.6–1.7 MeV)+p (12.6–17.5 MeV)]. This is shown in Fig. 7.8.

For DT fuel, the areal density is inferred by using magnetic recoil spectrometer (see Fig. 7.9) that measures the fraction of neutrons downscattered from fuel deuterons and tritons [28] (this fraction is directly proportional to the fuel ρR).

The main advantages in using charged-particle spectrometry to measure areal densities is that the peak in the reaction rate and peak fuel compression are not far apart (for Omega implosions they are separated by 20–30 ps with the peak in

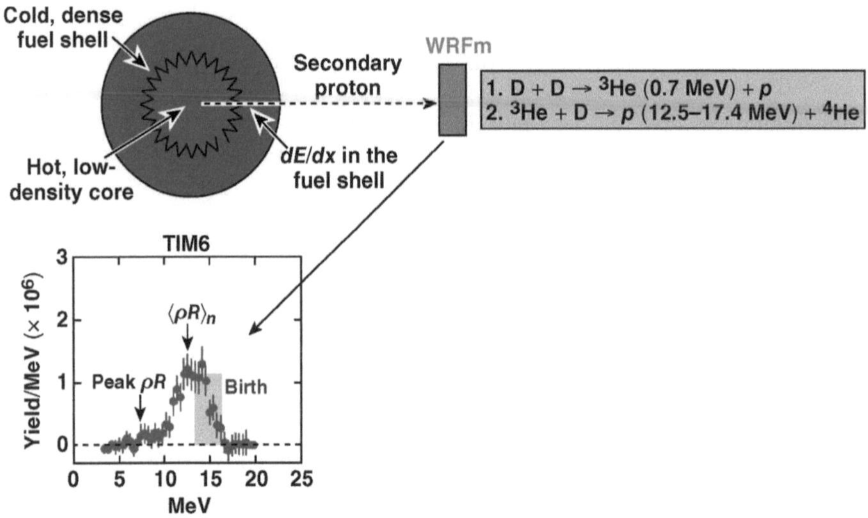

Fig. 7.8 Areal density in D_2 fuel is inferred from energy attenuation of secondary protons passing through the cold shell material

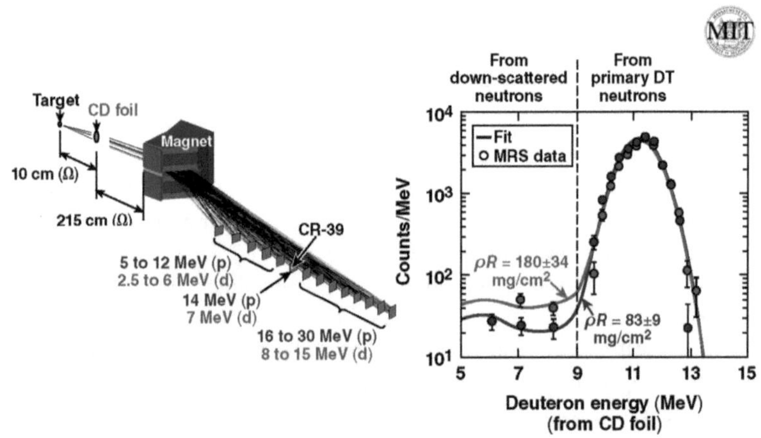

Fig. 7.9 Areal density in DT fuel is inferred from measurements of the down-scattered fraction of primary neutrons produced in a D+T reaction

neutron production being earlier), so the reaction products sample areal density close to its peak value. The fusion rates are affected, on the other hand, by the nonuniformity growth that reduces both the fuel ion temperature and fuel 'clean' volume where reactions take place. This changes timing and sampling of areal density by fusion-reaction products. The sensitivity of areal density measurement to neutron-production timing can be shown by noting that areal density evolve on

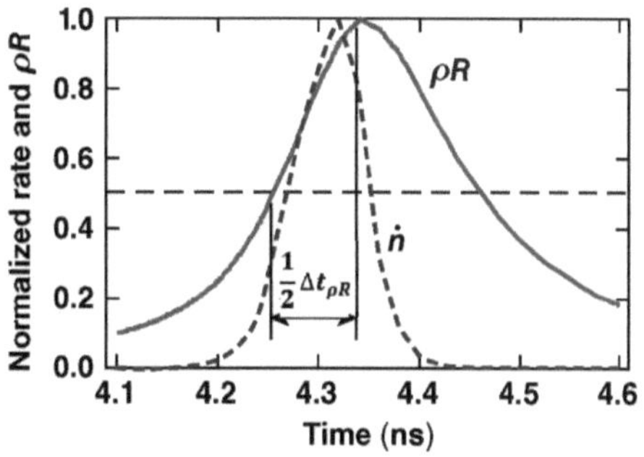

Fig. 7.10 Areal density and neutron-production–rate evolution for a typical cryogenic implosion on Omega

a time scale $\Delta t_{\rho R} \sim 2\Delta t$, where Δt is the confinement time defined in Eq. 7.9. For Omega-scale targets this gives

$$\Delta t_{\rho R} \sim 2\frac{R}{V_{\text{imp}}} \sim 2\frac{2\times 10^{-3}\text{cm}}{3\times 10^{7}\text{cm/s}} \simeq 130\text{ ps}, \tag{7.55}$$

while the temporal width of neutron production in a spherically symmetric implosion is twice less,

$$\Delta t_n \simeq \Delta t \simeq 70\text{ ps}. \tag{7.56}$$

The areal density and neutron production histories for a typical cryogenic-DT target are shown in Fig. 7.10.

Since the temporal scale of ρR evolution is short, the effect of perturbation growth on neutron-production timing and duration must be taken into account when comparing the experimentally inferred ρR values with the predictions.

7.3.2 Implosion Velocity

Implosion velocity is the key parameter that determines how much kinetic energy the fuel must acquire to ignite [see Eq. 7.1]. Shell velocity can be inferred from trajectory measurements using either time-resolved x-ray–backlit images [29] of an imploding shell or time-resolved self-emission images [25, 30]. The most accurate measurement (although indirect) of hydrocoupling efficiency in implosions on Omega is done by measuring the onset of neutron production. Temporal history of the neutron rate is measured on Omega using neutron temporal diagnostics (NTD) [31]. The absolute timing of NTD is calibrated to better than ±50 ps, which is equivalent to a spread in the implosion velocity of ±3.5 % for Omega-scale targets. Figure 7.11 illustrates the sensitivity of neutron-production timing to the variation in shell velocity. Here, the shell velocity (dashed lines) and neutron rate (solid lines) histories are calculated using two different laser-deposition models. The implosion velocity predicted with the less-efficient drive (thick lines) is 5 % lower than that

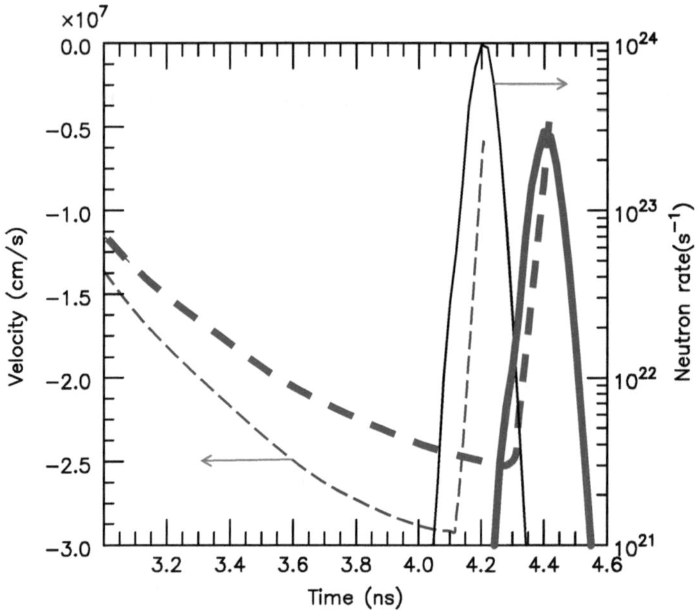

Fig. 7.11 Shell velocity (left axis, *dashed lines*) and neutron production rates (right axis, *solid lines*) calculated for an Omega cryogenic design using two different laser-deposition models. The less-efficient laser absorption model (*thick lines*) predicts smaller shell velocity and later neutron-production timing

predicted for higher-efficiency drive (thin lines), resulting in a 200-ps delay in neutron production. Such a delay is easily observed in an experiment since this time difference is well outside the measurement error bar.

7.3.3 Ion Temperature at Time of Peak Neutron Production

The fuel ion temperature at peak neutron production depends on the shell kinetic energy during the acceleration phase of implosion and on the growth of the hot-spot distortions while the shell decelerates. The ion temperature in an implosion is inferred by measuring temporal width of primary-neutron signal [32]. The thermal broadening of the neutron energy distribution $\Delta E_{\rm FWHM}$ is related to the local ion temperature T_i as [33]

$$\Delta E_{\rm FWHM} = 177\sqrt{T_i}, \tag{7.57}$$

where both $\Delta E_{\rm FWHM}$ and T_i are measured in keV. Then, measuring the neutron's time of flight (TOF) from the target to a detector, $\mathrm{TOF} = 72.3L/\sqrt{E_n}$, the neutron-averaged ion temperature is inferred relating TOF broadening $\Delta_{\rm TOF}$ with ΔE and using Eq. 7.57,

$$\langle T_\text{i} \rangle_\text{n,exp} = 68 \frac{\Delta^2_\text{TOF}}{L^2}, \tag{7.58}$$

where L is distance from detector to target in meters, $E_n = 14.1$ is the energy (in MeV) of primary neutrons in the D+T reaction, and TOF is measured in nanoseconds. Strictly speaking, the neutron spectral width is determined not only by thermal broadening, but also by gradients in the bulk fluid velocity of the reacting fuel. The latter contribution is not very important in a spherically symmetric implosion since the peak in neutron production occurs while the fuel is close to stagnation. When drive and target nonuniformities are taken into account, however, fuel flow caused by asymmetry growth can make a significant contribution to neutron spectral width. Thus, comparing $\langle T_\text{i} \rangle_\text{n,exp}$ with calculations, the bulk fluid motion needs to be taken into account in this case. To generalise Eq. 7.57, including the effect of bulk motion, we start with Eq. 29 of Ref. [33] and write the neutron kinetic energy as

$$E_n \simeq \frac{m_\alpha}{m_n + m_\alpha} Q + (\mathbf{V} \cdot \mathbf{e}_n) \sqrt{\frac{2 m_n m_\alpha Q}{m_n + m_\alpha}}, \tag{7.59}$$

where Q is nuclear energy released in a fusion reaction ($Q = 17.6$ MeV for D+T reaction), m_n and m_α are masses of reaction products (neutron and alpha-particle mass, respectively, for DT), $\mathbf{V}$ is the velocity of the centre of mass of reaction products, and $\mathbf{e}_n$ is a unit vector in the direction of neutron velocity (and direction to a neutron detector). If $\mathbf{V_f}$ is the fluid velocity, then averaging over thermal motion gives

$$\langle E_n \rangle \simeq E_0 + V_f \cos\theta_n \sqrt{2 m_n E_0}, \tag{7.60}$$

where $E_0 = m_\alpha/(m_n + m_\alpha) Q$ ($E_0 = 14.1$ MeV for DT), and θ_n is the angle between fluid flow and neutron velocity. Next, using Eq. 36 of Ref. [33], the neutron distribution at a particular location in a plasma with ion temperature T_i becomes

$$f_n(E) = e^{-\left(\frac{E - E_0}{\Delta E} - M_a \cos\theta_n\right)^2}, \quad \Delta E = 2\sqrt{\frac{m_n T_i E_0}{m_n + m_\alpha}}, \tag{7.61}$$

where $M_a = V_f / c_s$ is the flow Mach number, $c_s = \sqrt{T_\text{i}/m_\text{i}}$ is the ion sound speed, and $m_\text{i} = (m_n + m_\alpha)/2$ is the average fuel ion mass. According to Eq. 7.61, a fluid velocity, uniform in the direction of the neutron detector, affects only the position in the peak of the distribution function, but not its width. Averaging the distribution function over the fuel volume gives

$$\langle f_n(E) \rangle_V = \frac{\int dV n^2 \langle \sigma v \rangle e^{-[\alpha(E) - M_a \mu]^2}}{\int dV n^2 \langle \sigma v \rangle}, \tag{7.62}$$

where $\mu = \cos\theta$, $\langle\sigma v\rangle$ is reaction cross section, n is ion density, $\alpha(E) = (E - E_0)/\Delta E$. Taking the integral over the angles assuming spherical symmetry in Eq. 7.62 yields

$$\langle f_n(E)\rangle_V = \sqrt{\pi}\frac{\int_0^R drr^2n^2\langle\sigma v\rangle\{\mathrm{erf}[\alpha(E)+M_a]-\mathrm{erf}[\alpha(E)-M_a]\}}{4M_a\int_0^R drr^2n^2\langle\sigma v\rangle}, \quad (7.63)$$

where erf is the error function. Integrating Eq. 7.63 over the neutron-production time and fitting the result with a Gaussian with FWHM=ΔE_{fit},

$$\int dt\langle f_n(E)\rangle_V \overset{\mathrm{fit}}{\to} \exp\left[-4\ln 2\left(\frac{E-E_0}{\Delta E_{\mathrm{fit}}}\right)^2\right],$$

defines an effective temperature $\langle T_{\mathrm{i}}\rangle_{\mathrm{n,fit}} = (\Delta E_{\mathrm{fit}}/177)^2$ to be compared with the measurements [see Eq. 7.58]. A bulk flow with velocity distribution not pointing in the same direction broadens the neutron spectrum, leading to a higher effective ion temperature. This is illustrated by evaluating the angular integral in Eq. 7.62, assuming $M_a \ll 1$ and spherical symmetry,

$$\frac{1}{2}\int_{-1}^{1} d\mu e^{-\alpha^2+2\alpha M_a\mu} \simeq \frac{e^{-\alpha^2}}{2}\int_{-1}^{1} d\mu\left[1+2(\alpha M_a)^2\mu^2\right] \simeq e^{-\alpha^2/(1+2M_a^2/3)}. \quad (7.64)$$

Equation 7.64 gives

$$\langle T_{\mathrm{i}}\rangle_{\mathrm{fit}} = T_i\left(1+\frac{2}{3}M_a^2\right) = T_i + \frac{2}{3}m_{\mathrm{i}}V_f^2.$$

For a spherically symmetric flow, $\langle T_{\mathrm{i}}\rangle_{\mathrm{fit}}$ tracks T_i within a few percent since the fuel is close to stagnation at the neutron-production time. When significant asymmetries are present, bulk flow can lead to a significant contribution to $\langle T_{\mathrm{i}}\rangle_{\mathrm{fit}}$, making an inferred ion temperature larger than the actual thermodynamic value.

7.4 Early Experiments on Omega-24

The first experiments with layered DT targets were performed on the OMEGA-24 Laser System [34] in the late 1980s [35, 36]. The targets were spherical 3- to 5-μm-thick glass shells with outer radii of 100–150 μm. The cryogenic, 5–10 μm-thick solid DT layers were produced using a fast-freeze technique [37]. These targets were driven with 1–1.2 kJ of UV energy delivered with 650-ps Gaussian pulses (with a peak in drive intensity of up to 6×10^{14} W/cm^2). The target and drive pulse are shown in Fig. 7.12a. The predicted convergence ratios in these implosions

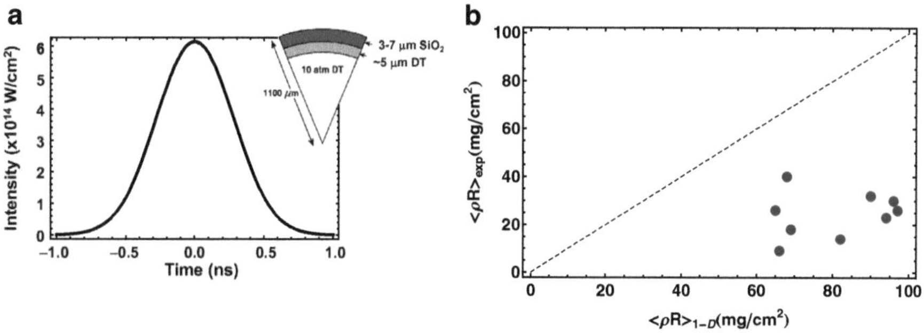

Fig. 7.12 (**a**) Pulse shape and target, and (**b**) predicted and inferred fuel areal densities for cryogenic implosions on the Omega-24 Laser System

were relatively high, $C_r \sim 20$ (C_r is defined as a ratio of the initial to the minimum radius of the fuel–glass interface) with a peak DT density of ~300 g/cm^3 and a peak fuel areal density of 150 mg/cm^2. For comparison, the all-DT ignition design described in Sect. 7.2.1 has $C_r = 27$. Targets were held inside the U-shaped cradle using three to five spider silks. These early designs were highly susceptible to the RT instability since peak of in-flight aspect ratio (IFAR) approached 70, a much higher value than currently considered to be acceptable for a robust design, IFAR < 40 (see Sect. 7.2.2.2).

The areal densities in these experiments were directly measured (the first such measurement performed in an ICF implosions at that time) by counting the down-scattered fraction of deuterium and tritium atoms [38]. Even though the inferred fuel areal density and mass density were the highest measured to date, they were lower than predictions by 40–60 %. Figure 7.12b plots the predicted value of fuel areal density using the 1-D hydrocode *LILAC* [39] and inferred areal densities using knock-on statistics. A significant deviation in the predicted value has occurred for an effective fuel adiabat $\alpha < 4$. This is not surprising considering the high IFAR of these shells. If a shell breaks up due to perturbation growth during acceleration, it creates a low-density precursor ahead of the imploding shell, which causes the shell to stagnate at a larger radius with a smaller peak areal density.

7.5 Cryogenic D_2 Implosions on the Omega Upgrade Laser System from 2001 Until Mid-2008

The fast-freezing technique employed to make cryogenic targets on Omega-24 could not be used to produce thicker fuel layers required for ignition-relevant Omega-scaled designs. Novel techniques for producing smooth DT and D_2 layers were introduced in 1980s and 1990s. A 'β-layering' was demonstrated to make uniform solid DT layers [40], and IR radiation was shown to produce layer smoothing in

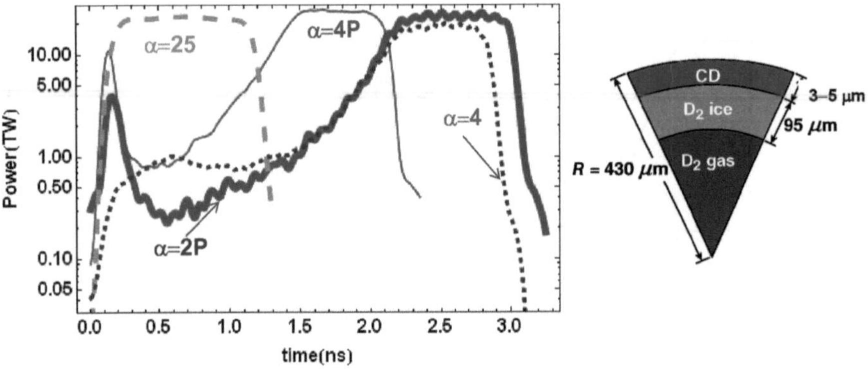

Fig. 7.13 Pulse shapes and target for $\alpha = 25$ (*dashed*), $\alpha = 4$ (*solid* and *dashed lines*) and $\alpha = 2$ (*thick solid line*) designs on Omega. The designs with the decaying-shock adiabat shaping are shown with *solid lines*

cryogenic D_2 fuel by exciting the vibration–rotation band [41]. The newly developed cryogenic system [42] on Omega Upgrade (30 kJ of UV energy, 60-beam system) [43] employed both these techniques for cryogenic target production. Cryogenic experiments on the new system started in 2000 by imploding D_2 targets [44]. DT was introduced in February 2006, after completion of an extensive system readiness review associated with the radiological impact of using tritium [45]. As target production was on a learning path to improving D_2-layer quality, the first implosions used a square laser drive pulse with laser energy ~23 kJ to set the cryogenic fuel on high adiabat $\alpha \sim 25$ (see dashed line in Fig. 7.13).

The acceleration phase in this design is very short so the impact of the RT growth on target performance is minimal. The yields, areal densities (30–60 mg/cm^2), and timing of neutron production were consistent with 1-D and 2-D hydrocode simulations [44, 46].

As the uniformity of ice layers has dramatically improved from $\sigma_{\rm rms}$ = 9–15 μm down to 1–3 μm in 2002, experiments began using designs that approached the Omega-scaled version of the all-DT ignition designs [47]. These were 3- to 5-μm-thick CD shells overcoated over 95- to 100-μm-thick D_2 ice layers driven at $I \sim 10^{15}$ W/cm^2 on $\alpha = 4$ adiabat (see dotted line in Fig. 7.13). These shells were somewhat thicker than required for hydrodynamic scaling (< 1 μm) since fill time was shorter and overall long-wavelength shell nonuniformities were smaller. By the middle of 2005, a large data set of these implosions was built sufficient to conclude that the measured areal densities were significantly lower than predicted, as shown in Fig. 7.14 with solid circles.

For the lowest adiabat (highest ρR) in this series, degradation in areal density was up to 50 %, which is equivalent to adiabat degradation [according to Eq. 7.46], by up to 70 %! The 2-D calculations using the hydrocode *DRACO* [48] and results of stability postprocessor [49] indicated that the shells in the low-adiabat implosions were sufficiently stable (the ratio of the mix width to the shell thickness did not

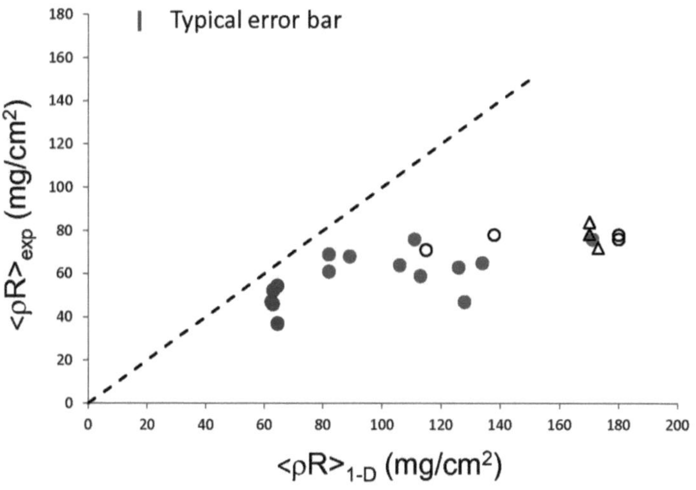

Fig. 7.14 Measured and predicted areal densities for cryogenic-D_2 implosions using the $\alpha = 4$ pulses shown in Fig. 7.13 (*solid circles*) and $\alpha = 4P$ (*open circles*)

exceed 50 %, where the short-scale mix at the ablation front, seeded mainly by laser imprint, is amplified by the RT instability). Measurements of the imprint efficiencies made earlier on planar targets [50], however, suggested that calculations could be underestimating imprint amplitude as much as by factor of 2, and the shell in low-adiabat implosions could be broken due to the imprint growth. Since shell stability was a main concern at that time, LLE was working on perturbation growth mitigation strategies. A novel technique for reducing the RT growth was proposed in 2002. The idea was to shape the adiabat through the shell (adiabat-shaping designs [21]). This can be accomplished either by launching a shock wave of decaying strength [Decaying-Shock (DS) design] through the shell [21] or by relaxing the shell material with a short-duration picket and recompressing it later with the shaped main pulse [adiabat shaping by relaxation (RX) design] [51]. This sets the outer part of the ablator on a higher adiabat, keeping the inner part of the shell on a lower adiabat. The higher adiabat at the ablation front increases the ablation velocity, mitigating the impact of the RT instability on target performance, as described in Sect. 7.2.2.1.

Pulse shapes, similar to ones shown in Fig. 7.13 with thin and thick solid lines, were used to implement adiabat-shaping designs on Omega. Calculations predicted a significant improvement in shell stability in designs with adiabat shaping in comparison with the original flat-foot designs (see Fig. 7.15).

The experiments, however, did not show any significant improvement in measured areal densities, which continued to saturate at ~80 mg/cm^2. These are shown as open circles in Fig. 7.14. To further support the conclusion that the short-scale mix due to the RT growth at the ablation front was not the main contributor to the observed ρR degradation, a series of implosions was performed with an enhanced

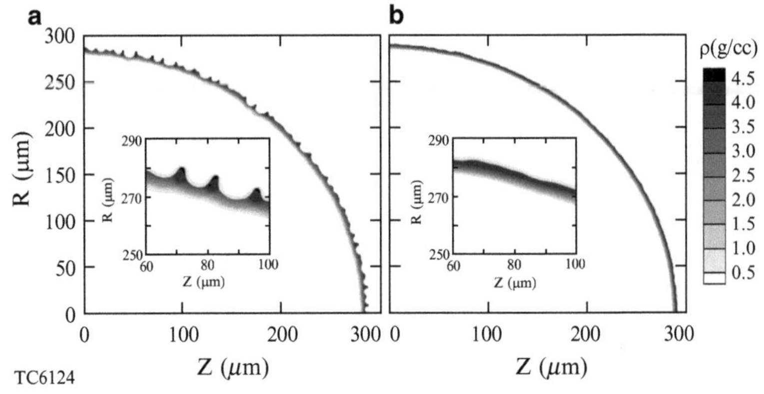

Fig. 7.15 In-flight shell density contours in designs (**a**) without and (**b**) with adiabat shaping

laser-imprint level by turning off the smoothing by spectral dispersion (SSD) [52]. The target yield dropped by a factor of 2 in these implosions, but the areal density remained unchanged (see open triangles in Fig. 7.14).

Since the source of excessive shell heating, not accounted for in a hydro simulation, was unknown at that time, several scenarios explaining the areal-density deficiency were considered: Excessive shell heating could have been due to (1) suprathermal electrons with $T_{\text{hot}} > 40$ keV, (2) radiation, or (3) shock waves. Next, we describe how each of these possibilities were addressed in Omega experiments.

7.5.1 Suprathermal Electrons

Suprathermal electrons are always present in a plasma because of high-energy tails in the electron distribution function. In addition, laser–plasma interaction processes, such as two-plasmon–decay (TPD) instability and Stimulated Raman Scattering (SRS) [53], can generate electrons with energies above 20 keV. These electrons can penetrate the ablator and fuel in Omega designs and deposit their energy close to the inner part of the fuel, degrading peak ρR. The electrons in the energetic tails of the distribution function will be addressed first.

7.5.1.1 Electron Distribution Tails and Nonlocal Thermal Transport

To model electron thermal transport in ICF experiments, a flux-limited model [54] is conventionally used in hydrocode simulations. Thermal conduction in such a model is calculated using Spitzer expression [55] q_{sp}, which is derived assuming that the electron mean free path is much shorter than the gradient scale-length of hydrodynamic variables [57]. In a narrow region, near the peak of the laser

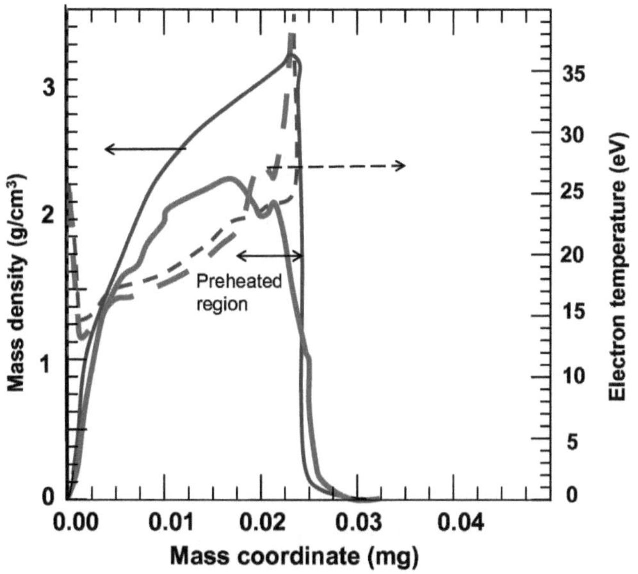

Fig. 7.16 In-flight shell density (*solid*) and electron temperature (*dashed*) with (*thin lines*) and without (*thick lines*) nonlocal effects in electron thermal conduction

deposition, the temperature profile is steep enough to break validity condition of Spitzer formula. The heat flux in this case is calculated as a fraction $f < 1$ of the free-stream conduction $q_{\mathrm{fs}} = nTv_{\mathrm{T}}$, where n and T are electron density and temperature, respectively, $v_{\mathrm{T}} = \sqrt{T/m}$ is the electron thermal velocity, and m is electron mass. The limiting factor f is referred to as 'flux limiter'. The flux limiter value of $f = 0.06$ is typically used to simulate direct-drive experiments.

Although it was successfully applied to simulate many experimental observables [58], the flux-limited thermal transport model neglects the effect of finite electron-stopping ranges and cannot be used to access the amount of shell preheat from the energetic electrons in plasma. To account for this effect, a simplified thermal transport model was developed and implemented in the 1-D hydrocode *LILAC*. The model used Krook-type approximation [56] to collisional operator to solve the Boltzmann equation without making the high collisionality approximation used in the 'classical' Chapman-Enskog method [57]. The modified energy-dependent Krook-type operator [58] conserves particles and energy by renormalising local electron density and electron temperature (which depend on gradients in hydro-dynamic profiles) in the symmetric part of the distribution function (Maxwellian modified to include effects of laser electric field [59]). When applied to the Omega experimental data, the nonlocal model showed no significant inner fuel preheat caused by the energetic electrons in the distribution tale (see Fig. 7.16). These electrons, instead, preheat the ablation front region [see how electron temperature in the calculation using the nonlocal model (thick dashed line in Fig. 7.16) increases towards the ablation front], leading to a greater ablative stabilisation of the RT

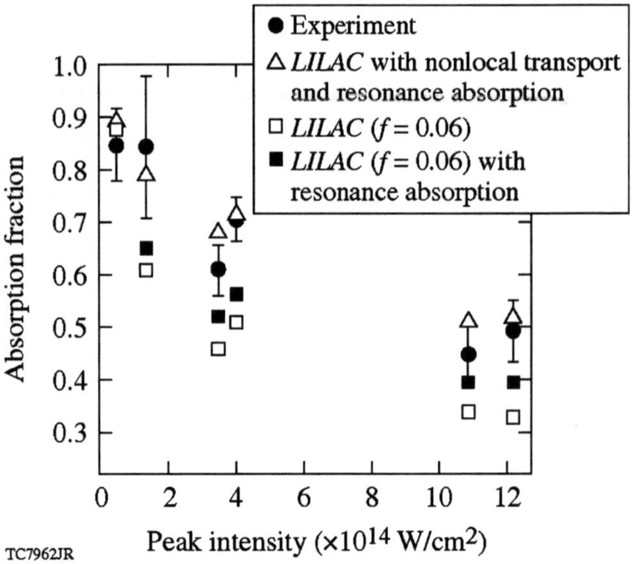

Fig. 7.17 Absorption fraction of the incident laser energy for a 20-μm-thick CH shell driven by a 200-ps Gaussian pulse at different peak intensities

growth. This preheat of the outer region of the shell can explain very little sensitivity of the measured ρR to variation in the source of short-scale perturbations described earlier in this section. Ablation front preheating due to the nonlocal electrons is also consistent with the short-wavelength stabilisation of the RT growth observed in experiments with accelerated planar foils [60].

In addition to the ablation region preheating, the strength of the first shock and a compression wave were significantly modified in calculations using the nonlocal electron-transport model [58]. At the beginning of the laser drive, where the hydrodynamic scale lengths are short, the shock strength predicted using the nonlocal model is larger compared to the results of the flux-limited model. This effectively leads to shock mistiming and an adiabat degradation prior to the shell acceleration. Experimental validation of the nonlocal model predictions by direct shock velocity measurement in spherical geometry was not available at that time (the experimental platform was developed in 2008). The existing shock velocity data in planar geometry, on the other hand, was not very sensitive to differences in predictions using the nonlocal and flux-limited models [58]. Measurements of early-time perturbation evolution (ablative Richtmyer–Meshkov instability [61]), however, clearly indicated that the higher heat fluxes, predicted by the nonlocal model at the beginning of the pulse, are consistent with the observations [62]. In addition, the absorption measurements of Gaussian pulses with FWHM of 200 ps and peak laser intensity varied from 5×10^{13} to 1.2×10^{15} W/cm^2 [63] were in much closer agreement with the results of the nonlocal heat-transfer model. These are shown in Fig. 7.17.

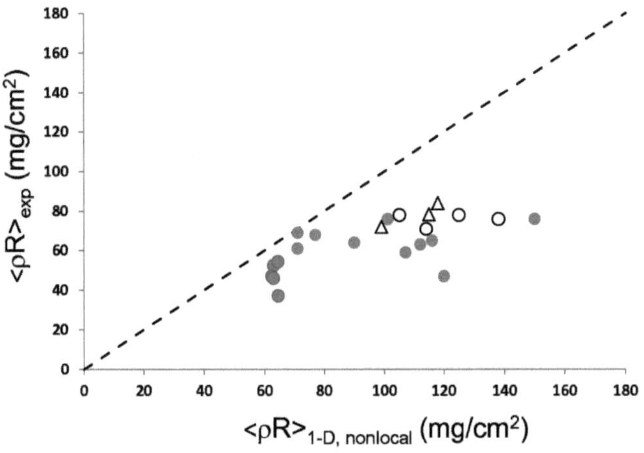

Fig. 7.18 Same as in Fig. 7.14, except these calculations are performed using the nonlocal thermal transport model

In addition to the inverse bremsstrahlung, the resonance absorption [53] resulting from tunnelling of the laser electric field from the turning point to the critical surface and exciting plasma waves was included in these simulations [58, 64]. The resonant absorption effects are important only early in the pulse when the density scale-length is short.

When the nonlocal model was used, the calculated areal densities became in closer agreement with the data compared to the results of the flux-limited model (see Fig. 7.18).

Nevertheless, some discrepancies in ρR, especially for implosions with the lowest adiabat, still remained.

The next step in the cryogenic campaign was to redesign the drive pulse design, taking into account modified coupling efficiency early in the pulse, as predicted by the new thermal transport model. Both the RX and DS designs driven at peak intensities of $\sim 6 \times 10^{14}$ W/cm^2 were used in this 'retuning' campaign. The experimental ρR values have marginally improved from 80 up to 100 mg/cm^2 (looking at this result with the knowledge that we have now, this 20 % increase in areal density was mainly due to a reduction in peak intensity from 9 to 6 $\times 10^{14}$ W/cm^2, which also reduced strength of secondary hydrodynamic waves launched by the pulse) but fell short of predicted values that were in the range of 150–170 mg/cm^2. Even though this campaign did not succeed in significantly increasing areal densities, it revealed a very interesting trend: the measured areal densities showed very strong dependence on CD shell thickness. These results are plotted in Fig. 7.19.

Such a dependence was not predicted in hydrocode simulations. Among the hypotheses explaining this trend are radiation preheat due to mix at the CD-D_2 interface (as discussed in Sect. 7.5.1.3), increased preheat due to suprathermal electron generation by the TPD instability, or short-scale magnetic field generation

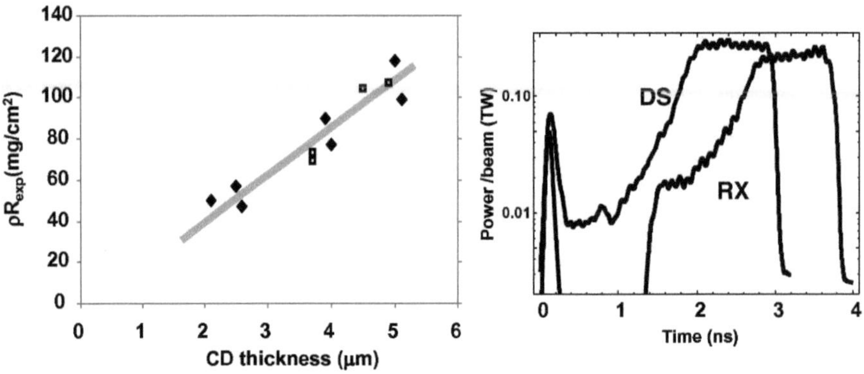

Fig. 7.19 Measured areal densities as function of CD shell thickness and pulse shapes used in this series

at the CH-D_2 interface as the latter travels through the ablation front and conduction zone. None of these hypotheses, however, could account for a factor of 2.5 reduction in areal density when the CD thickness decreased from 5 to 2.5 μm. The true explanation of this observation is still not found.

7.5.1.2 Suprathermal Electrons Generated by (TPD) Instability

In parallel to the study of the effect of nonlocal thermal transport on implosion performance, a different cryogenic design was proposed and used on Omega experiments to address a possible preheat issue caused by the suprathermal electrons created by the TPD instability. The threshold factor for the absolute TPD instability [65] is

$$\eta = \frac{I_{14} L_n(\mu\text{m})}{230 T_{\text{keV}}}. \tag{7.65}$$

It exceeds unity in direct-drive implosions on Omega when drive intensities are above $\sim 3 \times 10^{14}$ W/cm^2. Here, I_{14} is the laser intensity at quarter-critical surface in units of 10^{14} W/cm^2, L_n is the electron-density scale length in microns, and T is the electron temperature in keV. At these intensities, hard x-ray Bremsstrahlung radiation, emitted by suprathermal electrons as they slow down in plasma, is observed in Omega implosions [66] (see dotted line in Fig. 7.20). To prove that the preheat signal has its origin in the TPD instability, the measured hard-x-ray signal must correlate with $3/2\omega$ and $\omega/2$ emission [63]. An example of such correlation in a cryogenic implosion with a 5 μm CD shell is shown in Fig. 7.20. Here, $\omega/2$ signal is shown with thick solid line marked with '$\omega/2$'. Both signals are observed when calculated threshold parameter η (shown with the dashed line marked "Threshold η") exceeds unity.

The scale length for Omega spherical implosions, $L_n \simeq 150$ μm, is set by the target size. Thus, the main parameter that controls the TPD instability in an

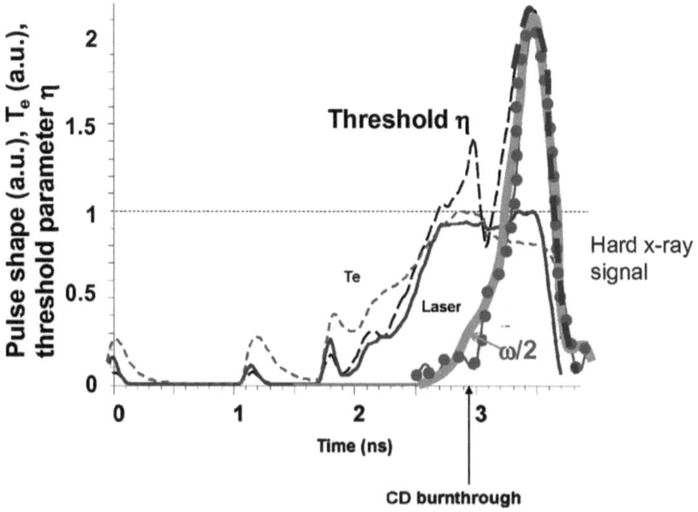

Fig. 7.20 Laser pulse shape (*thin solid line*), electron temperature at quarter-critical surface (*thin dashed line*), TPD instability threshold factor (*long-dashed line*), measured hard x-ray signal (*dotted line*), and measured ω/2 signal (*thick solid line*) for a cryogenic implosion with a 5 µm-thick CD shell

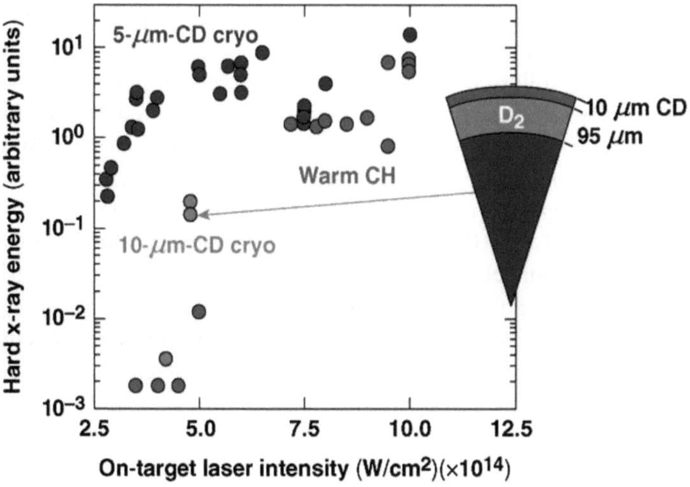

Fig. 7.21 Measured energy of hard x rays (above 40 keV) as a function of incident laser intensity for a variety of Omega implosions

experiment is the laser intensity. Since the hard x-ray emission increases with laser intensity [66], as plotted in Fig. 7.21, a 'low-intensity' series of cryogenic implosions was designed with peakz laser intensity reduced to below 3×10^{14} W/cm^2. Lowering drive intensity eliminates a possibility of fuel preheating caused by the

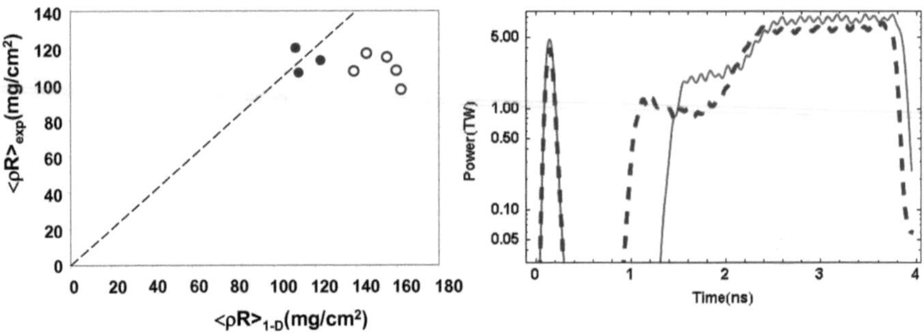

Fig. 7.22 Measured and predicted areal densities (*left panel*) for low-intensity RX drive pulses (*right panel*). *Solid symbols* correspond to the design shown with *solid line* on *right panel*. The open symbols correspond to lower-adiabat design shown with *dashed line* on the *right panel*

suprathermal electrons [67]. The first results of this campaign, shown in Fig. 7.22 by three solid circles, were very encouraging: for the first time the areal density measured in a low-adiabat ($\alpha \sim 3$) cryogenic implosion agreed with the simulation result! This initial success in ability to accurately predict fuel compression in a cryogenic implosion, however, was short lived. With the goal of increasing areal density in a low-drive design, the first picket energy was reduced and the intensity foot was reduced and extended in time (see dashed line on right panel in Fig. 7.22). The measurements, however, did not show any areal density increase predicted in simulations (see open circles in Fig. 7.22).

Instead, the data followed the same trend observed in higher-intensity implosions: areal density saturated to a value independent of the predicted adiabat.

Additional evidence supporting the conclusion that the suprathermal electrons alone cannot explain the areal density degradation (as shown in Fig. 7.18) was obtained using a 'dropping-intensity' design where the drive intensity was reduced from its peak value of 5×10^{14} down to $3 \times 10^{14}\,\mathrm{W/cm^2}$ starting from the time of onset of the suprathermal electron generation. This design and its comparison with the original flat-top design are shown in Fig. 7.23.

While the suprathermal electron preheat signal was substantially reduced, the dropping-intensity design has also failed to achieve areal densities above the saturation value of 80–100 $\mathrm{mg/cm^2}$.

7.5.1.3 Radiation Preheat

In addressing the second scenario for ρR degradation, excessive radiation preheating of the main fuel, the radiation x-ray power from plasma corona was measured using Dante [68]. Figure 7.24 (left panel) shows the total radiated x-ray power as a function of time for cryogenic implosion with 5-μm-thick CD shell.

Also plotted is the result of a *LILAC* simulation. The measured radiation power starts to deviate from the predictions at 3 ns. X-ray radiation spectrum, plotted on

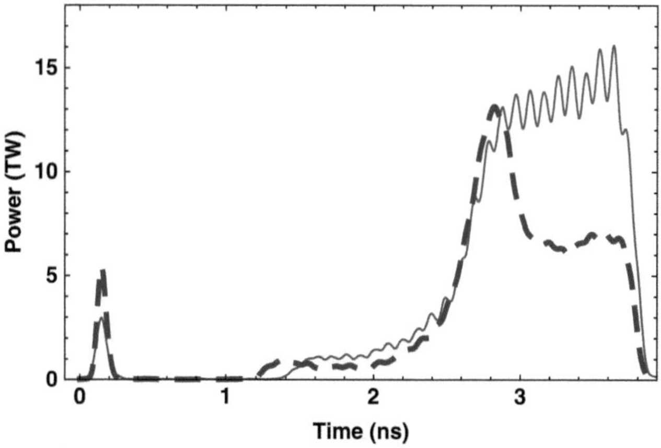

Fig. 7.23 The original (*solid line*) and modified (*dashed line*) design to reduce suprathermal electron production

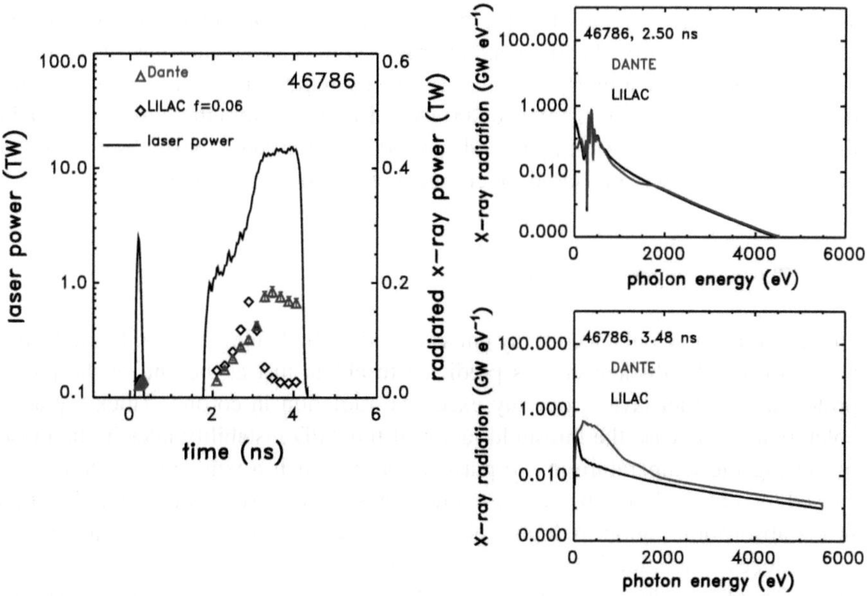

Fig. 7.24 Total radiated x-ray power and pulse shape (*left panel*) and x-ray radiation spectrum as measured using DANTE at $t = 2.5$ ns (*right panel, upper graph*) and $t = 3.48$ ns (*right panel, lower graph*)

the right panel in Fig. 7.24, also shows agreement with calculations early in the pulse. The spectrum deviates from calculations at $t = 3.48$ ns in the energy range from 100 eV to 1 keV. The plastic shell is totally ablated by that time, and the CD–D_2 interface starts to move into the plasma corona. Radiation in the hydrocode

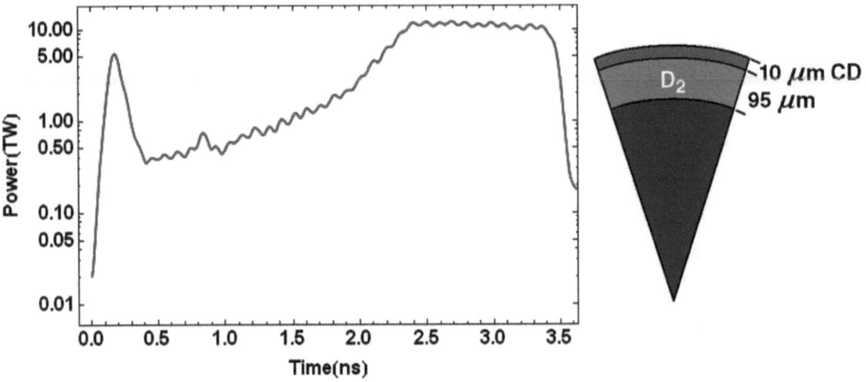

Fig. 7.25 Pulse shape and target dimensions for the thick plastic cryogenic design

calculation diminishes at this time because a higher-Z carbon is replaced by a lower-Z hydrogen in the x-ray–emitting region. Experimental data, on the other hand, showed a persistent signal after the burnthrough time. One plausible explanation of this effect is the mix of carbon and hydrogen at the CD–D_2 interface. This would cause carbon to stay longer at the higher-density region and significantly enhance the radiated x-ray power. Estimated 200 J was irradiated from the plasma corona in this experiment in excess of hydrocode predictions. Based on these observations, a new target design was proposed for cryogenic implosions on Omega.

Thick Plastic Cryogenic Designs

Observations of an enhanced x-ray emission led to thickening of the CD shell from 5 to 10 μm. The thicker shell is predicted to ablate just at the end of the pulse, protecting the fuel layer from any excessive radiation in corona. Thicker plastic ablators also increase the threshold factor of the TPD instability later in the pulse by raising the temperature in the plasma corona. Such a temperature increase is caused by a larger laser absorption fraction caused by presence of higher-Z carbon in the absorption region. A higher absorption fraction farther away in the corona also reduces irradiation intensity that reaches a quarter-critical surface. Both these effects lead to a reduction in η [see Eq. 7.65]. The cryogenic design with a 10-μm-thick CD ablator driven at $\sim 5 \times 10^{14}\,\mathrm{W/cm^2}$ is shown in Fig. 7.25.

Four shots with this design produced areal densities $\sim 200\,\mathrm{mg/cm^2}$, matching code predictions [58, 69]. Figure 7.26 shows predicted and measured spectra of downscattered secondary protons, confirming prediction accuracy.

The areal densities and fuel compression in these implosions were the highest ever achieved in an ICF implosion. As expected, both the hard x-ray signal (see points marked '10-μm-CD cryo' in Fig. 7.21) and x ray energy below 1 keV, emitted in excess to the predicted value, were significantly reduced in these experiments.

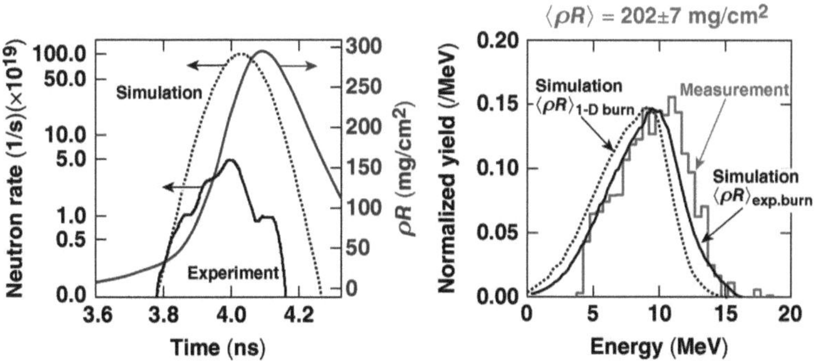

Fig. 7.26 The neutron-production history measured (*solid line*) and predicted (*dotted line*) for the design shown in Fig. 7.25. The ρR evolution calculated using 1-D code *LILAC* (*dashed line*, right axis) is also shown. *Right panel*: Measured secondary-proton spectrum (*solid line*). The *dotted line* shows the calculated spectrum averaged over the predicted 1-D neutron production, and the dotted line represents the calculated spectrum averaged over the experimental neutron-production history

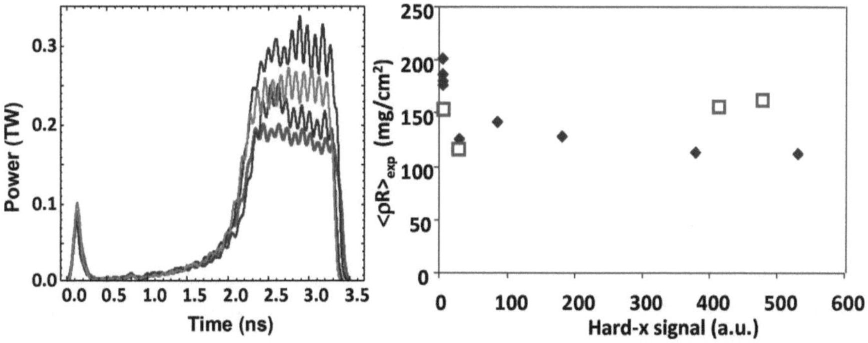

Fig. 7.27 Variation to the target design shown in Fig. 7.25 (*left*) and resulting measured areal densities for designs with 95 μm (*solid symbols*) and 80 μm (*open symbols*) D_2 ice thickness

Even though the designs with a thicker ablator demonstrated high compression, the drive intensity and implosion velocity $V_{\mathrm{imp}} \sim 2.2 \times 10^7$ cm/s were smaller than required for a robust direct-drive-ignition design, $I \sim 8 \times 10^{14}$ W/cm^2 and $V_{\mathrm{imp}} > 3.5 \times 10^7$ cm/s, respectively (see Sect. 7.1.1). The next step was to increase both the drive intensity and the implosion velocity (by reducing the shell mass). This turned out to be a very challenging task. Figure 7.27 (left panel) shows modifications made to the pulse shape in attempt to increase the drive intensity. Raising the intensity also increases the electron preheat signal. Right panel in Fig. 7.27 (solid symbols) shows measured areal densities as function of the preheat signal.

The measured areal density decreased dramatically even for minor variations in the laser pulse with very little or no sensitivity to the preheat signal. Reducing the thickness of the frozen D_2 layer from 95 to 80 μm also resulted in a decreased measured areal density (the predictions were ∼200 mg/cm^2 for all cases). This is

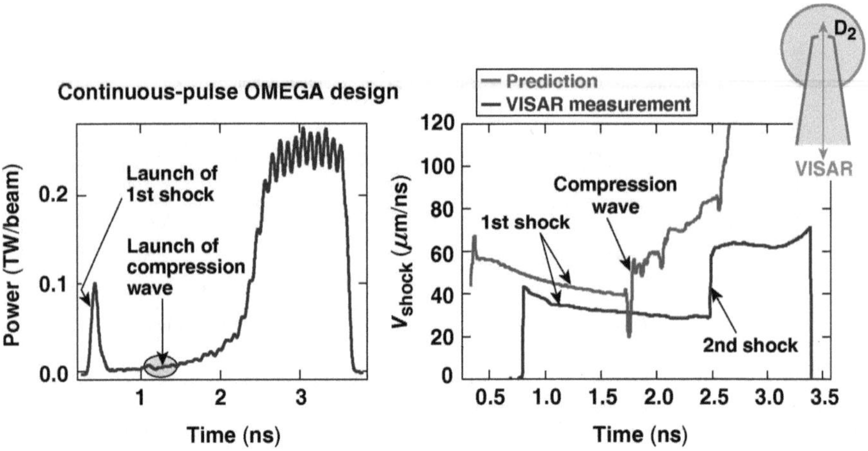

Fig. 7.28 Pulse shape (*left panel*) and leading shock velocity (*right panel*) measured (*lower curve*) and predicted (*upper curve*)

shown in Fig. 7.27 (right panel, open symbols). These results demonstrated that the continuous pulse designs cannot be easily extended to the ignition-relevant drive intensities and implosion velocities.

7.5.1.4 Shock Heating

The breakthrough in understanding cryogenic target performance came in 2008 when the shock-velocity measurement technique matured enough to give information on the formation of shock and compression waves in spherical geometry [70]. These measurements addressed the third scenario for explaining areal-density degradation, excessive shock heating. Accuracy of shock timing was verified by measuring the velocity of the leading shock wave using the velocity interferometry system for any reflector (VISAR) [71]. The targets in these experiments are spherical 5- or 10-μm-thick CD shells fitted with a diagnostic cone. The shell and cone are filled with liquid deuterium. An example of VISAR measurement performed using the continuous pulse design is shown in Fig. 7.28.

The measured shock velocity, as a function of time, is compared with 1-D predictions obtained using a *LILAC* simulation. An intensity picket at the beginning of the drive pulse sends a shock wave of decaying strength. As the drive intensity starts to rise from its minimum value, a compression wave is launched into the ablator at $t \simeq 1$ ns. After the head of the compression catches up with the first shock, strength and velocity of the leading shock increase gradually in time. The measured velocity history, however, shows a much steeper velocity increase that takes place later in the pulse, indicating that the compression wave turns into a shock prior to its coalescence with the first shock. Such a transition from adiabatic

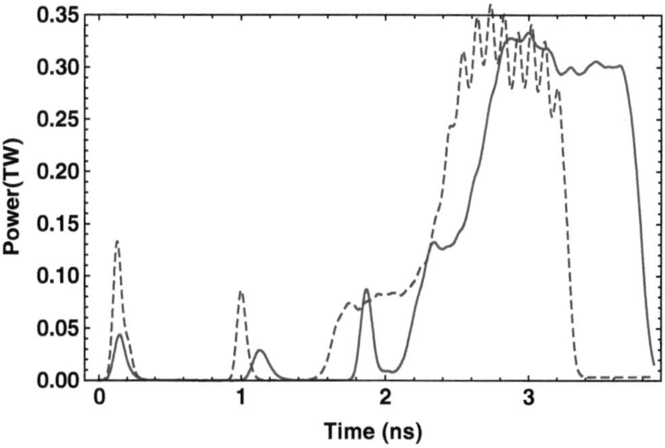

Fig. 7.29 Two- (*dashed line*) and three-picket (*solid line*) cryogenic Omega target designs

to shock compression raises the fuel adiabat at the inner part of the shell, limiting the final target convergence and peak fuel ρR. Since the effect of the compression wave steepening into a shock, not predicted by a simulation, is exacerbated by increasing peak drive pulse or changing the shell thickness, difficulty in tuning continuous-pulse designs can be explained by excessive shock heating.

After obtaining the VISAR results, the cryogenic program at LLE quickly moved to multiple-picket designs [72] by introducing double-picket, and later, triple-picket pulses (see Fig. 7.29).

To set the fuel on a low adiabat $\alpha \sim 1$–3, the double-picket design still requires a moderate-intensity foot ($1/4$–$1/3$ of peak intensity) and a gradual intensity increase to compress the fuel adiabatically (dashed line in Fig. 7.29). The triple-picket design (see solid line in Fig. 7.29), on the other hand, does not rely on an adiabatic compression and requires a short step at the beginning of the main pulse to control strength of the main shock.

7.6 Current Triple-Picket Cryogenic-DT Implosions

The main advantage in using multiple-picket designs is the ability to control all hydrodynamic waves launched by the drive pulse [72]. As described in Sect. 7.2.1, designs with continuous pulses rely on adiabatic fuel compression while the drive pressure increases by factor of 50 or more. The observed premature steepening of the adiabatic compression wave into a shock inside the shell makes it impractical to experimentally tune the shell adiabat in these designs. In a multiple-picket designs shown in Fig. 7.29, the required increase in drive pressure from a few Mbar to $\sim$100 Mbar is accomplished by launching a sequence of shocks that can be well controlled by adjusting the timing and energy of each individual intensity picket.

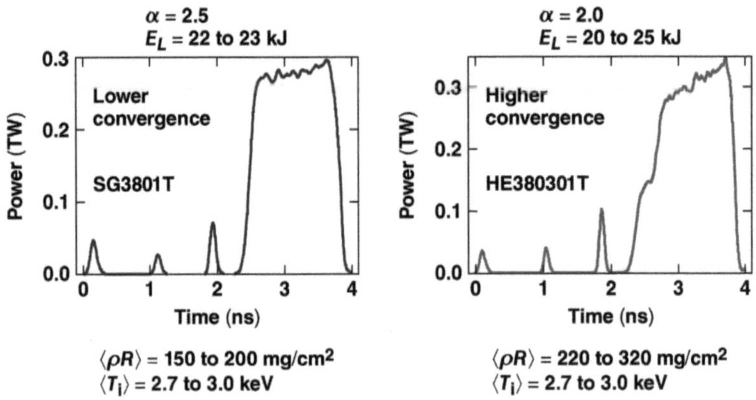

Fig. 7.30 Two triple-picket target designs used in cryogenic implosions Omega

There are two types of the triple-picket pulse shapes used in current cryogenic implosions on Omega. The laser power in the first design, shown on the left panel of Fig. 7.30, consists of three pickets and the main drive in the form of a square pulse. To control strength of the main shock, a short intensity step is introduced at the beginning of the main drive in the second design [shown on the right panel of Fig. 7.30b]. The stronger main shock launched in the first design sets the fuel on $\alpha = 2.5 - 3$. A weaker shock in the second design reduces adiabat to $\alpha = 2 - 2.5$.

Next, we describe how shock tuning was accomplished in these designs using Omega experiments.

7.6.1 Shock Tuning

Accuracy in predicting shock timing is verified by measuring the velocity of the leading shock wave using the VISAR. The targets in these experiments are spherical 5- or 10-μm-thick CD shells fitted with a diagnostic cone [73]. The shell and cone are filled with liquid deuterium. For an optimised design [72], all shocks should coalesce within 100 ps, soon after they break out of the shell. For the purpose of code validation, the time separation between shock coalescence events was increased in these experiments to accurately infer leading shock velocity after each coalescence. An example of such measurement is shown in Fig. 7.31.

Because of radiation precursor, the shock is not visible to the VISAR early in time while it travels through the plastic layer. Then, at $t \sim 300$ ps, the shock breaks out of CD into D_2 with the velocity of ~60 μm/ns. The shock is not supported by the laser at this time (picket duration is ~80 ps). Thus, the shock strength and its velocity decrease with time. Then, the second shock is launched at $t = 1.1$ ns. It travels through the relaxed density and pressure profiles established by the first blast wave. At $t = 2.5$ ns the second shock catches up with the first, resulting in a jump in

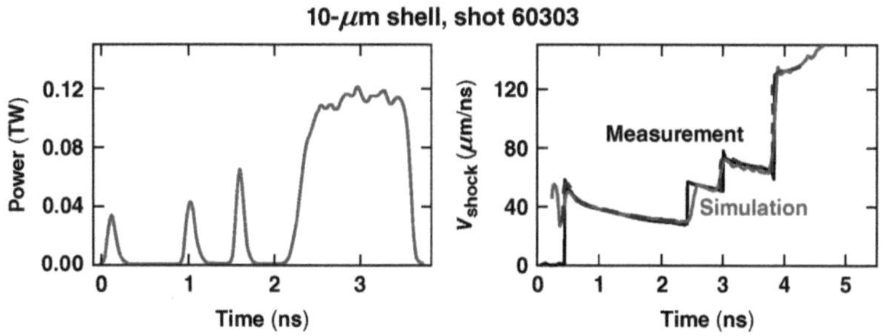

Fig. 7.31 Measured (*dashed line, right panel*) and predicted (*solid line, right panel*) leading shock velocity in triple-picket design (*left panel*)

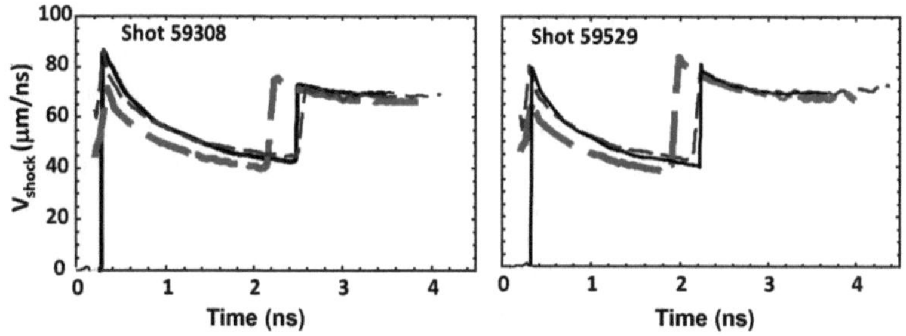

Fig. 7.32 Measured (*solid lines*) and predicted leading shock velocity using the flux-limited (*thick dashed lines*) and nonlocal (*thin dashed lines*) electron thermal transport models

leading shock velocity from 35 up to 60 μm/ns. The third picket and the main pulse launch two additional shocks that coalesce with the leading shock at $t = 3.0$ and 3.9 ns, respectively.

Matching both shock velocities and coalescence times is a good test of a thermal-conduction model used in a hydrocode simulation. The thermal conduction affects hydrodynamic profiles that determine energy coupling. The flux-limited model with $f = 0.06$ predicts a lower laser-absorption fraction than that calculated using the nonlocal thermal transport model, leading to a slower shock. The difference between two transport models increases with the energy in the first picket. The comparison between models predictions and experimental data is shown in Fig. 7.32.

As seen on this figure, agreement between predictions and measurements improves when the nonlocal thermal-transport model is used in the simulations.

Matching the predicted and measured shock velocities and coalescence times ensures that the shock heating is properly modelled. The in-flight shell adiabat, set by the shocks, can be degraded during the implosion by electron or radiation preheat as well as by secondary shock waves. As described in Sect. 7.3.1, the in-flight adiabat can be inferred from areal-density measurements if no significant shell decompression is induced by prolonged coasting phase [see discussion after Eq. 7.53]. The extended coasting phase could result from a loss in hydro-efficiency during shell acceleration. The latter would reduce shell implosion velocity and delay the time of neutron production. Thus, to connect any observable degradation in areal density with fuel preheat or any other effects that enhance in-flight adiabat, one must verify that hydrodynamic efficiency is accurately modelled and no extended coasting phase is present in the implosion. This will be addressed in the next subsection.

7.6.2 Laser Coupling and Hydrodynamic Efficiency

Accurate modelling of hydrodynamic efficiency of an imploding shell (defined as the ratio of the peak in shell kinetic energy to the total laser energy) is crucial for optimising high-convergence target designs, since a loss in the shell implosion velocity and kinetic energy leads to shell coasting after the laser drive turns off. During such coasting, both shell density and pressure drop. This reduces ρR [see Eq. 7.53] and gives a lower fuel ion temperature at the time of neutron production. One of the diagnostics that is most sensitive to deviations in the shell implosion velocity is a measurement of timing and temporal shape of primary neutrons produced as a result of fusion reactions. This is accomplished by using NTD (see discussion in Sect. 7.3.2). Currently, NTD is calibrated on Omega to $\sim$50 ps absolute timing accuracy with $\sim$10 ps shot-to-shot timing variation. In addition to the neutron-production timing, the laser-absorption measurement is performed using two full-aperture backscattering stations (FABS) [63]. Time-resolved scattered-light spectroscopy and time-integrated calorimetry in these stations are used to infer the absorption of laser light. Laser absorption, however, is not a direct measurement of hydrodynamic efficiency, since only a small fraction of the incident laser energy ($\sim$5 %) is converted (through the mass ablation) into the shell kinetic energy and the majority of the absorbed energy goes into heating the underdense plasma corona. Also, some fraction of laser energy can be deposited into plasma waves that accelerate suprathermal electrons and do not directly contribute to the drive.

Figure 7.33 compares the measured scattered laser light [(a) and (b)] and neutron production history (c) with the predictions (solid lines marked with 'without CBET') for an $\alpha = 2.5$ design.

As seen on Fig. 7.33b, calculations are in very good agreement with the measured scattered light data (dotted line) for the picket portion of the pulse. At the main drive, however, the predicted laser absorption overestimates the data, especially at the beginning of the drive. Higher predicted laser coupling results in an earlier bang time, as shown in Fig. 7.33c. On average, the rise of the neutron rate is earlier in simulations by 200 ps. Since calculations fail to accurately reproduce the laser-absorption fraction and neutron-production timing, an additional mechanism explaining a reduced laser coupling must be present in the experiments.

Such a mechanism, as discussed in a recent publication [74], is due to the Cross-Beam Energy Transfer (CBET) [75]. In the geometric optics approximation where each laser beam is subdivided into rays, the incoming ray in the central part of the

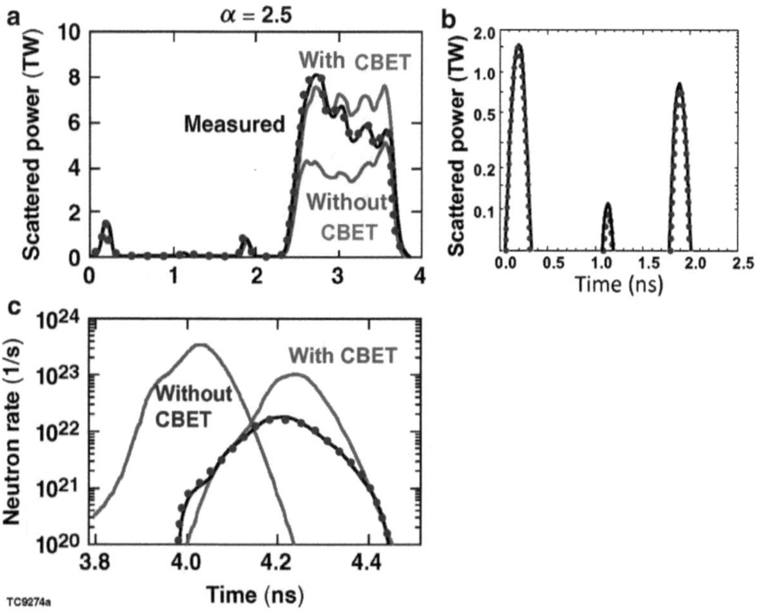

Fig. 7.33 Measured (*dotted lines*) and predicted (*solid lines*) scattered light data (**a**, **b**) and neutron-production history (**c**) for $\alpha = 2.5$ design shown on the *left panel* of Fig. 7.30

Fig. 7.34 The incoming ray in the central part of the beam (*2*) interacts with the outgoing ray on the outer edge of the beam (*1*), transferring its energy to that ray

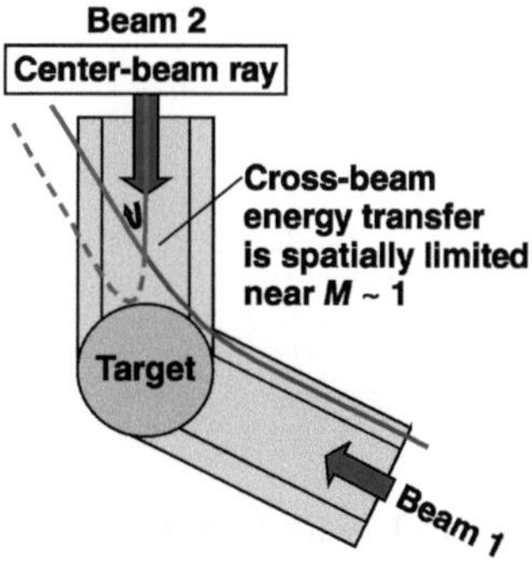

beam interacts (through the ion-acoustic waves) with the outgoing ray on the outer edge of the beam, transferring its energy to that ray. This is illustrated in Fig. 7.34.

Since the central part of the beam propagates closest to the target, CBET reduces the fraction of the beam energy that reaches the higher-density plasma corona, decreasing overall laser absorption. Because CBET reduces the total laser absorption, and, furthermore, the absorbed energy is deposited in corona farther away from the target surface, the hydro-efficiency of laser drive in directly driven implosions is degraded by 15–20 % in Omega implosions. When implemented into the hydrocode *LILAC*, a CBET model predicts a 10–15 % reduction in the absorbed energy, in agreement with experimental data. Shown in Fig. 7.33 with solid lines

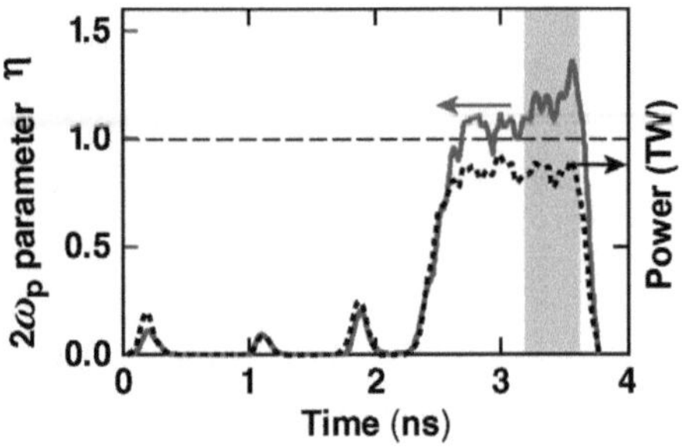

Fig. 7.35 Pulse shape (*dotted line*) and threshold parameter η of TPD instability

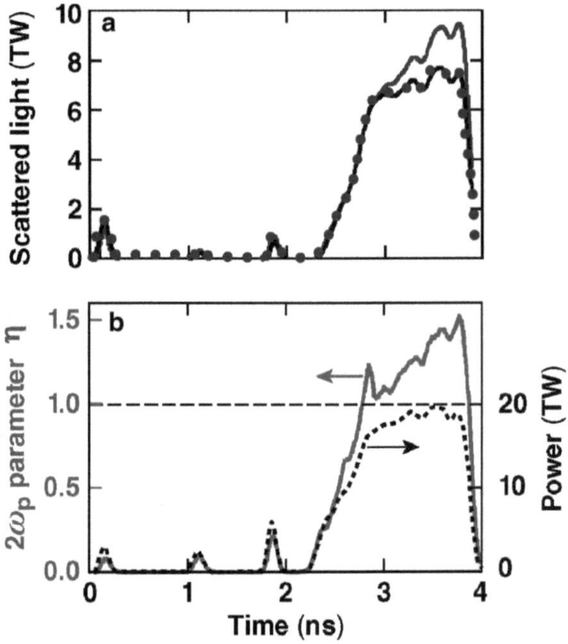

Fig. 7.36 (**a**) Measured (*dotted line*) and predicted scattered power. (**b**) Pulse shape (*dotted line*) and threshold parameter η of TPD instability (*solid line*)

marked 'with CBET' are the scattered light (a) and neutron-production rate (c) calculated using a combination of the nonlocal thermal transport and CBET models. The neutron-production timing matches data very well. The scattered-light power, however, deviates from the measurements at later times. This late-time discrepancy is likely due to extra absorption of laser energy by plasma waves excited by the TPD instability [14]. Figure 7.35 shows the drive pulses and threshold parameters for an $\alpha = 2.5$ designs. The threshold parameter exceeds unity at $t \sim 3.2$–3.3 ns, which matches the time when the experimental scattered light starts deviating from the predictions.

To further support the assertion that the observable fraction of laser energy being deposited into plasma waves, the scattered-light measurement and prediction are plotted in Fig. 7.36 for an implosion at a slightly higher drive intensity where the TPD instability threshold is exceeded at the beginning of the main drive [see Fig. 7.36b]. The calculated scattered-light power starts deviating from measurements earlier in this case, which is consistent with timing of η exceeding unity.

Incorporating the CBET model into hydrocode simulations shows only a marginal reduction (on average by $\sim$5 %) in neutron-averaged areal densities. This confirms that the areal densities in cryogenic implosions on Omega are affected mainly by the in-flight shell adiabat and the effect of shell decompression during the coasting phase is small.

7.6.3 Areal Densities in Triple-Picket Cryogenic Implosions

In this section we compare the calculated neutron-averaged areal density $\langle \rho R \rangle_n$ with the measurements. Since the predicted $\langle \rho R \rangle_n \sim 150$–$200\,\mathrm{mg/cm^2}$ for $\alpha = 2.5$ and $\langle \rho R \rangle_n \sim 220$–$300\,\mathrm{mg/cm^2}$ for $\alpha = 2$, the areal density is currently inferred using a single-view measurement with a magnetic recoil spectrometer (MRS) [28]. The MRS measures the number of primary neutrons and the number of neutrons scattered in the dense DT fuel. The ratio of these two is proportional to the fuel areal density during the neutron production. Two charged-particle spectrometers (CPS) were also used to measure the spectrum of knock-on deuterons, elastically scattered by primary DT neutrons. These measurements, however, are insensitive to $\langle \rho R \rangle_n > 180\,\mathrm{mg/cm^2}$ and were used to assess low-l-mode ρR asymmetries for implosions where areal density along CPS's line of sight is below 180 $\mathrm{mg/cm^2}$. Such asymmetries arise from errors in target positioning (offset) and ice roughness amplified during shell implosion. Since only a single-view MRS measurement is used for ρR analysis, it is important to take long-wavelength asymmetries into account when comparing the simulated and measured areal densities for high-convergence implosions. Strictly speaking, even a single MRS measurement averages fuel ρR over a solid angle of $\sim 1.5\pi$ since the downscattered neutrons have a finite spectral width, and neutrons with different energies sample different parts of the shell (see Fig. 7.37).

The scattering angle θ of a primary neutron (marked with 'n in Fig. 7.37) depends on downscattered neutron' ('n′') energy. MRS is sensitive to 8- to 13 MeV neutrons. The minimum scattering angle $\theta_{\mathrm{min}} = 29°$ and $23°$ correspond to 13-MeV neutrons scattered by tritons and deuterons, respectively. The maximum angle $\theta_{\mathrm{max}} = 80°$ and $62°$ corresponds to 8-MeV neutrons. The dark shell region in Fig. 7.37 corresponds to a region sampled by the downscattered neutrons in a single-view MRS measurement on Omega. Taking into account such averaging, Fig. 7.38 plots calculated variation in areal density as would be observed by the MRS in a single-view measurement taken along a different direction with respect to the target offset.

The results are shown for the offset values of $\delta_{\mathrm{offset}} = 10\ \mu\mathrm{m}$ (black line) and $30\,\mu\mathrm{m}$ (gray line). The calculations are performed by post-processing results of 2-D *DRACO* simulations [76] using Monte Carlo–based particle transport code *IRIS*.

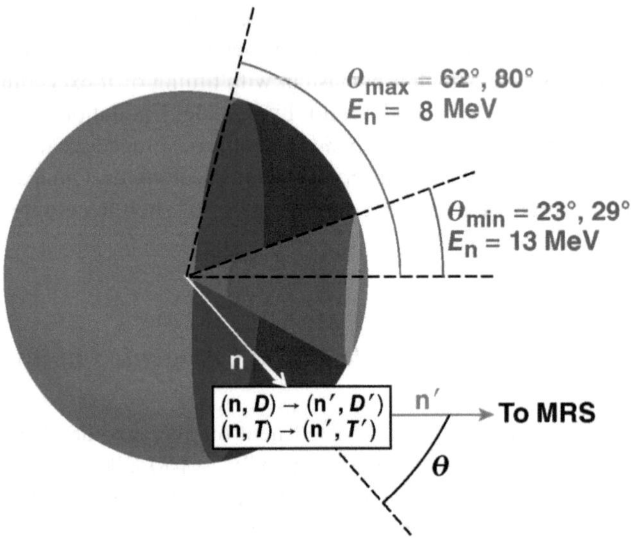

Fig. 7.37 The scattering angle θ of a primary neutron (marked with "n") depends on downscattered neutron ("n′") energy. MRS is sensitive to neutrons with energies between 8- and 13-MeV. The minimum scattering angle $\theta_{min} = 29°$ and 23° correspond to 13-MeV neutrons scattered by tritons and deuterons, respectively. The maximum angle $\theta_{max} = 80°$ and 62° corresponds to 8-MeV neutrons. The dark shell region corresponds to a region sampled by the downscattered neutrons in a single-view MRS measurement on Omega

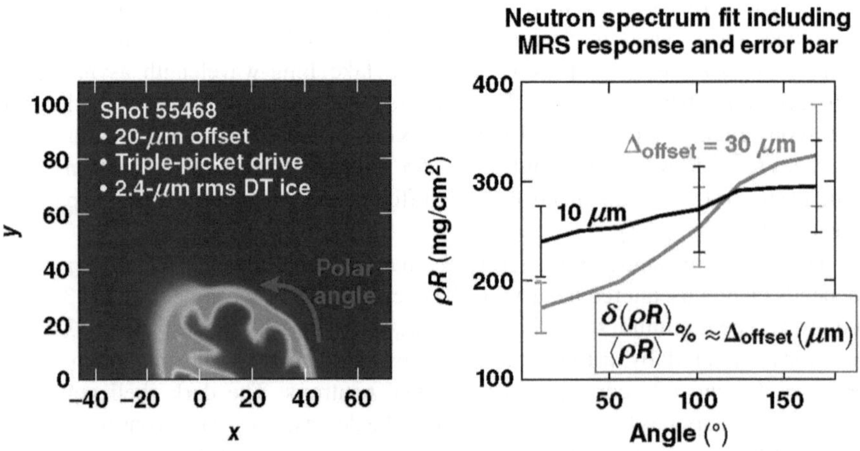

Fig. 7.38 Density contour of 2-D *DRACO* simulation of the cryogenic implosion on Omega (shot 55468) with the target offset of 20 μm (*left panel*). Predicted variation in areal density as would be observed by the MRS in a single-view measurement taken along a different direction with respect to the target offset (*right panel*)

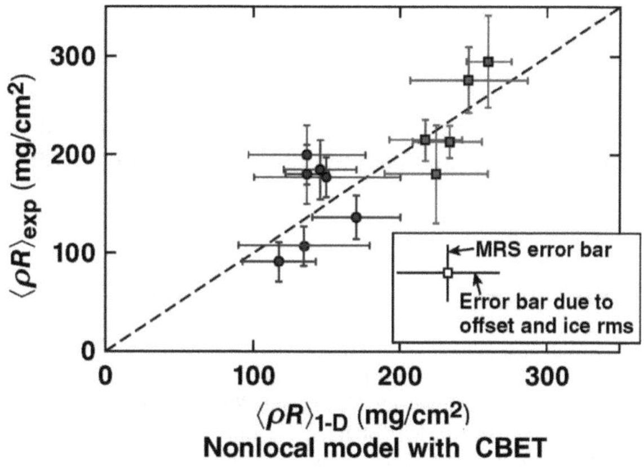

Fig. 7.39 Measured (symbols) versus predicted areal densities for triple-picket cryogenic implosions on Omega. *Squares* and *circles* correspond to $\alpha = 2$ and $\alpha = 2.5$ designs, respectively

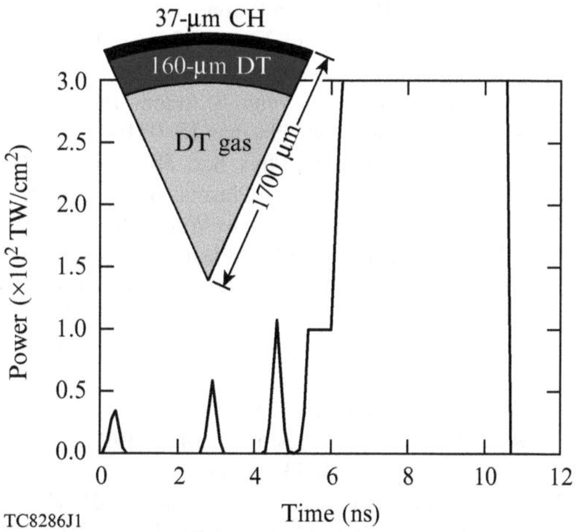

Fig. 7.40 Triple-picket, symmetric direct-drive design for the NIF

The error bars in Fig. 7.38 represent counting statistics errors in a typical cryogenic implosion on Omega. These calculations show that the $\langle \rho R \rangle_{\rm n}$ variation across the target can be approximated by a linear function of the offset,

$$\frac{\max\langle \rho R \rangle_{\rm n} - \min\langle \rho R \rangle_{\rm n}}{\langle \rho R \rangle_{\rm n}} \% \simeq \delta_{\rm offset}(\mu{\rm m}). \tag{7.66}$$

In addition to the target offset, the low-l modes ($l \leq 2$) seeded by ice roughness also lead to an azimuthal variation in the measured areal density. In plotting the predicted $\langle \rho R \rangle_{\rm n}$, we assign the error bar for each point taking into account ρR variation due to target offset and low-mode ice roughness measured for each target. The result is shown in Fig. 7.39 (see also Ref. [72]), where squares and circles correspond to $\alpha = 2$ and $\alpha = 2.5$ designs, respectively. In general, there is a good agreement between the experimental data and calculations. This confirms that adiabat is modelled accurately in low-adiabat, cryogenic implosions on Omega using the triple-picket designs.

Based on the good performance of the triple-picket design on Omega, this design was extended to a 1.5-MJ direct-drive–ignition design [72] for the National Ignition Facility (see Fig. 7.40). Driven at a peak intensity of 8×10^{14} W/cm^2, the shell reaches $V_{\mathrm{imp}} = 3.5$ to 4×10^7 cm/s, depending on the thickness of the fuel layer. This design is predicted to ignite with a gain $G = 48$.

References

1. J.D. Lindl, *Inertial Confinement Fusion* (Springer, New York, 1998)
2. S. Atzeni, J. Meyer-ter-Vehn, *The Physics of Inertial Fusion* (Clarendon press, Oxford, 2004)
3. M.C. Herrmann, M. Tabak, J.D. Lindl, Phys. Plasmas **8**, 2296 (2001)
4. R. Betti et al., Phys. Plasmas **9**, 2277 (2002)
5. V.N. Goncharov in *Laser-Plasma Interactions*, ed. by D.A. Jaroszynski, R. Bingham, R.A. Cairns (CRC Press, Boca Raton, 2009), p. 409
6. J. Meyer-ter-Vehn, Nucl. Fusion **22**, 561 (1982)
7. R. Betti et al., Phys. Plasmas **17**, 058102 (2010)
8. A. Kemp, J. Meyer-ter-Vehn, S. Atzeni, Phys. Rev. Lett. **15**, 3336 (2001)
9. S.W. Haan et al., Phys. Plasmas **18**, 051001 (2011)
10. C.P. Verdon, Bull. Am. Phys. Soc. **38**, 2010 (1993)
11. P.W. McKenty et al., Phys. Plasmas **8**, 2315 (2001)
12. J.A. Paisner et al., Laser Focus World, **30**, 75 (1994)
13. V.N. Goncharov et al., Phys. Plasmas **7**, 2062 (2000)
14. W.L. Kruer, *The Physics of Laser-Plasma Interactions, Frontiers in Physics*, vol. 73, ed. by D. Pines (Addison-Wesley, Redwood City, 1988), Chap. 4, p. 81
15. S. Chandrasekhar, *Hydrodynamic and Hydromagnetic Stability* (Claredon, Oxford, 1961), p. 428
16. J. Sanz, Phys. Rev. Lett. **73**, 2700 (1994); V.N. Goncharov et al., Phys. Plasmas **3**, 1402 (1996)
17. R. Betti et al., Phys. Plasmas **5**, 1446 (1998)
18. J. Sanz, Phys. Rev. E **53**, 4026 (1996)
19. V.N. Goncharov et al., Phys. Plasmas **13**, 012702 (2006)
20. C.D. Zhou, R. Betti, Phys. Plasmas **14**, 072703 (2007)
21. V.N. Goncharov et al., Phys. Plasmas **10**, 1906 (2003)
22. H. Sawada et al., Phys. Plasmas **14**, 122703 (2007)
23. A.L. Kritcher et al., Phys. Rev. Lett. **107**, 015002 (2011)
24. H. Sawada et al., Phys. Plasmas **16**, 052702 (2009)
25. F.J. Marshall et al., Phys. Rev. Lett. **102**, 185004 (2009)
26. R. Tommasini et al., Phys. Plasmas **18**, 056309 (2011)
27. F. Seguin et al., Phys. Plasmas **9**, 2725 (2002)
28. J.A. Frenje et al., Phys. Plasmas **16**, 042704 (2009); J.A. Frenje et al., Phys. Plasmas **17**, 056311 (2010)
29. D.G. Hicks et al., Phys. Plasmas **17**, 102703 (2010)
30. J.P. Knauer et al., Bull. Am. Phys. Soc. **50**, 133 (2005)
31. C. Stoeckl et al., Rev. Sci. Instrum. **74**, 1713 (2003)
32. T.J. Murphy et al., Rev. Sci. Instrum. **66**, 930 (1995)
33. H. Brysk, Plasma Phys. **15**, 611 (1973)
34. J.M. Soures et al., in *Proceedings of the 10th Symposium on Fusion Engineering, Philadelphia, 1983* (IEEE, New York, 1983), p. 1392
35. F.J. Marshall et al., Phys. Rev. A **40**, 2547 (1989)
36. R.L. McCrory et al., Nature **335**, 225 (1988)
37. D.L. Musinski et al., J. Appl. Phys. **51**, 1394 (1980)

38. S. Kacenjar et al., J. Appl. Phys. **56**, 2027 (1984); S. Skupsky, S. Kacenjar, J. Appl. Phys. **52** , 2608 (1981)
39. J. Delettrez et al., Phys. Rev. A **36**, 3926 (1987)
40. J.K. Hoffer, L.R. Foreman, Phys. Rev. Lett. **60**, 1310 (1988); A.J. Martin, R.J. Simms, R.B. Jacobs, J. Vac. Sci. Technol. A **6**, 1885 (1988)
41. G.W. Collins et al., J. Vac. Sci. Technol. A **14** , 2897 (1996)
42. See National Technical Information Service Document No. DOE/SF/19460-335 [Laboratory for Laser Energetics LLE Review **81**, 6 (1999)]. Copies may be obtained from the National Technical Information Service, Springfield, VA 22161
43. T.R. Boehly et al., Opt. Commun. **133**, 495 (1997)
44. C. Stoeckl et al., Phys. Plasmas **9**, 2195 (2002)
45. T.C. Sangster et al., Phys. Plasmas **14**, 058101 (2007)
46. P.W. McKenty et al., Phys. Plasmas **11**, 2790 (2004)
47. F.J. Marshall et al., Phys. Plasmas **12**, 056302 (2005)
48. P.B. Radha et al., Phys. Plasmas **12**, 032702 (2005)
49. V.N. Goncharov et al., Phys. Plasmas **7**, 5118 (2000)
50. T.R. Boehly et al., Phys. Plasmas **8**, 2331 (2001)
51. K. Anderson, R. Betti, Phys. Plasmas **11**, 5 (2004)
52. S. Skupsky et al., J. Appl. Phys. **66**, 3456 (1989)
53. W. Kruer, *The Physics of Laser Plasma Interactions* (Addison-Wesley, Redwood City, 1988)
54. R.C. Malone, R.L. McCrory, R.L. Morse, Phys. Rev. Lett. **34**, 721 (1975)
55. L. Spitzer, R. Harm, Phys. Rev. **89**, 977 (1953)
56. N.A. Krall, A.W. Trivelpiece, *Principles of Plasma Physics* (San Francisco Press, San Francisco, 1986)
57. S. Chapman, T.G. Cowling, *The Mathematical Theory of Non-uniform Gases* (Cambridge University Press, Cambridge, 1970)
58. V.N. Goncharov et al., Phys. Plasmas **15**, 056310 (2008)
59. V.N. Goncharov, G. Li, Phys. Plasmas **11**, 5680 (2004)
60. V.A. Smalyuk et al., Phys. Rev. Lett. **101**, 025002 (2008)
61. V.N. Goncharov, Phys. Rev. Lett. **82**, 2091 (1999)
62. O.V. Gotchev et al., Phys. Rev. Lett. **96**, 115005 (2006)
63. W. Seka et al., Phys. Plasmas. **15**, 056312 (2008)
64. I.V. Igumenshchev et al., Phys. Plasmas, **14**, 092701 (2007)
65. A. Simon et al., Phys. Fluids **26**, 3107 (1983)
66. C. Stoeckl et al., Rev. Sci. Instrum. **72**, 1197 (2001)
67. V.A. Smalyuk et al., Phys. Rev. Lett. **100**, 185005 (2008)
68. H.N. Kornblum, R.L. Kauffman, J.A. Smith, Rev. Sci. Instrum. **57**, 2179 (1986); K.M. Campbell et al., Rev. Sci. Instrum. **75**, 3768 (2004)
69. T.C. Sangster et al., Phys. Rev. Lett. **100**, 185006 (2008)
70. T.R. Boehly et al., Phys. Plasmas **16**, 056302 (2009)
71. L.M. Baker, R.E. Hollenbach, J. App. Phys. **43**, 4669 (1972); P.M. Celliers et al., Appl. Phys. Lett. **73**, 1320 (1998)
72. V.N. Goncharov et al., Phys. Rev. Lett. **104**, 165001 (2010)
73. T.R. Boehly et al., Phys. Rev. Lett. **106**, 195005 (2011)
74. I.V. Igumenshchev et al., Phys. Plasmas **17**, 122708 (2010)
75. C.J. Randall, J.R. Albritton, J.J. Thomson, Phys. Fluids **24**, 1474 (1981)
76. S.X. Hu et al., Phys. Plasmas **17**, 102706 (2010)

Chapter 8
Indirect Drive at the NIF Scale

Mordecai D. ('Mordy') Rosen

Abstract We review the basics of ICF ignition, and analytically derive the expected gains for the NIF scale and the reactor scale. We also analytically derive the radiation drive temperature for an empty NIF scale hohlraum. We describe improved physics models that better describe experiments at the NIF scale. We compare those improved models for NIF hohlraum data from 2009. We briefly review the status of NIF ignition experiments of 2011, including shock timing tuning, and the measurement of the implosion velocity.

8.1 Fundamentals of Indirect Drive ICF

8.1.1 Introduction

It has been a long-term goal to bring an imploded inertial confinement fusion (ICF) capsule to the proper conditions of density and temperature such that it will ignite its deuterium-tritium (DT) fuel via thermonuclear fusion reactions. These implosions can be driven either directly by a laser, or indirectly by having that laser impinge on the gold walls of a cylinder (a hohlraum) surrounding the capsule. The laser produces x-rays from the gold walls, which bathe the capsule in a quasi-black body radiation field. It is this bath of x-rays that implode the capsule. The general theory of this indirect drive ICF approach was covered in quite a bit of depth in our 2005 SUSSP60 St. Andrews lectures [1]. In Sect. 8.1 we will review some of those basics. We will also present material that describes improvements in our

M.D. Rosen (✉)
Lawrence Livermore National Laboratory, 7000 East Avenue, L-039, Livermore, CA 9450, USA
e-mail: rosen2@llnl.gov

P. McKenna et al. (eds.), *Laser-Plasma Interactions and Applications*,
Scottish Graduate Series, DOI 10.1007/978-3-319-00038-1_8,

physics understanding that have occurred in the interim. In Sect. 8.2 we will present a 'snapshot' in time (late 2011) of the progress of the National Ignition Campaign (NIC). Due to its many institutional partners, this campaign is truly national, and in fact, international in scope. The laser light it uses operates at a wavelength of 0.35 μm, and its energy/power of 1.8 MJ/500 TW is provided by the National Ignition Facility (NIF), located at the Lawrence Livermore National Laboratory (LLNL) in Livermore, CA.

In these lectures we will discuss the unique issues brought about by attempting ignition in large-scale hohlraums. The energy scale of the NIF is some 40 times larger than any other laser experiment previously attempted. The spatial scale for a hot (roughly 300 eV black body drive) hohlraums is about a factor of 4 larger than previous hohlraums that achieved radiation temperatures (T_r) of that order (actually more typically, a bit less, around 250 eV). Extrapolating from the previous database to this regime is challenging, especially in light of the very stringent specifications on accuracy and precision, in laser performance, target quality, and implosion performance, required to achieve ignition [2].

8.1.2 General Requirements for Ignition

In this section we will briefly summarise the requirements for ICF ignition and gain. A lengthier discussion and derivation of these requirements can be found in [3], which is meant to be a more complete tutorial on the subject. For even more depth, we refer the students to the books on ICF by Lindl [4] and by Atzeni and Meyer-ter-Vehn [5].

An ICF 1 GW reactor, with a 10 % efficient driver (supplying 6 MJ pulses @5 Hz to the target chamber) must have a target with gain $G > 100$. The inertial confinement time of an assembled fuel of radius R is given by $R/4C_S$. The burn-up fraction f_B of the fuel is given by $\rho R/(\rho R + 70)$ in MKS units. The numerator's ρR dependence can be thought of as burn fraction which should scale as the product of the burn rate $\sim$ density $\sim \rho$, and confinement time $\sim R$. The denominator is basically a constant, but the extra ρR factor therein assures a burn up fraction no larger than unity. We expect a target to operate optimally near $f_B = 1/3$ or a ρR of 30. With that f_B fixed, and using the energy per unit mass released by the fusing of DT, $Q_{DT} = 3.4 \times 10^{14}$ J/kg, we can see that to have the expected and containable output of 600 MJ we must compress DT 1000 fold so that its mass will be about 5×10^{-6} kg. The 600 MJ at 5 Hz will produce 3 GW, which when converted to electricity, will be able to send 1 GW out to the grid, and use the rest to power the input laser. This exploding 5 mg target has a momentum of 80 kg m/s or the impact an average person would have walking into a wall at average walking speed, which is obviously quite containable.

A spherical implosion is the least stressing way to compress matter that much. A hohlraum allows for good symmetry for such an implosion due to geometric smoothing. Two points, Alice and Bob, nearby each other on the outer surface of the capsule look out towards the hohlraum walls into their respective 2π skies. With a hohlraum wall typically a factor of 4 larger in radius than the initial capsule radius, they see nearly identical 'skies'. It doesn't matter, that due to non-uniformities of laser and plasma, they see rather complex radiant skies - the key point is that Alice and Bob see the *same* complex sky. Thus they are driven the same and thus these short wavelength (nearby location) asymmetries are smoothed by this geometric effect. Longer wavelength (e.g. Alice is at the pole of the capsule and Bob is at the equator) must be addressed by other means, as we will describe in Sect. 8.2.

The method of implosion is basically a rocket. Thus there are two coupling efficiencies that get us from driver energy E_D to the thermal energy of the assembled fuel: η_C the coupling efficiency of the driver to thermal energy on the surface of the capsule. That hot gas is the exhaust of the rocket, which delivers a radially inward moving payload at efficiency η_H. That kinetic energy of the imploding payload is reconverted to the thermal energy of the assembled self-stagnating fuel.

If we had to heat that entire fuel assembly to 10 keV to start the fusion process in earnest, it would require (1.5) (10 keV) (4) / 5 AMU = 10^{12} J/kg, a factor $1/340$ less than the fusion pay-off Q_{DT} quoted above. The ('4') of the above equation is due to the need to heat the deuteron, the triton, and the electron that comes with each. However, we only burn $1/3$ of that, and for a reactor scale hohlraum typical coupling efficiencies are $\eta_C = 0.2$ and $\eta_H = 0.2$, thus the gain is only $G = (340)(0.2)(0.2)(1/3) = 5$, far too low for the reactor of gain 100. The secret to high gain is to only heat a small central hot spot to 10 keV, and then let the alpha particle produced by the DT reaction stop within the surrounding fuel and do the heating of the bulk of the fuel. Typically that hot spot will have a density of 10^5 kg/m^3 (= 100 gm/cc in cgs), which is typically 0.1 of the density of the surrounding dense cold fuel shell. Moreover, the hot spot radius is typically at about $R/2$. Thus the hot spot will have a negligible 0.01 of the mass of the fuel.

The requirements of a hot spot then, are to be 10 keV, and have a ρR of 3 (kg/m^2). That is the range of an alpha particle in a 10 keV plasma. In fact because of that, the alphas can self-heat the hot spot. Thus, we only need to heat the hot spot (by non-fusion means such as the above-described 'rocket payload recompression' which equivalently can be thought of as hydrodynamic PdV compression heating) to about 5 keV. The alpha heating will take it to 10 keV. Moreover, the f_B formula implies that the hot spot will burn 5 % of its fuel. That is precisely the amount needed to supply enough alphas to the first thin shell (a layer inside of the dense fuel which also has a ρR of 3 kg/m^2) surrounding the hot spot to get it up to 10 keV, and thus launch the propagating thermonuclear burn wave.

Energy is also required to compress the cold dense fuel to its high-density state in the shell that surrounds the hot spot. This energy is required, as work must be done to counter the quantum pressure that resists compression. It is Fermi Degenerate matter, so its pressure will be $P_{FD} = 2.2 \times 10^6\, \alpha\rho^{5/3}$ (Pa) and the energy required is $E_{FD}(\mathrm{J}) = 3.3 \times 10^6 \alpha\rho^{2/3} M_{DT}$ (kg). Here α measures how far off the isentrope we are. We control that parameter by carefully pulse shaping the laser/x-ray drive.

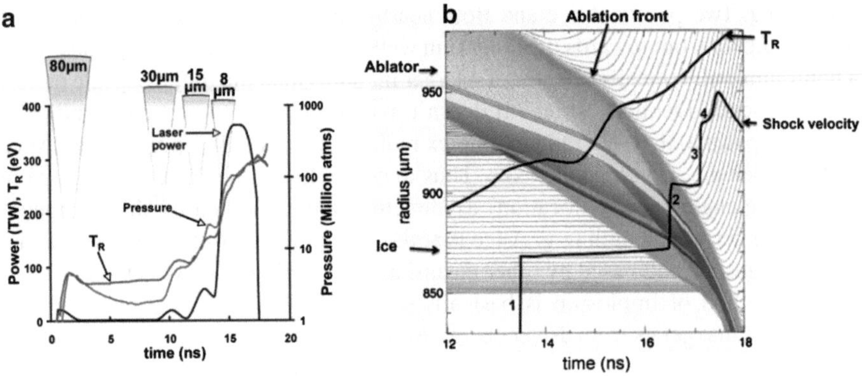

Fig. 8.1 (**a**) Time history of drive and schematic of shell position. (**b**) Detailed view of shock coalescence

To address pulse shaping we must recognise the following problem. Typically we must push on the main fuel shell with a peak pressure of about 100 Mbar, (1bar=10^6 in cgs), for several ns to accelerate it up to the required implosion velocity (to be discussed below) of about 3.6×10^5 m/s. The problem is we cannot apply that pressure to the original capsule! Consider the Fermi Degenerate (FD) isentrope of solid DT. $P_{FD} = 2 \times 10^6 \rho^{5/3} = 1.4 \times 10^{10}$ Pa for $\rho_0 = 200$ kg/m^3. Thus 100 Mbar (= 10^{13} Pa) is way off the isentrope and would make it energetically very difficult to proceed with the implosion. Hence there is a need for pulse shaping.

If we compare the Hugoniot relations for shocks vs. isentropic compression we learn a valuable lesson. For a jump of pressure $Y = P_1/P_0$ there will be a shocked density jump $X = \rho_1/\rho_0$. The Hugoniot relations tell us that $X = (4Y+1)/(Y+4)$. Let us compare that to $X_{isen} = Y^{1/\gamma}$ for $\gamma = 5/3$. For $Y = 1$ they are equal. For $Y = 2, X = 1.5$ and $X_{isen} = 1.51$. For $Y = 4$, $X = 2.13$ and $X_{isen} = 2.3$. For $Y = 8$, $X = 2.75$ and $X_{isen} = 3.5$. For infinite Y, $X = 4$ and X_{isen} is infinite. So as long as the pressure jumps are less than 4, the drift off of the isentrope via the sequential shock method will be less than 10 %.

Figure 8.1 illustrates the shock timing strategy. The first shock will necessarily be a strong one, hence $X = 4$, hence the post-shock density will be 800 kg/m^3. Thus the P_{FD} for that density will be 1.4×10^{11} Pa = 1.4 Mb, and that should precisely be the magnitude of our first shock, to match that and stay on the FD isentrope. After that we launch 3 more shocks each 4 times bigger than the previous one. Our final pressure will be $(4)^3$ (1.4 Mb) = 90 Mb as required, and now the shell has compressed to the proper high density to remain on the FD isentrope as we push on it at 90 Mb and accelerate it to the requisite implosion velocity of 3.6×10^5 m/s on its way to a successful thermonuclear implosion. We must of course carefully time those shocks so that all the shocks coalesce at very near the inside of the frozen DT shell so that most of the fuel will remain cold Fermi Degenerate fuel.

With these basics in hand, we are now in a position to calculate the gain.

8.1.3 Gain Calculation: NIF Ignition Capsule vs. High Gain Reactor Capsule

We begin with our current experimental campaign plans for a NIF ignition target. The current capsule is a CH ablator of initial radius about 1 mm. Inside that CH layer is a frozen DT shell with a mass of 170 mg. The current incident pulse has a main drive component of about $E_D = 1.3$ MJ. With the capsule coupling and rocket efficiencies each of about 0.12, this leaves us with an $E_F = 0.017$ MJ in the assembled fuel. As a further detail, the current conservative design leaves some part of the CH ablator layer, un-ablated, and thus that portion of the CH is part of the payload too. It has about 6 of those 17 kJ, leaving us with an energy of 11 kJ in the total DT (hot-spot surrounded by cold dense fuel) assembly. Let us assume the hot spot radius is 25 μm, namely $R_{HS} = 2.5 \times 10^{-5}$ m. This turns out to be near optimal. Since the hot spot requires ρR of 3, then ρ_{HS} must $= 1.2 \times 10^5$ kg/m^3. Then the mass of the hot spot is easily calculated to be 8×10^{-9} kg, and its thermal energy (at the required 5 keV) is $E_{HS} = 4 \times 10^3$ J. Its pressure $P_{HS} = 2n_i kT$ (the 2 because of the electrons) is 4.5×10^{16} Pa.

This self-stagnated assembly is isobaric, the hot spot pressure stops the cold fuel shell from imploding further, and pressures equilibrate $P_{HS} = P_C = P_{FD}$, which can then tells us what the cold density is: $\rho_C = 1.5 \times 10^6$ kg/m^3, which is an astounding 8000 fold denser than solid DT. We can plug that value of the cold density into the formula for the energy available for the cold dense shell: $E_{FD} = E_C = E_F - E_{HS} = 11\,\text{kJ} - 4\,\text{kJ} = 7\,\text{kJ}$, to obtain the mass of the shell, $M_C = 1.7 \times 10^{-7}$ kg, exactly in line with our assumption for the initial mass of the frozen DT shell. We can then set $4\pi R_{HS}^2 \rho_C(\Delta R) = M_C$ to find $\rho\Delta R = 22$ kg/m^2 thus $f_B = 22/(22+70) = 1/4$. Putting that altogether yields

$$G = \frac{f_B M_C Q_{DT}}{E_D} \tag{8.1}$$

$= (1/4)(1.7 \times 10^{-7}\,\text{kg})(3.4 \times 10^{14}\,\text{J/kg})/(1.3 \times 10^6\,\text{J}) = 10$. Thus we expect a successful NIF target to ignite and to have a gain of about 10. All of these values are in reasonable agreement with detailed simulations [2] for the NIF ignition point design, which use the Lasnex simulation code [6].

For completeness, and to stress the key idea that once NIF ignition is demonstrated, we can assume successful reactor scale performance with some confidence, let us consider a reactor scale performance. We begin with an incident pulse of $E_D = 6$ MJ. The coupling and rocket efficiencies are somewhat better at the reactor scale. This is due to better hohlraum coupling at longer pulse length, and the use of better wall materials such as cocktails (= mixtures of wall materials that optimise hohlraum performance, as discussed in Refs. [1,3]. As a result a coupling efficiency improvement from the NIF scale 0.1 to a reactor value of 0.2 is entirely reasonable. A hydro ('rocket') efficiency increase from the NIF scale of 0.1 to a reactor scale of about 0.15 is also to be expected, by using a more confident implosion design with less un-ablated CH. Thus $E_F = (0.2)(0.15)(6) = 0.18$ MJ is in the assembled fuel. Let us assume the hot spot radius is 50 μm, namely $R_{HS} = 510^{-5}$ m. This turns out to be near optimal. Since the hot spot requires ρR of 3, then ρ_{HS} must $= 6 \times 10^4$ kg/m^3. Then the mass of the hot spot is easily calculated to be 3×10^{-8} kg, and its thermal energy (at the required 5 keV) is $E_{HS} = 1.8 \times 10^4$ J. Its pressure $P_{HS} = 2n_ikT$ (the 2 because of the electrons) is 2.3×10^{16} Pa.

This self-stagnated assembly is isobaric, the hot spot pressure stops the cold fuel shell from imploding further, and pressures equilibrate $P_{HS} = P_C = P_{FD}$, which can then tells us what the cold density is: $\rho_C = 1.0 \times 10^6$ kg/m^3. We can plug that into the formula for the energy available for the cold dense shell: $E_{FD} = E_C = E_F - E_{HS} = 1.8 \times 10^5 - 1.8 \times 10^4 = 1.6 \times 10^5$ J, to obtain the mass of the shell, $M_C = 4.8 \times 10^{-6}$ kg, in line with our standard assumption for the 5 mg mass of a reactor target. We can then set $(4/3)\pi\rho_C(R^3 - R_{HS}^3) = M_C$ to find $R = 1.08 \times 10^{-4}$ m, thus $\Delta R = R - R_{HS} = 5.810^{-5}$ m, thus $\rho\Delta R = 58$ kg/m^2, and thus $f_B = 58/(58+70) = 0.45$. Putting that altogether yields $G = f_B M_C Q_{DT}/E_D = (0.45)(4.810^{-6}\text{kg})(3.410^{14}\text{J/kg})/(610^6\text{J}) = 123$. Thus this ICF target can the necessary gain of 100 needed to sustain a 1 GW power reactor.

Returning to our NIF hot spot gain example we can analyse it further and reach some important conclusions. We can ask: what was the required kinetic energy of the dense shell as it imploding in order to supply the 11 kJ of thermal energy of the fuel when it stagnated and fully assembled? We set $(1/2)M_cV_{imp}^2$ equal to 11 kJ. This then gives $V_{imp} = 3.6 \times 10^5$ m/s.

Other 'external' requirements flow from this: Since the final fuel radius was about 35 µm, and a typical convergence ratio to get to those high densities is of order 30, we get an initial radius of the capsule to be 1 mm. Then an implosion time would be $R/v_{imp} = 3$ ns, and thus a power requirement of about 450 TW. All of these estimates are close to the detailed calculated requirements. We now proceed to provide an update of some of the detailed physics models that go into a sophisticated numerical simulation and design of these indirectly driven targets.

8.1.4 An Improved Physics Model

One of the key ingredients of the simulation model is the choice of non-local-thermodynamic-equilibrium (NLTE) atomic physics model. This is necessary as the intense laser heats low density Au, blown off from the interior hohlraum wall, to high temperatures. Low densities, high temperatures and short time scales lead to NLTE conditions. Our standard model uses the XSN package [7]. It has done so for several decades based on analysis of gold disk emission data [8]. A second key ingredient is the choice of electron thermal flux limiter, which we will discuss in detail below. It was chosen to be $f = 0.05$, again based on that same Au disk data analysis. In general, hohlraum data prior to NIF have been matched rather well by using this standard model [9, 10].

As we shall describe in Sect. 8.2, in the Spring of 2010, we deployed a hohlraum simulation model that has several improvements over that of the standard model, including a more complete atomic physics description. We called it the 'high flux model' (HFM), because, when compared to the standard model, it produces a higher flux of x-ray emission and a higher flux of electron heat from a given laser heated high-Z (such as Au) plasma. A more detailed description of the high flux model, its historical antecedents, and its application to NIC results, is recorded in Rosen et al. [11]. What follows below is a shorter version of the story.

The HFM uses a detailed configuration accounting (DCA) NLTE atomic physics package [12] with many tens of levels while accounting for tens of iso-electronic ionisation states. The levels and transitions considered include $\Delta n = 0$ transitions, and dielectronic/auto-ionising processes. This is in contradistinction to the standard model's use of an XSN, 10 level, average atom NLTE model, which does not allow for $\Delta n = 0$ transitions, and which, in its default mode of operation, does not include dielectronic processes. The name 'detailed configuration accounting' is a misnomer here: the model used in hohlraum simulations is based on super-configurations described by principal quantum numbers and is considered to be highly averaged, except when compared to XSN.

The DCA model has been benchmarked extensively against even more detailed codes such as SCRAM [12]. For a given high Z ion, in a plasma at a fixed electron density and temperature, the DCA-predicted emissivity is greater than that predicted by the standard model. For example, Au at a temperature T of 2 keV and a mass density ρ of 0.01 gm/cc has an emissivity of 7.4 TW/cc according to SCRAM, but only 3.1 TW/cc according to XSN [13]. The DCA value is 7.9 TW/cc, quite close to SCRAM. In general, for a given hot, high Z plasma, the higher emissivity of DCA will more rapidly radiatively cool the plasma faster than the lower emissivity standard XSN model.

The second key element of the HFM is a more liberal electron heat flux limiter. A hot plasma with a steep temperature gradient violates the basic assumption of a local, Fick's Law form of heat transport that is in the hydrodynamic codes. The basic assumption is that the heat-carrying electron's mean-free-path is short compare to the gradient length. Quite often it is decidedly not [14]. To avoid non-physical results, the heat flux is limited to a fraction, f, of the free streaming heat flux, nvT, where n is the electron density and v is the thermal velocity. The HFM uses a relatively generous electron conduction flux limiter ($f = 0.15$), because that choice agrees favourably with the results obtained with a more physically motivated non-local transport model [15]. The HFM thus has more conduction cooling when compared to the standard model's choice of a relatively more restrictive $f = 0.05$.

The two key changes, DCA, and $f = 0.15$, each contribute directly to radiatively and conductively cooling a hot plasma faster than the standard model. Moreover, the cooler plasma due to more electron heat conduction places the ion in a somewhat cooler state with more electrons in 'active' atomic levels, and thus they do more radiative cooling. Similarly, the dielectronic processes also accomplish that. Together, the DCA and the $f = 0.15$ reinforce each other, and lead to a prediction of a hohlraum plasma that is substantially cooler than the standard simulation model.

For typical NIC ignition hohlraums, the difference in T is ~ 4.3 keV for the standard model vs. ~ 2.6 keV in the HFM. This difference proved to be a key element in solving the 'mysteries' discussed in Sect. 8.2.

Prior to NIC, experiments continued to be analysed via the standard model. Suter pointed out [13] that a key issue, as lasers and targets progressed upward in scale size, is the contribution of the volumetric laser heated Au coronal energy (and the emission there-from) to the general hohlraum energy balance. Whereas it was ~ 10 % on Nova scale, ~ 20 % on Omega scale [16] it exceeds 30 % on NIF scale. While hohlraum energetics are generally dominated by wall loss [1, 8] which scales with hohlraum area, as we progress to larger scales the coronal terms can be important: Volume / Area $\sim$ scale size. As such it is only of late, in the NIF era, that it was absolutely crucial to accurately calculate (and measure) the coronal x-ray emission, and thus to truly need a detailed, full physics model such as DCA.

Quite analogous to the atomic physics issues are the electron transport issues. There are numerous reasons why an effective flux limiter could be the restrictive value of 0.05, including finite spot effects, and the non-uniform two or even three dimensional issues of the cooler area and volume that surrounds the spot. Hohlraums such as those shot on Nova and Omega had their walls illuminated by tight laser spots. In that context, modelling with an $f = 0.05$ seemed to work reasonably well. In contrast, NIF's 192 large spot size beams more uniformly fill the hohlraum wall area. The uniformity may be a reason for the less restrictive, 'classical', flux limiter of 0.15 being operative now. Consistent with this view is the experience of R. London et al. [17] on Omega. Hohlraums were irradiated to study laser plasma interactions (LPI). Important for this study was the creation of a uniform density hot hohlraum fill, since that is the medium that undergoes the LPI under study. Initial experiments with the 'traditional' tightly focused Omega beams led to non-uniformities. A re-design with broad beams that covered the hohlraum walls much more uniformly led to a more uniform fill density. Importantly, with the uniform illumination, London et al. [17] found a much better fit to the LPI scattered light data (spectrum vs. time) when using an $f = 0.1$ model.

In a closely related way, was our experience with Au spheres illuminated uniformly by the Omega laser at the University of Rochester's Laboratory for Laser Energetics (URLLE). These experiments, and their analysis [18, 19] are, in essence, the modern analog of the gold disk experiments of the 1970s [8]. Just as those old gold disks set the stage for the development of the standard model, the Au spheres set the stage for the *new* model, namely the HFM. In particular, the spherical illumination uniformity in the Au sphere experiments may be the key ingredient needed to explain why the absorption and x-ray emission in those experiments are best matched with $f = 0.15$. Again, in those experiments, the non-local electron transport package supports that f= 0.15 result.

A model with $f = 0.05$ simply applied to these Omega Au sphere experiments has too hot a corona. This leads to less inverse bremsstrahlung absorption, as well as less efficient conversion of absorbed laser power to x-rays, since less energy is transported to higher density plasma where the x-rays are more efficiently created.

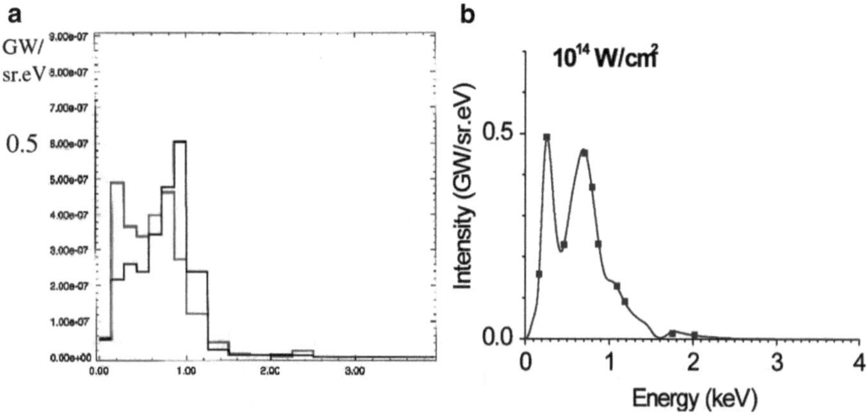

Fig. 8.2 (**a**) Spectrum from DCA (double humped curve) and XSN simulations vs. (**b**) data for Au 10kJ, 3 ns, 10^{14} W/cm^2 at $t = 2.9$ ns

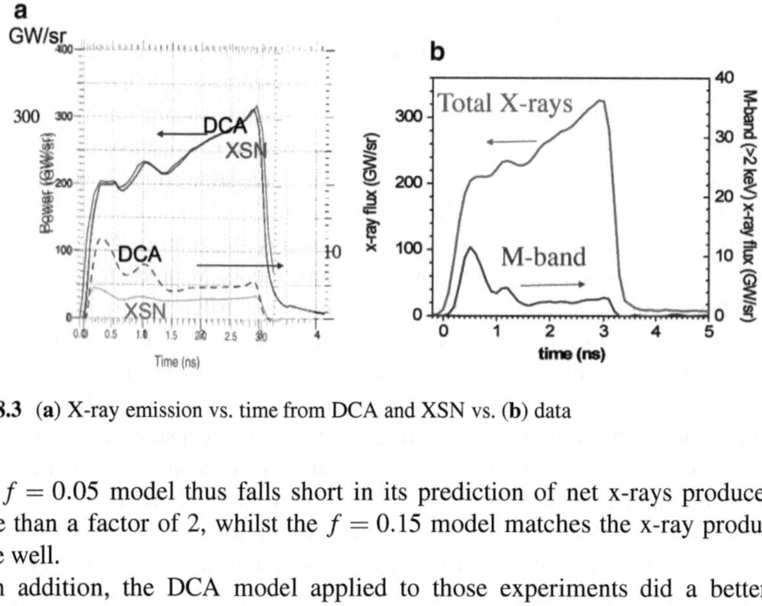

Fig. 8.3 (**a**) X-ray emission vs. time from DCA and XSN vs. (**b**) data

The $f = 0.05$ model thus falls short in its prediction of net x-rays produced by more than a factor of 2, whilst the $f = 0.15$ model matches the x-ray production quite well.

In addition, the DCA model applied to those experiments did a better job reproducing the spectral shape of the emission than XSN [19]. For example, for a 3 ns 10^{14} W/cm^2 illumination of the Au sphere, the observed spectrum [18] at the time of peak emission (just before the end of the 3 ns pulse) the DCA model reproduced the spectral shape of two equal height peaks of the correct width and photon energy central positions, while the XSN model does not. This is shown in Fig. 8.2.

Similarly the time history of the M-band x-rays is much better matched by the DCA model as seen in Fig. 8.3. The "bumpiness vs. time is not due to atomic physics but due to the incident laser having its ups and downs. Analogous to this

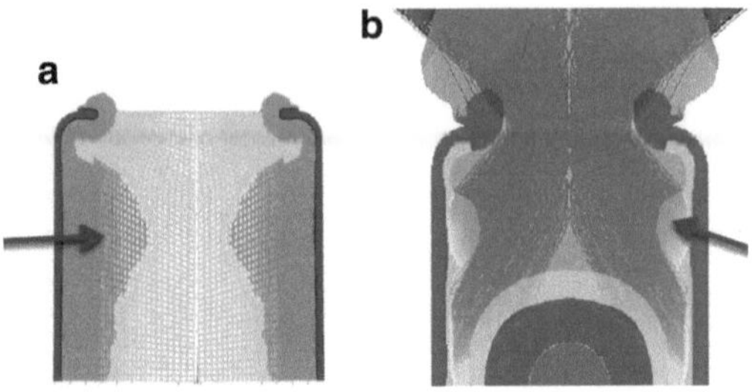

Fig. 8.4 (**a**) Empty hohlraum with large coronal emission region vs. (**b**) Ignition hohlraum with restricted emission region

low intensity data, is the high intensity case of 1 ns and 10^{15} W/cm^2 illumination of the Au sphere, where the observed spectrum and time history is also better matched by the DCA model. In a similar vein, in a study of large volume radiation sources, Colvin et al. [20, 21] independently chose to use a model consisting of DCA NLTE and $f = 0.2$ in order to better explain their data.

In the lead-up to the NIF experiments, Suter [13] predicted that the NIF hohlraum's coronal emission, calculated by DCA, would be important as a drive enhancer. Indeed, the first experiments, in the summer of 2009, were empty hohlraums. The drive emitted from those hohlraums exceeded the standard model's prediction by $\sim$ 25–30 %! [22,23]. The HFM explained this high drive level exactly. The higher flux limit of the full HFM allows for greater absorption of the laser in this empty hohlraum, and even more coronal emission (as discussed above). The large coronal emission at NIF scale came to the fore, showed the standard model was inadequate, and required the HFM to explain it.

In contradistinction to the 30 % effect in empty hohlraums, for the full NIF gas-filled, capsule-containing ignition hohlraums, the HFM predicted only a 10 % higher drive than the standard model. We believe this difference is due to the size of the re-emitting gold corona in the hohlraum interior. A 2-D simulation of an empty hohlraum has the hot laser- heated Au corona fill most of the volume in a semi-uniform way. The very same 2-D methodology applied to a gas-filled capsule-containing ignition hohlraum tells a different story. The capsule blow-off and the gas-fill conspire to severely limit the laser heated Au coronal blow-off to a much smaller volume than it had in the empty hohlraum. The volume is restricted both radially and especially axially: the inner beam absorption and the capsule blow-off severely restrict the size of the corona near the hohlraum waist (see Fig. 8.4).

Before leaving the subject of the HFM and its success in explaining the NIC empty hohlraum high drive results, we will engage in an analytic exercise in analysing that data.

8.1.5 An Analytic Model for Empty Hohlraum Performance

Before leaving this subject, we will do, as is our tradition, a 'hand calculation', namely, a purely analytic treatment, an extended 'back of the envelope' calculation,

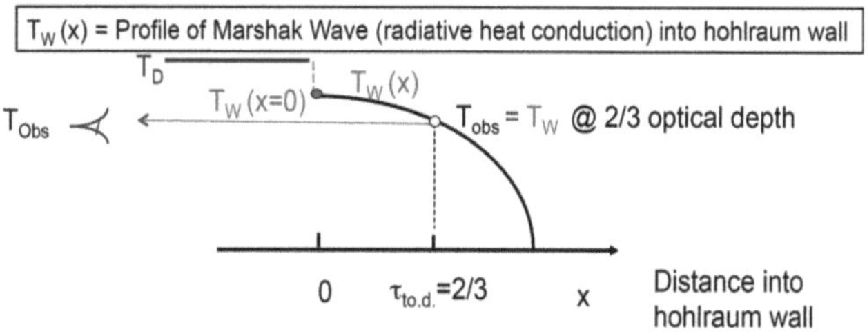

Fig. 8.5 Schematic of the three Temperatures discussed in the text

of the NIC empty hohlraum results. The advantage of doing so is that it then supports, rather explicitly, the notion of the extraordinarily high conversion of laser to x-rays. We have done similar calculations in our previous Sumer School lecture [1], applied to smaller energy scale experiments. What is different here is its application to NIF scale.

In addition, the detailed discussion that follows below, is to an important degree, an 'upgraded' version of the analytic theory, as it incorporates more fully and self consistently some of the lessons taught in that 2005 lecture [1]. In particular, this new and improved treatment does not treat the hohlraum drive as a single temperature, but makes the fine distinction that there are *three* temperatures to consider-though they are all causally and calculably related. These three temperatures are depicted schematically in Fig. 8.5.

The 3 T_r's are related in the following manner: The first temperature to consider is what we consider the basic drive temperature, T_D, which is the 'virtual drive' that leads to a wall surface at $T_W(x=0)$. These two are related by the Milne condition: $T_D^4 = T_W^4 + F/2$ where $F = F(T_W)$ is the absorbed flux per unit area. This absorbed flux, $F(T_W)$, in terms of T_W, is calculated in the theory of Hammer and Rosen [24]. This difference in temperatures, between T_D and T_W, as discussed in detail in [1], is due to the following: for a wall to have a surface temperature T_W, it must be driven at a drive higher than that, since the wall re-radiates some of the drive incident upon it. Considering that very notion of re-radiation, brings us to the third temperature: T_{Obs}, which is the observed drive when looking at the wall whose surface is at T_W. They are related by the albedo, $\alpha : T_{Obs}^4 = \alpha T_W^4$ where $\alpha = 1-(F/T_D^4)$. This latter equation is also derived in detail in [1]. Briefly, the T_{Obs} is lower than $T_W(x=0)$ because the observed drive is not emitted from the surface, but actually emitted from about two-third of an optical depth deeper into the wall from the surface, and the temperature is lower there, since T drops as we proceed deeper and deeper into the wall.

We consider a NIC empty hohlraum illuminated by a laser of energy E_L, and power P_L. It enters the hohlraum (usually made of a high Z material such as Au) and is absorbed along the inner walls where it is aimed. The hot plasma that ensues is a copious source of x-rays. We parametrise this process by a conversion efficiency η_{CE}. Thus we assume that η_{CE} E_L worth of x-rays now floods the hohlraum and uniformly bathes the wall areas of all that it sees. Some of the x-rays leave the hohlraum through the laser entrance holes (LEH) necessary to get the laser into the hohlraum in the first place. The LEH (which is the ultimate in energy sinks - a vacuum) absorbs all the flux σT_D^4 that impinges on it. Thus we know immediately how to calculate this energy loss channel: σT_D^4 times the area of the LEH, integrated over time.

Our major challenge is to calculate the wall loss, since the preponderance of wall area still make the wall the principal sink of energy in ICF hohlraums. Again we refer the reader to [1] and to [24] for the details of calculating the radiation diffusion wave, the so-called 'Marshak wave' [25] the non-linear radiative heat conduction wave that propagates into the gold wall.

To summarise the sources and sinks: The energy produced is:

$$E_X(t) = \int \eta_{abs}(t')\eta_{c.e.}(t')P_L(t')dt' \tag{8.2}$$

The energy lost into the wall is:

$$E_W = A_W \int F dt \tag{8.3}$$

where $F = F(T_W(t))$ is given by Hammer & Rosen in [24].

The energy lost out of the LEH is:

$$E_{LEH} = A_{LEH} \int T_D^4 dt \tag{8.4}$$

So, following the method of Brian Thomas [26], putting everything in terms of T_w:

$$E_X = [A_W + (A_{LEH}/2)] \int F dt + A_{LEH} \int T_W^4 dt \tag{8.5}$$

We use a convenient set of units, which we call Radiation Hohlraum Units (rhu), whose units are measured in: mm, ns, heV, hJ (yes that is hecta-joules or hundreds of joules and hecta-volts, hundreds of eV). In these units, familiar quantities are: Power is in hJ / ns = 100 GW = 10^{11} W, Irradiance is in 10^{13} W/cm^2 and delightfully, σ of "σT^4" fame, = 1.03. For the NIC example we will consider a 'Scale 0.7' empty NIF hohlraum, a right circular gold walled cylinder with a length of 0.64 cm, a diameter of 0.355 cm and an LEH diameter on each end cap of 0.265 cm. With those dimensions we find that $A_W = 80$ mm^2 and $A_{LEH} = 11$ mm^2. In these experiments the laser energy was 635 kJ, or, in rhu: $E_L = 6350$ hJ in a flat-top 2 ns pulse.

The key point in all of this is that we will 'test' the 'High flux model' by using a very high conversion efficiency:

$$\eta_{CE}(t) = 0.87t^{0.2} \tag{8.6}$$

with t in ns. This conversion efficiency is much higher than the $\eta_{CE} = 0.7$ used in smaller scale hohlraums and used in Ref. [1]. Then, with the assumed time dependence of $T_w(t) = T_0 t^{0.14}$, again with t in ns, we obtain, for gold, from [24]:

$$
\begin{aligned}
F(t) &= 0.42 T_w(t)^{3.3} t^{-0.41} \quad \text{hJ/ns} - \text{mm}^2 \\
E_W(t) &= A_W 0.43 T_0^{3.3} t^{1.05} \quad \text{hJ} \\
E_{LEH}(t) &= A_{LEH} 0.66 T_0^4 t^{1.56} \quad \text{hJ}
\end{aligned}
\tag{8.7}
$$

There is an additional complication/ model upgrade that we also consider here, namely LEH closure. As the hohlraum heats up the edges of the LEH expand inward towards the axis. This has a three-fold effect. From an energetics point of view it lowers LEH vacuum loss, but by the same token it raises wall loss, as now there is effectively more wall area. In addition it directly affects the radiant intensity (GW/Sr) seen by the Dante x-ray detector, since it collects its signal coming out of the hohlraum through this closing LEH aperture. We model the LEH radius vs. time as: $R_{LEH} = R_0 - C_s t$ where $C_s = 0.03\, T_0^{3/4} t^{1.15}$ mm/ns. This will modify the equations above, since now: $A = A(t)$ too!

Let us consider the situation at a time of 2 ns, the end of the drive pulse. Hole closure subtracts 1.3 mm^2 from A_{LEH} and adds it to A_W. Equating sources to sinks: $6350 = 71.5 T_0^{3.3} + 18.6 T_0^4$. This is solved by: $T_0 = 3.37$ heV. Then: $T_W = 3.71$ and $F = 24$. Then $T_D = 3.77$ and $\alpha = 0.88$. We then can find $T_{obs-no.h.c} = 3.65$ which has no hole closure correction. Then, since the Dante broadband soft x-ray detector looks through the LEH, we must consider the 'T' observed with hole closure correction (~ 20 % correction @ 2 ns), and find:

$$
T_{obs-wi.h.c} = T_{obs} (A_{LEH(@2ns)}/A_0)^{1/4} = 3.46 \tag{8.8}
$$

We can also ask more directly what radiant intensity the Dante would see, which is the GW/sr at 37.4° out of one

$$
\begin{aligned}
LEH :&= T^4_{obs-no.h.c} \cos\theta\, A_{LEH}(@2\,\text{ns})/\pi \\
&= 26.2 T^4_{obs-no.h.c} (0.81)(5.5) \\
&= 20640\,\text{GW/sr}
\end{aligned}
\tag{8.9}
$$

We can do a similar calculation early in time, at 0.5 ns. Hole closure changes A_{LEH} by 4 %. Equating sources to sinks:$1202 = 16.5 T_0^{3.3} + 2.4 T_0^4$. This is solved by: $T_0 = 3.36$ heV (Note that this is consistent with our 2 ns calculation's T_0!). Then: $T_W = 3.05$ and $F = 22$ and $T_D = 3.14$, and $\alpha = 0.77$. We find $T_{obs-no.h.c} = 2.95$ (no hole closure correction) and with hole closure correction (~ 4 % correction @ 0.5 ns):

$$
T_{obs-wi.h.c} = T_{obs} (A_{LEH(@2ns)}/A_0)^{1/4} = 2.92 \tag{8.10}
$$

Also, we find the GW/sr at 37.4° out of $1 LEH := T^4_{obs-no.h.c} \cos\theta\, A_{LEH}(@2ns)/\pi = 26.2 T^4_{obs-no.h.c} (0.99)(5.5) = 10470\,\text{GW/sr}$.

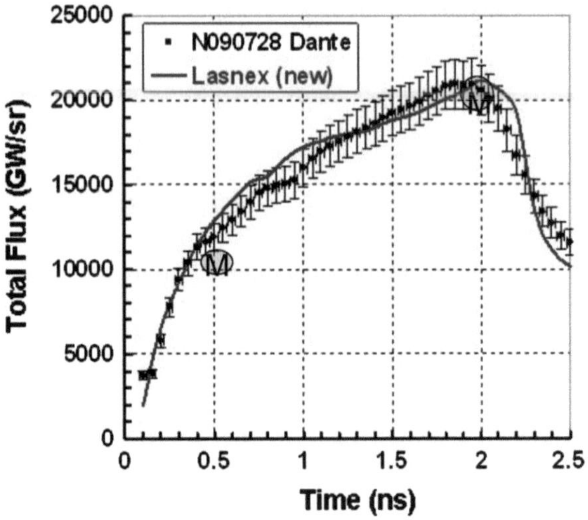

Fig. 8.6 Empty NIC hohlraum data and HFM simulations compared with 'M', analytic model predictions.

These values are in quite reasonable agreement with the data and simulation [22, 27] as shown in Fig. 8.6. The early time model can be improved by assuming a self consistently steeper temporal behaviour to $T(t)$. We omit that calculation here, but it is described in [28]. Thus simulation and analytic theory show that a HFM-like high conversion efficiency is key to getting agreement with the empty hohlraum data.

Thus, the HFM was ready to be applied to the 2009 NIC ignition hohlraum energetics campaign. Due to real time uncertainty in the astounding empty hohlraum data (which, in the end turned out to be true) and due to a desire to be conservative (at least with respect to drive expectations) the standard model was still the tool of choice going into the campaign. As we shall see in Sect. 8.2, in hindsight, the choice of the standard model turned out to *not* be conservative with respect to LPI and coupling issues.

So let us now turn to the ensuing NIC ignition campaign.

8.2 Current Status of the NIC Campaign

8.2.1 Introduction

Before we delve into the specifics of the NIC campaign, we must put across a key concept. The energy scale of the NIF is some 40 times larger than any other laser experiment previously attempted and the spatial scale for a hot hohlraums is about a factor of 4 larger than previous hohlraums. See Fig. 8.7 for details of the hohlraum. Extrapolating from the previous data-base to this regime is challenging, especially in light of the very stringent specifications on accuracy and precision, in

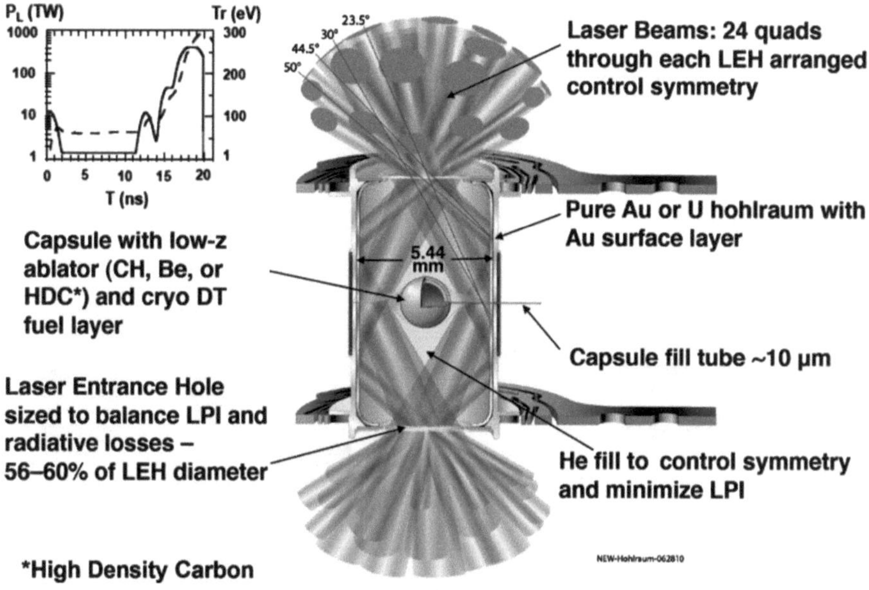

Fig. 8.7 The NIC hohlraum

laser performance, target quality, and implosion performance, required to achieve ignition [2]. While the NIF laser has performed admirably up to this stringent task, and target fabrication efforts continue to make great strides as well, it is the uncertainties in target physics that simply will not allow us to take a 'point design' and hope to have it ignite, right from the first shot. We have always planned a semi-empirical 'tuning campaign', to be sure, informed by simulations, to lead us to ignition.

The tuning campaign [29] is organised and formulated as follows: A multi-variable sensitivity study (MVSS) is done to place the ignition campaign into a unified context. What emerges is an ignition threshold factor (ITF) [2] that if greater than 1, implies a very high probability of reaching ignition. Margin, as defined in [30, 31] is equal to ITF-1. The ITF involves four terms. Achieving a low adiabat implosion, a sufficiently high implosion velocity, a sufficiently clean (namely, only partially mixed) fuel, and a sufficiently round, symmetric, implosion. See Fig. 8.8 for a schematic of methods to measure these quantities and to carry out the tuning campaign described below. We have formulated an experimentally observable form of ITF, called ITFX. It is a metric of how close to the ignition / Lawson criteria (of the density, temperature and confinement time required for ignition) we have come [32, 33].

The adiabat is achieved via tuning of the height and timing of the laser pulse that leads to four shocks. The velocity is achieved via tuning the height and duration of the last part of the pulse, where most of the laser power and energy resides, as well as by tuning the capsule ablator thickness. The mix problem is controlled by

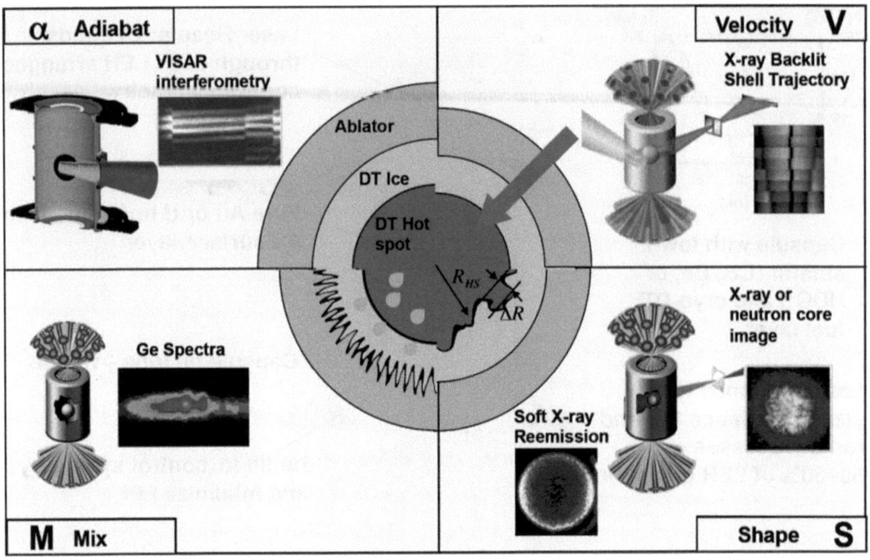

Fig. 8.8 Typical data involved in the process of measuring and tuning the four key parameters that go into the ITF

varying the high Z dopant levels in the ablator, as well as by varying the ablator thickness. Symmetry is achieved by varying the pointing of the laser beams as well as by controlling the time variation of the power balance between the "inner beams' that hit the waist of the hohlraum vs. the "outer beams" that hit the hohlraum wall closer to the laser entrance hole.

After the hohlraum drive is 'certified', and laser plasma interaction (LPI) issues under sufficient control, the NIC strategy is to do the shock timing and symmetry campaigns in a somewhat interlaced mode, as we proceed from first through fourth shock. We then do a campaign called THD [34] wherein the D is minimised and replaced by H (T is still needed for beta layering smoothness of the DT ice) in order to cut down on neutron flux, so that it is easier for x-ray diagnostics to inform us of adiabat, mix and shape issues in the final fuel assembly. The THD campaign will be followed by attempts at DT ignition.

To stress the point that this semi-empirical tuning approach was always the strategy, it is worth emphasising that to test this methodology, a virtual campaign was performed well before NIF was complete [35]. We formed two teams. The Blue Team executed this virtual NIC campaign as if it were real. They were handed 'data' and made decisions, based on that data, of what the next shots will be, by specifying target and laser parameters. The Red Team had two roles and two phases of operation. In phase one the red team determined a physics/data model that is reasonable yet different from the one currently implemented in the hydro-codes. The point design failed under this new mode, but the Red Team re-designed one that ignited.

In phase two the Red Team acted as a virtual facility, simulating the irradiation of the targets 'ordered' by the Blue Team, by a laser pulse, also specified by the Blue Team, and using their 'Red Team Physics Model'. The results of the virtual experiment were reported out to the Blue Team in terms of virtual diagnostic outputs, e.g. scope voltages vs. time, x-ray pinhole camera images, etc., all convolved with spatial and temporal blurring response functions as well as noise. The Blue team was kept in the dark as to exactly what target and laser was used (which vary shot to shot within the NIC specifications). They were also not told how the Red Team Physics Model differed from their own. A referee was put in place to ensure the efficacy of this entire exercise by monitoring the virtual campaign and maintaining separation of Red Team knowledge from the Blue team.

And so, the virtual campaign began. The Red Team varied (via a one-time throw of the dice) the equations of state and opacities of all capsule and hohlraum materials, within the 1σ (typically 10–20 %) uncertainties that exist. Far more uncertainty ($\sim$ factor of 2) exists for NLTE collisional excitation rates and for electron thermal conduction. Those were also varied accordingly. With this new model the point design failed. A re-tuned target/laser combination restored ignition to nominal robustness. As this was the first such exercise we simplified it by legislating that the hohlraum drive / LPI issues were near nominal. We also did not explore 3-D issues at this point, nor include mix issues at the ablator DT ice interface. Our main concern was to see if the shock timing / symmetry interlaced campaigns would converge in a reasonable number of virtual shots, and not enter an endless do-loop. The schematic of this exercise is shown in Fig. 8.9.

The first 'experiments' surprised the Blue Team in that the x-ray drive on the first foot of the pulse came in low. They ordered the next shot to have a higher laser power in the foot to establish a slope of drive vs. laser power, and by so doing could then specify the correct laser power to achieve the required first shock intensity. This is the empirical approach of the NIC strategy and it worked quite well. The Blue team attempted to formulate a new model (via an Au hohlraum wall opacity model) though the 'reality' was that the discrepancy was much more due to the electron conduction changes in the Red Team model. The main point is that the tuning was done successfully, empirically.

Similar discrepancies occurred in the symmetry campaign. There too, a slope of symmetry vs. some specific symmetry tuning parameter (in this case the colour separation between inner and outer beams that controls cross beam transfer [39]) was established empirically. It was parallel to the Blue Team's original expectations, but offset. Again, the main point is the ability to tune empirically via this slope method.

After about 30 'virtual shots' the Blue Team was ready to specify an ignition target. Three targets were 'shot', first in THD mode, and then, to save time, the very same were then shot in DT mode. In THD mode, the x-ray self-emission implied a hot, warm and very hot result, respectively. The backlit images implied, respectively, a dense, fluffy, and very dense core. (In DT mode those three targets gave 4, 0.1, and 14 MJ yields respectively, thus achieving ignition on two of the three shots. In retrospect the middle, failure shot, had target parameters out of

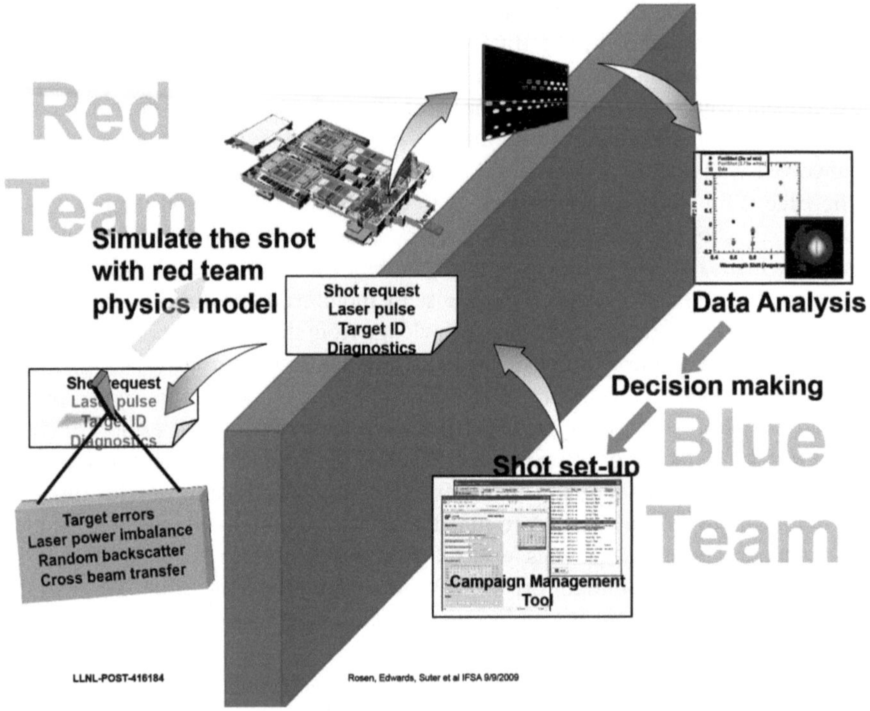

Fig. 8.9 Schematic of the virtual tuning campaign discussed in the text

the allowed specifications. The corresponding ITFs and margins of these three shots are consistent with their performance. Thus, in this virtual exercise, ignition was achieved via this empirical tuning campaign, despite the blue team using, by construction, the 'wrong physics' model.

8.2.2 The 2009 Energetics Campaign

NIF began full 192-beam performance in the summer of 2009. As mentioned in Sect. 8.1, the campaign then involved empty hohlraums. In the fall of 2009 the NIC began to explore ignition scale hohlraums with surrogate targets, culminating in a 1 MJ shot.

The goal of the campaign was to understand the energy balance in such a capsule/hohlraum/laser-pointing configuration. In general this campaign showed very good laser-hohlraum coupling [36] reasonably high drive [37] and good implosion symmetry control via the technique of cross beam transfer [38] to be discussed in further detail below. To date, based on extensive data analysis, all of these very positive conclusions remain essentially unchanged.

However, there were discrepancies between several types of data and the expectations based on both pre- and post-shot simulations using the standard simulation modelling methodology. It took the HFM to 'solve' most of those puzzling discrepancies. Let us proceed to first discuss the results, then the discrepancies, and finally the reasons why the HFM helped resolve most of those issues.

8.2.2.1 Experimental Results from the 2009 NIC Energetics Campaign

The NIC 2009 energetics campaign progressed from sub MJ laser energies incident into smaller hohlraums, and culminated, on Dec. 4, 2009 when the NIF illuminated a full ignition scale cylindrical gold hohlraum, of length 1.0 cm and a diameter of 0.544 cm with over 1 MJ of laser light. At the hohlraum centre was a 2 mm diameter capsule. The capsule was composed of a 180 μm thick Ge-doped plastic shell filled with a predominantly He gas that contained some deuterium to produce neutron signals. This capsule served as a surrogate to an ignition capsule that would have a frozen shell of equi-molar Deuterium / Tritium (DT) just inside the plastic shell. The hohlraum itself was filled with He gas to help hold back the inward expansion of the hohlraum gold walls which are heated to several keV temperatures by the incident beams.

The pulse shape was a typical ignition "shaped pulse" [2] namely a series of three pickets, which produce three shocks, followed by a main pulse of 3 ns in duration that occurs 16 ns after the first picket, and which provides most of the drive. The NIF laser has one-third of its beams entering the vertical hohlraum (equally split to enter the top and bottom laser entrance holes (LEHs)) at 50° with respect to the hohlraum's vertical rotational axis. Another one-third enter at 44.5°. These two sets of beams, comprising two-third of the NIF power and energy are called "outer beams" because they intersect the walls of the hohlraum at an axial position roughly midway between the hohlraum mid-plane and the LEH end-caps. The remaining one-third of the NIF beams are split equally between 30° and 23.5° beams that are called 'inner beams' because they intersect the hohlraum wall at an axial position very near its mid-plane, directly surrounding the capsule's equator, situated at the hohlraum centre. The NIF beams come in a cluster of 4 unit called a 'quad', with 32 outer quads and 16 inner quads.

The NIF was designed with the flexibility to change the "colours" of the laser beams in anticipation of the possibility that these colour differences ("$\Delta\lambda$") could help control the transfer of energy between beams. When the beams overlap near the LEH, the local velocity field can provide a resonance which can facilitate Brillouin side scattering processes that transfer the energy from one beam to another [39].

This 2009 NIC gas-filled /capsule-imploding hohlraum energetics campaign showed good laser-hohlraum coupling [36]. Nearly 90 % of the incident laser was absorbed by the hohlraum. That 90 % level is high enough to mitigate stress on the laser system and help it to routinely provide sufficient incident power needed for ignition. The $\sim$ 10 % loss is due to scattered light produced by laser plasma instabilities (LPI). These include the Brillouin process in which the incident laser light stimulates ion waves, which then act as a grating/mirror to scatter that incident light, called stimulated Brillouin scattering ('SBS'). In a similar way the Raman process stimulates electron plasma waves that scatter the light via a process called stimulated Raman scattering ('SRS). The scattered light's power level and spectrum vs. time was measured on one 50° outer beam quad and on one 30° inner beam quad. In general the loss was due to SRS on the inner beams.

In addition, the SRS-induced plasma waves eventually 'break' and can accelerate electrons to high energy. These 'hot electrons' can pre-heat the capsule, making ignition more difficult. The hot electron temperature ('T_{hot-e}') and level ('f_{hot-e}') of hot electrons produced are inferred by the hard x-ray bremsstrahlung created as the hot electrons stop in the gold wall of the hohlraum.

The campaign also showed reasonably high radiation drive [37] of nearly 300 eV. This is measured by 'Dante', which is a multi-broad-band-channel x-ray detector, with a coverage from 0.1-several keV photons looking into the hohlraum through the LEH at an angle of 37.5°. The 'brightness temperature' is determined by the observed x-ray power integrated over the entire spectrum of emission, accounting for the fact that the LEH closes during the observation time. We post-process the radiation hydrodynamic simulation in order to simulate the detector, and thus compare the radiant intensity (W/Sr) emitted from the hohlraum at the given viewing angle. The temporal signal peak comes at 19 ns, at the end of the main drive pulse. Since the spectrum is close to a sub-keV Planckian in shape, the colour temperature and brightness temperatures are quite similar. We also monitor the emission from the M-band of laser heated Au, which occurs at 1–3 keV. This M-band radiation can also preheat the fusion capsule, and the Ge doping of the plastic ablator shell is adjusted to mitigate this issue. The 300 eV level of radiation drive temperature is the drive required to implode the ignition capsule to sufficiently high velocity such that, upon stagnation, the hot spot temperature will be sufficient for ignition [3].

The capsule implosion symmetry is another important parameter. Hohlraums naturally control short wavelength drive asymmetries due to geometric, 'view factor' considerations [1,3]. However long wavelength ('$P2$' and '$P4$') asymmetries are controlled by beam placement along the hohlraum walls and the relative power in the inner and outer beams. Here $P2$ and $P4$ refer to a Legendre polynomial decomposition of the imploded capsule symmetry pattern. The outer beams make hot-spots on the hohlraum wall whose x-rays tend to push on the poles of the capsule, aligned with the vertical hohlraum axis. The inner beams counteract that push, by creating a hot source near the midplane ('equator') of the capsule, since they propagate to the wall at the midplane of the hohlraum.

The NIC 2009 campaign showed that we were able to control symmetry via cross beam transfer [38]. The implosions converged about a factor of 10 and their symmetry was monitored by measuring the shape of their ~ 5 keV x-ray emission upon stagnation. The images clearly went from severely 'pancaked', implying ineffective inner beam drive, to round as $\Delta\lambda$ was increased, and power was transferred from outer to inner beams. Another metric of the increase of cross beam transfer from outer to inner beams as $\Delta\lambda$ was increased, was the decrease in x-ray brightness of the spots at the positions where the outer beams hit the hohlraum wall. After optimising, by using $\Delta\lambda$ variation, the images were within about 10 % of round, using $P2$ and $P4$ as the figure of merit, implying of order 1 % in the time integrated drive symmetry. Actual ignition capsules will converge further, but we are only now (Autumn 2011) in the midst of a full campaign that monitors symmetry and adjusts beam power in a time dependent manner to ensure even better time-integrated symmetry. Nonetheless, these initial results are encouraging as they show that the $\Delta\lambda$ technique acts in a reproducible and controllable way to transfer energy between beams to help achieve good symmetry.

While all of these results are very positive and very promising with regards to achieving hohlraum conditions conducive to driving targets to ignition, there were, however, a number of unresolved questions that remained. Achieving a fuller understanding of the plasma conditions in the hohlraum and arriving at a more fully self consistent picture of the physics at play here, could lead to more optimised hohlraums. We discuss those discrepancies in detail, in the next section.

8.2.2.2 Discrepancies Between the Data and the Simulation Model's Predictions

Our expectations from any given shot during the campaign are formed by the following procedure. We use the radiation-hydrodynamic two-dimensional simulation code LASNEX [6]. We input the measured laser power but subtract from it the estimated SRS and SBS losses. The estimate takes the measured value of SBS and SRS detected on a single 50° outer beam quad, assumes this loss happens for all of the outer beam quads equally, and thus multiplies that observed value by 32. Similarly the losses measured on the single 30° inner beam quad are multiplied by 16.

For each experiment there was a conscious choice of the $\Delta\lambda$ between inner and outer beams. The procedure by which we predict how much transfer of power occurs from outer to inner beams is described in detail elsewhere [16]. With all of these ingredients, the simulation is performed and then post processed to mimic the diagnostics that provide the data from the shot. A summary of the surprises are listed below, and then shown schematically in Fig. 8.10.

While the results are broadly excellent from the point of view of the ignition campaign goals, subjecting the results to detailed analysis revealed disagreements with our expectations based on the methodology described above:

1. The level of the Stimulated Raman Scatter (SRS) light, detected as it leaves the hohlraum, was higher than expected.
2. Its spectrum was below 580 nm, when 650 nm was expected.
3. The hot electron fraction, f_{hot-e}, inferred from the hard x-rays, did <u>not</u> track the SRS levels.
4. The slope of the hard x-ray spectrum, T_{hot-e}, was 30 keV, when, based on the (surprising) SRS spectra we would have expected 18 keV.

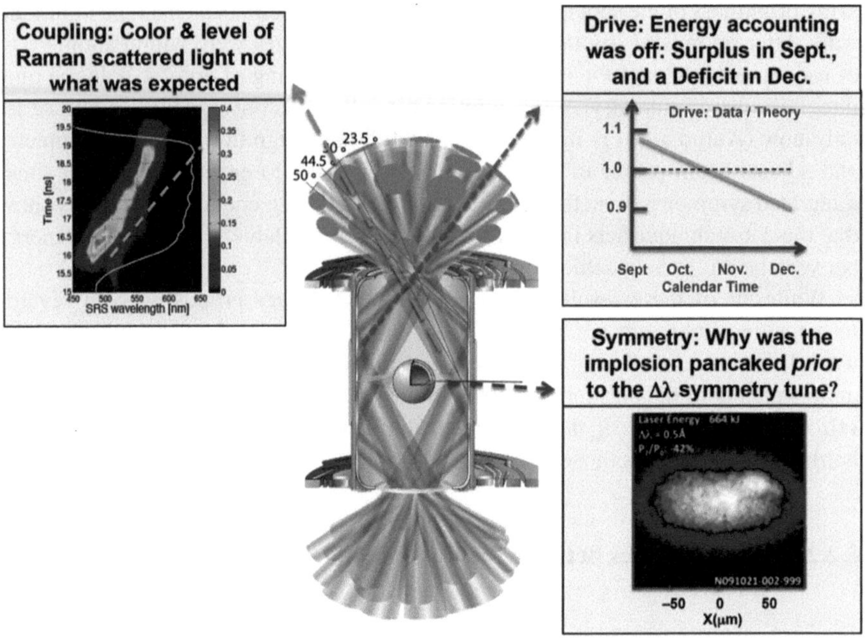

Fig. 8.10 Pictorial representation of the major surprises encountered in the first NIC energetics campaign (2009)

5. The drive of the September '09 shots was more than predictions, and those of the Nov. '09 shots less than predictions.
6. The implosion, before applying the '$\Delta\lambda$ technique", was pancaked, when it was designed (of course) to be round.

Other questions regarding capsule performance, yield, peak x-ray brightness times, hydrodynamic instabilities, etc. will be addressed later in this lecture, as they were beyond the original scope of this energetics campaign. What would prove key in unlocking the mystery of these discrepancies listed above would be a better physics model, namely the High Flux Model (HFM) described in Sect. 8.1.

8.2.2.3 The High Flux Model Helps Resolve the Discrepancies

As noted at the close of Sect. 8.1, despite the HFM's many past successes in correctly modelling high radiative fluxes seen in the Omega Laser Au sphere data [18, 19] and in explaining the surprisingly high drive seen in NIC empty-hohlraums [22, 23], it was not initially applied to the NIC 2009 campaign. Given that this improved model has led to an overall understanding of the hohlraum performance, by virtue of its consistency with a great variety of observations described below, the HFM has become the preferred hohlraum model for going forward towards ignition.

The Raman Spectrum and Levels

The first breakthrough in explaining the discrepancies and inconsistencies that were delineated above, came in a creative attempt to understand the observed SRS spectrum. In general, the gain of SRS will increase with both laser intensity I_L,and electron density n, and it will decrease with temperature T. The standard model predicted a rather high T in the fill gas that occupies most of the volume of the hohlraum.

In particular, consider the 'mid-point of the road' position, which is at about the midway point of the path of the inner laser beam as it moves from outside the hohlraum, through the LEH (the 'beginning-of-the-road', point) and eventually hits the hohlraum mid-plane above the capsule (the 'end-of-the-road' point). In the standard model the 'mid-point of the road' position was deemed too hot to have much SRS gain there as it is ~ 4.5 keV near the end of laser peak power at ~ 19 ns. The plasma is even hotter at the 'beginning-of-the-road' near the LEH where the beams overlap, and in addition, the plasma is less dense since the plasma flows out of the hohlraum there. Thus on several counts, the SRS was even less likely at the LEH. On the other hand, SRS was *most* likely near the cooler, denser region near the hohlraum waist, the 'end-of-the-road' position.

As time progresses during the pulse, the density throughout the hohlraum rises, as does the plasma frequency. As such, the SRS scattered light shifts downward in frequency (upward in wavelength) throughout the pulse. The standard model thus predicted a spectral shift characteristic of the highest density, which occurs at the 'end-of-the-road' density), and thus a large wavelength shift. That was not what was observed.

D. Hinkel and E. Williams [40] invoked two insights into getting theory to agree much better with the spectral data. One was to hypothesise a T lower than that of the simulations. This inspired 'guess' was taken purely in order to match the data, as it would then allow SRS to happen at the 'middle of the road' at a lower density. The second insight was to realise that, with SRS now allowed to happen at 'the mid-point of the road' location, three-dimensional effects would now play a role. The nearest-neighbour inner-beams overlap in an azimuthal sense, and thus, are effectively more intense. They progress from complete overlap ($3\times$ the single beam intensity) at the LEH, to partial overlap ($2\times$ the single beam intensity at the 'mid-point of the road' position) as they propagate, in an axial sense, about halfway into the hohlraum. By the time they are at the hohlraum midplane (the 'end-of-the-road' position), the beams have all separated azimuthally. The combination of lower T and $2\times$ the intensity at the 'mid-point of the road' position now allows the peak SRS gain to occur there, at a lower density than previously thought, and thus to much better match the SRS spectrum vs. time.

Upon hearing of this result, we suggested [11] the use of the HFM, since it naturally gives an appropriately low T_e, due to its high radiative and electron flux cooling of the corona, as discussed above. The HFM gives a T of about 2.6 keV, (vs. the standard model's 4.3 keV) thus obviating the previous need to artificially lower the T when previously using the standard model simulation. The SRS spectrum was thus matched in a physical, understandable manner.

In addition, the HFM's cooler plasma leads to less Landau Damping and thus predicts [40] higher levels of SRS. That means a higher fraction of incident energy back reflected by SRS, than the standard model does with its hotter T_e. This higher level of SRS approximately agrees with observations.

The Capsule Implosion Symmetry

Upon applying the HFM to the 2009 NIC hohlraum energetics campaign in support of the efforts to better match the SRS spectrum, as described just above in Sect. 8.2.2.3, we discovered [11] a delightful bonus. We found that it was immediately clear that the HFM would match the observed implosion symmetry behaviour. Relative to the standard model, the outer beams, with their higher electron conduction, convert laser light to x-rays more efficiently. These x-rays shine on the poles of the capsule driving it towards a natural 'pancaking' shape upon implosion. In addition, the cooler plasma of the HFM inhibits the propagation of the inner beams deeper into the hohlraum via inverse bremsstrahlung absorption, thus preventing them from reaching the midplane of the hohlraum wall surrounding the capsule waist. If these beams cannot reach the midplane, they cannot provide the drive on the waist to supply that which is needed to counter the outer beams' drive on the pole. They cannot efficiently produce a 'sausaging - counter-force' to the outer beams 'pancaking -force'. A balance of forces would produce a round implosion. The standard model with its hotter plasma produces such balance. The HFM with its cooler plasma inhibits the 'sausaging' force, resulting in an imbalance that produces a net "pancaking".

Thus, the enhanced outer beam drive, and the absorption of the inner beams before getting to the hohlraum midplane, together give a natural 'pancaking' to an implosion, as observed. It takes crossbeam transfer, via $\Delta\lambda$, to make the capsule implosion round, as observed. The detailed modelling of the symmetry vs. $\Delta\lambda$, using the HFM, and its very successful matching of the data are discussed in detail by R. Town and M. Rosen et al. [16].

Since most of the reportage to date involves symmetry vs. $\Delta\lambda$ [16, 36, 38] we present here an additional successful result of the HFM in its matching of symmetry data. Here we consider the change in implosion symmetry at fixed hohlraum and laser conditions, including, at a fixed $\Delta\lambda$. Instead, here we vary the capsule's CH ablator thickness: from the nominal 180 μm to a thinner 155 μm. The experiment went from round for the nominal case, to 40 % $P2$ sausage for the thinner ablator. A rather large $\Delta\lambda$ was used in both shots. See Fig. 8.11.

The standard model, used incorrectly with no beam transfer despite the high $\Delta\lambda$ used in the experiment to obtain a round shape for the nominal case, predicts a less than 20 % $P2$ sausage for the thinner ablator. The HFM correctly gets the nominal capsule round and more significantly, gets the correct result for the thinner ablator: a +40 % P_2 sausage. This result required using a 65 % enhancement of incident inner beam energy due to the transfer of energy from the outer beams to the inner beams, as is reasonable for the experimental value of $\Delta\lambda$.

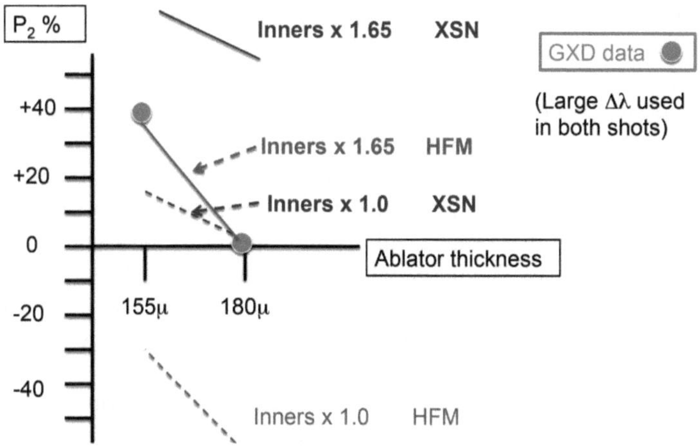

Fig. 8.11 Symmetry vs. ablator thickness. The HFM matches the data. The standard model is very far from it

The physics of this difference is clear. The standard model's hot corona did not put much of a roadblock in front of the inner beam's propagation path to the wall at the hohlraum mid-plane. So a thinner ablator, which fills the hohlraum with less plasma, simply made an 'easy job easier'. The HFM however, made life very difficult for the inner beams to propagate to the wall at the hohlraum midplane. The thinner ablator put less plasma out into the hohlraum and thus made a 'difficult job much easier'. Hence the standard model predicted a very small change in the symmetry. The HFM predicted a very large change, and indeed it was that very large change that was observed. This is all depicted in Fig. 8.11.

Energy Balance

Having succeeded in explaining the SRS spectrum and level, as well as the surprising pan-cake symmetry image when little $\Delta\lambda$ is applied, and the rapid sausaging of the symmetry image when the ablator thickness is diminished, it remained to be seen how well the HFM would explain the measured drive. For a given laser input (after subtracting off the measured LPI losses, the model should correctly match the observed drive if energy balance was intact.

The HFM immediately solved the problematic energy balance prediction of too much drive for shots early in the campaign, when using the standard model. The HFM, with its higher emissivity, naturally gives more drive than the standard model, for a given laser input. Early in the campaign the power was relatively low compared to the MJ class experiments at the end of the campaign, so LPI coupling issues were small. In general, early in the campaign, $\Delta\lambda$ tended to be small as well. The disposable debris shields were rather pristine early in the campaign as well. Thus all of the possible losses were accounted for. In essence, the early-in-the-campaign high drive 'discrepancy' was merely Mother Nature's way of hinting to us that we should have been using the HFM, and not the less emissive standard model.

An important challenge was for the HFM to demonstrate energy balance for the 'pay-off' 1 MJ shot late in the campaign. However, there was already a 'missing energy' problem. Even with the low-emitting, standard model, the observed drive was lower than expectations, which accounted for the known losses at that time. Applying the HFM to these shots, with its high emissivity, only made the 'missing energy' problem worse! The observed drive was now much lower than the HFM based expectations. It seemed as if the HFM had failed its most important test, since the 1 MJ shot was the culmination of the entire 2009 campaign, and the HFM made the drive discrepancy problem worse, not better.

We considered this challenge of 'missing energy' to be an opportunity for the HFM to not only, in a post-shot fashion, successfully model experimental data, but for it to boldly make a prediction. The prediction [11] had to be that there were more losses, and lower coupling to the hohlraum than had been assumed as of the March 2010 time frame.

The prediction of more losses had several components. The first was rather obvious: that the disposable debris shields (DDSs) were ageing by collecting debris. They had not been replaced throughout the campaign, so it was quite plausible that, by late in the campaign, the built-up debris and damage sites might be scattering incident light into larger angles that would lead to some fraction of the incident light to not enter the LEH in the first place. With the campaign over, these DDSs were assessed. They were deemed to be scattering $\sim 5\ \%$ of the incident laser light into angles that would miss the LEH.

The second component of the prediction was less obvious: that the level of the SRS losses was higher than presumed at the time. As the SRS level was only measured on the 30° inner beams, in order to restore energy balance for the HFM model via the route of postulating increased the level of losses, we made the bold assertion that there was more SRS on the un-monitored 23.5° inner beams.

A breakthrough in confirming this prediction came when L. Divol and P. Michel et al. [41] re-interpreted the hard x-ray spectrum, not in terms of a single 30 keV T_{hot-e} and f_{hot-e}, but rather as a two temperature distribution:

1. A dominant f_{warm}, with an18 keV T_{warm}. This is the value of T_{warm} that was expected from the observed, and now understood, SRS spectrum. The expectation is based on a $T_{warm} \sim (1/2)mv_{phase}^2$ argument, where v_{phase} is the phase velocity of the plasma wave that both scatters the incident light out of the hohlraum and then breaks to create hot electrons.
2. A much smaller f_{hotter}, with a 60 keV T_{hotter}. This hotter component may be due to SRS happening at higher density 'end-of-the-road' position above the capsule waist, whose reflected light would refract and be trapped within the hohlraum, or perhaps another LPI issue- the $2\omega_p$ instability in which the laser light decays directly into two plasma waves.

Following through on this two-temperature insight, from that f_{warm}, they inferred a total SRS level. Since we only observe SRS light from the 30° inner beams, subtracting that 30° data from the newly inferred total SRS tells us how much SRS is coming out of the 23.5° inner beams. They found that for the larger incident laser energy shots, with larger $\Delta\lambda$, such as the 1 MJ shot, there was indeed substantially more SRS on the 23.5° inner beams. These surprising results were in line with our SRS loss predictions inspired by the need for the HFM to conserve energy balance.

In summary, late in the campaign losses were larger, as the higher laser powers and the higher values of $\Delta\lambda$ that were used, led to SRS in the un-monitored 23.5° beams. Those losses, as well as the DDS losses, were initially un-accounted for. Therefore, initially, the drive predictions were above the drive data for the Nov.-Dec. '09 shots. Now, with all the losses accounted for, the HFM matched the observed radiant intensity emerging from the hohlraum. Again, as with the symmetry, R. Town, M.D. Rosen et al. describe this agreement of the HFM drive with the data in great detail [16].

There was one question remaining. The total SRS loss, and the extra SRS loss occurring in the unmonitored 23.5° inner beams, as just discussed, was all based on a string of inferences from the hard x-ray spectrum. It would be far more convincing to actually measure the SRS on the 23.5° inner beams. Thus, the credibility of the HFM hung in the balance for nearly half a year as a diagnostic was prepared for a 23.5° inner beam line to do exactly that. In the latter half of 2010 the measurement was made [16] and the SRS levels directly observed agreed well with the SRS amounts that previously, were only inferred. The HFM model had withstood the test.

Despite these successes, an annoying discrepancy remained. The implosion times, inferred from the time of peak x-ray emission brightness from the capsule, were later than predicted by the HFM. We will return to this issue in Sect. 8.2.4.

8.2.2.4 Lessons Learnt from the Energetics Campaign

Reaching this understanding of the ignition scale hohlraums, based on finding consistency with the wide variety of data from the NIC '09 energetics campaign, allowed us to project into the future and to invent new schemes for achieving even more optimal hohlraum conditions.

For example, incorporating a suggestion by E. Moses, P. Michel [41] calculated a $\Delta\lambda_{30-23.5}$ that transfers laser power from the 23.5° inner beams, which have proven to be more prone to SRS, to the more benign 30° inner beams. This is a possible method of reducing SRS losses and reducing the level of hot electrons that they create. This experiment has been done, and the proof of principle been demonstrated [42].

Another example of lessons learnt comes from the follow up work of D. Callahan [46]. She made HFM based design changes to hohlraum geometry. A somewhat shorter and somewhat wider hohlraum allows the inner beams better access to the waist. The cylindrical aspect ratio (length/diameter) has dropped from $(10.01/5.44 = 1.84)$ to $(9.41/5.75 = 1.64)$, close to the 'golden ratio'! Due to its diameter being 5.75 mm, this improved hohlraum is simply referred to as 'the 575'.

Other names for this new and improved hohlraum include 'golraum' due to its golden ratio aspect ratio, and 'haiku hohlraum'. A haiku is a 17 syllable, 3-line poem, distributed as 5, 7,and 5 syllables respectively. A haiku about this hohlraum reads: "Hohlraum width is new. Five, seven, five like haiku! Too good to be true?". As we will discuss below, this new hohlraum has already helped in accomplishing good symmetry in the present (Autumn 2011) campaign.

8.2.3 The 2010 Cryogenic DT and THD Capsule Campaign

In 2010 the NIF facility was prepared for ignition experiments, with a great deal of concrete shielding added. In addition, in the latter part of 2010 the first frozen DT shell experiments were performed. Initial experiments had low amounts of D, replaced by H, in order to cut down on neutron damage to x-ray instrumentation. These experiments extended into early 2011 [34]. There were a number of technological issues to overcome. Ice would form on the hohlraum window. The window was there to hold in the He gas that filled the hohlraum, which helps keep the high Z wall expansion into the interior of the hohlraum in check. The ice on the window delayed the first picket of the 4-step pulse shape. A double window ('storm window') had to be implemented to avoid this problem. With those and other technological cryogenic related issues solved, the program was ready to actually begin the 'real' ignition tuning campaign.

8.2.4 The 2011 Ignition Tuning Campaign

8.2.4.1 Introduction

We are about 6 months into the actual tuning campaign. As such, results are still in their preliminary and pre-publication phase. Because of that we will not 'publish' any of this preliminary data here, but only describe what we think the situation is, in general terms. In due time all of the data will be scrutinised, quality checked, and then published.

We will describe experiments roughly in the order they were performed. Typical types of platforms for tuning the various quantities, and typical data are depicted back in Fig. 8.8.

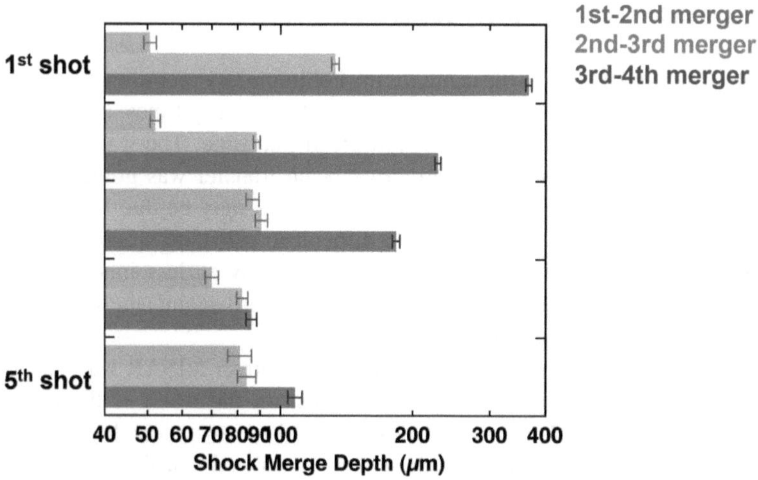

Fig. 8.12 The shocks were tuned to coalesce within five attempts

8.2.4.2 The First Shock Tuning Experiments

Using a re-entrant cone that allowed a laser velocity interferometry ('VISAR') set up to probe shocks [43, 44] emerging from the CH ablator into liquid D_2 at the centre of the hohlraum, we were able to observed the four shock coalescence, or should we say, lack of coalescence. The results of the first shot, which showed a lack of coalescence is yet another clear example of how unlikely it would have been for ignition to occur 'on the first shot'. Here, on the first tuning shot, we saw the lack of good shock timing. This is due in part, we believe to the need for a better equation of state on release, as the shock leaves the denser CH and enters the D_2, which is less dense.

However by tuning (yes, there's that key word again!) the laser pickets in height and in timing, within five shots the four shocks coalesced as required. This is shown in Fig. 8.12. The fourth shock was weaker than expected, and we will return to discuss this issue in more depth later in this lecture. When applying this new tuned pulse (which ended up about 1 ns longer than before) to a full DT or THD capsule, we increased the measured ρR product of the imploded assembly. This quantity is measured by observing the 'down scatter ratio' (dsr) which is the ratio of energy-down-scattered neutrons compared to the number of neutrons at their original 'birth energy' as they emerge from the fusion process. After the shock-tuned pulse was applied, the 14 MeV neutrons encountered a denser path out of the capsule to the detector by about a factor of two over pre-tuned implosions.

The newly shock-tuned pulse, which, as mentioned is longer in duration, had made achieving symmetric implosions more difficult. Luckily, the '575' hohlraums appeared just in time. They have provided more 'head room' for symmetry control via the $\Delta\lambda$ technique, since they are naturally shifted a bit towards less pancaked implosions.

8.2.4.3 Early Time Symmetry Experiments

We employed a Bi sphere at the centre of the hohlraum. By recording its re-emission of x-rays [45] from this non-imploded sphere, at very early times, (time of the first picket) we were able to assess the symmetry at that time. The waist was brighter than the pole. This told us that more cross beam transfer was going on from the outer beams to the inner beams than expected. This may be due to subtle effects of the plasma conditions early on in the difficult to calculate vicinity of the LEH. Nonetheless, by tuning the pulse in time, namely having less inner beam power early, compared to the pre-shot, we tuned (yes that word again!) the symmetry early in time to be within ignition specifications.

In a tour de force of the VISAR technique, a mirror VISAR was set up. This allowed us to watch shock coalescence not just on the waist of the capsule, as reported in Sect. 8.2.4.2 above, but also allowed us to watch the pole position shock breakout behaviour. This measurement confirmed that the first picket early time symmetry was indeed fixed by the lower inner beam retune. However, it did show that pickets 2 and 3 were not breaking out symmetrically (pole vs. waist), so that too needed to be retuned. This was indeed recently accomplished. In this way the symmetry through the shock-timing phase of the drive was corrected empirically [46].

8.2.4.4 Implosion Trajectory / Velocity Experiments

By backlighting an imploding capsule [47] we have been able to measure the trajectory of the dense shell. This gives us not only position vs. time but also, via its slope, the implosion velocity vs. time.

For the simulations to match that trajectory for the original design which had Ge doping in the CH ablator (for M-band shielding purposes) we have had to make two ad-hoc adjustments to the incident laser pulse [48]. The first, is to delay the rise of the fourth, main pulse, by about 200–300 ps. Doing so does not particularly match the Dante x-ray temporal rise. The second ad hoc correction is to lower the peak laser drive by about 15 %.

Subsequently a target design change used Si doping instead of Ge. The Si shields the M band due to its K edge at 1.8 keV. The Ge did so via an L edge, which is at 1 keV, precisely at the peak of the 300 eV Planckian drive. The Si doped capsule did implode faster, and only required a main pulse ad-hoc reduction of 8 %. This implies that there is a problem in our understanding of the Ge opacity, and how it affects the transport of the thermal x-rays through the Ge doped CH ablator. Experiments were done at Omega that showed good agreement with Ge doped CH [27]. It is possible we are simply having a 'scale size' issue. A small, undiagnosed error, say 5 % in the Omega scale, can be magnified to a 20 % effect at scale size that is a factor of 4 larger. An alternative explanation is that the Omega experiments were done at 240 eV, with a drive peak at 670 eV, so the 1.0 KeV Ge L edge did not play as crucial a role there.

With the velocity lower than required for ignition, we needed to increase it. The next section discusses experiments that have done so and future plans.

8.2.4.5 Experiments to Increase the Implosion Velocity

As the previous section alluded to, it appears as if the targets do not immediately respond to the rise of the main pulse. As such they have effectively less of a main drive duration than they are 'supposed' to have, in order to be accelerated up to the required thermonuclear ignition implosion velocity of 3.6×10^5 m/s. As they are at a relatively larger radius than they are 'supposed' to be at when the laser goes off, then they will coast on the way in, decompress and be less effective an implosion.

These observations lead to an obvious, but energy costly 'fix', which is to increase the duration of the main pulse. Then the acceleration of the shell inward will continue the requisite time. Experiments were recently done to do precisely that, and indeed the velocity increased. To further increase the velocity we will have to have the peak power during the main pulse increased. Plans are underway to do so.

Another way to effectively increase the main pulse drive, without increasing the laser power is to use a more efficient hohlraum. In [1] this was a central theme. In particular we discussed cocktail-wall hohlraums there, namely a wall made of an optimised mixture of materials. There we derived the basic Marshak wave scaling of wall loss. We showed that the wall loss scales as $e^{0.7}/\kappa^{0.4}$, where e is the specific heat coefficient and κ is the opacity coefficient. The specific heat scales as $(Z_B+1)/A_N$ where is the ionisation state. Cocktails work best if they contain U since then the wall loss numerator will decrease due to uranium's higher A_N (compared to Au). Cocktails really work well since the wall loss denominator also increases. They are optimised, via their mixture, to increase κ, since one element's opacity valley gets filled-in by another's peaks. Cocktail hohlraums have been difficult to fabricate with any throughput. As a compromise, U walled hohlraums at least have the numerator decreased.

Very recently U hohlraums have been shot and indeed the velocity did increase, compared to Au hohlraums shot with the same laser pulse [46]. The *U* had a thin over-coating of Au to prevent oxidation. This technological necessity was confirmed in cocktail walled hohlraums (in which U was a component of the cocktail) on Omega [49, 50]. Prior to this fix, the wall was oxidised. This oxygen, low Z component, increased 'e' because of its (relative to Au) higher $(Z_B+1)/A_N$ factor. When oxygen was carefully eliminated in the production process and 'sealed' with the thin Au overcoat, the cocktail hohlraum immediately showed the enhanced performance predicted by theory. Thus the success at NIC of the *U* hohlraum is, in some sense, a 'double no surprise'. Both the physics and the technological issues were previously studied and proven at Omega.

8.2.5 Some Physics Issues that Remain

A current campaign is busy optimising both $P2$ symmetry and $m = 4$ symmetry. When that is done we can more fully assess what is contributing to lower than expected yields. If 3-D mix is an issue, we need to separate it out from the 3-D $P2$ and $m = 4$ components of the imploded configuration.

There is also the possibility of a 1-D issue. We have been calling it the 'fifth shock hypothesis', after the suggestions of P. Springer and colleagues. Since it appears that the fourth shock rise (= main pulse rise) of the drive is delayed, there may be a fifth shock that eventually occurs in the implosion. It can add more energy and entropy into the hot spot, which would result in an implosion with a decreased compression, lower pressure etc. Clearly a priority is to fix the delay of the fourth, main pulse, and that will be a subject of a separate campaign.

Ideally we would like to understand the issue better, in order to more optimally fix it. It could be NLTE issues of the ablator as it adjusts to the rapidly rising main pulse. It could be internal LPI, whose signatures simply do not get out of the hohlraum for us to detect them. The internal LPI may direct light to parts of the hohlraum, like the central fill gas, rather than to the walls, resulting in less drive. We have simulated such a scenario, by forcing Raman backscatter near the mid-plane of the hohlraum during the rise of the pulse. The back-scattered light, coming from the higher density plasma there is of sufficiently long wavelength so as to significantly refract on its way back out of the hohlraum. As a result it does not exit the LEH but is absorbed in lower density hohlraum fill gas, (rather than near the Au wall) and does not lead to much drive. As such the capsule is not driven properly during the rise and the peak x-ray brightness time is indeed significantly delayed. For this hypothesis to be taken more seriously, we would need to understand why the Dante diagnostic does see the drive rise, while the capsule somehow does not.

Another hypothesis involves the high Z wall somewhat mixing with the fill gas, and preventing beam propagation due to high Z inverse bremsstrahlung, especially during the colder conditions at the rise of the main pulse. The issue with Dante just mentioned applies to this hypothesis as well. Nonetheless, it is conceivable that somehow the Dante sees hot hi-Z / low Z mix near the LEH and 'reports' a signal, whereas deeper in the hohlraum the laser rays have not 'burned through' this 'inverse bremsstrahlung high Z barrier' and thus denies the capsule a view of the drive.

Yet another hypothesis involves the complicated time history of the closure of the LEH. It is difficult to model. This is especially true near the rise of the main pulse when the LEH actually 'opens' due to the blown off plasma along the outer edges of the LEH heating up and going transparent to x-rays. Again, a detailed scenario that explains the Dante observation as well as the delayed coupling to the capsule must be constructed, and then tested.

Another modelling issue is the need to drop the main power by 8 % in order to match the velocity data. In some sense this is less crucial, as, in principle, an 8 %

increase in laser power should fix that. The theoretical reason for the discrepancy could be as follows. While we 'match' the Dante, it is a product of drive, T^4, times the LEH area. Meezan [51] has suggested that we may be making compensating errors- the theoretical LEH is smaller than reality and the theoretical T^4 is larger than reality, but their product matches the data. It is easy enough to make these compensating errors. The LEH dynamics are very difficult to model accurately. The in-line NLTE DCA model is necessarily a compromise vs. better models which are too complex to run in-line, and thus DCA could be in error. The numerical resolution to properly resolve the narrow region that converts laser light to x-rays is also not easy to implement in a full 2-D simulation. All of these issues are under active scrutiny and investigation.

The next year of experiments will address the issues presented here. In less than half a year of actual experiments we have increased the ITFX by a factor of 50. We did this by first, significantly improving the shock timing, and thus lowering the 'off the isentrope parameter', α, from a value greater than 3 to a value of about 1.7, as described above. Then ITFX was increased further by improving the symmetry both of $P2$ and $m = 4$. After that, we improved it still further by finding ways to increase the implosion velocity, also as described above. We still have another factor of 10 to go in ITFX. This will be the focus of the experiments to come. This is the most exciting time in ICF history.

Acknowledgements We are very grateful to the very large team of our colleagues who were responsible for the work presented here. For the development and application of the HFM we thank H.A. Scott, D.E. Hinkel, E.A. Williams, D.A. Callahan, R.P.J. Town, L. Divol, P.A. Michel, W.L. Kruer, L.J. Suter, R. Olson, J. Hammer, S. MacLaren, J.A. Harte, and G.B. Zimmerman. The parallel efforts in, and results from, the three-dimensional design code, Hydra are also to be recognised, in particular the efforts of M. Marinak, N. Meezan, R.A. London, O. Jones, M. Patel, J. Milovich and G. Kerbel. Other design issues discussed here have benefitted from the work and thoughts of S. Haan, S. Weber, P. Springer, C. Cerjan, M. Key, D. Clark, J. Salmanson, P. Amendt, B. Hammel, C. Thomas, R. Berger, D. Strozzi, J. Castor, D. Munro, S. Hatchett, D. Braun, S. Prisbey, and B. Spears. The collaborative efforts of S. Hansen and P. Springer in launching a renewed NLTE modelling initiative are greatly appreciated.

The 2009 energetics campaign, of which we reported in detail, had N. Meezan, principal designer, and S. Glenzer principal experimentalist, and we thank them for it. Many other experimentalists, whose work informs our discussions here, include J. Kline, J. Moody, E. Dewald, M. Schneider, D. Hicks, H. Robey, P. Celliers, G. Kyrala, S. Dixit, J. Ralph, K. Widmann, D. Bradley, S. Regan, T. Doppner, A. MacKinnon, A. Moore, R. Scott, and the MIT group led by R. Petrasso. The NIF laser has performed spectacularly well, and we thank R. Patterson, B. Van Wonterghem, C. Haynam and their entire team. Target fabrication continues to perform miracles, and we are grateful to A. Nikroo, H. Wilkins and the GA team, and to A. Hamza, E. Dzentitis, and the LLNL team.

We thank J. Lindl, J. Edwards, N. Landen, J. Kilkenny, B. MacGowan, J. Atherton, J. Nuckolls, and E. Moses, as well as D. Pilkington, C. Verdon, B. Goodwin, and G. Miller for their support and encouragement. We thank P. McKenna and his staff for their wonderful organisation of a very successful summer school in Glasgow, our fellow faculty members for their illuminating lectures, and the students for their astute observations and questions. Finally, a very big thank you goes to R. Rosen for her constant support in all things.

This work was performed by LLNL for the DoE under contract #DE-AC52-07NA27344.

References

1. M.D. Rosen, in *Laser-Plasma Interactions*, ed. by D.A. Jaroszynski, R. Bingham, R.A. Cairns, Scottish Universities Summer School in Physics 2005 (CRC Press, Boca Raton, 2009), pp. 325–353
2. S.W. Haan, J.D. Lindl, D.A. Callahan et al., Phys. Plasmas **18**, 051001 (2011)
3. M.D. Rosen, Phys. Plasmas **6**, 1690 (1999)
4. J.D. Lindl, *Inertial Confinement Fusion; The Quest for Ignition and Energy Gain Using Indirect Drive* (Springer-Verlag, New York, 1998)
5. S. Atzeni, J. Meyer-ter-Vehn, *The Physics of Inertial Fusion: Beam Plasma Interaction, Hydrodynamics, Hot Dense Matter* (Oxford University Press, New York, 2004)
6. G.B. Zimmerman, W.L. Kruer, Comments Plasma Phys. Contr. Fusion **2**, 85 (1975)
7. W.A. Lokke, H.W. Grasberger, National Technical Information Service Document No. UCRL52276 Copies may be ordered from the National Technical Information Service, Springfield
8. M.D. Rosen, D.W. Phillion, V.C. Rupert et al., Phys. Fluids **22**, 2020 (1979)
9. L.J. Suter, R.L. Kauffman, C.B. Darrow et al., Phys. Plasmas **3**(5), 2057 (1996)
10. M.D. Rosen, J.D. Lindl, J.D. Kilkenny, J. Fusion Energy **13**(2), 155 (1994)
11. M.D. Rosen, H.A. Scott, D.E. Hinkel et al., High Energy Density Phys. (2011)
12. H.A. Scott, S.B. Hansen, High Energy Density Phys. **6**(1), 39 (2010)
13. L.J. Suter, S. Hansen, M.D. Rosen, P. Springer, S.Haan, AIP Conf. Proc. **1161**, 3 (2009)
14. D. Shvarts, J. Delettrez, R.L. McCrory, C.P. Verdon, Phys. Rev. Lett **47**(4), 247 (1981)
15. G. P. Schurtz, X. Nicolai, and M. Busquet, Phys. Plasmas **7**, 4238 (2000)
16. R.P.J. Town, M.D. Rosen, P.A. Michel et al., Phys. Plasmas **18**, 056302 (2011)
17. R.A. London, D.H. Froula, R.L. Berger et al., Bull. Am. Phys. Soc. **DPP.NO4.5** (2008)
18. E.L. Dewald, M.D. Rosen, S.H. Glenzer et al., Phys. Plasmas **15**, 072706 (2008)
19. M.D. Rosen, H.A. Scott, L. Suter, S. Hansen, Bull. Am. Phys. Soc. **DPP.BP6.28** (2008)
20. J.D. Colvin, K.B. Fournier, M.J. May, H.A. Scott, Phys. Plasmas **17**, 073111 (2010)
21. J. Colvin, K. Fournier, J. Kane, M. May, High Energy Density Phys. **7** (2011)
22. J.L. Kline, K. Widmann, A. Warrick et al., Rev. Sci. Instrum. **81**, 10E321 (2010)
23. R.E. Olson, L.J. Suter, J.L. Kline et al., J. Phys.: Conf. Ser. **244**, 032057 (2010)
24. J.H. Hammer, M.D. Rosen, Phys. Plasmas **10**, 1829 (2003)
25. R.E. Marshak, Phys. Fluids **1**, 24 (1958)
26. B. Thomas, J. Phys.: Conf. Ser. **112**, 012006 (2008)
27. R.E. Olson, G.A. Rochau, O.L. Landen, R.J. Leeper, Phys. Plasmas **18**, 032706 (2011)
28. M.D. Rosen, Bull. Am. Phys. Soc. **DPP.CP8.00082** (2009)
29. O.L. Landen, J. Edwards, S.W. Haan et al., Phys. Plasmas **18**, 051002 (2011)
30. S.W. Haan, M.C. Herrmann, T.R. Dittrich et al., Phys. Plasmas **12**, 056316 (2005)
31. D.S. Clark, S.W. Haan, J.D. Salmonson, Phys. Plasmas **15**, 056305 (2008)
32. M.J. Edwards, J.D. Lindl, B.K. Spears et al., Phys. Plasmas **18**, 051003 (2011)
33. R.S. Betti, P.Y. Chang, B.K. Spears et al., Phys. Plasmas **17**, 058102 (2010)
34. J.D. Lindl, L.J. Atherton, P.A. Amednt et al., Nucl. Fusion **51**, 094024 (2011)
35. M.D. Rosen, M.J. Edwards, and L.J. Suter, in *The National Ignition Campaign Blue team/Red Team Simulated Campaigns*, ed. by B. Hammel, C. Laubanne, H. Azechi. Inertial Fusion Sciences and Applications 2009. J. Phys., IOP Publishers (2009).
36. S.H. Glenzer, B.J. MacGowan, P. Michel et al., Science **327**(5970), 1228 (2010)
37. N.B. Meezan, L.J. Atherton, D.A. Callahan et al., Phys. Plasmas **17**, 056304 (2010)
38. P. Michel, S.H. Glenzer, L. Divol et al., Phys. Plasmas **17**, 056305 (2010)
39. P. Michel, L. Divol, E. Williams et al., Phys. Rev. Lett **102**(2), 25004 (2009)
40. D.E. Hinkel, M.D. Rosen, E.A. Williams et al., Phys. Plasmas **18**, 056312 (2011)
41. P. Michel, L. Divol, R. Town et al., Phys. Rev. E **83**(4), 046409 (2011)
42. J.D. Moody, P. Michel, L. Divol et al., Nature Phys. **8**, 344 (2012)
43. D.H. Munro, P.M. Celliers, G.W. Collins et al., Phys. Plasmas **8**, 2245 (2001)

44. P.M. Celliers, D.K. Bradley, G.W. Collins et al., Rev. Sci. Instrum. **75**, 4916 (2004)
45. E.L. Dewald, J. Milovich, C. Thomas et al., Phys. Plasmas **18**, 092703 (2011)
46. D.A. Callahan, N.B. Meezan, S.H. Glenzer et al., Phys. Plasmas **19**, 056305 (2012)
47. D.G. Hicks, B.K. Spears, D.G. Braun et al., Phys. Plasmas **17**, 102703 (2010)
48. R.E. Olson, D.A. Callahan, D.G. Hicks et al., J. Phys.: Conf. Ser. (2012, submitted)
49. J. Schein, O. Jones, M.D. Rosen et al., Phys. Rev. Lett **98**(17), 175003 (2007)
50. O.S. Jones, J. Schein, M.D. Rosen et al., Phys. Plasmas **14**, 056311 (2007)
51. N.B. Meezan, D.A. Callahan, R.P.J. Town et al., J. Phys.: Conf. Ser. (2011, At Press)

[illegible]

Chapter 9
Laser-Plasma Coupling with Ignition-Scale Targets: New Regimes and Frontiers on the National Ignition Facility

William L. Kruer

Abstract It is very exciting that the National Ignition Facility (NIF) is now operational and being used to irradiate ignition-scale hohlraums. As discussed in the last Scottish Universities Summer School in Physics on the topic of laser-plasma interactions in 2005, laser plasma physics faces its biggest challenge ever on NIF. Excellent temporal and spatial control of the laser energy deposition is required for highly convergent implosions, while many new regimes are being accessed. These new regimes include much larger plasmas and laser beam spots, many crossing laser beams, and interaction physics in highly shaped laser pulses. Furthermore, it is key that a broad range of plasma conditions be accurately modelled in the design codes in order that the laser plasma interactions be correctly modelled. Indeed, all these new regimes have proved to be significant challenges and to require new and ongoing understanding. There are many fascinating trade-offs. For example, crossing beam effects both provide a convenient tool to quickly provide the laser beam balance needed for more symmetric implosions but also enhance the stimulated scattering of the incident laser light. The impact of these new coupling regimes in recent NIF experiments is discussed, and the ongoing understanding outlined.

9.1 Challenges of Inertial Confinement Fusion: Importance of Laser-Plasma Coupling

The quest for inertial confinement fusion [1] has continued and culminated in the National Ignition Facility (NIF) [2], which is now operational and being used for ignition-scale experiments. As shown in Fig. 9.1, the NIF laser system is a 192 beam

W.L. Kruer (✉)
Department of Applied Science, University of California, Davis, One Shields Avenue, Davis, CA 95616-8254, USA
e-mail: wlkruer@ucdavis.edu

P. McKenna et al. (eds.), *Laser-Plasma Interactions and Applications*, Scottish Graduate Series, DOI 10.1007/978-3-319-00038-1_9,

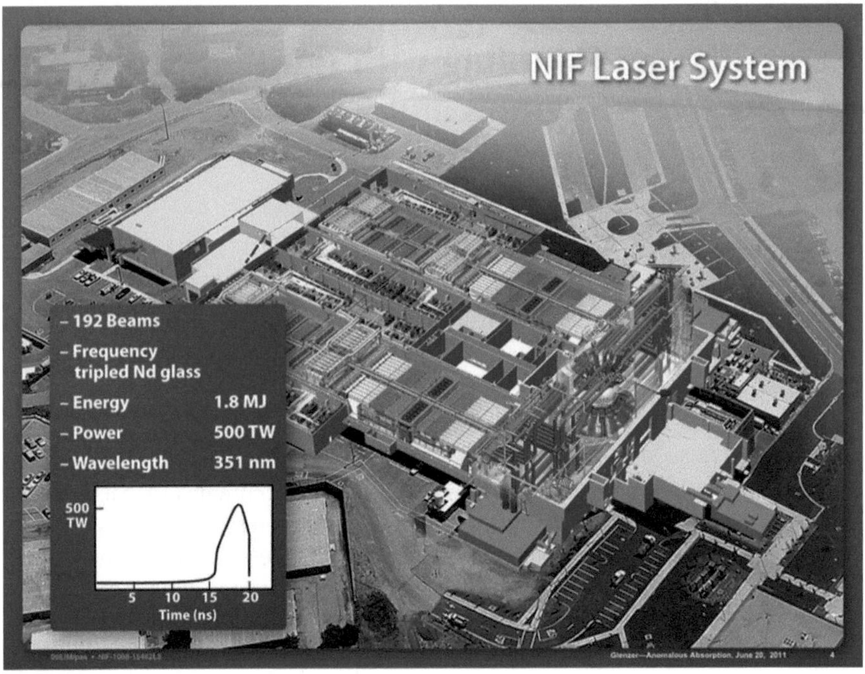

Fig. 9.1 A schematic illustrating the NIF laser system and its capabilities (Courtesy of Lawrence Livermore National Laboratory, Livermore)

Nd glass laser which is frequency-tripled to a wavelength of 351 nm. Its peak energy is 1.8 MJ and peak power is 500 TW. NIF routinely delivers precisely shaped laser pulses with pulse lengths in the range of 10–20 ns. NIF is a remarkable engineering feat. The energy of lasers in a football-sized stadium ends up concentrated in a volume much less than a centimeter cubed.

Although the basic idea is straightforward, inertial confinement fusion is very challenging. An incident energy flux heats and blows off the outer portion of a capsule (the so-called ablator which might be CH or Be), creating an ablation pressure which compresses the DT fuel inside the capsule. To minimise the pressure of the fuel being compressed, the compression must be with sufficient precision that the fuel remains on a low adiabat; hence a carefully shaped laser pulse is needed. A hotspot of DT gas in the centre of the capsule is heated by the compression to a thermonuclear temperature ($\approx$ 5–10 keV), launching a burn wave throughout the compressed fuel.

A few simple calculations illustrate the magnitude of the challenge [3]. The compressed fuel must burn before it flies apart. The burn fraction f_b can be shown to be

$$f_b \simeq \frac{\rho R}{\rho R + 6} \tag{9.1}$$

where ρ is the mass density and R the radius of the compressed fuel. For reasonably efficient burn let's choose $f_b = 1/3$, which then requires that $\rho R \simeq 3$ g/cm^2. How much DT fuel mass can one use? The fusion energy output E_{fusion} is readily calculated to be

$$E_{fusion} = 3.3 \times 10^{11} M_b \tag{9.2}$$

where M_b is the DT mass burned. Taking $f_b = 1/3$ and $E_{fusion} = 10^7$ J (appropriate for ignition experiments), the DT fuel mass $M = 10^{-4}$ g.

To obtain $rR = 3$ g/cm^2 with such a small mass requires a large compression. Noting that the mass can be expressed as

$$M = \frac{4\pi}{3} \frac{(\rho R)^3}{\rho^2}, \tag{9.3}$$

a compression to $\rho = 10^3$ g/cm^3 is seen to be needed. The challenge is apparent. Compressing DT fuel from $\rho = 0.21$ g/cm^3 (liquid density) to 10^3 g/cm^3 entails a volume compression greater than 10^4, which corresponds to a convergence ratio of ~ 30 for the fuel capsule.

To achieve these large compressions, one must drive the capsule very symmetrically. In particular,

$$\frac{\delta I}{I} \leq \frac{1}{4C_r} \approx 1\,\% \tag{9.4}$$

Here I is the energy flux, δI its variation around the capsule, and C_r is the convergence ratio (~ 30). To bring the fuel to such high density with reasonable drive pressure, one must keep the fuel on a low adiabat (nearly Fermi degenerate). This means that the shock timing and the preheat due to high energy electrons and x-rays must be well controlled. For example, the energy in suprathermal electrons reaching and heating the fuel (those with energy greater than about 170 keV for a typical ignition scale indirect drive capsule) should be less than about 50–100 J late in the shaped pulse and much less early in the pulse when the fuel is less compressed. This was estimated by simply requiring that the change in fuel pressure due to preheat δp be much less than (~ 0.1) the Fermi pressure p_{fermi}; i.e.,

$$\delta p \ll p_{fermi} \cong 2.1 \rho^{5/3} \,\text{Mbar} \tag{9.5}$$

Finally the implosion is hydrodynamically unstable, since one is in effect accelerating a heavy fluid (the fuel) with a lighter fluid (the low density ablating plasma). One must then compress the capsule quickly enough to limit the growth of the Rayleigh-Taylor instability. As an example, Fig. 9.2 illustrates how the operating regime for ICF target design is constrained by the combination of hydrodynamic and laser-plasma instabilities. The operating regime for ignition target design in laser power versus energy space is denoted by the shaded area [4]. The boundary at low power is determined by hydrodynamic instabilities; i.e., compress the capsule

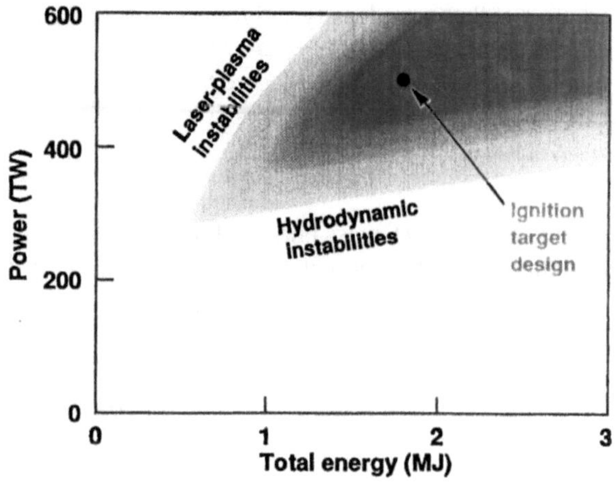

Fig. 9.2 A schematic illustrating the operating regime [4] for ignition target design as constrained by hydrodynamic and laser plasma instabilities

too slowly, and there's time for too much Rayleigh-Taylor instability growth. The high power boundary is determined by laser plasma instabilities; i.e., drive the capsule too strongly, and laser plasma instabilities generate too much stimulated scattering and too many suprathermal electrons. Of course, the operating space can be expanded as one learns how to better control these instabilities.

Let's now get more specific so that we can proceed to the ignition-scale experiments on NIF. There is a direct and an indirect drive approach to inertial confinement fusion. In direct drive [5], the capsule is directly driven by the laser light. In indirect drive [5], the capsule is driven by x-rays generated by irradiating the walls of a hohlraum (a high *Z*, usually cylindrical enclosure with holes at either end to admit the laser light). These are complementary approaches; each has pluses and minuses. For direct drive, the targets are simpler, the irradiated plasmas smaller, and the overall energy transfer to the fuel more efficient. On the other hand, the ablator is thinner, and so the implosions are more sensitive to hot electron preheat and hydrodynamic instability. For indirect drive, the x-rays are symmetrising, the ablator thicker and less sensitive to hot electron preheat, the hydrodynamic efficiency greater, and there are fewer holes in the target chamber to admit the laser light. On the other hand, the overall efficiency is less due to x-ray losses in the walls and out the laser entrance holes, and the irradiated plasma is much larger and so more susceptible to stimulated scattering. Both direct and indirect drive continue to be pursued.

We now focus on indirect drive, which is the principal approach to ignition on NIF. As indicated in Fig. 9.3, it is instructive to summarise the challenges [6] of achieving ignition in terms of four parameters: velocity, shape, entropy and mix. One must accelerate the dense fuel layer to a sufficiently high velocity while keeping the fuel on a low adiabat; i.e., nearly Fermi degenerate. To achieve the large compressions, the x-ray drive must by quite symmetric, so that the shape of the capsule remains reasonably round. Finally the Rayleigh-Taylor growth at the ablator fuel interface must be controlled to avoid too much mixing of other

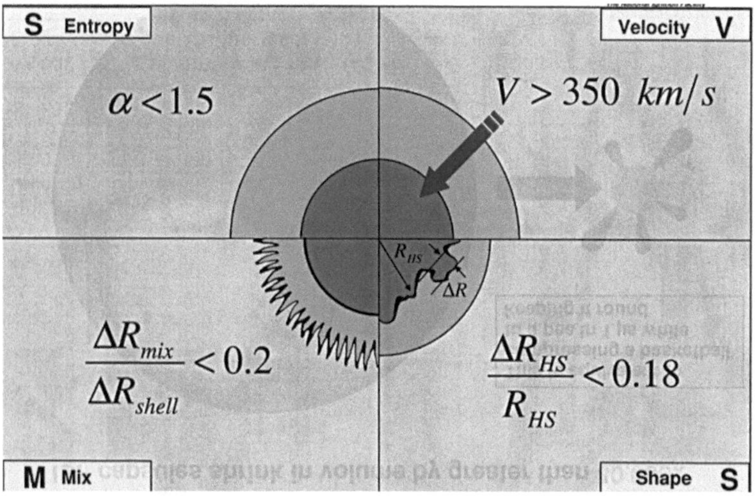

Fig. 9.3 A schematic illustrating four key variables which must be controlled to achieve ignition [6]

material into the fuel. Each of these four crucial parameters has a dependence on the laser plasma coupling. For example, the velocity can be reduced by stimulated scattering of laser light out of the hohlraum and the entropy increased by preheat due to high energy electrons generated by laser plasma instabilities. The shape can be modified by laser plasma effects on the temporal and spatial deposition of laser energy onto the hohlraum walls, and the mix enhanced if the overall laser plasma coupling requires one to drive the capsule more slowly.

9.2 New Regimes of Laser-Plasma Coupling in Ignition-Scale Experiments

Having considered the challenges of ignition and how laser plasma coupling matters, let's now proceed to discuss the coupling in NIF experiments. To appreciate the new features, consider a representative experiment [7] as shown in Fig. 9.4. A hohlraum with a length of 1 cm and a radius of 2.72 mm is irradiated with 192 laser beams in a 20+ns shaped laser pulse with an energy of 1.3 MJ. The capsule has a radius of about 1 mm and a CH ablator, and the hohlraum is filled with a low density plasma to slow down the Au wall motion. The laser beams are grouped into inner and outer beams. The outer beams are in cones which are incident at 44.5° and 50° with respect to the hohlraum axis and are aimed at the wall nearer the laser entrance holes. The inner beams are in cones incident at 23.5° and 30° and are aimed at the hohlraum wall over the capsule. The individual beams are grouped into so-call quads of four neighbouring beams which largely overlap one another within the hohlraum. There are twice as many quads in an outer cone as in an inner

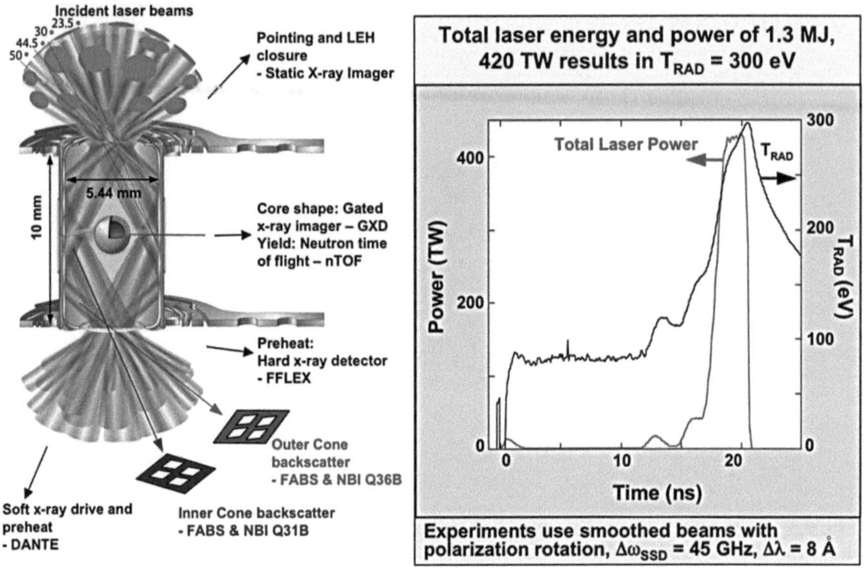

Fig. 9.4 A representative indirect drive NIF experiment [7] in which a hohlraum is irradiated by 192 beams in a 20+ns shaped pulse with an energy of 1.3 MJ

cone. By choosing the relative power in the inner and outer cones (and its phasing in time), the symmetry of the x-ray drive can be tuned. Some of the numerous diagnostics are also indicated in Fig. 9.4. These include diagnostics to monitor the laser light scattered out of the hohlraum (for a few representative beams), the x-ray flux generated, the neutron yield, the very high energy x-rays, and the shape of the imploded capsule.

The new challenges for the laser plasma coupling are apparent. This hohlraum is over 3 times larger than previous hohlraums irradiated with the NOVA and Omega lasers. Hence the irradiated plasma is much larger. The plasma is irradiated by many overlapping laser beams with large spot sizes ($>$ 1 mm). Furthermore, the plasma is irradiated with a long carefully shaped laser pulse, meaning that time-dependent coupling can matter. In addition, the preheat due to high energy electrons must be kept quite low. And of course, an accurate modelling of these much larger plasmas is needed to confidently assess the coupling.

9.2.1 The Challenge of Much Larger Plasmas

Let's first consider the challenges associated with much larger plasmas. Clearly laser plasma instabilities can be a greater threat in the larger plasmas which allow for more growth lengths. As a very simple example the gain exponent for stimulated

Raman back scattering by a damped electron plasma wave in a plasma with density n and length L is:

$$G = \frac{k_p^2 L}{8k_s}\left(\frac{v_{os}}{v_e}\right)^2 / \left(\frac{\gamma_e}{\omega_e}\right) \propto IL\lambda_0 / \left(\frac{\gamma_e}{\omega_e}\right) \tag{9.6}$$

Here k_p (k_s) is the wave number of the plasma wave (scattered light wave), v_{os} (v_e) is the oscillatory velocity (thermal velocity) of an electron, $\gamma_e(\omega_e)$ the plasma wave damping (frequency), and I (λ_0) the intensity (wavelength) of the laser light.

An attempt to avoid this threat was made in the following way. Employ as many control techniques as possible. Many of these control techniques were discussed [8] at the previous laser-plasma SUSSP in 2005. Keep the electron and ion waves heavily damped either via plasma composition or high electron temperature. Use beam smoothing techniques, including smoothing by spectral dispersion and polarisation smoothing. Finally, monitor the instability gain and design targets to keep them modest. There are many trade-offs possible. For example, the laser beam intensity could be reduced by the use of a larger hohlraum with lower x-ray drive temperature.

A postprocessor code [9] (LIP) has been used to assess the instability gain coefficients for stimulated Raman and Brillouin backscatter. This code assumes a single quad laser beam (no speckles) and uses plasmas profiles from the target design codes. Gain coefficients are evaluated in the convective, heavily damped approximation using linear Landau damping rates on Maxwellian velocity distributions. The coefficients are calculated along ray paths representing different portions of the beam using a pre-assigned intensity for each ray path; i.e., no intensification due to refraction or beam overlap effects. The LIP calculations are used to guide the choice of where and when to look with more detailed calculations.

The more detailed calculations use a code (PF3D) which solves the coupled wave equations in the paraxial approximation for a realistic laser beam (again a quad) and for convectively amplified backscattered waves due to stimulated Raman and Brillouin scattering [10]. The plasma waves associated with the scattering are assumed to be linear and to be heavily damped by Landau damping on Maxwellian velocity distributions. The zeroth order plasma conditions are taken from target design codes. Intensity changes due to refraction are now included but not those due to crossing beam effects. PF3D essentially calculates the linear theory of both Raman and Brillouin backscattering (and ponderomotive filamentation) including pump deletion, a realistic laser beam, and the detailed profiles of plasma conditions. The calculations can follow an entire quad for $\approx$ 100 ps. They are very computer intensive and in practise limited to a few selected time windows.

Of course, the biggest risk to this strategy is that the plasma conditions have not been modelled or resolved in the design codes with sufficient accuracy. Plasma instability gains can be quite sensitive to the plasma conditions and thus to uncertainties in the models for the plasma emissivity, the electron heat transport, and the hydrodynamics (W.L. Kruer, 14–16 January 2009, Presentation to the

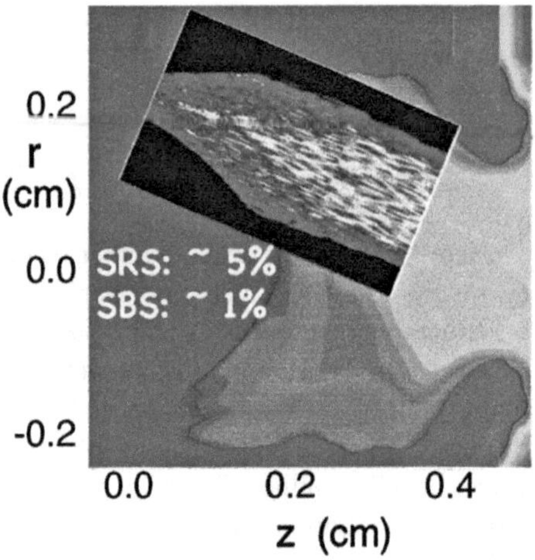

Fig. 9.5 An illustration showing the results of a PF3D calculation [11] of stimulated scattering at peak power in a NIF hohlraum using plasma conditions in the original design code

Jason review of NIC, La Jolla, unpublished). As an example, consider stimulated Raman backscatter of laser light in a plasma with a density $n = 0.12n_{\text{cr}}$ (the critical density) and an electron temperature $T_e = 2.5\,\text{keV}$. The gain coefficient G is inversely proportional to γ_L, the linear Landau damping coefficient for the unstable electron plasma wave. The damping rate depends on the slope of the electron distribution function at the phase velocity v_p of the wave and can change greatly as v_p varies relative to the electron thermal velocity. Let G_o be the gain coefficient for $n = 0.12n_{\text{cr}}$ and $T_e = 2.5\,\text{keV}$. If T_e is held fixed and the density varied by plus or minus 20 %, G will change from about $0.36G_o$ to about $3.7G_o$! Likewise, if the density is held fixed and T_e is varied by plus or minus 20 %, G changes from about $0.54G_o$ to about $2.7G_o$! Remember that G is a gain exponent. The gain exponent is not always this sensitive to the plasma conditions, especially for stimulated Brillouin backscattering in a mixed species plasma. However the need for accurate modelling of the plasma conditions is apparent.

There are other limitations to the instability gain analysis. The gain coefficients are proportional to the laser light intensity. As will be discussed shortly, overlapping beam effects can raise the effective intensity in several ways. In addition, the Landau damping of an electrostatic wave can be strongly reduced by particle trapping, which onsets at a remarkably low amplitude as discussed at the last summer school.

The PF3D calculations predicted low to modest Raman and Brillouin scattering in NIF ignition-scale experiments. The insert in Fig. 9.5 illustrates the results of a PF3D calculation [11] of the stimulated scattering of an inner quad (incident at 30°) at peak power in a hohlraum designed for a peak radiation temperature of 300 eV. The speckle structure in the laser beam is apparent, especially in the scattered light. In this example, the stimulated Raman scattering was ≈ 5 %, and the stimulated Brillouin scattering was ~ 1 %.

Enough preamble! What has actually happened in recent NIF ignition-scale experiments? As discussed by Rosen at this summer school, measurements of both

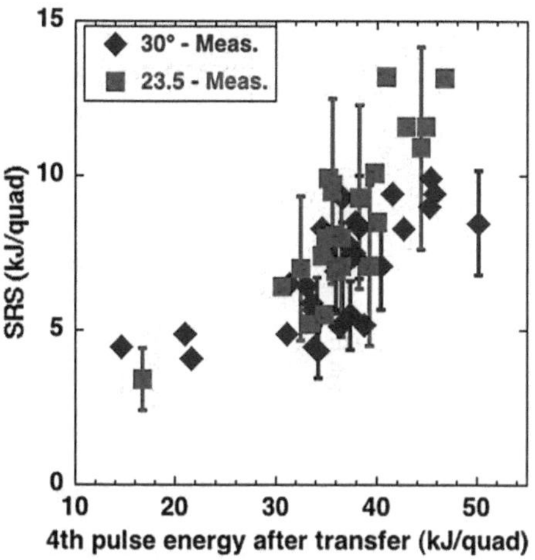

Fig. 9.6 The measured energy [14] in Raman scattered light in a 23.5° and a 30° inner quad versus the calculated energy in the inner quads after cross beam energy transfer

the x-ray flux and the spectrum of the Raman-scattered light led to significant changes in the design codes. The radiation physics and the heat transport models both needed to be significantly improved [12], leading to an irradiated plasma which is more emissive and which transports heat more efficiently. Consequently the plasma within the hohlraum is now calculated to be significantly cooler than had been predicted with the default design codes. For example, in a typical experiment T_e at density $\lesssim 0.1 n_{cr}$ in the main body of the hohlraum is not about 5 keV but rather about 2.5 keV! Not surprisingly, the measured SRS of the inner beams is quite a bit larger than had been predicted and occurs at lower plasma densities.

In the experiments the scattering of an 30° inner quad was measured [13] both back into the beam line (using a so-called FABS) and at nearby angles (using a scatter plate). In Fig. 9.6, the energy in Raman-scattered light measured in a number of experiments with input laser energy from about 0.8 MJ to about 1.3 MJ is plotted [14] versus the energy inferred to be in this quad after cross beam energy transfer. (This will be explained shortly.) For the 30° quad (the diamonds), the time averaged reflectivity due to stimulated Raman scattering is inferred to be ≈ 20 % Time-resolved measurements indicate peak SRS reflectivity's up to about 50 %! The squares denote measurements of the stimulated Raman scattering of a 23.5° quad using a less accurate cross scatter plate. This quad shows a slightly greater Raman reflectivity of ≈ 25 %. Similar measurements for the Brillouin scattered light indicate a time-averaged reflectivity of about 1–8 %, increasing with input laser energy. To date, the outer quads show only a small amount of stimulated Brillouin scattering.

These reflectivity's are consistent with those previously measured in large strongly driven plasmas. Figure 9.7 shows an illustrative plot [15] of the peak reflectivity as a function of the estimated density scale length of the plasma in units of the wave length of the incident laser light. The targets included disks and thin foils irradiated with 0.53 μm laser light with the Novette laser as well as thin

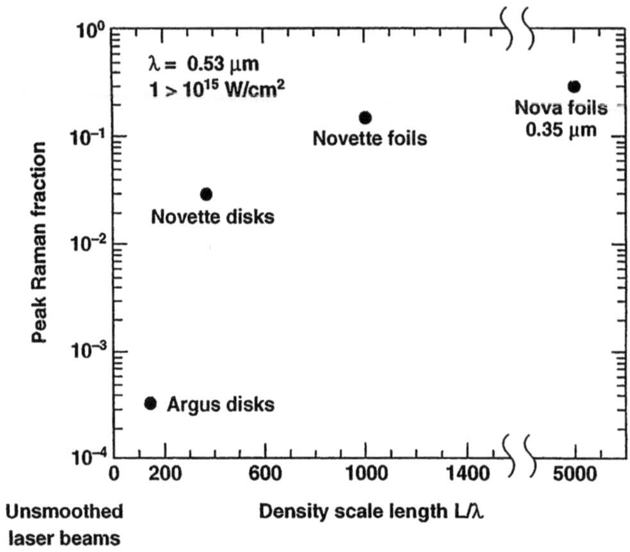

Fig. 9.7 The peak reflectivity due to stimulated Raman scattering versus estimated density scale length (size) as measured in a variety of strongly driven experiments [15]

foils irradiated with 0.35 μm laser light on the Nova laser. The laser beams were unsmoothed. As the size of the irradiated plasmas increased, the reflectivity rapidly increased and tended to saturate at levels of 20–30 %. It should be noted that a series of strongly driven hohlraums [16] irradiated with the NOVA laser also showed $r_{SRS} \approx 20$–30 %.

9.2.2 NIF and the New Frontier of Many Overlapping Laser Beams with Large Spots

In NIF hohlraums, 24 quads overlap one another near each laser entrance hole. These quads have large elliptical spots of various sizes. For example, the major spot diameters are >1 mm, and the minor spot diameters are ~ 0.6–0.7 as large. Not surprisingly, there are very significant overlapping beam effects, both favourable and unfavourable. Energy transfer among the crossing laser beams can change the power balance between the inner and outer cones and so be used to tune the implosion symmetry. However, large cross beam energy transfer significantly enhances the intensity of the inner beams and exacerbates their stimulated scattering. Furthermore, having the inner-outer beam balance and its phasing in time be mostly determined by a laser plasma process operating at high levels may not be prudent for highly convergent implosions. An additional unfavourable effect is the possibility of cooperative amplification [17] of stimulated scattered light by the overlapped laser beams.

Cross beam energy transfer was first predicted [18] for NIF hohlraums about 15 years ago and was discussed at the last summer school. Two crossing beams with different frequencies can readily transfer energy by stimulated Brillouin forward scattering at an angle. The energy transfer can be a strong effect since the crossing laser beams have comparable intensity; i.e. the noise level is quite high and little gain is needed. Assuming a heavily damped ion wave with energy damping rate ν, the spatial gain rate (g_s) for this instability is:

$$g_s = \frac{\gamma_o^2}{2\nu v_{gz}} I, \tag{9.7}$$

where

$$I = \frac{1}{1+4\left(\frac{\Delta\omega_D}{\nu}\right)^2}. \tag{9.8}$$

Here γ_o is the instability growth rate in a spatially homogeneous plasma, v_{gz} the group of the scattered light wave in the direction of the pump wave, and $\Delta\omega_D$ is the Doppler-shifted frequency mismatch. Clearly the energy transfer can be made more or less by adjusting the frequencies of the laser beams; i.e., by making the process more or less resonant. Even if the crossing beams have the same frequency, some transfer can still occur due to the Doppler shift of the ion wave by plasma flow.

Experiments [19] in which thin foils were irradiated confirmed the energy transfer between laser beams. Generally the measured transfer was less than predicted. This reduced transfer was attributed to the effect of long wavelength modulations in the flow velocity of the plasma. Calculations [20, 21] of the cross beam energy transfer in NIF hohlraums have become much more sophisticated. Twenty-four quads cross and exchange energy with one another on each side of the hohlraum. Over a hundred different ion waves are generated in the beam crossing regions. Interestingly, the theory continues to significantly over predict the actual transfer and needs to be corrected to match experiments by the use of ad hoc limiters on the density fluctuations of the ion waves.

Cross beam energy transfer has proven to be an extremely useful tool for adjusting the power and energy balance between the inner and outer beams. Of course, this balance determines the symmetry of the x-ray drive and so the symmetry of the imploded capsule. The original strategy [5] was to configure the laser to give the correct power balance (predicted to be about one third in the inner beams and two thirds in the outer beams) and to use beam pointing to fine-tune the symmetry. Hence there are twice as many quads in the outer beams than in the inner beams. Alas, the hohlraum plasma has proved to be more emissive and to transport heat more efficiently than predicted. Since the plasma is cooler, the inner beams undergo more collisional absorption and more stimulated scattering, so more power needs to be given to the inner beams to achieve a symmetrical implosion. The required power balance now becomes roughly one half in the outer beams and one half in the inner beams. Fortunately this was quickly achieved by intentionally transferring significant energy from the outer beams into the inner beams [22]. As shown in

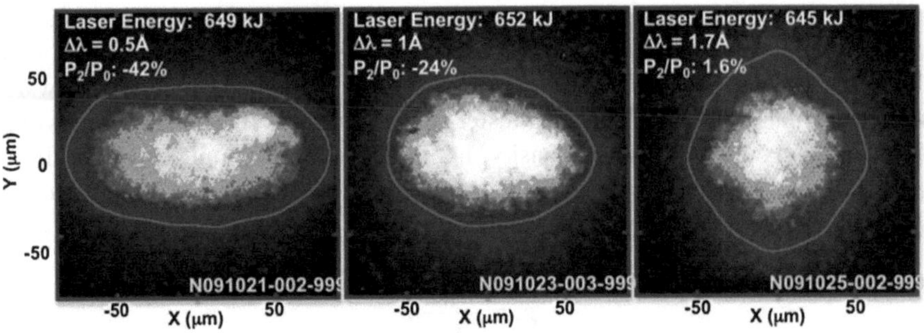

Fig. 9.8 X-ray images showing the symmetry of an imploded capsule [22]. The shape is changed from pancaked to nearly round by adjusting the wave length separation between the inner and outer beams

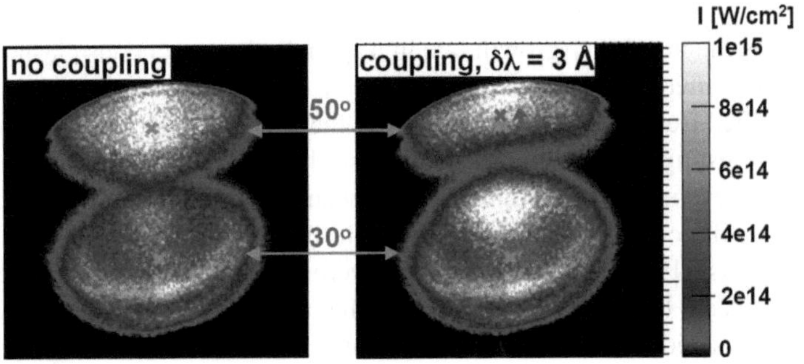

Fig. 9.9 The calculated intensity distributions in an outer and an inner quad before (on the *left*) and after (on the *right*) cross beam energy transfer [23]. The inner quad is on the bottom

Fig. 9.8, the symmetry of an imploded capsule was successfully tuned from a very pancaked implosion to a nearly round one by adjusting the wave length separation between the inner and outer beams. In this example the input laser energy was about 650 kJ. As the length of the shaped pulse (and the input laser energy) has increased in recent experiments, a larger wave length separation (up to ≈ 8 Å) corresponding to greater energy transfer has been needed.

Crossing beam effects also enhance stimulated scattering of the inner beams. The energy transfer increases the power in the inner beams and hence their intensity. With the default hohlraum, the power is typically enhanced by a factor greater than 1.6. Actually the intensity increase is even larger ($\approx 2-3$ times), since the energy transfer is greater to the upper part of the inner beams. An example of this change of the intensity distribution of a 30° beam (quad) is illustrated in Fig. 9.9, which shows the calculated intensity distribution before and after the energy transfer [23]. Of course, these intensity enhancements can also increase the role of other instabilities, such as the laser beam filamentation and two plasmon decay instabilities.

Crossing beams can also introduce cooperative scattering [17], which occurs when many laser beams drive a common wave. As a concrete example,

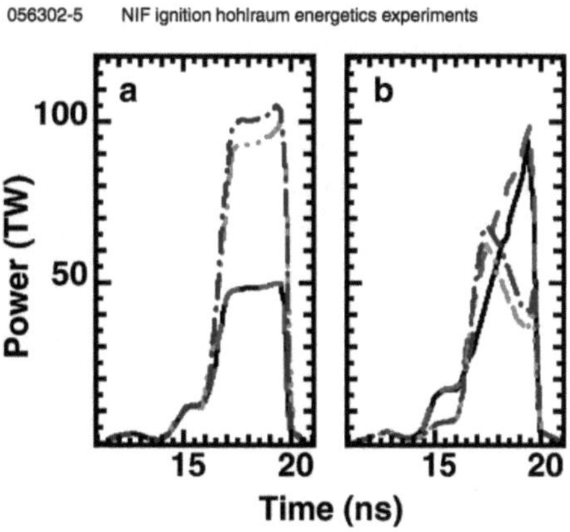

Fig. 9.10 The power balance (**a**) in the incident inner and outer beam cones and (**b**) as modified by calculations of cross beam energy transfer and measurements of time-dependent stimulated scattering [26]

Raman-backscattered light in one beam can be further amplified by other crossing beams, becoming backward scattering at an angle for those beams [24]. Very recently, the overlap of two nearest neighbour quads as well as the enhanced intensity due to cross beam energy transfer has been included in PF3D simulations [25]. The stimulated Raman reflectivity is indeed found to be significantly increased and now clearly requires that nonlinear effects be added to the PF3D model. The understanding of cooperative scattering is at an early stage.

Finally it should be noted that using too much cross beam energy transfer may not be prudent for highly convergent implosions. As shown in Fig. 9.10, the transfer strongly changes the time-dependent beam balance [26]. Calculations of the cross beam transfer rely on nonlinear limiters to approximately match the measured time-averaged implosion symmetry in experiments. In addition, the laser beam spots on the wall are also significantly modified as already discussed. Of course, it would be best to use cross beam energy transfer to fine-tune the beam balance, not help determine it to zeroth order.

9.2.3 NIF and the New Frontier of Laser-Plasma Coupling in Highly Shaped Laser Pulses

Compressions of a capsule by factors of greater than 10^4 are obtained by using a carefully shaped laser pulse as illustrated in Fig. 9.4. In NIF experiments there are four precisely timed pickets. These pickets launch shocks which must coalesce in the fuel in tight sequence to achieve the requisite drive pressure. For example, if the shocks coalesce too late, the material decompresses between the shocks, leading to fuel with a high adiabat. Clearly stimulated scattering on the rising parts of the highly shaped input laser pulse can complicate shock timing.

Stimulated scattering can in fact become important on the leading edge of the final (high power) laser picket. A simple model suffices to illustrate one possibility.

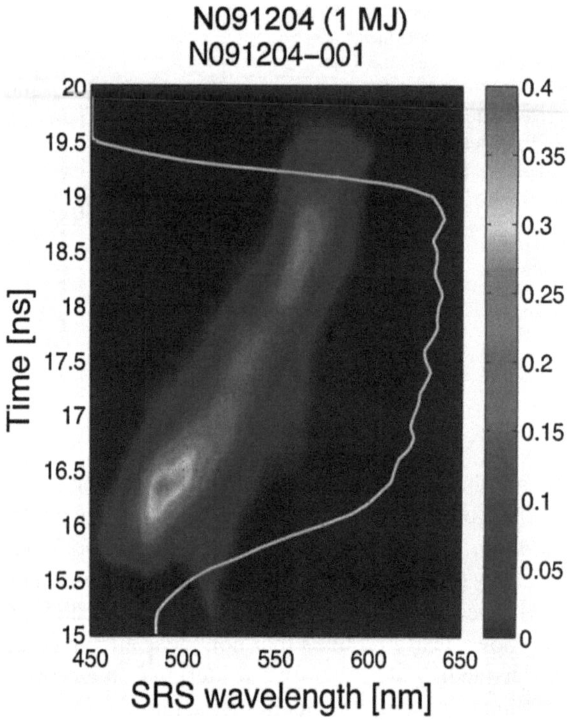

Fig. 9.11 The time-dependent spectrum of the Raman scattered light [13, 14] from a 30° inner quad during the fourth picket in a NIF experiment with an input laser energy of 1.05 MJ

Let the gain coefficient for stimulated Raman backscatter at peak power be G_o. For a modest decrease in laser intensity, the gain coefficient can increase! The gain coefficient is proportional to I/γ_L, where I is the laser intensity and γ_L is the Landau damping rate for the electron plasma wave associated with the backscattering. As previously noted, γ_L can be a very sensitive function of the electron temperature T_e. Let the intensity be I_o at peak power and take $n = 0.12n_{\rm cr}$, $T_e = 2.5\,\text{keV}$, and $G = G_o$. Simply estimate the intensity dependence of T_e by using flux-limited heat flow (whether the flux limiter is 0.05 or 0.15 does not matter), which give $T_e \propto I^{1/3}$. Then when $I = I_o/2$, $T_e \approx 2\,\text{keV}$, and $G \geq G_o$.

Another possibility is that cross beam energy transfer to the inner quads peaks earlier in time than calculated by using ad hoc clamps to approximately match the time-averaged energy transfer needed to achieve the observed implosion symmetry. The energy transfer could initially be the significantly larger value calculated without limiters which then reduces in time as the self-consistent energy and momentum deposition modifies the plasma conditions in the beam transfer region on the nanosecond time scale.

Strong stimulated Raman scattering on the rising part of the fourth picket has indeed been observed [13, 14]. Figure 9.11 shows the time-dependent spectrum of the Raman-scattered light measured for a 30° beam (quad) in a NIF experiment with an incident laser energy of 1.05 MJ. The line denotes the time dependence of the fourth laser picket. Note that the scattering onsets and indeed peaks in this example on the rising portion of the input laser pulse. The Raman scattering has a peak reflectivity of $\sim 40\,\%$ in this experiment and wave lengths appropriate to scattering from a plasma with density $\lesssim 0.1 n_{cr}$. The spectrum of the Raman scattered light is quite different than had been predicted using the default design code, which was one of the first indications that the plasma was actually much cooler [12]. In other experiments with higher input laser energy, it is sometimes stimulated Brillouin scattering which first onsets on the rising part of the high intensity picket.

9.2.4 NIF and the Challenge of Suprathermal Electron Preheat.

As already discussed, in order to achieve a high compression the fuel must be kept on a low adiabat; i.e., nearly Fermi degenerate. It is then very important to control preheating of the fuel by suprathermal electrons. Both stimulated Raman scattering and the two plasmon decay can produce electron plasma waves with a high phase velocity v_p. These waves grow until they ultimately nonlinearly transfer their energy to the electrons. Since an electrostatic wave interacts most strongly with electrons near the phase velocity, a suprathermal electron tail is created on the electron velocity distribution function. Based on early strongly driven computer simulations [27], a useful but rather crude approximation for the effective 'temperature' of this suprathermal tail is

$$T_{hot} \approx \frac{mv_p^2}{2} \tag{9.9}$$

Figure 9.12 displays this T_{hot} for Raman backscattering versus the plasma density normalised to the critical density. Note the dependence on the background electron temperature due to thermal corrections to the frequency of the electron plasma waves. For moderate T_e (a few keV), this T_{hot} varies from about 20 keV at $0.1n_{cr}$ to about 100 keV at $0.25n_{cr}$. Similar arguments give a $T_{hot} \sim 90$–100 keV due to the strongly-driven two plasmon decay instability. Of course, much larger and more weakly driven PIC simulations are needed to better understand the expected suprathermal electron distributions functions. Little attention has been given to this issue for a long time.

Suprathermal electron preheat depends on several factors. First, only some energetic electrons reach the fuel. This geometric dilution factor is usually assumed to be $\sim$ 10–20, but could be significantly less if the suprathermal electrons are generated near or beamed towards the capsule. Secondly, only some of the energetic electrons actually have sufficient energy to penetrate the ablator and reach the fuel. Taking $\rho_{abl}\Delta r \sim 0.02$ (a typical value in recent NIF experiments) and computing the electron range (which is proportional to the square of the electron energy E_e), one finds that E_e must be greater than or about equal to 170 keV to penetrate the ablator and reach the fuel. Clearly the fraction of the suprathermal electrons which can reach and heat the fuel is very dependent on their effective T_{hot} since the fraction

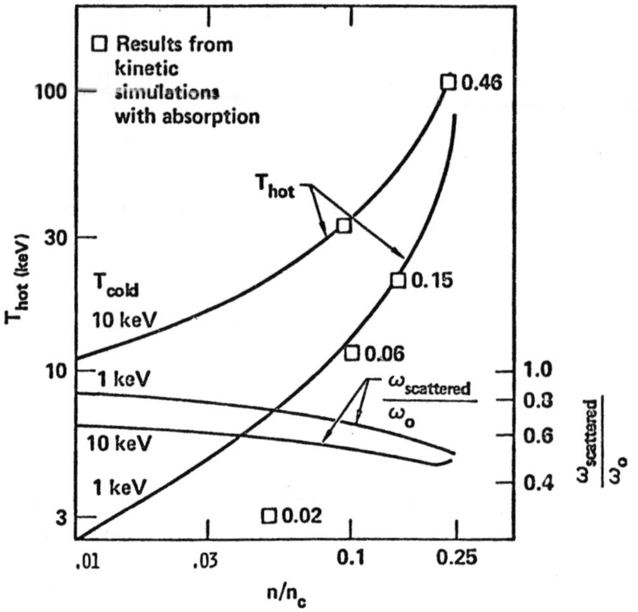

Fig. 9.12 The estimated T_{hot} versus the plasma density normalised to the critical density [27]. Some results of strongly driven kinetic simulations are included

of the electron energy with $E_e > 170$ keV is a strong function of T_{hot}. Finally, the timing of the preheating electrons also matters. If they come near the end of the laser pulse, the fuel is more compressed and so less sensitive to a given amount of preheat.

The estimated suprathermal electron temperature depends on the high frequency instability which generates the suprathermals and on its location. If strongly driven, stimulated Raman backscatter would produce $T_{hot} \lesssim 20$ keV at $n \leq 0.1n_{cr}$, $T_{hot} \sim$ 30–50 keV at $n \sim 0.15n_{cr}$, and $T_{hot} \sim 90$ keV at $n \sim 0.25n_{cr}$. Similarly, strongly driven two plasmon decay is estimated to produce $T_{hot} \sim 90$ keV at $n \sim 0.25n_{cr}$. (This instability can actually occur down to $n \sim 0.22n_{cr}$ when thermal effects are accounted for.) This all suggests a very simplified but useful three temperature description of suprathermal electrons in NIF hohlraums: not-so-hot electrons ($T_{hot} \sim$ 10–20 keV), hot electrons ($T_{hot} \sim$ 30–50 keV), and superhot electrons ($T_{hot} \sim 90$ keV). Clearly the superhot electrons are a special preheat danger.

As discussed, a substantial Raman reflectivity of $\sim$ 20–25 % of the inner beam energy is measured in recent NIF hohlraum experiments with input laser energy > 1 MJ. If we assume a collisionless plasma, the energy deposited into electron plasma waves in the Raman process ultimately Landau damps, producing a suprathermal tail on the electron distribution function. The Manley-Rowe relations then predict that

$$f_{hot} = \frac{\omega_e}{\omega_{sc}} r_{SRS} \tag{9.10}$$

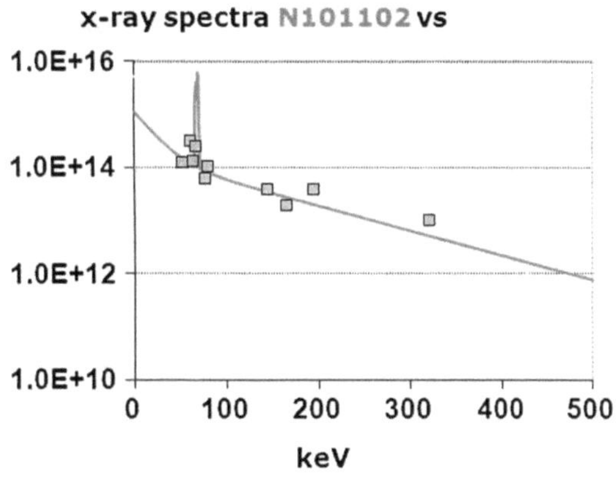

Fig. 9.13 The hard x-ray spectrum (intensity versus energy) measured in a NIF experiment [31] with an input laser energy of 1.3 MJ

where f_{hot} is the fraction of the laser energy ending up in suprathermal electrons, ω_e (ω_{sc}) the frequency of the electron plasma wave (scattered light wave), and r_{SRS} is the Raman reflectivity. Taking $r_{SRS} \sim 20$ % gives $f_{hot} \sim 10$ % of the inner beam energy. After cross beam energy transfer about 40–50 % of the incident laser energy is in the inner beams. Hence about 5 % of the incident laser energy has been transformed into suprathermal electrons! Fortunately most of the Raman scattering is occurring at $n \lesssim 0.1 n_{cr}$ and generating [28] not-so-hot electrons with $T_{hot} \lesssim 20$ keV. There are relatively few electrons with energy above 170 keV in this not-so-hot component.

A component of superhot electrons has been predicted to occur in the high intensity, fourth picket in NIF hohlraums due to excitation of the two plasmon decay instability [17]. Sean Regan subsequently discovered superhot electrons at *early* times in experiments [29] in which hohlraums were irradiated with the Omega laser. These electrons were correlated with the two plasmon decay instability in the window plasma, where a $0.25 n_{cr}$ surface is present as the window blows apart under irradiation by the first picket. This early time excitation of this instability is avoided in NIF experiments by suitably limiting the intensity of the laser light in the first picket.

A component of superhot electrons in the main picket in NIF-irradiated hohlraums has recently been inferred from FFLEX measurements of the hard x-rays [30, 31]. Figure 9.13 shows the spectrum of these x-rays from an experiment in which the input laser energy was about 1.3 MJ. This spectrum is consistent with roughly 1–1.5 kJ in superhot electrons with a $T_{hot} \sim 93$ keV. In addition, these superhot electrons appear to come late in the fourth picket.

To estimate the energy at risk , the laser energy striking the $0.25 n_{cr}$ surface with sufficient intensity to be above threshold for the two plasmon decay instability is monitored by following the rays in a design code. This energy at risk diagnostic was first applied using a gradient threshold to better understand the superhot electron generation in the window plasma at early times. Later a collisional threshold was added so that the diagnostic could be applied to the interior of the hohlraum in the high intensity fourth picket. An early calculation [32] of the energy at risk in a 1 MJ

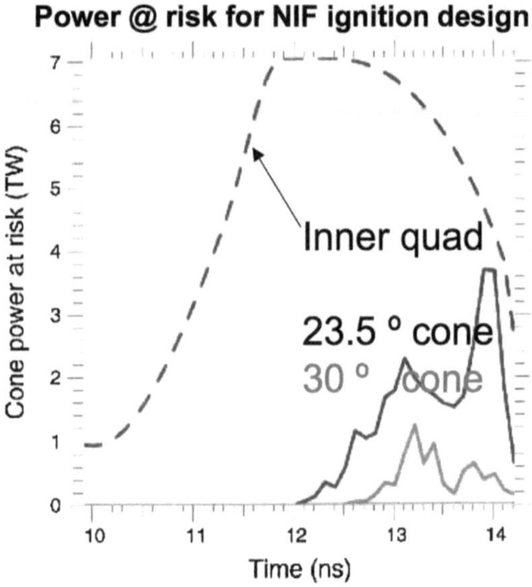

Fig. 9.14 The power at risk to the two plasmon decay instability in the fourth picket of a 1 MJ hohlraum design using a capsule with a Be ablator [32]

design using a Be capsule is shown in Fig. 9.14. Note that the energy (power) at risk was in the inner beams and came late in the fourth picket. In this example only about 4 kJ of laser energy was at risk, but this early calculation did not include the large cross beam energy transfer from the outer to the inner beams. This is expected to enhance the energy at risk at least several fold.

Based on these early calculations, let's estimate how the energy at risk might scale with input laser energy. Let's assume an intensity distribution appropriate to a speckled laser beam:

$$P(I) = \frac{\exp(-I/I_0)}{I_0} \tag{9.11}$$

where P is the probability for intensity I in a beam with average intensity I_o. Then

$$\frac{E_{risk}}{E_{inner}} = \left(1 + \frac{I_T}{I_o}\right) \exp\left(-\frac{I_T}{I_o}\right) \tag{9.12}$$

where I_T is the threshold intensity, whose variation with modest changes in input energy has been neglected. (So this is likely an underestimate for the scaling with input laser energy.) To get symmetry in the cooler hohlraums one needs roughly an equal balance of power in the inner and outer beams, rather than the $1/3 - 2/3$ balance for which the laser was configured based on the original design codes. Furthermore, the inner beam energy reaching the wall is reduced by its reflectivity due to stimulated scattering at lower density, which is taken to vary from $r \sim 0.2$ at 1 MJ to $r \sim 0.35$ at 1.8 MJ. Taking these changes to the inner beam energy (intensity) into account, the energy at risk is estimated to be ~ 31 kJ when $E_L = 1.3$ MJ and ~ 67 kJ when $E_L = 1.8$ MJ. Again this is likely an underestimate. Clearly the efficiency with which the $2\omega_{pe}$ instability produces superhot electrons is a significant issue. Even an efficiency of a few percent could become problematic. Improved estimates for the energy at risk in recent NIF experiments are being made using the detailed plasma and inner beam intensity profiles from the improved design codes (E.A. Williams et al., 2011, private communication).

A potential way to reduce superhot electrons is to point the inner beams so that they strike the $0.25n_{cr}$ surface in the high Z wall plasma (as much as possible). Then less energy is at risk for at least two reasons: the collisional threshold scales as Z^2, and there's more collisional absorption in the higher Z plasma. One must also be careful not to design hohlraums which fill with near $0.25n_{cr}$ density plasma in the region of the hohlraum over the capsule. Some hohlraums seem dangerously close to having this happen or to having significant portions of the inner beams intercepted by Au blowing off the walls, particularly if the pulse length is further increased.

9.3 Summary and Ongoing Challenges

Laser plasma coupling continues to play a key role in the quest for inertial confinement fusion. In NIF hohlraums, much larger plasmas are irradiated with many overlapping large spot laser beams in a long shaped laser pulse. The recent NIF experiments have driven significant improvements in the radiation and electron heat transport models in the target design codes which have led to a much improved modelling of the plasma conditions within the hohlraum. Consistent with the cooler, more emissive plasma now calculated, significant levels of stimulated Raman scattering are observed and the power balance between inner and outer beams needed for symmetric implosions significantly changed. Fortunately changing this power balance by cross beam energy transfer has enabled the implosion symmetry to be rapidly tuned in current NIF experiments.

To date, the coupling has been 'good enough' to enable a first pass at fielding implosion experiments with cryogenic capsules and testing an impressive range of sophisticated diagnostics (for shock timing, preheat, implosion symmetry, etc.). While significant, the stimulated scattering out of the hohlraum has been less than $\sim 18\ \%$ of the laser energy, giving a net absorption of $> 82\ \%$. The energy in suprathermal electrons is potentially significant ($\sim 5\ \%$ of the laser energy). Fortunately most of this energy seems to be characterised by a low hot electron temperature, which lessens the preheat danger. A component of superhot electrons is being carefully monitored. The implosion symmetry has been conveniently controlled by using large amounts of cross beam energy transfer. This nonlinear, time dependent transfer may become problematic for highly convergent implosions.

There are many options for improving and more precisely controlling the coupling. Re-design of the hohlraums with the improved design codes is ongoing, seeking to reduce the need for so much cross beam transfer and to reduce the stimulated scattering. Techniques to reduce stimulated scattering include changing the laser beam spot sizes and pointing, intentionally enhancing the temperature in the fill gas, and using larger hohlraums which fill with lower density plasma. Designs using capsules with different ablators (such as Be) and different pulse lengths and radiation temperatures will be explored. The program is early on the learning curve.

Hohlraums are clearly rather complex environments in which to carry out energy poor, highly convergent implosions. An improved understanding of the plasma conditions has been obtained but they likely need to be better modelled and characterised. Laser light propagation within the hohlraum depends upon how the laser entrance hole fills with plasma, how the high Z walls expand into the hohlraum interior, how the low Z gas fill impedes this expansion, and how the ablator plasma fills the hohlraum, especially over the capsule. All these issues need to be better understood. For example, there seems to be a delicate trade-off in the choice of the initial density of the low Z fill gas in the hohlraum, too little density and the Au walls blow in too quickly; too much density and hydrodynamic coupling perturbs the capsule.

The codes which model stimulated scattering clearly need to include nonlinear effects on the electron plasma and ion sound waves. Complementary codes which do not restrict attention to convective amplification of backscattering in single (or even nearby quads) are needed. With large laser beam spots, scattering either sideward and or at large angles can be important (and sometimes instabilities are absolute). Indeed in past experiments [33] with large spots, large backward scattering was found to be accompanied by significant sideward scattering (which would change the distribution of the laser light on the hohlraum walls). Given that the large plasmas are indeed producing very significant levels of stimulated scattering as well as showing an important interplay between stimulated Raman and Brillouin scattering [34], codes with the broader physics capabilities will be important to complement PF3D calculations. The time-dependent nonlinear cross beam energy transfer is now adjusted using ad hoc limiters to approximately match the time averaged symmetry found in the experiments; this needs to be better understood given the critical role this transfer is playing in current hohlraums. As discussed above, its time dependence likely depends on how the transfer is nonlinearly limited. It would be very instructive to carry out comparison experiments in which the cross beam energy transfer is minimised and the requisite beam balance obtained directly by using precision laser pulses. Given the power constraints on the laser, such experiments seem possible with input laser energies up to about 1 MJ. Kinetic simulations are needed to both address suprathermal electron distribution functions and guide the inclusion of kinetic nonlinearities in the reduced models. Such simulations can also show how efficiently the two plasmon decay instability generates superhot electrons.

These are challenging and exciting times. We have entered important new regimes involving very large plasmas, many crossing laser beams with large spots, and highly shaped laser pulses. It's not surprising that there are a number of issues to better understand. The ideas of the community continue to be important.

Acknowledgements This paper is based on lectures given at the 68th Scottish Universities Summer School on Laser Plasmas. I am grateful to the organisers (P. McKenna, R. Bingham, D. Neely, D. Jaroszynski, and A. Andreev) for putting together such a stimulating summer school with about 100 bright students from all over the world working on many exciting applications of laser plasmas. In preparing this paper I am very grateful for many discussions and interactions with J. Nuckolls, M. Rosen, J. Moody, E. Dewald, S. Glenzer, R. Kirkwood, J. Klein, H. Robey, R. Olson, O. Landen, P. Michel, L. Divol, R. Town, C. Thomas, J. Salmonson, S. Wilks, K. Estabrook, P. Amendt, J. Hammer, N. Meezan, D. Callahan, D. Strozzi, E.Williams, D. Hinkel, B. Langdon, S. Haan, S. Dixit, B. MacGowan, J. Edwards, L. Suter, J. Lindl, and E. Moses. The entire NIF team has done an outstanding job delivering precision laser pulses, fielding sophisticated targets, and carrying out complex experiments with state of the art diagnostics. This work was performed under the auspicies of the U.S. Department of Energy by the Lawrence Livermore National Laboratory under Contract DE-AC52-07NA27344.

References

1. J. Nuckolls, L. Wood, A. Thiessen, G. Zimmerman, Nature **239**(139), 139–142 (1972)
2. E. Moses, C. Wuest, Fusion Sci. Technol. **47**(3), 314–322 (2005)
3. M. Rosen, in *Laser-Plasma Interactions*, ed. by D.A. Jaroszynski, R. Bingham, R.A. Cairns (CRC Press, Boca Raton, 2009), pp. 325–350
4. S. Haan, S. Pollaine, J. Lindl et al., Phys. Plasmas **2**, 2480 (1995)
5. J. Lindl, *Inertial Confinement Fusion* (Springer, New York, 1998)
6. O.L. Landen, T.R. Boehly, D.K. Bradley et al., Phys. Plasmas **17**, 056301 (2010)
7. S.H. Glenzer, B.J. MacGowan, N.B. Meezan et al., Phys. Rev. Lett **106**(8), 85004 (2011)
8. W.L. Kruer, in *Laser-Plasma Interactions*, ed. by D.A. Jaroszynski, R. Bingham, R.A. Cairns (CRC Press, Boca Raton, 2009), pp. 1–18
9. R.L. Berger, E.A. Williams, A. Simon, Phys. Fluids B **1**(414), 414–421 (1989)
10. R.L. Berger, C.H. Still, E.A. Williams, A.B. Langdon, Phys. Plasmas **5**, 4337 (1998)
11. D.E. Hinkel, D.A. Callahan, A.B. Langdon et al., Phys. Plasmas **15**, 056314 (2008)
12. M. Rosen, H. Scott, D. Hinkel et al., High Energy Density Phys. **7**(3), 180 (2011)
13. J.D. Moody, P. Datte, K. Krauter et al., Rev. Sci. Instrum. **81**, 10D921 (2010)
14. J.D. Moody, P. Michel, L. Divol et al., Nature Phys. **8**, 344 (2012)
15. H.A. Baldis, E.M. Campbell, W.L. Kruer, Handbook of Plasma Physics (North-Holland, Amsterdam, 1991), Vol. 3, (Elsevier Science Publishing, B.V, 1991), pp. 361–434
16. J. Fernandez, B. Bauer, J. Cobble et al., Phys. Plasmas **4**, 1849 (1997)
17. W. Kruer, Bull. Am. Phys. Soc. **52**, BP8.56 (2007)
18. W. Kruer, S. Wilks, B. Afeyan, R. Kirkwood, Phys. Plasmas **3**, 382 (1996)
19. R. Kirkwood, B. Afeyan, W. Kruer et al., Phys. Rev. Lett **76**(12), 2065 (1996)
20. P. Michel, L. Divol, E. Williams et al., Phys. Rev. Lett **102**(2), 25004 (2009)
21. P. Michel, S.H. Glenzer, L. Divol et al., Phys. Plasmas **17**, 056305 (2010)
22. S.H. Glenzer, B.J. MacGowan, P. Michel et al., Science **327**(5970), 1228 (2010)
23. P. Michel, L. Divol, E.A. Williams et al., Phys. Plasmas **16**, 042702 (2009)
24. R.K. Kirkwood, P. Michel, R. London et al., Phys. Plasmas **18**, 056311 (2011)
25. D.E. Hinkel, M.D. Rosen, E.A. Williams et al., Phys. Plasmas **18**, 056312 (2011)
26. R.P.J. Town, M.D. Rosen, P.A. Michel et al., Phys. Plasmas **18**, 056302 (2011)
27. K. Estabrook, W.L. Kruer, B.F. Lasinski, Phys. Rev. Lett **45**(17), 1399 (1980)
28. P. Michel, L. Divol, R.P.J. Town et al., Phys. Rev. E **83**(4), 046409 (2011)
29. S.P. Regan, N.B. Meezan, L.J. Suter et al., Phys. Plasmas **17**, 020703 (2010)

30. E.L. Dewald, C. Thomas, S. Hunter et al., Rev. Sci. Instrum. **81**, 10D938 (2010)
31. T. Döppner, G. A. Thomas, L. Divol et al., Phys. Rev. Lett. **108**, 135006 (2012)
32. W.L. Kruer, N. Meezan, S.P. Regan et al., J. Phys.: Conf. Ser. **244**, 022020 (2010)
33. M.D. Rosen, D.W. Phillion, V.C. Rupert et al., Phys. Fluids **22**(10), 2020 (1979)
34. D. Montgomery, B.B. Afeyan, J.A. Cobble et al., Phys. Plasmas **5**, 1973 (1998)

Chapter 10
Inertial Confinement Fusion with Advanced Ignition Schemes: Fast Ignition and Shock Ignition

Stefano Atzeni

Abstract Essential ingredients of inertial confinement fusion (ICF) are fuel compression to very high density and hot spot ignition. In the conventional approach to ICF both fuel compression and hot spot formation are produced by the implosion of a suitable target driven by a time-tailored pulse of laser light or X-rays. This scheme requires an implosion velocity of 350–400 km/s. In advanced ignition schemes, instead, the stages of compression and hot spot heating are separated. First, implosion at somewhat smaller velocity produces a compressed fuel assembly. The hot spot is then generated by a separate mechanism in the pre-compressed fuel. The reduced implosion velocity relaxes issues concerning hydrodynamic instabilities, laser-plasma instabilities and preheat control. In addition, it can lead to higher target energy gain (ratio of fusion energy to driver energy). Fast ignition and shock ignition are promising advanced ignition schemes. In fast ignition the hot spot is created by either relativistic electrons or multi-MeV protons or light-ions, produced by a tightly focused ultra-intense laser beam. In shock ignition, intense laser pulses drive a converging shock wave that helps creating a hot spot at the centre of the fuel. These advanced schemes are illustrated in the present chapter. Motivation, potential advantages and issues are described. Research needs and perspective are also briefly discussed.

10.1 Introduction

In inertial confinement fusion (ICF) energy is released from deuterium-tritium (DT) reactions occurring in highly compressed and extremely hot fuel elements. The pressure of the burning fuel is so high (hundred–thousand of gigabars) that no

S. Atzeni (✉)
Dipartimento SBAI, Università di Roma *La Sapienza* and CNISM,
Via A. Scarpa 14, 00161 Roma, Italy
e-mail: stefano.atzeni@uniroma1.it

P. McKenna et al. (eds.), *Laser-Plasma Interactions and Applications*,
Scottish Graduate Series, DOI 10.1007/978-3-319-00038-1_10,

means can be exploited to contain the fuel. It only remains confined, by its own inertia, for a time of the order of the ratio of the fuel linear dimension to the speed of sound, hence the name inertial confinement. For comprehensive presentations of ICF see, e.g., Refs. [1–3]. ICF is an intrinsically pulsed, explosive process. Future Inertial Fusion electrical Energy (IFE) power plants will operate cyclically. At each cycle a *driver* delivers a pulse of energy E_d to a fuel element (target), and brings it to fusion conditions. The released energy E_{fus} is conveyed to a suitable fluid, converted into mechanical energy, and then into electricity. Part of this energy is used by the driver and the remaining fraction is sent to the grid. As presently envisaged, an IFE reactor releasing a power of 1–2 GW to the electric grid will employ a driver delivering pulses of a few MJ of energy at rate of 5–20 Hz. Each pulse induces compression, ignition and burn of a target containing a few mg of DT fuel mixture. Economic operation requires that the energy gain of the target, $G = E_{fus}/E_d$, satisfies $G > 10/\eta_d$, where η_d is the efficiency of the driver. For $\eta_d = 10$ %, $G > 100$ is needed [4]. Another essential requirement concerns the cost of the targets: targets releasing 300 MJ and containing 2–3 mg of DT fuel, should cost at most 30 Eurocent. In this chapter we will only consider ICF driven by powerful multi-beam lasers. However, ICF experiments are also performed using Z-pinches [5, 6], and heavy-ion accelerators (see, e.g., Sect. 9.5.1 of [1] and [7]) are studied as possible IFE drivers.

Essential ingredients of any ICF/IFE scheme are strong fuel compression and hot spot ignition. In the standard ICF scheme [8], a single, time-shaped laser pulse drives the implosion of a fuel capsule and generates a compressed fuel assembly with a central hot spot. Laser irradiation can be either direct or indirect (see Sect. 10.2.3). The National Ignition Campaign (NIC) [9, 10], currently ongoing at the U.S. National Ignition Facility (NIF) [11], Livermore, California, is testing this scheme using indirect-drive. Its first goals are fuel ignition and energy gain of 10–20 [9]. Subsequent optimisation is expected to increase the gain [12]. However, substantially higher gain can potentially be achieved by employing direct-drive and, in addition, using schemes separating fuel compression from hot spot generation. Predicted gain as a function of laser energy is shown in Fig. 10.1 for different ICF schemes.

In this chapter we illustrate and discuss such advanced ignition schemes. We first briefly review the ICF principles (Sect. 10.2), emphasising the main issues that constrain target design (Sect. 10.3). In particular, we show how such issues are related to implosion velocity. The advanced ignition schemes of fast ignition [13] and shock ignition [14], introduced in Sect. 10.4, reduce instability risks by imploding the fuel at lower velocity. On the other hand, they require additional more intense pulses to generate the hot spot, and involve new regimes of laser-plasma interaction. General ignition requirements and gain potentials of advanced ignition schemes are discussed in Sect. 10.5 using simple models. The following Sects. 10.6 and 10.7 are then devoted to fast ignition and shock ignition, respectively. Conclusions are drawn in Sect. 10.8.

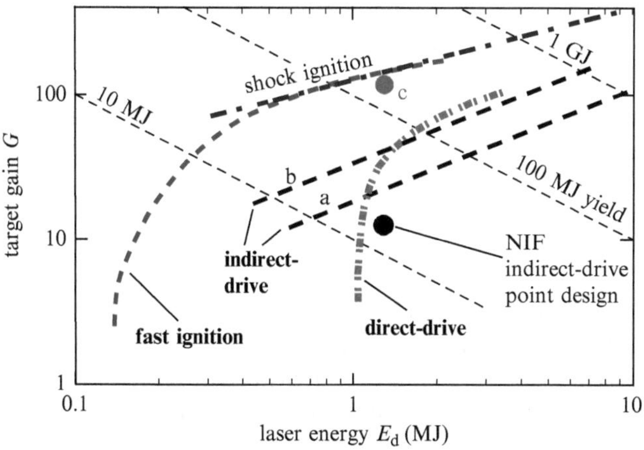

Fig. 10.1 Gain curves for different ICF schemes. The potentials of advanced ignition schemes are apparent. Curves and points refer to: NIF indirect-drive point design [9]; indirect drive limiting gain assuming the same coupling efficiency as for the NIF point design (a) [2]; indirect-drive limiting gain assuming improved coupling efficiency (b); direct-drive (from Fig. 7 of [2]); advanced direct-drive [15] (point c); fast ignition [16] and shock ignition [17]. The wavelength of the compression laser is $\lambda = 0.25$ μm for point (c), and $\lambda = 0.35$ μm for all other points and curves (Adapted from Atzeni [16])

10.2 ICF Principles

Advantages and issues of advanced ICF schemes are best appreciated by analysing general features of ICF. In this section, we briefly review the essential ICF requirements and describe the standard ICF scheme.

10.2.1 Essential Ingredients: Compression and Hot Spot Ignition

ICF has two basic ingredients: fuel compression to very high density, $\rho \geq 200$ g/cm^3, and hot spot ignition. They follow from simple considerations on the target gain. Fusion energy is released by the burn of a fraction Φ of the fuel mass m contained in the target. We then have

$$G = \frac{m\Phi Y_{DT}}{E_d} = \eta\Phi\frac{Y_{DT}}{\varepsilon}, \tag{10.1}$$

where $Y_{DT} = 340$ GJ/g is the specific DT yield (i.e. the energy released by the full reaction of a unit mass of DT fuel), $\varepsilon = \eta E_d/m$ is the specific energy delivered by the driver to the fuel and η is the beam-to-fuel coupling efficiency. Typically, $\eta = 5$–$10\ \%$. It is apparent that large gain requires efficient burn (e.g. $\Phi = 30\ \%$) and relatively small ε.

High density is required to burn a substantial fraction of the fuel. Indeed the burn fraction of a homogeneous spherical fuel of radius R is well approximated by [1–3]

$$\Phi = \frac{\rho R}{\rho R + H_B}, \tag{10.2}$$

with $H_B = 7\ \mathrm{g/cm^2}$ the approximate value of $H(T) = 8c_s m_i / <\sigma v>$ at a burn temperature $T = 30$–100 keV. Here m_i is the average ion mass, $<\sigma v>$ is the temperature dependent Maxwellian-averaged DT reactivity and $c_s = 2.7 \times 10^7 \sqrt{T(\mathrm{keV})}$ cm/s is the sound speed. Notice that the ratio $\rho R/H$ is just equal to the ratio τ_c/τ_r of the fuel confinement time $\tau_c = R/4c_s$ to the reaction time $\tau_r = n/(dn/dt) \approx 2/n <\sigma v>$ of an equimolar DT fuel with ion density n. Equation 10.2 shows that $\rho R > 2$–3 g/cm^2 is required to achieve the burn fraction $\Phi > 0.2$–0.3 needed for high gain. On the other hand, the fuel mass m has to be limited to a few mg in order to contain the explosive energy release: 1 mg of DT releases the same energy as 85 kg of high explosive! The need for compression follows immediately from the relationship between mass and confinement parameter: $\rho^2 = (4\pi/3)(\rho R)^3/m$ for a sphere.

DT reactions (with a Q-value of 17.6 MeV) occur at high rate and can self-sustain themselves at temperatures above 5 keV. However, uniform heating of the fuel to 5 keV costs 30 keV for each DT pair ($3kT/2$ for each nucleus and electron), leading to unacceptably small gain $G = \eta\Phi 17{,}600/30 = 17.6(\eta/0.1)(\Phi/0.3)$. This limitation is overcome by relying on *hot spot ignition*. This consists in heating to a temperature of 5–10 keV only a small fraction of fuel, with mass $m_h \ll m$, but still capable of first self-heating due to the power released by the fusion alpha-particles, and then triggering a burn wave reaching the whole fuel. According to detailed studies [1] (see also Sect. 10.5 below) the hot spot temperature T_h, density ρ_h and radius R_h must satisfy $T_h > 5$–10 keV and $\rho_h R_h > 0.2$–0.5 g/cm^2.

10.2.2 *ICF by Central Ignition*

The standard inertial fusion scheme [8] is based on beam-driven spherical implosion and central hot spot ignition. It is illustrated in Fig. 10.2, referring to laser direct-drive. A hollow shell target, containing an inner layer of cryogenic DT fuel, is irradiated symmetrically by powerful beams. Its outer layers absorb radiation, heat and evaporate. As a reaction to the sudden outward expansion of the ablated material, the remaining solid shell is accelerated inward, reaching a velocity u_{imp}. The imploding shell therefore behaves as a spherical rocket, driven by the ejection of the ablated material. Equivalently, one can say that the ablated material exerts an *ablation pressure* p_a on the shell. At implosion stagnation the shell kinetic energy

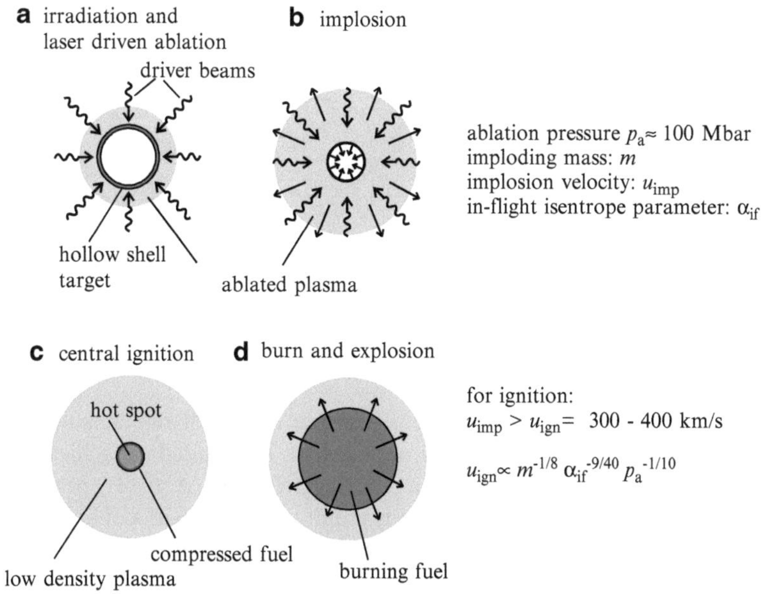

Fig. 10.2 The standard scheme of laser inertial confinement fusion by spherical implosion and central hot spot ignition (Adapted from Atzeni and Meyer-ter-Vehn [1])

is converted into internal energy, and the remaining shell material, mostly fuel, with mass m, is compressed and heated. A central hot spot is formed and the fuel ignites and burn.

Central ignition occurs for implosion velocity exceeding a threshold [18]

$$u_{imp} > u_{ign} \propto m^{-1/8} \alpha_{if}^{9/40} p_a^{1/10}, \tag{10.3}$$

where α_{if} is the in-flight isentrope parameter. This quantity will be defined and discussed below in Sect. 10.3.2. For the parameters of the current NIC campaign [9], $m = 0.2$ mg, $\alpha_{if} \simeq 1.45$, $p_a \simeq 100$ Mbar, the required implosion velocity is $u_{ign} \simeq 360$ km/s.

The implosion velocity depends on laser and target parameters. A particularly important quantity is the required ablation pressure. A simple estimate is obtained as follows [2]. Consider a hollow shell, with initial radius R_0, thickness ΔR_0 and mass $m \simeq 4\pi \rho_{DT} R_0^2 \Delta R_0$, where $\rho_{DT} = 0.25$ g/cm^3 is the density of the solid DT fuel. A constant pressure p is applied at the shell outer surface until the radius shrinks by 50 %, and the shell achieves velocity u_{imp}. Then, by equating the shell kinetic energy to the pressure work we have

$$p = \frac{12}{7} \rho_{DT} \frac{u_{imp}^2}{R_0/\Delta R_0}, \tag{10.4}$$

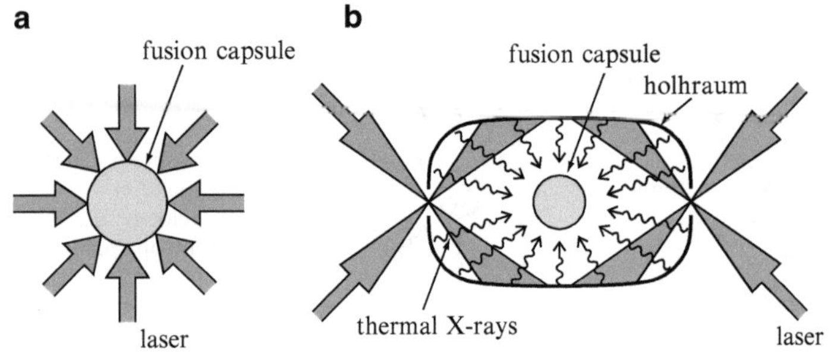

Fig. 10.3 (**a**) Direct and (**b**) indirect drive

which makes apparent the advantage of using a thin shell rather than a sphere. According to Eq. 10.4 a pressure $p = 56$ Mbar is required to achieve $u_{\rm imp} =$ 360 km/s, for a shell with aspect ratio $A = R_0/\Delta R_0 = 10$. Actually, peak ablation pressure higher by about a factor of two is required, since the model neglects target mass variation due to ablation and assumes a constant pressure.

10.2.3 Direct- and Indirect-Drive

ICF targets can be driven either directly or indirectly. In direct drive (Fig. 10.3a), many laser beams irradiate the target outer surface uniformly. In indirect drive (Fig. 10.3b), laser beams irradiate the inner walls of a cavity (a hohlraum), creating a plasma that generates thermal radiation with a temperature of about 300 eV [2, 3, 19]. Such a radiation, confined within the cavity, in turn irradiates the fusion capsule and drives its ablative implosion. Indirect-drive is pursued because it can relax the symmetry and stability issues that will be discussed later. In particular, (i) the thermal radiation field is very smooth on short scales and (ii) Rayleigh-Taylor instability (RTI, see Sect. 10.3.4 below, Ch. 8 of [1], Sect. VI of [2]) is milder at a radiation driven front than at a laser driven front. On the other hand, the coupling efficiency η is reduced by conversion of laser energy into x-rays, and by the additional loss of the significant fraction of the radiation energy that is absorbed in the hohlraum wall. In the following, we shall only consider direct-drive schemes, given their potentials for higher gain. However, it has to be recalled that direct-drive schemes require excellent control of any mechanism that can seed RTI, in particular small-scale laser nonuniformities, which can *imprint* corrugations on the target surface.

10.3 ICF Issues: The Role of Implosion Velocity

The scheme illustrated above cleverly compresses the fuel and concentrates energy into the hot spot. However, the achievement of ignition with affordable driver energy presents four main issues [20, 21]:

1. efficient coupling of laser energy to the target, to achieve adequate implosion velocity;
2. efficient use of the coupled energy to compress the fuel;
3. maintaining nearly spherical symmetry, to create a small, central hot spot;
4. limiting the dangerous effects of hydrodynamic instabilities, RTI in particular.

We now briefly discuss each of these issues.

10.3.1 Coupling Laser Energy to the Target

The laser pulse has to generate the required ablation pressure efficiently. While the ablation pressure increases with absorbed laser intensity, $p_a \propto (I/\lambda)^{2/3}$ (see, e.g. Sect. V of [2]), absorption of the laser light worsen with intensity [22]. Moreover, as shown in Fig. 10.4, absorption efficiency and ablation pressure decrease with increasing laser light wavelength [22]. In addition, laser intensity has to be limited to avoid the occurrence of parametric instabilities, which both degrade absorption and generate undesired hot electrons. As a rule of thumb, $I[\mathrm{W/cm^2}]\{\lambda[\mu\mathrm{m}]\}^2 < 10^{14}$ (see, e.g., [23]). It is apparent from Fig. 10.4b that the pressure required for direct-drive central ignition can only be obtained with UV light (i.e. $\lambda = 0.35\ \mu$m or smaller) and relatively thin hollow shells with $R_0/\Delta R_0 \simeq 10$.

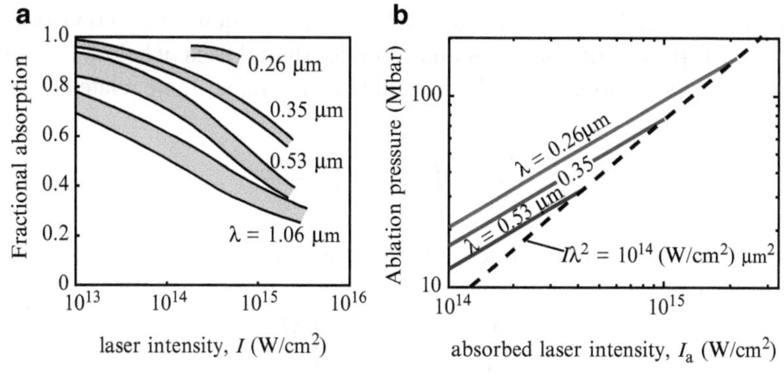

Fig. 10.4 (**a**) Absorption efficiency vs laser intensity for different laser light wavelengths [22]. (**b**) Ablation pressure vs absorbed laser intensity for different wavelengths. A low atomic number absorber is assumed

10.3.2 Efficient Compression

Compressing matter has an energy cost, which has to be kept as small as possible in order to achieve large energy gain. Since an energy $dE = p(\rho,T)\rho^{-2}d\rho$ is required to increase the density of a given lump of matter by an amount $d\rho$, the lower the pressure, i.e. the lower the temperature (at given ρ), the more efficient the compression. One has therefore to limit *preheating* of the shell due penetrating fast particles (hence the limitations discussed in the previous subsection), as well as preheating due to shock waves. Indeed, sudden application of the peak ablation pressure would just compress the fuel by a factor about four [24], and cause substantial heating. Instead, the pressure pulse has to be tuned (in jargon, *time-shaped*) as to reach peak pressure gradually, and to produce a sequence of not-too-strong shocks, approximating an adiabatic compression.

Since the minimum pressure of a material at a given density ρ is that of a 0 K Fermi electron gas, $p_F[\text{Mbar}] = 2.13\{\rho[\text{gcm}^{-3}]\}^{5/3}$ for DT, the quality of compression is often measured by the *isentrope parameter* $\alpha = p(T,\rho)/p_F(\rho) \geq 1$. The lower α the more efficient the compression. In a typical ICF target, the isentrope of the shell during acceleration, called *in-flight-isentrope* α_{if} is determined by the first shock crossing the shell. If preheating is negligible and the subsequent increase of the pressure is well timed, the isentrope stays constants during the implosion. At implosion stagnation, additional entropy is generated and the final entropy of the cold portion of the fuel is $\alpha_c > \alpha_{if}$ (see, e.g., Sect. 5.4.2 of [1] and Refs. [18, 25]).

10.3.3 Implosion Symmetry

Creation of a small hot spot at the centre of the compressed fuel, with spot radius R_h much smaller than the initial shell radius, $R_h \simeq R_0/30$, requires very good irradiation symmetry. Typically, irradiation nonuniformities should be kept below 1 %. Indeed, relative hot spot deformations, which should be $\delta R_h/R_h \ll 1$, are related to velocity nonuniformities,

$$\frac{\delta R_h}{R_h} \simeq \left(\frac{R_0}{R_h} - 1\right)\frac{\delta u_{imp}}{u_{imp}}. \tag{10.5}$$

In turn, velocity non uniformities are related to pressure and irradiation intensity nonuniformity by

$$\frac{\delta u_{imp}}{u_{imp}} \simeq \frac{\delta p_a}{p_a} \simeq \frac{2}{3}\frac{\delta I}{I}, \tag{10.6}$$

where any perturbation amplification by instabilities has been neglected. This shows that even under optimistic assumptions $\delta I/I \ll (3/2)(R_h/R_0)$.

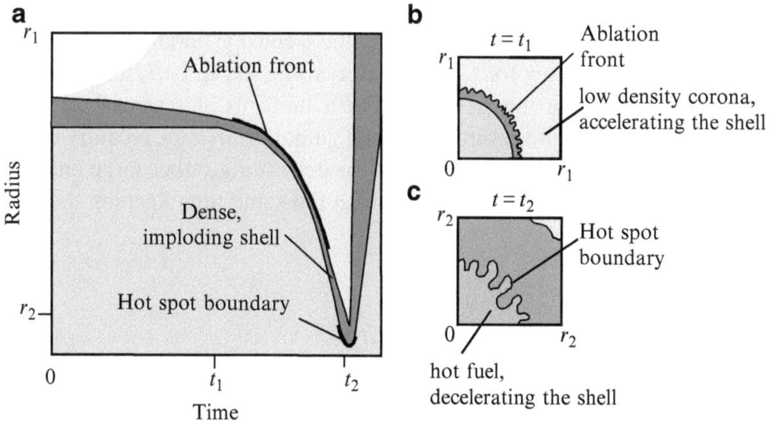

Fig. 10.5 (**a**) Qualitative implosion chart (radius vs time evolution) of a laser-driven ICF shell. Rayleigh-Taylor instability occurs at the ablation front during ablative acceleration (frame **b**), and at the shell inner surface at implosion stagnation (frame **c**). The first instability may cause shell rupture, the second one is responsible for material mixing or cold-hot fuel mixing (Adapted from Atzeni and Meyer-ter-Vehn [1])

10.3.4 Stability

The fourth, and probably the most serious, issue concerns fluid instabilities, in particular Rayleigh-Taylor instability (RTI), which occurs when a lighter fuel accelerates a heavier one [26]. In ICF, RTI at the shell outer surface (ablation-front RTI) threatens integrity of the shell, while RTI at the shell inner surface during shell deceleration (deceleration-phase RTI) hinders formation of the hot spot; see Fig. 10.5. Pedagogical discussions of RTI in ICF are presented in Sect. VI of [2] and in Ch. 8 of [1]. Ablation-front RTI amplifies seeds due to target imperfections or generated by laser imprint. Perturbations with wavenumber k and initial amplitude $\xi_0 \ll 1/k$ grow exponentially, $\xi = \xi_0 \exp[\Gamma(k)]$ at the end of the acceleration stage. Of course, the final perturbation amplitude can be limited by reducing ξ_0, i.e. short scale irradiation nonuniformities and target defects. However, this may not be sufficient, particularly for large implosion velocity. Indeed, RTI growth factors Γ increase with implosion velocity. This is simply explained as follows. For perturbations at the surface of an accelerated layer, Γ increases with the ratio of layer displacement to layer thickness, i.e., in our case, with the in-flight-aspect-ratio of the shell (ratio of radius to in-flight thickness). Accelerating a shell to higher velocity in turn requires higher driving pressure (see Eq. 10.4), causing stronger in-flight compression [since $\rho \propto (p/\alpha_{if})^{3/5}$], which means higher in-flight-aspect ratio. A more quantitative analysis [2] shows that the largest growth factor $\Gamma_{max} = \max_k[\Gamma(k)]$ is given by

$$\Gamma_{max} \simeq \frac{8.5}{\alpha_{af}^{2/5}} I_{15}^{1/15} \left(\frac{u_{imp}}{3 \times 10^7 \text{ cm/s}} \right)^{1.4}, \qquad (10.7)$$

for a shell irradiated by laser light with wavelength $\lambda = 0.35\ \mu\text{m}$. Here I_{15} is the laser intensity in units of $10^{15}\ \text{W/cm}^2$ and α_{af} is the isentrope parameter at the ablation front. Typical designs allow for $\Gamma_{\max} < 7$. According to Eq. 10.7, RTI amplification is reduced by increasing the isentrope, but a uniform increase of the isentrope throughout the target would result in reduced gain. Fortunately, recently introduced adiabat shaping techniques [27, 28] allow for generating rather large entropy (e.g. $\alpha_{\text{af}} \simeq 4$) in the material to be ablated while at the same time keeping the in-flight-fuel adiabat $\alpha_{\text{if}} \simeq 1$.

10.3.5 Issues vs Implosion Velocity

The issues we have just discussed are mainly related to the implosion velocity and to creation of the central hot spot. Indeed, issues 1 (coupling efficiency), 2 (control of preheat) and 4 (RTI) are all made worst by increasing the implosion velocity. Higher implosion velocity requires higher pressure and laser intensity, and consequently increases risks related to plasma instabilities and preheating. It also requires better pulse shaping, and leads to larger growth of ablation front RTI. In addition, creating the central hot spot also demands a high degree of symmetry (issue 3) and control of mixing caused by deceleration phase RTI (issue 4). One then wonders whether alternate ICF schemes can be conceived, which operate al lower velocity and, possibly, also relax symmetry constraints.

10.4 Decoupling Compression and Ignition: Advanced Ignition Schemes

For a large class of targets, the peak average density of the compressed fuel and the peak confinement parameter are well approximated by [29]

$$<\rho>_{\max} \simeq 0.6\rho_{\text{peak}} \simeq \frac{500}{\alpha_{\text{if}}} \left(\frac{I_{15}}{\lambda_\mu / 0.35} \right)^{0.13} \left(\frac{u_{\text{imp}}}{300\ \text{km/s}} \right)^{0.96} \text{g/cm}^3, \quad (10.8)$$

$$<\rho R>_{\max} \simeq \frac{1.46}{(\lambda_\mu / 0.35)^{1/4} \alpha_{\text{if}}^{0.55}} \left(\frac{\eta_{\text{a}} E_{\text{L-c}}}{100\ \text{kJ}} \right)^{0.33} \left(\frac{u_{\text{imp}}}{300\ \text{km/s}} \right)^{0.06} \text{g/cm}^2, (10.9)$$

where λ_μ is the laser wavelength in units of μm and $E_{\text{L-c}}$ is the energy of the compression pulse. These equations show that an implosion velocity of 250–300 km/s, i.e. somewhat below the ignition threshold, still leads to strong compression and good confinement, if the isentrope is kept sufficiently small. A separate mechanism is however needed to achieve ignition. Advanced ignition schemes indeed rely on different mechanisms (and in some cases even different drivers) to compress the fuel and generate the ignition hot spot, respectively.

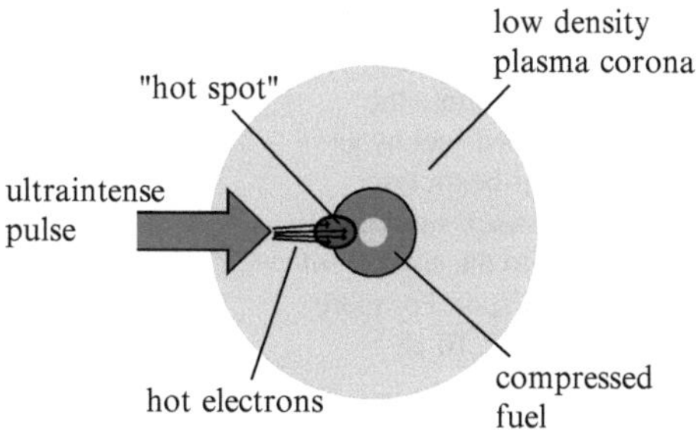

Fig. 10.6 Original fast ignition scheme. An ultraintense laser pulse is focused onto a tiny spot and generates a beam of hot electrons (with one-to-few MeV energy). The hot electrons reach the pre-compressed plasma, where they are stopped and create the hot spot

Before discussing specific schemes, let us assume that compression and ignition are driven by laser pulses of energy E_{L-c} and E_{L-ig}, respectively. The target gain can then be written as

$$G = \frac{m_{DT}\Phi Q_{DT}}{E_{L-c} + E_{L-ig}} = \frac{\Phi Q_{DT}}{\frac{u_{imp}^2}{2\eta} + \frac{E_{L-ig}}{m_{DT}}}. \qquad (10.10)$$

This equation shows that the gain increases with decreasing implosion velocity, if ignition requires (much) less energy than compression (i.e. $E_{L-ig} \ll E_{L-c}$) and Φ and η depend weakly on implosion velocity.

It is then apparent that ICF schemes separating compression and ignition can both relax stability issues and allow for the achievement of higher gain. However, this occurs at the cost of an additional heating mechanism, as it will be discussed in the next subsection.

10.4.1 Fast Ignition and Shock Ignition: A Preview

Fast ignition and shock ignition are two advanced ignition schemes, in which the hot spot is created in a previously compressed fuel. However, they differ largely both concerning the mechanism of hot spot generation and the parameters of the required laser pulse.

In fast ignition, the ignition hot spot is generated by fast particles, in turn produced by the interaction of an ultraintense laser beam [13]. Different igniting particles have been considered (see Sect. 10.6). In the original and most studied scheme an ultraintense laser beam is focused onto a pre-compressed target. Interaction of the light with the plasma corona leads to the generation of a beam of forward directed electrons with energy in the MeV range. Such hot electrons are stopped in the compressed plasma, where they create the hot spot (Fig. 10.6). Fast ignition thus relaxes compression symmetry and stability constraints, because implosion velocity is smaller than in standard ICF and no central hot spot is required. On the other hand energy has to be delivered into the (very small) hot spot volume, in the (very short)

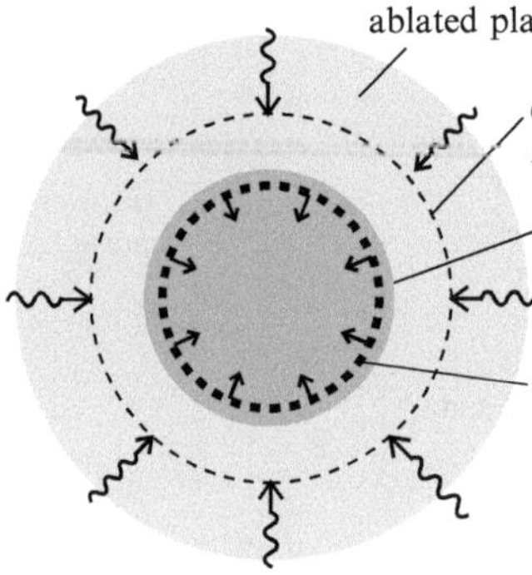

Fig. 10.7 Shock ignition: a strong, nearly spherical shock, is driven into the pre-compressed fuel by an intense multi-beam laser pulse. The shock substantially contributes to the creation of a central hot spot. For more details, see Fig. 10.18

time the fuel remains confined. As we shall see in the next section, this involves laser pulses with energy about 100 kJ and power of a few PW, focused onto spots of radius of 10–30 microns, so that the intensity is about 10^{20} W/cm^2.

In shock ignition [14], just as in fast ignition, the target is first imploded in the usual way. At a time close to implosion stagnation the target is irradiated by intense beams which generate an ablation pressure of about 300 Mbar, driving a spherical imploding strong shock wave (Fig. 10.7). Collision of this shock wave with the shock wave bouncing from the target centre leads to the formation of the igniting hot spot. In this case the laser beam power is of the order of a few hundred TW, distributed over the whole solid angle, and the corresponding intensity in the range $(0.2\text{–}1)\times 10^{16}$ W/cm^2.

10.4.2 Regimes of Laser Interaction

Examples of laser pulses for fast and shock ignition are shown in Fig. 10.8 [30–33]. They refer to the HiPER baseline target (HBT), designed at the beginning of the HiPER study to test ignition with total laser energy below 400 kJ [30]. Targets for IFE will be bigger, and pulses more energetic, but with the same qualitative features. Figure 10.9 shows the HiPER baseline target and lists parameters characterising implosion and compression [31, 33], as well as relations used for scaling the pulse when the linear dimensions of the target are scaled by a factor s.

Representative points in the intensity–wavelength plane are shown in Fig. 10.10. The compression pulse interacts in the classical regime (with absorption taking place by inverse bremsstrahlung), somewhat below the threshold for significant parametric instabilities. The shock ignition spike is absorbed mainly collisionally, but involves important laser-plasma parametric processes. They may either degrade or increase absorption, and can generate hot electrons. The fast ignition spike, instead, involves relativistic plasma physics. Most of the absorbed energy is delivered to relativistic electrons with a wide energy spectrum, with average energy of one to a few MeV.

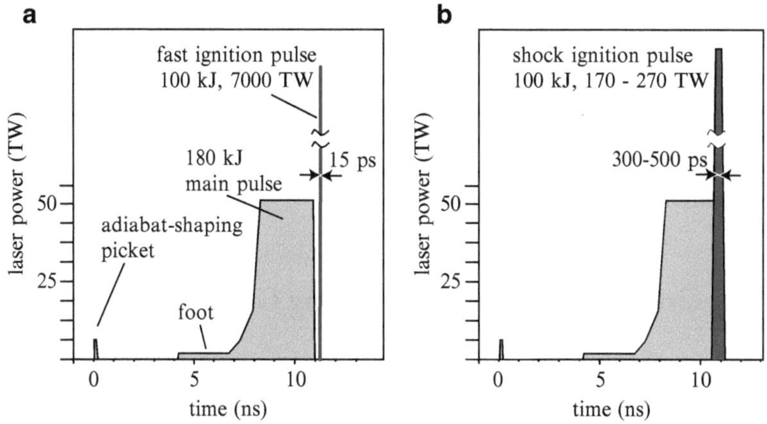

Fig. 10.8 Typical laser pulses for (**a**) fast ignition and (**b**) shock ignition: power vs time. The numbers refer to the HiPER baseline target (see Fig. 10.9), designed to test the feasibility of advanced ignition schemes at laser energy of a few hundred kJ. The wavelength of the compression pulse is $\lambda = 0.35$ μm

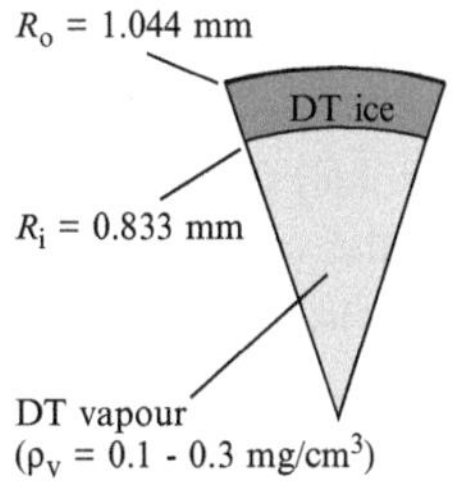

compression pulse and implosion performance	
foot power:	1.7 TW
flat-top power:	42-46 TW
absorption efficiency:	70%
imploding mass:	0.28 mg
implosion velocity:	280 - 292 km/s
in-flight isentrope:	1.2
peak <ρR>:	1.4 - 1.5 g/cm^2

scaling with size at fixed u_{imp}	
radius:	s
mass:	s^3
time:	s
energy:	s^3
power:	s^2
intensity:	s^0

Fig. 10.9 The HiPER baseline target: target section; main implosion parameters; scaling with size

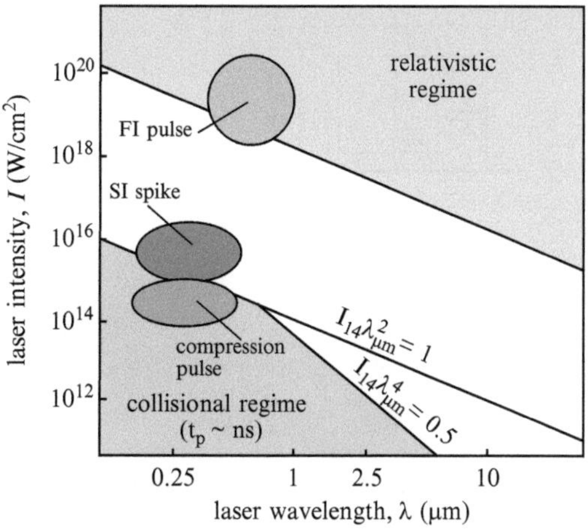

Fig. 10.10 Typical interaction conditions in the wavelength-intensity plane, for compression, shock ignition and fast ignition

It is then apparent that advanced ignition schemes have great potential advantages (lower susceptibility to RTI, allowance for low-entropy direct-drive, higher gain), but involve new aspects (in particular concerning laser interaction), that are at a preliminary stage of investigation. Some of these new issues will be briefly discussed in the following Sects. 10.6 and 10.7. In the next section, instead, we focus on the ignition process, and highlight the differences between conventional central ignition and advanced ignition.

10.5 Ignition Conditions and Limiting Gain Curves

Ignition in ICF is the process by which the hot spot, created either by hydrodynamic processes or by electron or ion heating, self-heats due to the power deposited by fusion reaction products and then drives a burn wave that propagates through the surrounding fuel [8]. This requires that the power released by the DT fusion reactions and deposited by the 3.5 MeV alpha-particles in the hot spot overcomes the losses due to thermal conduction, bremsstrahlung and mechanical work [1,2,34].

In the standard ICF scheme the igniting fuel is nearly at rest and quasi-isobaric: the central hot spot is surrounded by much denser and colder fuel, so that the pressure is nearly uniform over most of the fuel (Fig. 10.11a), and power losses by mechanical work are negligible. In fast ignition, instead, the fuel is nearly isochoric, and the pressure is much higher in the hot spot than in the colder fuel (Fig. 10.11b). The case of shock ignition is intermediate (Fig. 10.11c): the cold fuel is denser than the hot spot, but its pressure p_c is lower than that of the hot spot; typically, $\beta = p_h/p_c = 2$–4.

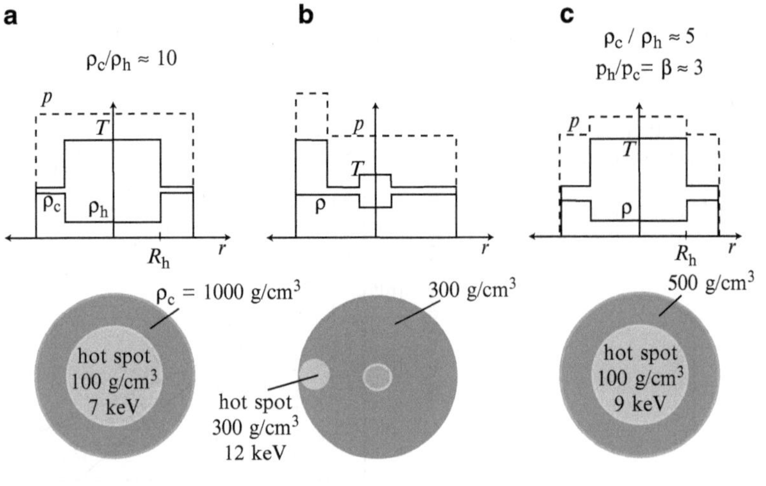

Fig. 10.11 Schematic fuel conditions at ignition. (**a**) Standard central ignition; (**b**) fast ignition and (**c**) shock ignition

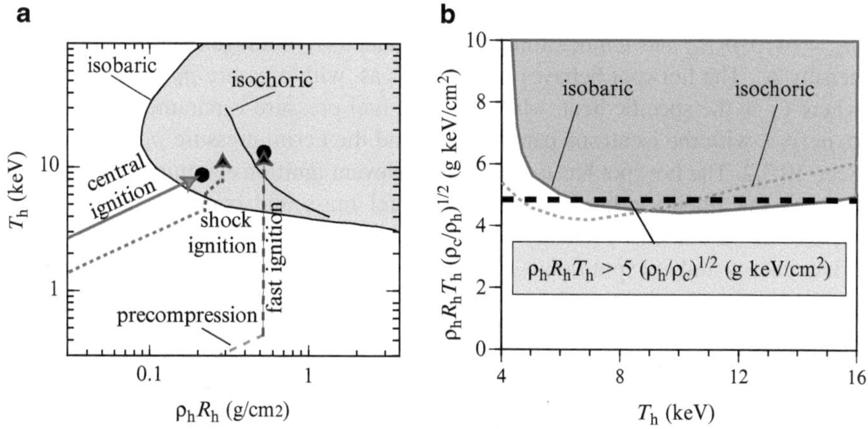

Fig. 10.12 (**a**) Curves delimiting the ignition region in the plane of hot spot confinement parameter $\rho_h R_h$ and hot spot temperature T_h, for isobaric and isochoric initial conditions. The filled circles mark optimal ignition points. (**b**) Ignition curves, $\rho_h R_h T_h \sqrt{\rho_c/\rho_h}$ vs T_h. In the temperature range 6–15 keV the thick dashed lines approximates both ignition curves

10.5.1 Ignition Conditions

Ignition conditions obtained by numerical simulations [34] and well approximated by simple models [1, 34] are shown in Fig. 10.12a in the plane $\rho_h R_h$–T_h of hot spot confinement parameter and hot spot temperature, for isobaric initial conditions and isochoric initial conditions. These curves are analogous to the Lawson curves in $n\tau$–T plane [35], often considered in magnetic fusion research [36]. The ignition condition is more severe for the isochoric case, because alpha-particle heating has to compensate large losses due to mechanical work [34]. The figure also shows the route followed to create the hot spot in the different ignition schemes.

An alternative presentation of the ignition condition is provided by Fig. 10.12b, showing the ignition domain in the plane of the temperature and of the parameter $\rho_h R_h T_h (\rho_c/\rho_h)^{1/2}$. The figure also shows a simple ignition condition [1] that approximates simulation results in the temperature interval 6–15 keV.

10.5.2 Limiting Gain Curves

Important information on the potentials of different ICF schemes is provided by a simple model, giving the target gain as a function of a small number of driver and target parameters. We summarise here its main features, while details are given both in the original publications [37,38] and in pedagogical presentations [1,21]. We assume that laser energy is delivered to the fuel with overall efficiency η. At ignition the fuel consists of a homogeneous hot spot [with radius R_h, density ρ_h, mass $m_h = (4/3)\rho_h R_h^3$ and temperature T_h] surrounded by homogeneous colder fuel at density ρ_c. The hot spot behaves as an ideal gas, with pressure $p_h = (2/3)C_v\rho_h T_h$, where C_v is the specific heat, while the cold fuel pressure is parametrised as $p_c = \alpha_c p_F(\rho_c)$, with the isentrope parameter α_c and the Fermi pressure p_F defined as in Sect. 10.3.2. The hot spot has to satisfy the relevant ignition condition (represented, e.g., by the filled circles in Fig. 10.12). Fuel mass and confinement parameter $<\rho R>$ are obtained by energy and mass conservation, for given values of the driver energy and hot spot radius. The gain $G(E_d, R_h, \alpha_c, \eta)$ is then computed using Eq. 10.1, with the fractional burn-up given by Eq. 10.2. For each choice of the parameters α_c and η one can then draw a family of gain curves $G(E_d)|_{R_h}$ at different values of R_h. [Analogously, one can generate gain curves $G(E_d)|_m$, or $G(E_d)|_{p_h}$]. It turns out that the envelope of this family of curves, i.e. the curve giving the maximum gain, or *limiting gain* that can be achieved for a given driver energy is accurately approximated by a simple expression that can be derived analytically. For isobaric and isochoric igniting fuel configurations, i.e. the configurations relevant to conventional central ignition and fast ignition, respectively, we have

$$G_{\rm lim}^{\rm isobaric} = 6{,}000\,\eta \left[\frac{\eta E_d\,[\rm MJ]}{\alpha_c^3}\right]^{0.3}, \qquad (10.11)$$

$$G_{\rm lim}^{\rm isochoric} = 19{,}000\,\overline{\eta} \left[\frac{\overline{\eta} E_d\,[\rm MJ]}{\alpha_c^3}\right]^{7/18}, \qquad (10.12)$$

where $\overline{\eta} = \eta_{L-ig}^{4/25}\eta^{21/25}$, with η_{L-ig} the coupling efficiency of the fast ignition beam [39]. It is apparent that fast ignition has potentials for achieving ignition at lower total energy, and higher gain at given energy than standard central ignition. This is because in central ignition the fuel is isobaric, with the bulk cold fuel at very high density to match the hot spot pressure. In fast ignition the cold fuel is instead at lower pressure than the hot spot, and hence has lower specific energy than the cold fuel in the isobaric case. This advantage by far out-weights the drawback of the more severe ignition condition.

For a fuel assembly as sketched in Fig. 10.11c, relevant to shock ignition, the model gives

$$G_{\rm lim} = 6{,}000\,\beta^{0.34}\eta \left[\frac{\eta E_d\,[\rm MJ]}{\alpha_c^3}\right]^{0.3}, \qquad (10.13)$$

assuming the same ignition condition as for the isobaric case and the same coupling efficiency for compression and ignition beam. We see that the gain is between the conventional case and fast ignition (Fig. 10.13).

Equations 10.11 and 10.13 clearly show the gain potentials of advanced ignition schemes. They also indicate that in any case the gain depends strongly on coupling efficiency and isentrope parameter. It is worth remarking that the model above does not consider symmetry and stability issues. Their inclusion would substantially

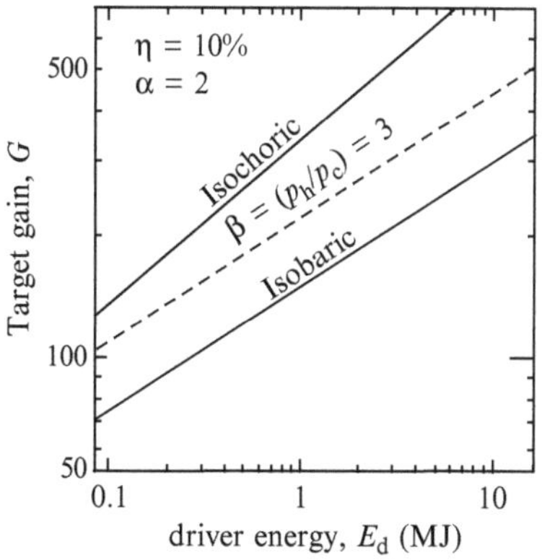

Fig. 10.13 Limiting gain vs fuel energy for initial fuel model configurations relevant to different ignition schemes: (**a**) isobaric initial conditions (relevant to conventional central ignition); (**b**) isochoric initial conditions (relevant to fast ignition); (**c**) hot spot with pressure $\beta = 3$ times larger than the cold fuel pressure (relevant to shock ignition)

modify the behaviour of the gain curves at small driver energy. Indeed, according to the model, the hot spot pressure increases and the hot spot radius decreases as the energy decreases. This implies higher implosion velocity (hence stronger RTI) and more severe symmetry requirements. In fact, symmetry and stability constraints set minimum thresholds to laser ignition energy (see Fig. 10.1).

10.6 Fast Ignition

Several fast ignition schemes have been proposed, as shown in Fig. 10.14. In all cases an ultraintense laser pulse generates a beam of fast particles, which create the ignition hot spot in the pre-compressed fuel. In this section we first discuss general beam requirements, then consider specific aspects of the different schemes.

10.6.1 General Beam Requirements

The main parameters of the energetic particle beams can in all cases be estimated by referring to the optimal ignition condition for an isochoric fuel (see Fig. 10.12a). We then have to create a hot spot with $\rho_h R_h = 0.5\ \mathrm{g/cm^2}$ and $T_h = 12$ keV, by delivering a pulsed beam onto a spot of radius $r_b = R_h$ in a time shorter than the hot spot confinement time $R_h/c_s(T_h)$. This requires an energy $E_{ig} = C_v m_h T_h = 9/\hat{\rho}^2$ kJ, where $\hat{\rho} = \rho_h/(300\ \mathrm{g/cm^3})$, i.e. the density normalised to a reference value of 300 g/cm^3. Of course, this energy will coincide with the particle beam energy only if the heating particles deliver all their energy within a depth $\approx R_h$, i.e. if their mass penetration depth $\mathscr{R} \leq 1\ \mathrm{g/cm^2}$. In this case, beam power, beam intensity, beam focus and pulse duration scale with fuel density as $W_{ig} \propto 1/\rho_h$, $I_{ig} \propto \rho_h$, $r_b \propto 1/\rho_h$ and $t_b \propto 1/\rho_h$, respectively.

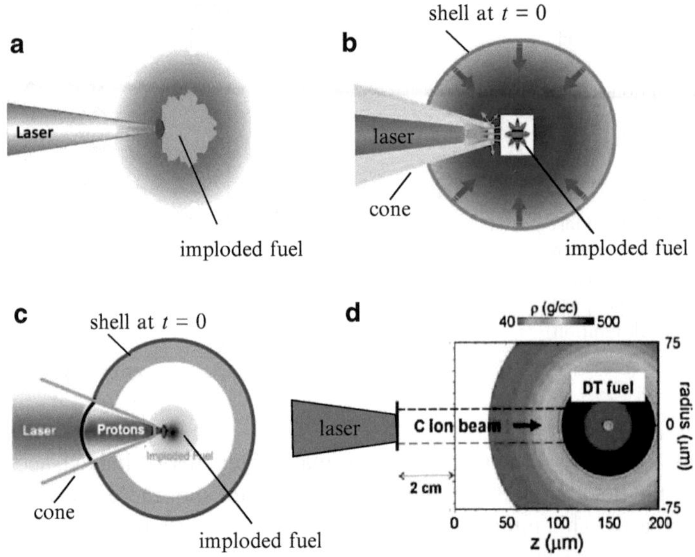

Fig. 10.14 Fast ignition schemes: (**a**) by laser accelerated electrons produced in the coronal plasma; (**b**) by electrons produced by laser interaction in the tip of a cone inserted in a spherical target; (**c**) by laser accelerated proton beams; (**d**) by laser accelerated (Carbon) ions (Courtesy of R. R. Freeman, Ohio State University, with frame (**d**) from Fernandez et al. [40])

Numerical hydrodynamic simulations nearly confirm the above dependencies on the density, but with a significant difference in the front factors. Indeed simulations considering a parallel beam of particles, with constant stopping power and assigned range, injected into an initially homogeneous DT sphere, show that the required delivered beam energy, power and intensity are [39, 41]

$$E_{\rm ig} = 18\, h(\mathscr{R})\, \hat{\rho}^{-1.85} \ \text{kJ}, \tag{10.14}$$

$$W_{\rm ig} = 0.9 \times 10^{15}\, h(\mathscr{R})\, \hat{\rho}^{-1} \ \text{W}, \tag{10.15}$$

$$I_{\rm ig} = 7.2 \times 10^{19}\, h(\mathscr{R})\, \hat{\rho}^{0.95} \ \text{W/cm}^2, \tag{10.16}$$

where the factor $h(\mathscr{R}) = \max(1, \mathscr{R}/\mathscr{R}_0)$, with $\mathscr{R}_0 = 1.2\ \text{g/cm}^2$, accounts for the energy penalty to be paid if the particle range exceeds the optimal hot spot dimension. The previous expressions apply to beams focused onto a spot of radius $r_{\rm b} \le 20\hat{\rho}^{-0.97}$ μm and to pulses of duration $t_{\rm p} \le 20\hat{\rho}^{-0.85}$ ps. Further corrections to Eqs. 10.14–10.16 apply in case of non optimally focused beams [42]. According to Eq. 10.14 the ignition energy decreases strongly with increasing density; however, this implies reducing the beam focus and increasing beam intensity. As a reference density for fast ignition study we can take $\rho = 300\ \text{g/cm}^3$, since it is unlikely that a single laser beam delivering pulses of tens of kJ can be focused onto a spot with radius smaller than 15–20 μm, or many less energetic beams can be overlapped within less than 15–20 μm.

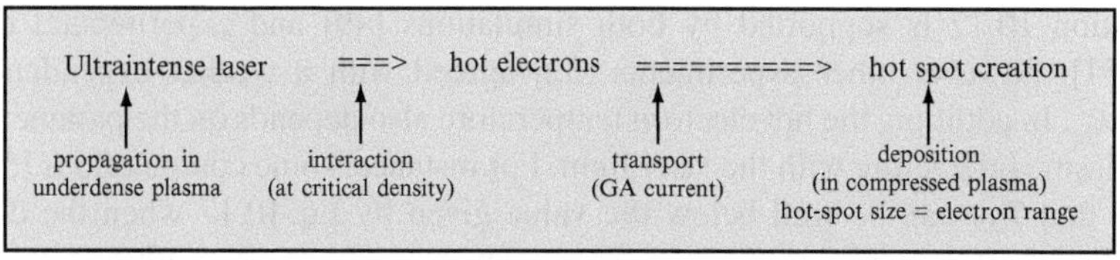

Fig. 10.15 Energy flow in the conventional fast ignition scheme, and relevant physics processes

Equation 10.14 refers to the energy E_{ig} deposited into the compressed fuel. The energy of the igniting laser pulse should be $E_{L-ig} = E_{ig}/\eta_{L-ig}$. Here the coupling efficiency η_{L-ig} accounts for laser absorption, conversion to fast particles and transport from the particle source to the compressed fuel. Assuming $\eta_{L-ig} = 20\ \%$ and the reference density of 300 g/cm^3, we see that fast ignition requires pulses of about 100 kJ and 5 PW, delivered onto a spot of 15–20 μm radius. Such values are rather extreme, but in principle achievable with chirped-pulse-amplified solid-state lasers [43, 44].

10.6.2 *Fast Ignition by Hot Electrons*

In the most studied fast-ignition (FI) scheme an ultraintense laser beam is used to generate relativistic electrons. Such electrons, accelerated around critical density, have to reach the compressed fuel, i.e. they have first to cross the distance separating the birth point from the fuel core, without too much energy loss, scattering, or defocusing. In addition, their range must match the desired hot spot size. The involved processes and main issues are schematically shown in Fig. 10.15. Space limitations do not allow us to cover all these aspects; we only deal with a few of them, mainly concerning the overall energetics. Other topics are widely discussed in the literature and nicely summarised in several review papers. See, in particular, [45] and the collection of papers in [46].

Laser absorption and hot electron production have been reviewed in [47]. Subsequent experiments for parameters relevant to fast ignition (intensity $I_{L-ig} > 10^{18}\,\mathrm{W/cm^2}$, laser wavelength of 1.06 μm and pulse duration of 1–10 ps) report conversion efficiency of 20 % ±10 %, independent of pulse duration [48].

Hot electrons are produced with a nearly 1D Maxwellian spectrum, characterised by a temperature (average kinetic energy) T_{hot}. Scaling of this temperature with laser parameters is controversial. Most target studies assume that T_{hot} is related to laser intensity and wavelength λ_{ig} by the so-called ponderomotive scaling [49]

$$T_{hot} \simeq \left[\frac{I_{L-ig}}{1.2 \times 10^{19}\mathrm{W/cm^2}} \left(\frac{\lambda_{ig}}{1.06\ \mu\mathrm{m}} \right)^2 \right]^{\frac{1}{2}} \mathrm{MeV}. \tag{10.17}$$

Equation 10.17 is supported by both simulations [49] and experimental results [50, 51]; however other experiments [52] agreed with a weaker dependence on $I_{L-ig}\lambda_{ig}^2$. In addition, the hot electron temperature also depends on the parameters of the plasma interacting with the laser light. For instance, some computations [53,54] show that T_{hot} can be well below the value given by Eq. 10.17 when the density scalelength at critical density is shorter than the light wavelength and the light interacts with an overdense plasma.

We have seen that fast ignition requires laser intensity $I_{L-ig} > 10^{20}\ \mathrm{W/cm^2}$. On the other hand we shall show in Sect. 10.6.2.1 that the hot electron temperature should be limited to 1 or perhaps 2 MeV. From the ponderomotive scaling, it follows that fast electron fast ignition either requires laser light with wavelength of $\frac{1}{2}$ μm or shorter [30] and/or the occurrence of processes leading to range shortening.

Propagation of the ultraintense laser beam in the underdense corona surrounding the imploding shell is problematic. To overcome this difficulty, in the original FI scheme (Fig. 10.14a) the igniting beam is preceded by an additional *hole boring* beam of intermediate intensity ($\approx 10^{18}\mathrm{W/cm^2}$), which drills a sort of hole in the plasma corona, due to the ponderomotive force. The FI beam then finds an open channel for propagation. An alternative scheme employs cone-inserted targets (see Fig. 10.14b). The FI beam propagates in the vacuum space inside the hollow cone. Hot electrons are produced at the interaction with the cone tip, close to the compressed fuel, and also the issue of transport from critical density to the dense fuel is therefore relaxed. The concept was successfully tested in experiments [55] with heating pulse of about 100 J. A coupling efficiency η_{L-ig} of 20–25 % was inferred. Recent experiments, with somewhat larger laser energy (300 J heating laser beam), have qualitatively confirmed the previous results, with somewhat smaller $\eta_{L-ig} = 10$–20 % [56]. However, experiments at larger scale are needed to prove that (i) the same efficiency can be achieved when multi-kJ laser pulses propagate inside the cone; (ii) the cone does not degrade target implosion symmetry [57] and (iii) mixing of cone material with the fuel is negligible.

10.6.2.1 Hot Electron Energy Deposition

Once the hot electrons reach the compressed fuel, they are slowed down mainly by Coulomb collisions. Monte Carlo simulations are used to compute energy deposition, including the effects of scattering and straggling (see Fig. 10.16). We find that in a uniform plasma of density ρ, electrons with initial energy $\mathscr{E}$ penetrate to a depth $\mathscr{R}$ (here defined as the depth at which they have deposited 90 % of their energy) [58]

$$\mathscr{R} \simeq \frac{\mathscr{E}}{1.4\ \mathrm{MeV}} \left(\frac{\rho}{300\ \mathrm{g/cm^3}} \right)^{0.06} \quad \mathrm{g/cm^2}, \tag{10.18}$$

for $1 \le \mathscr{E} \le 5$ MeV. The specific deposition does not vary substantially with depth (see Fig. 10.16b). However, we have seen that laser interaction produces electrons

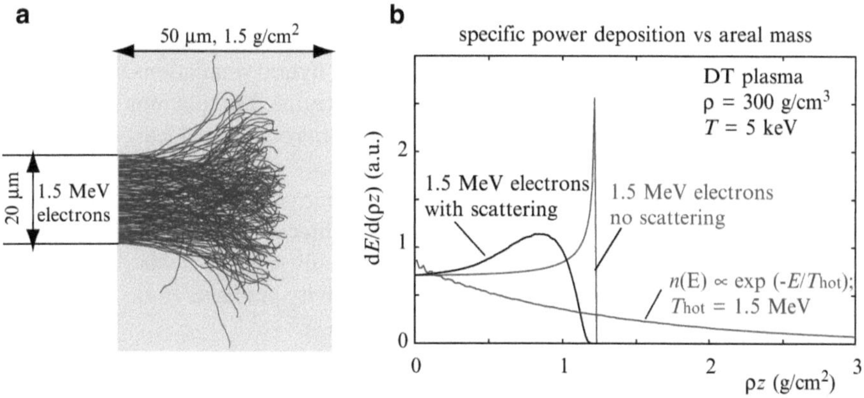

Fig. 10.16 Energy deposition by fast electron in a plasma with density $\rho = 300\ \text{g/cm}^3$ and temperature $T = 5$ keV. (**a**) Trajectories of 1.5 MeV electrons, showing the effects of straggling and scattering; (**b**) specific power deposition vs penetration for 1.5 MeV electrons and for a beam with exponential energy distribution, with temperature of 1.5 MeV (Adapted from Atzeni et al. [58])

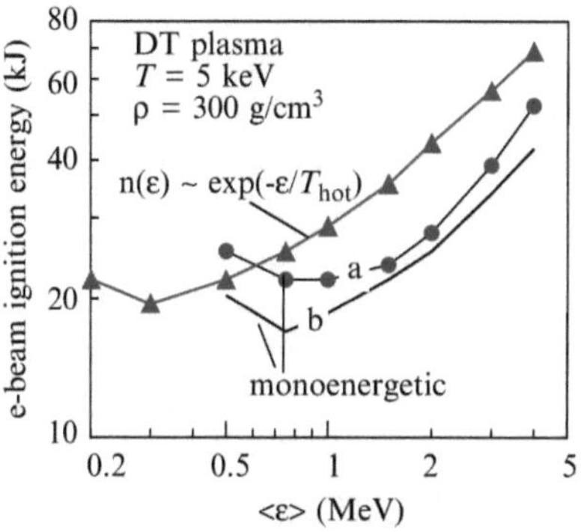

Fig. 10.17 Electron beam energy required for fast ignition vs electron kinetic energy. Results for monoenergetic beams (**a**: including scattering; **b**: neglecting scattering) and for beams with exponential energy distribution (scattering included) are presented. In all cases an initially parallel, optimally focused beam is considered (Adapted from Atzeni et al. [58])

with a nearly exponential energy spectrum. Simulations show that an electron beam with such a spectrum and temperature T_{hot} has longer average penetration than a monoenergetic beam with $\mathcal{E} = T_{\text{hot}}$ (see Fig. 10.16b).

Computations of fast ignition of pre-compressed fuels by electron beams with the above spectrum show that the beam energy required for ignition depends strongly on T_{hot}; it is close to that given by Eq. 10.14 with $h = 1$ only if $T_{\text{hot}} \leq 1$ MeV, and increases rapidly for larger T_{hot} [58, 59]; see Fig. 10.17.

All above computations assume a cylindrical electron beam. Actually, both experiments [51] and simulations [60] show large electron beam divergence.

Electrons are emitted in a cone of half-width of 30–50°, increasing with laser intensity. While earlier simulations [61] suggested that such electron beams could be pinched by self-generated magnetic fields, recent hybrid simulations with a better model of the electron source [60] and large-scale integrated simulations [62] did not confirm this effect. The minimum ignition energy turns out to be at least twice larger than previously estimated.

It is now agreed that limitation of beam divergence is essential for the feasibility of electron fast ignition [63]. To this purpose, two direction of research are pursued [63]. The first one concerns schemes with externally applied fields. The second relies on the generation of magnetic fields by resistivity gradients [64].

10.6.2.2 Gain Curves

Fast ignition research is still at a preliminary stage. The effectiveness of some of the basic processes has to be proven at full scale, before detailed target design can be performed. However, gain estimates based on simple models are useful to identify possible operating windows, potentials and issues. Parametric gain curves have been produced by several authors [30, 42, 65]. Here we refer to the curve shown in Fig. 10.1 [16]. It was obtained with a model ([30], similar to that originally proposed in [65]) that takes into account (i) laser ablation and rocket-like implosion, (ii) pressure, density and entropy multiplication at stagnation, (iii) ignition by hot electrons, (iv) ponderomotive scaling for the hot electron temperature; (v) fuel burn. The wavelengths of the compression laser λ_c and of the ignition laser λ_{ig}, the in-flight-isentrope parameter α_{if}, and the coupling efficiency of the ultraintense beam η_{ig} are free parameters of the model. Constraints are instead imposed to limit parametric instabilities ($I_{15}\lambda_\mu < 0.05$ at the compression stage) and Rayleigh-Taylor instability growth at the shell outer surface ($\Gamma_{max} < 6$). The focal spot radius is also constrained ($r_b \geq 20$ μm). The curve of Fig. 10.1 refers to $\lambda_c = 0.35$ μm, $\lambda_{ig} = 0.53$ μm, $\eta_{L-ig} = 25$ %, and $\alpha_{if} = 1.5$. The possibly of achieving $\alpha_{if} = 1.5$ and $\Gamma_{max} < 6$ simultaneously relies on the effectiveness of adiabat shaping. The above parameters are necessary to achieve significant energy gain with total laser energy below 500 kJ and ignition laser energy below 100 kJ. Following the discussion of the previous sections the most critical issues concern the coupling efficiency. In addition, the model assumes that the cone (if any) does not affect implosion symmetry and ignition conditions.

10.6.3 Fast Ignition by Proton or Light Ion Beams

Laser accelerated protons [66, 67] and light ions [40, 68] are also considered as possible igniting particles, following the experimental discovery of collimated beams of multi-MeV protons and ions from solid foils irradiated by ultraintense ps or sub-picosecond laser pulses. Indeed, protons with kinetic energy $\mathcal{E} = 5 - 15$ MeV (or Carbon ions with energy of 400–500 MeV) have range appropriate for fast ignition. When compared with electrons, such heavier particles would have the great advantage of much simpler and understood transport. However, there are still serious problems concerning laser-to-hot spot coupling efficiency.

The status of proton and ion generation by high-intensity laser irradiation of solids in 2006 was reviewed in [69]. The most studied proton acceleration scheme is the so-called target normal sheet acceleration (or TNSA); protons are emitted from the foil surface opposite to the laser irradiated one. They are extracted from the target by the electrostatic potential of the double layer which forms at the sharp solid-vacuum interface when laser accelerated hot electrons try to leave the target. The scheme produces, with efficiency up to 10 %, low emittance proton beams with a nearly exponential energy spectrum, $dn/d\mathcal{E} \propto \exp(-\mathcal{E}/T_p)$, with average energy $<\mathcal{E}> = T_p$. The target emitting surface must be protected from the radiation emitted by the corona of the fusion capsule and placed at some distance d from the compressed fuel (i.e. from the centre of the fuel capsule); see Fig. 10.14c. Due to velocity dispersion, when the beam impinges onto the fuel, the pulse length increases, $\tau \propto d/\sqrt{T_p}$ and pulse power decreases. It turns out that the minimum laser energy required for fast ignition increases significantly with d. For instance, ignition of a pre-compressed homogeneous DT sphere requires [70]

$$E_{L-ig} = \frac{220}{\eta_p/0.1}\left(\frac{d}{1\ \text{mm}}\right)^{0.7}\left(\frac{\rho}{300\ \text{g/cm}^3}\right)^{-1.3}\ \text{kJ} \qquad (10.19)$$

for distances d between 1 and 4 mm, and protons with optimal temperature $T_p = 5$–10 keV. Since ICF targets have radius of 1–2 mm, placing the TNSA target at some distance from the target (e.g., at $d = 4$–5 mm) leads to an unacceptable increase of laser energy. This severe limitation could be overcome by resorting to conically guided targets (Fig. 10.14c).

For heavier ions TNSA efficiency is definitely too low. Instead the recently discovered break-out afterburner (BOA) mechanism [71, 72] is promising. In separate small scale experiments, with laser energy of 80 J, it has been proved that BOA can produce Carbon ions with the energy (400–500 MeV), energy spread (lower than 20 %) and conversion efficiency (10 %) required for fast ignition [73]. However, considerable research is required to demonstrate the above features simultaneously at the energy and power levels relevant to fast ignition and in a realistic fusion environment. A BOA source could be employed in the fast ignition scheme shown in Fig. 10.14d [40].

10.6.4 Perspective

The gain curve for FI shown in Fig. 10.1 indicates that fast ignition has potentials for high gain at driver energy of about 1 MJ and for achieving ignition at sub-MJ energy. On the technology side, whatever the specific scheme, fast ignition requires multi-PW, hundred kJ pulses, focused onto spots of a few tens of microns. These features involve outstanding technology issues, probably including the conversion of the laser beam to the second harmonic. Regarding target physics, issues for electron fast ignition concern generation, transport and focusability of the electron beams, as well as the hydrodynamics of cone-inserted target. Probably the most critical aspect is the reduction of beam divergence, which has to rely on strong self-generated or externally applied magnetic fields. For ion fast ignition, the main problem is efficient generation of the ions from a practical target. All these aspects are currently addressed by relatively small-size experiments performed at several different institutions. Unfortunately, such main issues are hardly scalable, and no facility exists where full-scale tests can be performed. This means that target and laser specifications for a point design are difficult to establish at the moment.

10.7 Shock Ignition

In this section we consider shock ignition. We first describe the main features of the shock-assisted ignition process, and then discuss target studies.

10.7.1 Shock Assisted Ignition

The ignition condition is essentially a condition on the hot spot pressure. Indeed, the criterion

$$\rho_\mathrm{h} R_\mathrm{h} T_\mathrm{h} > 5\sqrt{\rho_h/\rho_c} \quad \mathrm{g\ keV/cm^2}, \tag{10.20}$$

(see Fig. 10.12b) can also be written as

$$p_\mathrm{h} > 520 \left[\frac{30\ \mu\mathrm{m}}{R_\mathrm{h}} \sqrt{\frac{6}{\rho_c/\rho_h}} \right] \quad \mathrm{Gbar}, \tag{10.21}$$

where the factor within brackets depends weakly on fuel mass (for a family of geometrically scaled targets, $R_h \propto m^{1/3}$) and on details of the processes leading to the formation of the hot spot. On the other hand, the hot spot pressure at stagnation is a strong function of the implosion velocity, approximately [18, 25, 74]

$$p_\mathrm{stagn} \propto u_\mathrm{imp}^3 \alpha_\mathrm{if}^{-9/10} p_\mathrm{a}^{2/5}. \tag{10.22}$$

This explains the implosion velocity threshold for central ignition discussed in Sect. 10.1 (see Eq. 10.3).

When the implosion velocity is below the ignition threshold a hot spot is still formed, but its pressure is insufficient to achieve ignition. In the shock ignition

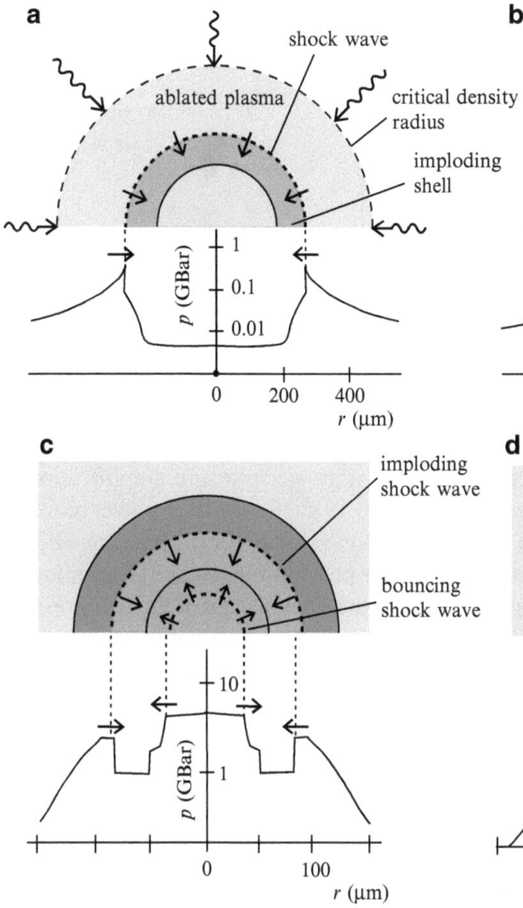

Fig. 10.18 Shock ignition: (**a**) the intense laser pulse drives an imploding shock wave; (**b**) this shock is amplified as it converges towards the centre; (**c**) a shock bounces from the centre; (**d**) the collision between the two shocks gives rise to stronger imploding and outgoing shocks

scheme the laser-produced converging shock increases the hot spot pressure to reach the ignition threshold [14, 17, 32]. The sequence of processes leading to hot spot pressure amplification is schematically shown in Fig. 10.18.

An intense laser pulse generates an ablation pressure of 200–300 Mbar, driving an imploding shock wave; this shock is amplified as it converges towards the centre (with pressure varying approximately as $p \propto 1/r$); while the imploding shock-wave progresses towards the centre, a shock bounces from the centre; the collision between the two shocks gives rise to stronger imploding and outgoing shocks. The imploding shock compresses and heats the hot spot. The final configuration at peak pressure (not shown in Fig. 10.18) is analogous to that already seen in Fig. 10.11c.

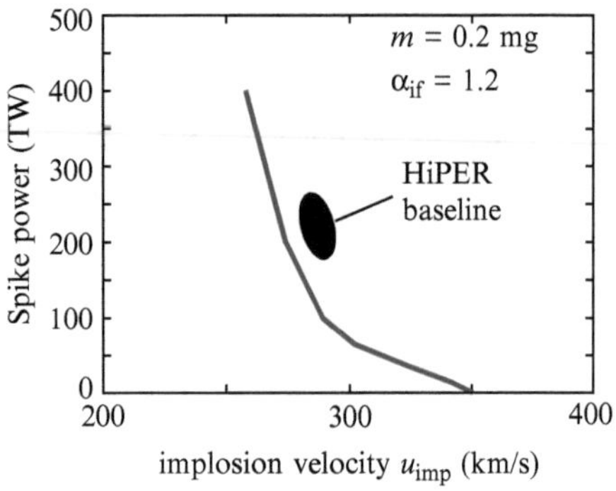

Fig. 10.19 (Additional) spike power required for ignition vs implosion velocity, for the HiPER baseline target of Figs. 10.8 and 10.9. The implosion velocity is varied by changing the flat-top power of the compression laser pulse

The required strength of the shock, and hence the intensity and of the laser pulse of course depends on the specific design. For a given target, the lower the implosion velocity the higher the laser intensity, because a greater pressure amplification is required to reach ignition. As an example, Fig. 10.19 shows the laser power required for ignition and propagating burn vs the implosion velocity for the baseline HiPER target. The target is first compressed by the laser pulse of Fig. 10.8b. The implosion velocity is varied simply changing the power of the final plateau of the compression pulse.When the implosion velocity $u_{imp} \geq 360$ km/s the target ignites without any additional pulse; this is just conventional central ignition. In the reference shock ignition HiPER design [33] the implosion velocity is about 290 km/s, and an ignition pulse of 150–250 TW is required. If the velocity is further reduced to 250 km/s, ignition is still possible, with a power as large as 400 TW.

10.7.2 A Shock Ignition Target Study

A few groups have designed shock ignition targets and analysed their performance, robustness, and scalability [14, 32, 33, 75–78]. As an example, in this subsection we summarise a few results of shock ignition studies concerning the HiPER baseline target.

10.7.2.1 One-Dimensional Simulation Results

We refer to the target shown in Fig. 10.9. The main target and pulse parameters and results of one-dimensional simulations [33] are summarised in Table 10.1. A typical implosion chart is presented in Fig. 10.20, where the trajectories of the spike-driven shock, of the bouncing shocks and of the post-collision shocks are clearly seen.

A few aspects are worth mentioning. The initial laser picket is essential to shape the adiabat and reduce RTI growth. Target compression is driven by 48 beams arranged as described in [79]. Each beam has Gaussian intensity profile

Table 10.1 Shock ignition of the HiPER baseline target (Fig. 10.9)

Laser wavelength	0.351 μm
Adiabat-shaping picket	
Peak power P_p	3.8–4.7 TW
Δt_p	250 ps
Compression pulse	
Spot width w_c	640 μm
Foot power P_c	1.7 TW
Flat-top power P_c	42–46 TW
Energy E_c	164–180 kJ
Absorption efficiency	75 %
In-flight-aspect-ratio (at $R/R_0 = 0.5$)	28
Implosion velocity	280–292 km/s
SI pulse	
Spot width w_s	400 μm
Peak power P_s	170–270 TW
Δt_s	300 ps
Energy E_s	80–135 kJ
Absorption efficiency	43 %
Time window for spike firing	120 ps at $P_s = 170$ TW 250 ps at $P_s = 270$ TW
Hot spot convergence ratio	35 for $\rho_v = 0.3$ g/cm^2 42 for $\rho_v = 0.1$ g/cm^2
1D fusion performance	
Fusion energy	$\simeq$ 20 MJ
Gain	65–80

Note: Pulse parameters and performance

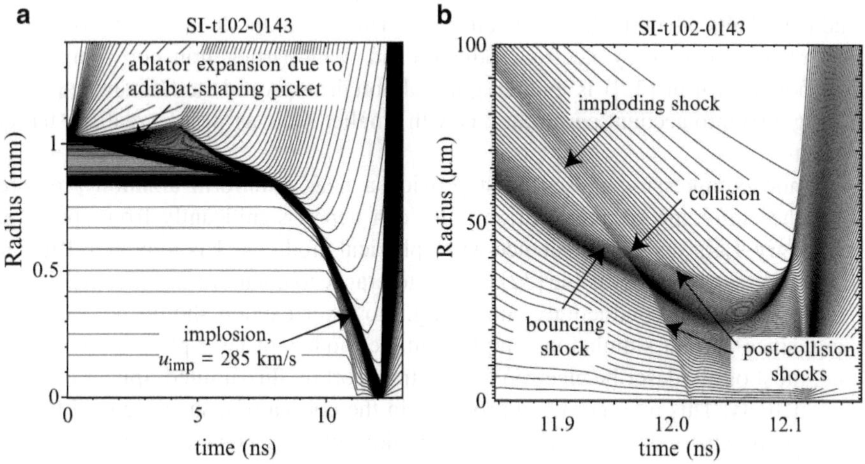

Fig. 10.20 Shock-ignited HiPER baseline target. Implosion charts (radius vs time evolution of selected Lagrangian elements: (**a**) whole evolution; (**b**) enlarged view showing shock collisions. Ignition occurs at $t = 12.1$ ns (Adapted from Atzeni et al. [33])

$\propto \exp(-r/w_c)^2$, with $w_c = 640$ μm. This scheme guarantees very good irradiation uniformity and acceptable absorption efficiency $\eta_a = 75$ %. The igniting spike beams are instead focused onto smaller spots, (with width $w_s = 400$ μm) since they hit an already shrunk shell [32]. Synchronisation of the igniting pulse with the compression pulse, required to provide proper shock amplification, is not critical. A 150–250 ps spike launch window is found, depending on the power of the final spike. A possible issue can instead come from the strong target convergence. Indeed, the hot spot convergence ratio (ratio of the initial shell radius R_0 to the hot spot radius R_h) is rather large ($R_0/R_h \geq 35$). Creation of the hot spot therefore requires very good control of implosion symmetry.

10.7.2.2 Robustness

Several aspects of target robustness, i.e. target performance sensitivity to deviation of parameters from their nominal values and/or from perfect spherical symmetry, have been studied by 2D numerical simulations, performed with the code DUED ([80] and refs. therein). The simulations [33] used a simplified description of laser irradiation scheme. The actual 48 laser beams are replaced by radial rays, with power adjusted to produce the correct 1D implosion, and with angular dependence of intensity corresponding to the initial (2D averaged) illumination spectrum generated by the actual irradiation scheme. Legendre modes up to 16 have been considered.

The simulations show that despite the very low level of irradiation nonuniformity (0.2 % rms) the compressed shell is significantly distorted, but this does not hinder ignition; see Fig. 10.21. Indeed, the growth of hot spot surface perturbations, seeded by the nonuniform irradiation, and amplified by deceleration-phase RTI, suddenly halts when the hot spot self-heats. This is mainly due to *fire polishing* of perturbations caused by electron, radiation and alpha-particle transport. This result (already shown in [32]) is encouraging, although improved studies are required, taking also into account perturbations with smaller scale as well as realistic target defects.

While compression pulses must provide a nearly uniform ablation pressure over the entire sphere, the ignition pulse can depart significantly from spherical symmetry. Large power imbalance with spherical mode $l = 1$ is tolerated. Bipolar irradiation seems also feasible [32]. Such a tolerance to nonuniform spike irradiation is due to thermal conductivity smoothing. However caution should be exercised here, since current simulations employ simple models of laser-plasma interaction and, most of all, describe electron energy transport by flux-limited Spitzer thermal conductivity. This may not be appropriate in the corona of a shock ignited target, with temperatures of a few keV. In some regions of the corona, the electron-mean-free-path becomes comparable with the electron temperature scale-length, thus invalidating classical local treatment of electron transport (see, e.g. [81]).

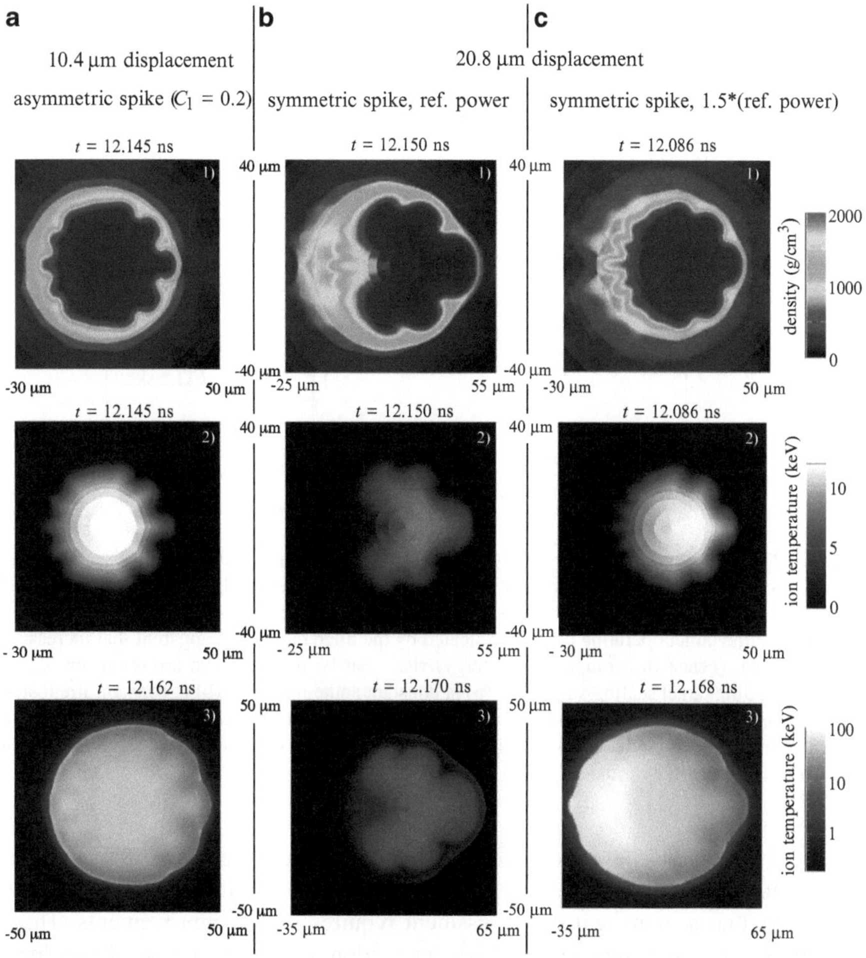

Fig. 10.21 Shock-ignited HiPER baseline target: effect of non-uniform irradiation and target mispositioning studied by 2D simulations. (**a**) Target displaced by 10.4 μm (1 % of the initial radius) and ignited by a spike with left-right power asymmetry (with $C_1 = 0.2$). (**b**) and (**c**) Target displaced by 20.8 μm and shocked by a perfectly uniform pulse, with nominal power (**b**), and with spike power increased by 50 % (**c**). In all cases the target is compressed by the nominal pulse (see main text and [33]). Cases (**a**) and (**c**) ignite and achieve more than 90 % of the 1D yield. Case (**b**) does not ignite

Spherical implosion symmetry also requires accurate positioning of the target with respect to the laser beams. Simulations show that the nominal HiPER targets tolerates displacements of no more than 15 μm, i.e. 1.5 % of the initial radius [33]. Increasing the spike power allows for somewhat larger displacement. Examples are shown in Fig. 10.21.

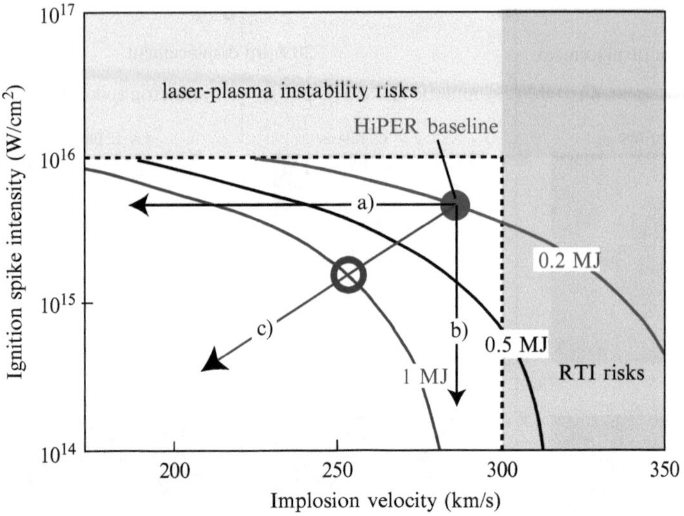

Fig. 10.22 Ignition spike intensity vs implosion velocity for different values of the laser compression incident energy. Laser spike intensity should be limited to reduce Laser-Plasma Instabilities (LPI), while the implosion velocity should be limited to reduce hydrodynamic instability. The HiPER baseline target operating point is indicated by the filled circle. It is apparent that increasing the target size (hence the compression energy) risks can be reduced. The arrows in the figure indicate possible target scaling strategies: (**a**) at constant spike intensity; (**b**) at constant implosion velocity; (**c**) by reducing both spike intensity and implosion velocity. E.g. by choosing the point indicated by the open circle at 1 MJ both LPI and RTI risks are considerably reduced (Modified version of a figure kindly provided by G. Schurtz)

The above studies have provided first indications about the robustness of shock-ignited targets, and suggested possible ways to increase it. At the same time, it appears that a more realistic assessment requires model improvements. These concern (i) the (3D) treatment of laser interaction, (ii) a description of non-local electron transport, (iii) high resolution hydrodynamics, with inclusion of small scale target-defects.

10.7.3 Target Scaling

Once a shock ignition target has been designed, scaling to larger/smaller energy (and mass) is relatively straightforward [17, 77]. Scaling to larger mass offers opportunities for more robust design. Let us call s the geometrical scale factor. At given laser wavelength and implosion velocity, target mass and compression laser energy scale as s^3, laser power as s^2 and the intensity does not change. However, since the ignition pressure depends on size (see Eq. 10.21) spike intensity decreases with s, resulting in reduced risks related to laser-plasma instabilities (see Fig. 10.22). If, instead, the spike intensity is kept constant, the implosion velocity decreases and RTI risks are reduced. Intermediate choices, as indicated by the arrow (c) in the figure, allow a simultaneous reduction of both implosion velocity and spike intensity, thus increasing the distance from both risk boundaries. In any case, increasing target size and driver energy results in increased robustness and larger gain (see the SI gain curve in Fig. 10.1).

10.7.4 Issues and Perspective

Shock ignition is a novel ICF scheme. It was proposed in 2006–2007 [14], and a first preliminary experiment was performed immediately afterwards [82]. Target studies show that the scheme is quite robust and allows for some degree of flexibility. When compared with conventional central ignition and with fast ignition, shock ignition presents a number of attractive features. Just as fast ignition it requires a reduced implosion velocity, but – differently from fast ignition – ignition relies on classical hydrodynamic processes and a single laser can probably be used for both compression and ignition.

A few issues have however to be addressed. As discussed in Sect. 10.4.2, the shock ignition spike involves a poorly explored laser-plasma interaction regime (with intensity I up to 10^{16} W/cm^2), where collisional absorption is not expected to be highly efficient and parametric instabilities can occur. First of all, it must be be demonstrated that shock pressures of 200–300 Mbar can be generated by laser pulses with the above intensity. Additional issues concern laser-plasma instabilities, which could degrade absorption and produce hot electrons causing fuel preheating. A recently emerged problem concerns energy transfer between overlapping laser beams [83]. Also electron transport in the corona [81], including possible reduction by self-generated magnetic fields requires experimental investigation. A hydrodynamic aspect deserving experimental research and simulations regards the propagation of the imploding shock wave in the inhomogeneous material of the deformed shell. Concerning LPI, recent studies based on particle-in-cell simulations [84] are encouraging. They show that after a short initial transient in which most on the incoming light is reflected by Stimulated Raman Scattering (SRS), a nearly steady state is achieved, characterised by absorption efficiency of about 70 % in cavities generated by SRS. Most of the energy is carried by electrons with energy about 30 keV, which do not penetrate up to the fuel and instead might even help the development of the shock-wave. Experiments are required to confirm these results. Results of first experiments on laser-plasma interaction under shock-ignition relevant conditions have been reported in [85].

It is important to notice that lasers producing pulses with the power and energy required for shock ignition are already available [75]. This should allow for first proving the feasibility of the strong laser-driven imploding shock. Subsequently, a roadmap for full scale demonstration at a large scale facility (NIF [11], or LMJ [86]) should be developed. Since these facilities have been designed primarily for indirect-drive ICF, they do not allow for straightforward spherically symmetric irradiation. Polar direct-drive schemes [87, 88] should be therefore designed and tested. An essential element of a roadmap for shock ignition will therefore be the development of a suitable polar direct-drive platform. Proof-of-principle scaled-down compression experiments could however be performed within a relatively short time.

10.8 Conclusions

In this chapter we have illustrated the main features of fast ignition and shock ignition ICF schemes. Such advanced ICF schemes attain the required fuel compression and hot spot creation by separate pulses. This allows to reduce the implosion velocity, thus relaxing instability issues, and potentially allowing for higher target energy gain. On the other hand, both shock ignition and fast ignition require intense (or even ultra-intense) ignition pulses. Preliminary, proof-of-principle experiments have yielded encouraging results for both schemes. However laser-plasma interaction occurs in conditions which have been not thoroughly investigated, and which at the moment cannot be accurately simulated under realistic target conditions. Extensive experiments are therefore required. The situation is however different for fast ignition and shock ignition.

Electron fast ignition relies on relativistic laser-plasma interaction, and on the generation of collimated beams of energetic hot electrons. This involves nonlinear plasma processes, which are intensively investigated at existing facilities. However, mainly due to the energy of the available ultraintense beams (limited to 1 kJ), experiments refer to conditions which cannot reproduce an ICF target environment, and scaled down experiments are not feasible. Problems related to the interaction of the ultraintense beam with underdense plasma can be partly overcome by the use of cone-inserted targets, which however make target irradiation and compression more difficult and can lead to fuel contamination. An outstanding critical physics issue concerns the excessive divergence of the laser-accelerated electron beam. Success of electron fast ignition requires substantial reduction of such a divergence; properly engineered magnetic fields are proposed to help beam collimation. Proton or ion beam fast ignition relies on much simpler particle transport. However, the overall suitability of the proposed schemes has to be proved, and, again, existing lasers only allow experiments at scales largely different from those of an ignition target. In conclusion, fast ignition has in principle great potential, but involves great uncertainties. It certainly deserves investigation, but a reliable assessment cannot be performed, and specifications for laser and target design cannot be established at present.

Shock ignition involves more conventional physics, and its main features can be addressed by scaled down experiments. Also in this case a few key aspects require experimental demonstration. However the main specific issues of the generation of the required shock pressure and limitation of parametric instabilities can be studied at existing facilities of intermediate scale. Once these have been satisfactorily addressed a point design can be performed, and a detailed roadmap for full scale demonstration can be defined. Other issues, such as control of implosion symmetry, limitation of RTI growth, reduction of laser imprint are common to any conventional direct-drive scheme. Shock ignition could then eventually be tested at full scale at NIF or LMJ, using a polar direct-drive scheme [75].

In concluding this chapter it is appropriate stressing that the advanced ignition schemes we have discussed are still ICF schemes; they have to satisfy the essential requirements discussed in Sect. 10.2.1 and have to deal with the general ICF issues illustrated in Sect. 10.3. Some of these issues are relaxed by the separation of compression and ignition, but new physics is involved. Both schemes are (much) less developed than conventional central ignition approaches, and therefore present larger risks. On the other hand, they offer challenging and fascinating opportunities to fusion physicists and engineers.

Acknowledgements Work partially supported by the Italian project PRIN 2009FCC9MS and by HiPER project and Preparatory Phase Funding Agencies (EC, MSMT and STFC). I thank A. Schiavi and A. Marocchino for continuous collaboration on several topics discussed in this paper. I also thank G. Schurtz for many discussions on advanced ignition schemes and, particularly, for communicating me a few results on shock ignition prior to publication.

References

1. S. Atzeni, J. Meyer-ter-Vehn, *The Physics of Inertial Fusion* (Clarendon, Oxford, 2004)
2. J.D. Lindl, Phys. Plasmas **2**, 3933 (1995)
3. J.D. Lindl, *Inertial Confinement Fusion: The Quest for Ignition and High Gain Using Indirect Drive* (Springer and AIP, New York, 1997)
4. J.W. Hogan (ed.), *Energy from Inertial Fusion* (International Atomic Energy Agency, Vienna, 1995)
5. M.E. Cuneo et al., Plasma Phys. Control. Fusion **48**, R1 (2006)
6. C. Olson et al., Fusion Sci. Technol. **47**, 633 (2005)
7. B.G. Logan, L.J. Perkins, J.J. Barnard, Phys. Plasmas **15**, 072701 (2008)
8. J.H. Nuckolls, L. Wood, A. Thiessen, G.B. Zimmermann, Nature **239**, 139 (1972)
9. S.W. Haan et al., Phys. Plasmas **18**, 051001 (2011)
10. J.D. Lindl, E.I. Moses, Phys. Plasmas **18**, 050901 (2011)
11. G.H. Miller, E.I. Moses, C.R. Wuest, Nucl. Fusion **44**, S228 (2004)
12. L.J. Suter et al., Phys. Plasmas **7**, 2092 (2000)
13. M. Tabak et al., Phys. Plasmas **1**, 1626 (1994)
14. R. Betti et al., Phys. Rev. Lett. **98**, 155001 (2007)
15. S.E. Bodner et al., Phys. Plasmas **7**, 2298 (2000)
16. S. Atzeni, Plasma Phys. Control. Fusion **51**, 124029 (2009)
17. M. Lafon, J. Ribeyre, G. Schurtz, Phys. Plasmas **17**, 052704 (2010)
18. M.C. Herrmann, M. Tabak, J.D. Lindl, Nucl. Fusion **41**, 99 (2001)
19. J.D. Lindl et al., Phys. Plasmas **11**, 339 (2004)
20. S.E. Bodner, J. Fusion Energy **1**, 221 (1981)
21. M.D. Rosen, Phys. Plasma **6**, 1690 (1999)
22. C. Garban-Labaune et al., Phys. Rev. Lett. **48**, 1018 (1982)
23. W.L. Kruer, *The Physics of Laser Plasma Interactions* (Addison Wesley, Redwood City, 1988)

24. Ya.B. Zel'dovich, Yu.P. Raizer, *Physics of Shock Waves and High Temperature Hydrodynamic Phenomena* (Academic, New York, 1967)
25. J. Meyer-ter-Vehn, C. Schalk, Z. Naturf. **37a**, 955 (1982)
26. G.I. Taylor, Proc. R. Soc. (London) **A201**, 159 (1950)
27. K. Anderson, R. Betti, Phys. Plasmas **11**, 5 (2004)
28. V.N. Goncharov et al., Phys. Plasmas **10**, 1906 (2003)
29. R. Betti, C. Zhou, Phys. Plasmas **12**, 110702 (2005)
30. S. Atzeni, A. Schiavi, C. Bellei, Phys. Plasmas **14**, 052702 (2007)
31. X. Ribeyre et al., Plasma Phys. Control. Fusion **50**, 025007 (2008)
32. X. Ribeyre et al., Plasma Phys. Control. Fusion **51**, 015013 (2009)
33. S. Atzeni, A. Schiavi, A. Marocchino, Plasma Phys. Control. Fusion **53**, 035010 (2011)
34. S. Atzeni, Jpn. J. Appl. Phys. **34**, 1980 (1995)
35. J.D. Lawson, Proc. R. Soc. (London) **70**, 6 (1957)
36. J. Wesson, *Tokamaks*, 3rd edn. (Clarendon, Oxford, 2003)
37. J. Meyer-ter-Vehn, Nucl. Fusion **22**, 561 (1982)
38. M.D. Rosen, J.D. Lindl, Laser Program Annual Report, Report UCRL-520021-83, 3.5 (1984)
39. S. Atzeni, Phys. Plasmas **6**, 3316 (1999)
40. J.C. Fernandez et al., Nucl. Fusion **49**, 065004 (2009)
41. S. Atzeni, M. Tabak, Plasma Phys. Control. Fusion **47**, B769 (2005)
42. M. Tabak et al., Fusion Sci. Technol. **49**, 254 (2006)
43. D. Strickland, G. Mourou, Opt. Commun. **55**, 447 (1985)
44. M. Perry, G. Mourou, Science **264**, 917 (1994)
45. M.H. Key, Phys. Plasmas **14**, 055502 (2007)
46. E.M. Campbell, R.R. Freeman, K.A. Tanaka (guest editors), Fusion Sci. Technol. **49**, 249 (2006)
47. J.R. Davies, Plasma Phys. Control. Fusion **51**, 014006 (2009)
48. P.M. Nilson et al., Phys. Rev. Lett **105**, 235001 (2010)
49. S.C. Wilks et al., Phys. Rev. Lett. **69**, 1383 (1992)
50. G. Malka, J.L. Miquel, Phys. Rev. Lett. **77**, 75 (1996)
51. J.S. Green et al., Phys. Rev. Lett. **100**, 0150032 (2010)
52. F.N. Beg et al., Phys. Plasmas **4**, 448 (1997)
53. S.C. Wilks, W.L. Kruer, IEEE J. Quantum Electron. **33**, 1954 (1997)
54. B. Chrisman, Y. Sentoku, A.J. Kemp, Phys. Plasmas **15**, 056309 (2008)
55. R. Kodama et al., Nature **418**, 933 (2002)
56. H. Shiraga et al., Plasma Phys. Control. Fusion **53**, 124029 (2011)
57. S.P. Hatchett et al., Fusion Sci. Technol. **49**, 327 (2006)
58. S. Atzeni, A. Schiavi, J.R. Davies, Plasma Phys. Control. Fusion **51**, 015016 (2009)
59. S. Atzeni et al., Phys. Plasmas **15**, 056311 (2008)
60. A. Debayle, J.J. Honrubia, E.D. D'Humieres, V.T. Tikhonchuk, Plasma Phys. Control. Fusion **53**, 124024 (2011)
61. J.J. Honrubia, J. Meyer-ter-Vehn, Plasma Phys. Control. Fusion **51**, 014008 (2009)
62. D.J. Strozzi et al., EPJ Web of Conf. Preprint: arXiv:1111.5089v1 (2012)
63. R.R. Freeman, Presentation at the National Academy Review on *Prospects for Inertial Confinement Fusion Energy Systems* (2011); fire.pppl.gov/IFE_NAS3_Fast_Ignite_Freeman.pdf
64. B. Ramakrishna et al., Phys. Rev. Lett. **105**, 135001 (2010)
65. M. Tabak, D. Callahan, Nucl. Instr. Meth. Phys. Res. **A544**, 48 (2005)
66. M. Roth et al., Phys. Rev. Lett. **86**, 436 (2001)
67. M.H. Key et al., Fusion Sci. Technol. **49**, 440 (2006)
68. V.Yu. Bychenkov et al., Plasma Phys. Rep. **27**, 1017 (2001)
69. M. Borghesi et al., Fusion Sci. Technol. **49**, 412 (2006)
70. S. Atzeni, M. Temporal, J.J. Honrubia, Nucl. Fusion **42**, L1 (2002)
71. L. Yin, B.J. Albright, B.M. Hegelich, J. Fernandez, Laser Part. Beams **24**, 291 (2006)
72. B.J. Albright et al., Phys. Plasmas **14**, 094502 (2007)

73. B.M. Hegelich et al., Nucl. Fusion **51**, 083011 (2011)
74. A. Kemp, J. Meyer-ter-Vehn, S. Atzeni, Phys. Rev. Lett **86**, 3336 (2001)
75. L.J. Perkins et al., Phys. Rev. Lett. **103**, 045004 (2009)
76. A.J. Schmitt et al., Phys. Plasmas **17**, 042701 (2010)
77. X. Ribeyre et al., Plasma Phys. Control. Fusion **51**, 124030 (2009)
78. S. Atzeni, G. Schurtz, Proc. SPIE **8080**, 808022 (2011)
79. S. Atzeni et al., Nucl. Fusion **49**, 055008 (2009)
80. S. Atzeni et al., Comput. Phys. Commun. **169**, 153 (2005)
81. A.R. Bell, M. Tzoufras, Plasma Phys. Control. Fusion **53**, 045010 (2011)
82. W. Theobald et al., Phys. Plasmas **15**, 056306 (2008)
83. W.L. Kruer et al., Phys. Plasmas **3**, 382 (1996)
84. O. Klimo et al., Plasma Phys. Control. Fusion **52**, 055013 (2010)
85. S. Depierreux et al., Plasma Phys. Control. Fusion **53**, 124034 (2011)
86. J. Ebrardt, J.M. Chaput, J. Phys. Conf. Ser. **244**, 032017 (2010)
87. S. Skupsky et al., Phys. Plasmas **11**, 2763 (2004)
88. J.A. Marozas et al., Phys. Plasmas **13**, 056311 (2006)

Part IV
Laser-Plasma Particle and Radiation Sources

Chapter 11
Laser Plasma Accelerators

Victor Malka

Abstract The continuing development of powerful laser systems has permitted to extend the interaction of laser beams with matter far into the relativistic domain, and to demonstrate new approaches for producing energetic particle beams. The extremely large electric fields, with amplitudes exceeding the TV/m level, that are produced in plasma medium are of relevance particle acceleration. Since the value of this longitudinal electric field, 10,000 times larger than those produced in conventional radio-frequency cavities, plasma accelerators appear to be very promising for the development of compact accelerators. The incredible progresses in the understanding of laser plasma interaction physic, allows an excellent control of electron injection and acceleration. Thanks to these recent achievements, laser plasma accelerators deliver today high quality beams of energetic radiation and particles. These beams have a number of interesting properties such as shortness, brightness and spatial quality, and could lend themselves to applications in many fields, including medicine, radio-biology, chemistry, physics and material science, security (material inspection), and of course in accelerator science.

11.1 Introduction

In 1979 Tajima and Dawson [1], on the basis of theoretical work and simulations, have shown that an intense electromagnetic pulse can create a weak of plasma oscillation through the non linear ponderomotive force, demonstrating how relativistic plasma can be suitable for the development of compact accelerators.

V. Malka (✉)
Laboratoire d'Optique Appliquée, ENSTA Paris Tech, CNRS,
Ecole Polytechnique, UMR 7639, 91761 Palaiseau, France
e-mail: victor.malka@ensta.fr

P. McKenna et al. (eds.), *Laser-Plasma Interactions and Applications*,
Scottish Graduate Series, DOI 10.1007/978-3-319-00038-1_11,

In the proposed schemes, relativistic electrons were injected externally and were accelerated through the very high electric field (GV/m) sustained by relativistic plasma waves driven by lasers. In this former article, the authors have proposed two schemes called the laser beat wave and the laser wakefield. Several experiments have been performed in the beginning of the 1990s following their idea, and injected electrons in the few MeV level have indeed been accelerated by electric fields in the GV/m range in a plasma medium using either the beat wave or the laser wakefield scheme. These first experiments have shown that it was possible to use plasma medium to accelerate electrons. With the development of more powerful lasers, much higher electric fields were achieved, from few GV/m to more than one TV/m. A major breakthrough, was obtained in 1994 at the Rutherford Appleton Laboratory, where electrons from the plasma itself were trapped and accelerated [2]. In this relativistic wave breaking limit, the amplitude of the plasma wave was so large, that copious number of electrons were trapped and accelerated in the laser direction, producing an energetic electron beam. Few hundreds of GV/m electric field were at this time measured. The corresponding mechanism is called the self-modulated laser wakefield, an extension of the forward Raman instability at relativistic intensities. In those experiments, the electron beam had a Maxwellian-like distribution as it is expected from random injection processes in relativistic plasma waves. These first beams did not compare well to beams produced by conventional accelerators. To control or to shape the electron beam distribution, one has to reduce electrons injection to a very small volume of the phase space. In practice, this means that injected electrons must have a duration much shorter that the plasma period, i.e. much less than ten femtoseconds. This was done in the breakthrough experiments in 2004 when three groups produced quite simultaneously electron beams with quasi-monoenergetic distributions [3–5] demonstrating experimentally the bubble/blowout regime. This bubble regime is reached when the laser power is high enough and when the laser pulse length and waist match with the plasma wavelength. When these conditions are met, the laser ponderomotive force expels radially the plasma electrons and leaves a cavitated region behind the pulse. Electrons are progressively injected at the back of this cavity forming a dense electron beam in the cavity. The increasing charge of the forming electron beam progressively reduces the electric value and the injection process eventually stops, leading to the formation of a quasi-monoenergetic electron beam. These results have had a significant impact in the accelerators community. Nevertheless, the shot to shot reproducibility was not so good and the control of the beam parameters was not possible. In 2006, stable and tunable quasi-monoenergetic electron beams were measured by using two laser beams in the colliding scheme with a counter propagating geometry. The use of two laser beams instead of one offers more flexibility and enables one to separate the injection from the acceleration process[6]. The first laser beam the pump beam (the injection beam) is used to heat electrons during its collision with the pump beam. As a consequence, electrons gain enough momentum to 'catch' the relativistic plasma wave to be efficiently accelerated.

11.2 Laser and Plasma Parameters

11.2.1 Laser Parameters

The first important parameter, that we consider, is the laser normalised vector potential a_0, which is related to the maximal intensity of the laser pulse I_0 by

$$a_0 = \left(\frac{e^2}{2\pi^2 \varepsilon_0 m_e^2 c^5} \lambda_0^2 I_0 \right)^{1/2} \tag{11.1}$$

where c is the celerity of light, ε_0 the permittivity of vacuum, e the electron charge and m_e its mass, and λ_0 the laser wavelength.

For a Gaussian laser beam in space and time, laser peak intensity is given by

$$I_0 = \frac{2P}{\pi \mathrm{w}_0^2} \tag{11.2}$$

with $P = 2\sqrt{\frac{\ln 2}{\pi}} \frac{E}{\tau_0} \sim \frac{E}{\tau_0}$, where E is the energy contained in the pulse and τ_0 is the pulse duration at full width at half maximum (FWHM), and w_0 is the waist of the focal spot (the radius at $1/e$ of the electric field).

When a_0 exceeds unity, the oscillations of an electron in the laser field become relativistic. In laser plasma accelerators the motion of the electrons is mostly relativistic. For a visible laser light intensity $I_0 = 3 \times 10^{18}$ W/cm^2, corresponds a $a_0 = 1.3$.

An electron submitted to an electromagnetic field oscillates at the laser frequency, and for finite laser pulse, in addition to this high frequency motion, electron moves away from the high intensity region. This motion results from the ponderomotive force. The mathematical expression of this force is deduced from the electron equation of motion in the laser field averaged over an optical cycle. This force repels electrons from region where laser intensity gradient is large. The ponderomotive force derives from a ponderomotive potential which is written as follows:

$$\phi_p = \frac{I}{2cn_c} = \frac{e^2 E^2}{4m_e \omega_0^2} \tag{11.3}$$

For an intensity $I_0 = 1 \times 10^{19}$ W/cm^2 and a wavelength 1 μm, one obtains a ponderomotive potential of $\phi_p = 1$ MeV.

11.2.2 Plasma Parameters

A plasma is a state of matter made of free electrons, totally or partially ionised ions and neutral atoms or molecules, the whole medium being globally neutral.

Let's assume an initially uniform, non-collisional plasma in which a slab of electron is displaced from the equilibrium position. The restoring force which applies on this electron slab, drives them towards the equilibrium position. For the time scale corresponding to the electron motion, one neglects the motion of the ions because of the inertia. This gives in the end oscillations around the equilibrium position at a frequency called the electron plasma frequency ω_{pe}:

$$\omega_{pe} = \sqrt{\frac{n_e e^2}{m_e \varepsilon_0}} \tag{11.4}$$

where n_e is the unperturbed electron density.

This frequency has to be compared to the laser frequency: if $\omega_{pe} < \omega_0$ then the characteristic time scale of the plasma is longer than the optical period of the incoming radiation. The medium can't stop the propagation of the electromagnetic wave. The medium is said to be transparent or under-dense. On the opposite, when $\omega_{pe} > \omega_0$ then the characteristic time scale of the electrons is fast enough to adapt to the incoming wave and to reflect totally or partially the radiation. The medium is said to be overdense.

These two domains are separated at frequency ω_0, which corresponds to the critical density, $n_c = \omega_0^2 m_e \varepsilon_0 / e^2$. For a wavelength $\lambda_0 = 820$ nm, one obtains $n_c = 1.7 \times 10^{21}$ cm^{-3}. The typical range of electron density of laser plasma accelerators, with current laser technology, is $[10^{17}$–10^{20} cm$^{-3}]$.

11.2.3 Electric Field of the Plasma Wave

For a periodic sinusoidal perturbation of the electron plasma density in a uniform ion layer, the density perturbation δn is written:

$$\delta n = \delta n_e \sin(k_p z - \omega_p t) \tag{11.5}$$

where ω_p and k_p are the angular frequency and the wave number of the plasma wave.

This density perturbation leads to a perturbation of the electric field $\delta \mathbf{E}$ via the Poisson equation:

$$\nabla . \delta \mathbf{E} = -\frac{\delta n \, e}{\varepsilon_0} \tag{11.6}$$

This gives

$$\delta \mathbf{E}(z,t) = \frac{\delta n_e \, e}{k_p \varepsilon_0} \cos(k_p z - \omega_p t) \mathbf{e_z} \tag{11.7}$$

To describe electron acceleration in a relativistic plasma wave, i.e. with a phase velocity close to the speed of light $v_p = \omega_p / k_p \sim c$. Let $E_0 = m_e c \omega_{pe} / e$. Let's consider an electric field:

$$\delta \mathbf{E}(z,t) = E_0 \frac{\delta n_e}{n_e} \cos(k_p z - \omega_p t)\mathbf{e_z} \tag{11.8}$$

One notice that the electric field is dephased by $-\pi/4$ with respect to the electron density.

11.2.4 Electron Energy Gain in the Plasma Wave

Let's now describe what happens to an electron placed in this electric field. The goal is to obtain the required conditions for trapping to occur. The following variables are introduced to describe the electron in the laboratory frame: z the position, t the associated time, β the velocity normalised to c, $\gamma = 1/\sqrt{1-\beta^2}$ the associated Lorentz's factor. In the frame of the plasma wave, let z', t', β' and γ' represent the equivalent quantities.

The frame linked to the plasma wave is in uniform constant translation at speed $v_p = \beta_p c$. One writes γ_p the Lorentz's factor associated to this velocity. The Lorentz's transform allows to switch from the laboratory frame to the wave frame :

$$\begin{cases} z' = \gamma_p(z - v_p t) \\ t' = \gamma_p(t - \frac{v_p}{c}x) \\ \gamma' = \gamma\gamma_p(1 - \beta.\beta_\mathbf{p}) \end{cases} \tag{11.9}$$

In this new frame, without magnetic field, the electric field remains unchanged $\delta \mathbf{E}'$

$$\delta \mathbf{E}'(z') = \delta \mathbf{E}(z,t) = E_0 \frac{\delta n_e}{n_e} \cos(k_p z'/\gamma_p)\mathbf{e_z} \tag{11.10}$$

Consequently, in terms of potential, the electric field is derived from potential Φ' defined by

$$\mathbf{F} = -e\delta\mathbf{E}' \equiv -\nabla'\Phi' \tag{11.11}$$

This leads to

$$\Phi'(z') = mc^2\gamma_p \frac{\delta n_e}{n_e} \sin(k_p z'/\gamma_p) \equiv mc^2\phi'(z') \tag{11.12}$$

Finally, one writes the total energy conservation for the particle in this frame compared to the initial energy at the injection time (labelled with subscript 0):

$$\gamma'(z') + \phi'(z') = \gamma_0'(z_0') + \phi_0'(z_0') \tag{11.13}$$

Equation 11.13 gives the relation between the electron energy and its position in the plasma wave. Finally, the reverse Lorentz's transform gives this energy in the laboratory frame.

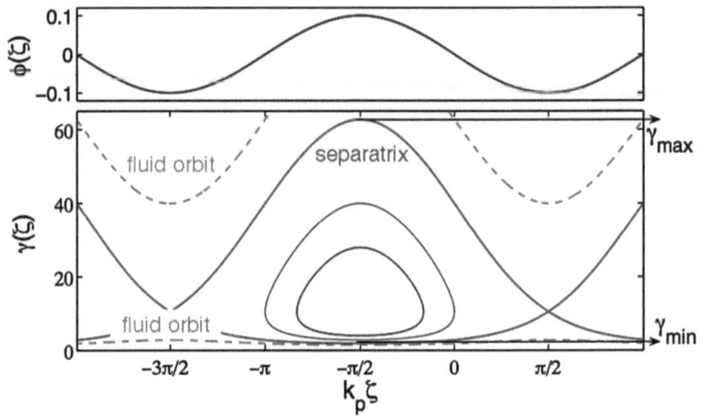

Fig. 11.1 *Up*: Potential in phase space. *Down*: Trajectory of an electron injected in the potential of the plasma wave in the frame of the wave with the fluid orbit (dashed line), the trapped orbit and, in between, the separatrix

For $\beta' > 0$, the scalar product in Eq. 11.9 is positive

$$\gamma = \gamma'\gamma_p + \sqrt{\gamma'^2 - 1}\sqrt{\gamma_p^2 - 1} \tag{11.14}$$

For $\beta' < 0$, scalar product in Eq. 11.9 is negative

$$\gamma = \gamma'\gamma_p - \sqrt{\gamma'^2 - 1}\sqrt{\gamma_p^2 - 1} \tag{11.15}$$

Figure 11.1 represents an example of electron trajectory in a plasma wave. In this phase space, the closed orbits correspond to trapped particles. Open orbits represent untrapped electrons, either because the initial velocity is too low, or to high. The curve which separates these two regions is called the separatrix. This separatrix gives the minimum and maximum energies for trapped particles. This is comparable to the hydrodynamic case, where a surfer has to crawl to gain velocity and to catch the wave. In terms of relativistic factor, γ has to belong to the interval $[\gamma_{min}; \gamma_{max}]$ with:

$$\begin{cases} \gamma_{min} = \gamma_p(1 + 2\gamma_p\delta) - \sqrt{\gamma_p^2 - 1}\sqrt{(1 + 2\gamma_p\delta)^2 - 1} \\ \gamma_{max} = \gamma_p(1 + 2\gamma_p\delta) + \sqrt{\gamma_p^2 - 1}\sqrt{(1 + 2\gamma_p\delta)^2 - 1} \end{cases} \tag{11.16}$$

where $\delta = \delta n_e / n_e$ is the relative amplitude of the density perturbation.

One deduces that the maximum energy gain ΔW_{max} for a trapped particle is reached for a closed orbit with maximum amplitude. This corresponds to the injection at γ_{min} on the separatrix and its extraction at γ_{max}. The maximum energy gain is then written

$$\Delta W_{max} = (\gamma_{max} - \gamma_{min})mc^2 \tag{11.17}$$

For an electron density much lower than the critical density $n_e \ll n_c$, one has $\gamma_p = \omega_0/\omega_p \gg 1$ and

$$\Delta W_{max} = 4\gamma_p^2 \frac{\delta n_e}{n_e} mc^2 \tag{11.18}$$

For electron travelling along the separatrix, the time necessary to reach maximal energy is infinite because there exists a stationary point at energy γ_p. On other closed orbits, the electron successively gains and looses energy during its rotation of the phase space. In order to design an experiment, one needs an estimation of the distance an electron travels before reaching maximal energy gain. This length, which is called the dephasing length L_{deph}, corresponds to a phase rotation of $\lambda_p/2$ in the phase space. In order to have a simple analytical estimation, one can assume that the energy gain is small compared to the initial energy of the particle and that the plasma wave is relativistic $\gamma_p \gg 1$, then the dephasing length is written:

$$L_{deph} \sim \gamma_p^2 \lambda_p \tag{11.19}$$

This concept of dephasing length in a 1D case can be refined in a bi-dimensional case. Indeed, if one also takes into account the transverse effects of the plasma wave, this one is focusing or defocusing for the electrons along their acceleration [7]. Because these transverse effects are shifted by $\lambda_p/4$ with respect to the pair acceleration/deceleration, the distance over which the plasma wave is both focusing and accelerating is restricted to a rotation of $\lambda_p/4$ in phase space, which decreases by a factor 2 the dephasing length L_{deph}.

$$L_{deph}^{2D} \sim \gamma_p^2 \lambda_p/2 \tag{11.20}$$

In these formulae, one has considered a unique test electron, which has no influence on the plasma wave. In reality, a massive trapping of particles modifies electric fields and distorts the plasma wave. This is called space-charge or beam loading effects (which results from the Coulomb repulsion force). Finally, this linear theory is difficult to apply to highly non-linear regimes which are explored experimentally.

11.3 Experimental Results

11.3.1 Linear Regime

The ponderomotive force of the laser excites a longitudinal electron plasma wave with a phase equal to the group velocity of the laser close to the speed of light. Two regimes have been proposed to excite relativistic electron plasma wave.

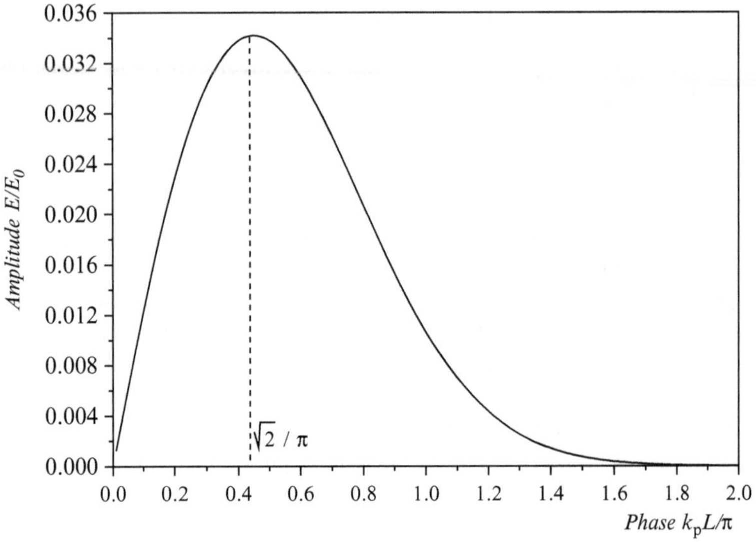

Fig. 11.2 Amplitude of the electric field as function of the length of a Gaussian laser pulse for a normalised vector potential $a_0 = 0.3$

11.3.1.1 Laser Wakefield

In the standard laser wakefield acceleration (LWFA) approach, a single short laser pulse excites the relativistic electron plasma wave. As the ponderomotive force associated with the longitudinal gradient of the laser intensity exerts two successive pushes in opposite directions on the plasma electrons, the excitation of the electron plasma wave is maximum when the laser pulse duration is of the order of $1/\omega_p$. For a linearly polarised laser pulse with full width at half maximum (FWHM) $\sqrt{2\ln 2}L$ (in intensity), the normalised vector potential, called also the force parameter of the laser beam, is written:

$$a(z,t) = a_0 \exp\left[-\left(\frac{k_0 z - \omega_0 t}{\sqrt{2}k_p L}\right)^2\right] \tag{11.21}$$

In the linear regime $a_0 \ll 1$, the electronic response obtained behind a Gaussian laser pulse can be easily calculated [8]. In this case, the longitudinal electric field is given by:

$$\mathbf{E}(z,t) = E_0 \frac{\sqrt{\pi}a_0^2}{4} k_p L \exp(-k_p^2 L^2/4)\cos(k_0 z - \omega_0 t)\mathbf{e_z} \tag{11.22}$$

Equation 11.22 explicitly shows the dependence of the amplitude of the wave with the length of the exciting pulse. In particular, the maximal value for the amplitude is obtained for a length $L = \sqrt{2}/k_p$ (see Fig. 11.2).

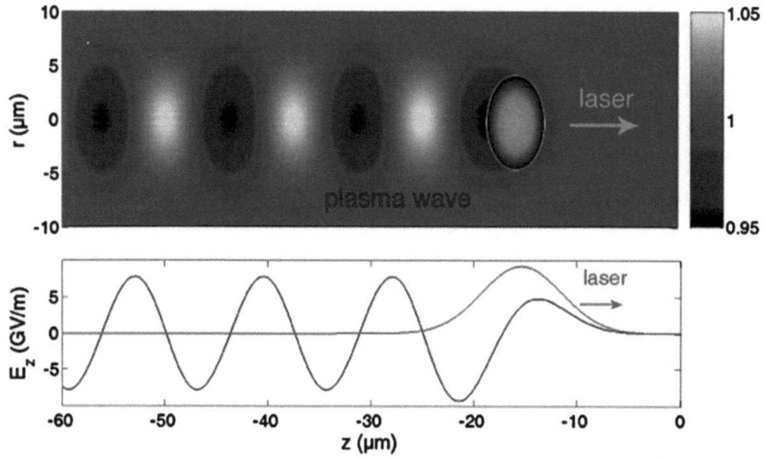

Fig. 11.3 Density perturbation (*top*) and electric field (*bottom*) produced in the linear regime

In Fig. 11.3 the density perturbation and the corresponding electric field produced by a 30 fs laser pulse at low intensity $I_{laser} = 3 \times 10^{17}$ W/cm^2 are shown. One can note that in the linear regime the electric field has sinusoidal shape and reach maximal values of a few GV/m. For example, for an electron density $n_e = 10^{19}$ cm^{-3}, the optimal pulse duration equals $L = 2.4$ µm (equivalent to a pulse duration $\tau = 8$ fs). For $a_0 = 0.3$, the maximal electric field is in the GV/m range.

In experiments carry out at LULI, relativistic plasma waves with 1 % amplitude were demonstrated. 3 MeV electron beam has been injected in this relativistic plasma wave and some of them were accelerated up to 4.6 MeV [9]. The electron spectra has a broad energy distribution with a Maxwellian like shape as expected when injected an electron beam with a duration much longer that the plasma period, and in this case with a duration much longer than the plasma wave live-time. Optical observation of radial plasma oscillation has been observed at LOA [10, 11] and in Austin [12] with a time resolution of less than the pulse duration by using spectroscopy in the time-frequency domain diagnostic. More recently, using the same technique but with a chirp probe laser pulse, has allowed in a single shot a complete visualisation of relativistic plasma wave, showing very interesting features such as the relativistic lengthening of the laser plasma wavelength [13].

11.3.1.2 Laser Beat Wave

Before the advent of short and intense laser pulses, physicists have used the beat wave of two long laser pulses (100 ps range) with frequencies (ω_1 and ω_2) close enough to generate modulation of the laser envelop in the low frequency domain of interest. Such co-propagating laser beams in a perfect homogeneous plasma at a density for which the plasma frequency satisfies exactly the matching condition,

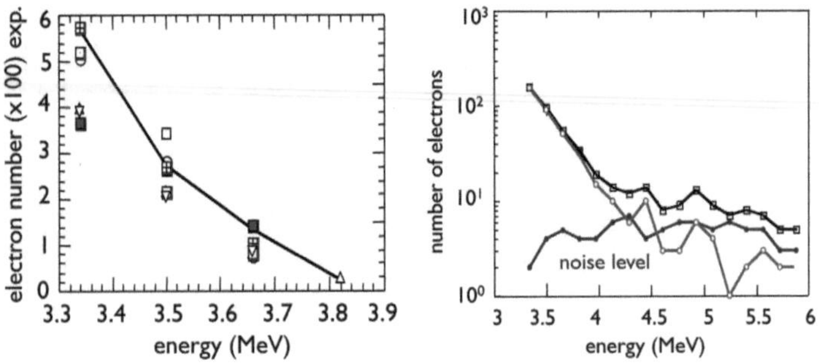

Fig. 11.4 Electrons spectra obtained at LULI. *Left*: in the laser beat wave scheme, *right*: in the laser wakefield scheme

$\omega_p = \omega_1 - \omega_2$ have been used to excite efficiently relativistic plasma waves. Without saturation mechanism the amplitude of the plasma wave grows linearly with time:

$$\frac{\delta n}{n_0}(t) = \frac{1}{4}\sqrt{a_1 a_2}\,\omega_p t \tag{11.23}$$

where a_1 and a_2 are the force parameters of the laser beams. Due to the sharp resonance of the beat wave scheme any changes of the electron density reduces the performance of this scheme. Limiting factors due either to relativistic effects or either to ion motion have been observed. Beat wave experiments using a CO_2 lasers at about 10 μm wavelengths or Nd lasers at about 1 μm wavelengths have shown electric fields in GV/m order. In the CO_2 lasers case the saturation of the electric field was attributed to the relativistic detuning that occurs when the oscillation velocity of the electrons is so high that the relativistic mass correction has to be taken into account and has to detune the plasma electrons from the pump beat wave term. In the Nd case, experiments have been performed in plasmas with a higher density ($n_e = 10^{17}$ cm^{-3} instead of 10^{16} cm^{-3}) that leads to a much faster coupling by modulational instability between electron waves and ion waves.

The first observation of relativistic plasma waves using Thomson scattered technique [14] was performed at UCLA. Acceleration of injected electrons at 2 MeV up to 9 MeV and up to 30 MeV has been demonstrated by the same UCLA group [15, 16]. Acceleration of electrons in a plasma beat wave experiments using 1 μm lasers [17] has also been performed. Electrons injected at an energy of 3 MeV have been accelerated up to 3.7 MeV in a plasma wave with a peak amplitude of 2 % corresponding to a peak electric field strength of 0.6 GV/m. Electron spectra obtained at LULI in the laser beat wave scheme and in the laser wakefield scheme are shown on Fig. 11.4

In order to reduce the coupling between electron waves and ion waves which was a limiting factor of previous experiment done with 100 ps Nd lasers [18], experiments done at Rutherford Appleton Laboratory, with a 3 ps laser pulse have shown excitation of higher amplitude relativistic plasma wave [19].

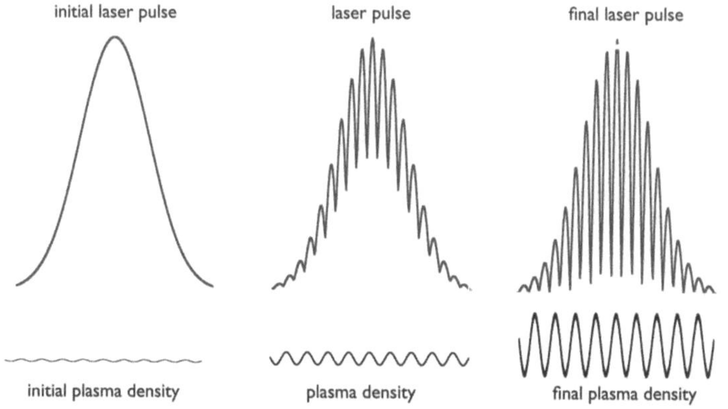

Fig. 11.5 Evolution of the laser pulse and plasma density in the self modulated laser wakefield regime

11.3.2 Non-Linear Regime

11.3.2.1 Self-Modulated Wakefield (SMWF)

Thanks to the development of powerful laser systems with short pulse duration (500 fs), new non linear effects in plasmas has been studied. The cumulative effects of the self-focusing and the self-modulation of the laser envelope by the initial perturbation of the electron density generates a train of laser pulses which becomes resonant with the plasma wave. These effects are described on Fig. 11.5. The self-modulated laser wakefield regime has been investigated theoretically [20–22]. Their works show that when the laser pulse duration exceeds the plasma period and when the power exceeds the critical power for self-focusing, a unique Gaussian laser pulse becomes modulated at the plasma wavelength during its propagation. This mechanism, which is close to Forward Raman Scattering Instability, can be described as the decomposition of an electromagnetic wave into a plasma wave at a frequency shifted by the plasma frequency, that gives finally modulations similar to those produced in the beat wave scheme [23]. In the experiment done at RAL, the relativistic plasma wave was exited by an intense laser ($> 5 \times 10^{18}$ W/cm^2), short duration (<1 ps), 1.054 μm wavelength laser pulse in the self modulated laser wakefield regime. This instability, induced by a noise level plasma wave of the strong electromagnetic pump wave of the laser (ω_0, k_0) into plasma wave (ω_p, k_p) and two forward propagating electromagnetic cascades at the Stokes $(\omega_0 - n\omega_p)$ and anti-Stokes $(\omega_0 + n\omega_p)$ frequencies. n being a positive integer, and ω and k being the angular frequency and the wavenumber respectively of the indicated waves. The spatial and temporal interference of these sidebands with the laser produces an electromagnetic beat pattern propagating synchronously with the plasma wave. The electromagnetic beat exerts a force on the plasma electrons, reinforcing the original noise level plasma wave which the scatters more sidebands, thus closing the feedback loop for

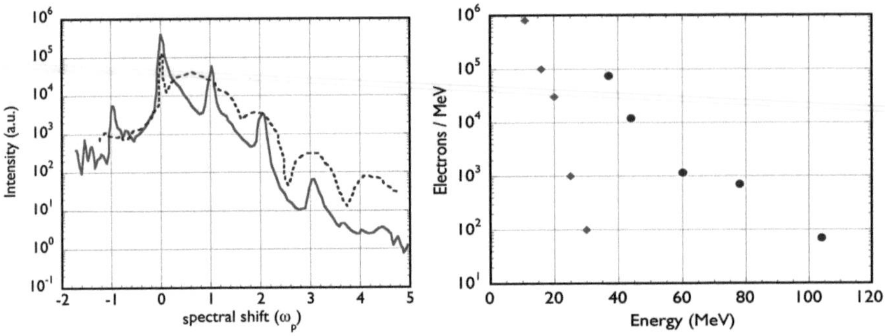

Fig. 11.6 Frequency and electron spectrum in the self modulated laser wakefield regime for two different electron plasma densities : 0.54×10^{18} cm^{-3} (solid line and diamond symbols) and 1.5×10^{19} cm^{-3} (dashed line and circle symbols)

the instability. The solid curve in Fig. 11.6 shows the electromagnetic frequency spectrum emerging form the plasma with a density of $> 5 \times 10^{18}$ cm^{-3} where the abscissa is the shift in frequency of the forward scattered light from the laser frequency in units of ω_p. The upshifted anti-Stokes and downshifted Stokes signals at $\Delta\omega/\omega_p = \pm 1$ are clearly visible as is the transmitted pump at $\Delta\omega/\omega_p = 0$ and the second and third anti-Stokes sidebands. These signals are sharply peaked, and their widths indicate that the plasma wave which generated these signals must have a coherence time of the order of the laser pulse. The dashed curve shows the spectrum when the density is increased to 1.5×10^{19} cm^{-3}. The most startling feature is the tremendous broadening of the individual anti-Stokes peaks at this higher density. This broadening corresponds to the wave-breaking and is mainly caused by the loss of coherence due to severe amplitude and phase modulation as the wave breaks. As wave breaking evolves, the laser light no longer scatters off a collective mode of the plasma but instead scatters off the trapped electrons which are still periodically deployed in space but having a range of momenta producing, therefore a range of scatter frequencies.

During experiments carried out in England in 1994 [2], the amplitude of the plasma waves reached the wave-breaking limit, where electrons initially belonging to the plasma wave are self-trapped and accelerated to high energies. The fact that the external injection of electrons in the wave is no longer necessary is a major improvement. Electron spectrum extending up to 44 MeV have been measured during this first campaign, and up to 104 MeV in the second campaign. This regime has also been reached for instance in the United States at CUOS [24], at NRL [25]. However, because of the heating of the plasma by these relatively ‘long’ pulses, the wave breaking occurred well before reaching the cold wave breaking limit, which limited the maximum electric field to a few 100 GV/m. The maximum amplitude of the plasma wave has also been measured to be in the range 20–60 % [26].

Experiments performed at LOA since 1999 have shown that electron beam can also be produced using a compact 10 Hz laser system [27]. Figure 11.7 shows two

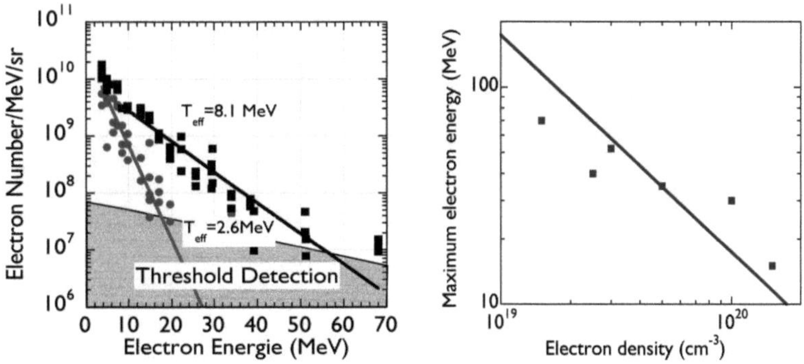

Fig. 11.7 *Left*: Typical electron spectra obtained at 5×10^{19} cm^{-3} (squares) and 1.5×10^{20} cm^{-3} (circles). The corresponding effective temperatures are 8.1 MeV (2.6 MeV) for electron density of 5×10^{19} cm^{-3} (1.5×10^{20} cm^{-3}). *Right*: Maximum electron energy as a function of the plasma electron density. Experimental data: squares, theoretical value: line

typical electron spectra obtained at 5×10^{19} cm^{-3} and 1.5×10^{20} cm^{-3}. The 0.6 J, 35 fs laser beam was focused tightly 6 μm focal spot leading a peak laser intensity of 2×10^{19} W/cm^2. Electron distribution with electron energy greater than 4 MeV is well fitted by an exponential function, characteristic of an effective temperature for the electron beam. These effective temperatures are 8.1 MeV (2.6 MeV) for electron density of 5×10^{19} cm^{-3} (1.5×10^{20} cm^{-3}), to which correspond typical values of 54 MeV (20 MeV) for the maximum electron energy. This maximum energy is defined by the intersection between the exponential fit and the detection threshold. One can observe an important decrease of the effective temperature and of the maximum electron energy for increasing the electron densities.

This is summarised in Fig. 11.7 where we present the maximum electron energy as a function of the electron density. It decreases from 70 to 15 MeV when the electron density increases from 1.5×10^{19} to 5×10^{20} cm^{-3}. Also presented in Fig. 11.7 is the theoretical value [28]:

$$W_{max} \approx 4\gamma_p^2 (E_z/E_0) mc^2 F_{NL} \tag{11.24}$$

Here the maximum electron energy is greater than the conventional one, given by the simple formula: $W_{max} \approx 2\gamma_p^2 (E_z/E_0) mc^2$, where γ_p is the plasma wave Lorentz factor (which is equal to the critical density to electron density ratio n_c/n_e) and E_z/E_0 is the electrostatic field normalised to E_0 ($E_0 = cm\omega_p/e$). The factor of two is due to self-channeling induced by space charge field which focuses accelerated electrons for all phases. The correction factor $F_{NL} \approx (\gamma_{\perp 0} n_0/n)^{3/2}$ corresponds to non linear correction due to relativistic pump effect and to self-channeling. In this formula n_0 is the initial electron density and n the effective one, $\gamma_{\perp 0}$ is the Lorentz factor associated to the laser intensity: $\gamma_{\perp 0} = (1 + a_0^2/2)^{1/2}$. The electron density depression is estimated by balancing the space-charge force and laser ponderomotive force, and evaluated by $\delta n/n = (a_0^2/2\pi^2)(1 + a_0^2/2)^{-1/2}(\lambda_P/w_0)^2$.

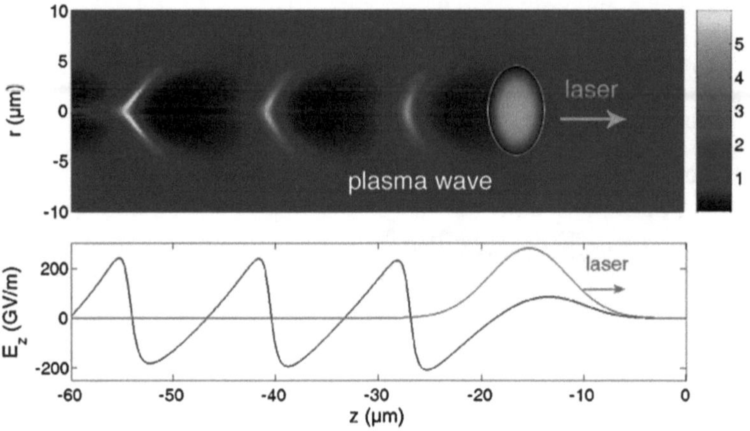

Fig. 11.8 Density perturbation (*top*) and electric field (*bottom*) produced in the non linear regime

In the lower electron density case the depression correction will introduce an important increase of the maximum energy gain which is multiplied by a factor of 2 at 1.5×10^{19} cm^{-3}. For density greater than 1.0×10^{20} cm^{-3}, the main contribution is due to relativistic pump effect as is outlined on the plot in Fig. 11.7. It is also crucial to notice that the fact that the electron maximum energy increases when the electron density decreases, demonstrates that electrons are mainly accelerated by relativistic plasma waves. The maximum electron energy calculates at lower density overestimated the experimental ones indicating that the dephasing length becomes shorter than the Rayleigh length. In order to solve this problem, experiments were performed at LOA using a longer off axis parabola, more energetic electrons have been measured, with a peak laser intensity ten times smaller that in this first experiment. Electron beams with Maxwellian spectral distributions, generated by compact high repetition rate ultra-short laser pulses, have been also at this time produced in many laboratories in the world: at LBNL [29], at NERL [30], and in Germany [31] for instance.

11.3.2.2 Forced Laser Wakefield (FLW)

The regime [32], is reached when the laser pulse duration is approximately equal to the plasma period and when the laser waist is about the plasma wavelength. This regime allows to reduce heating effects That is produced when the laser pulse interacts with trapped electrons. In this regime highly non linear plasma wave can be reached as one can see on Fig. 11.8.

The laser power needs also to be greater that the critical power for relativistic self focusing in order to the laser beam to be shrunk in time and in space. Due to self-focusing the pulse erosion can take place, which can allow efficient wake

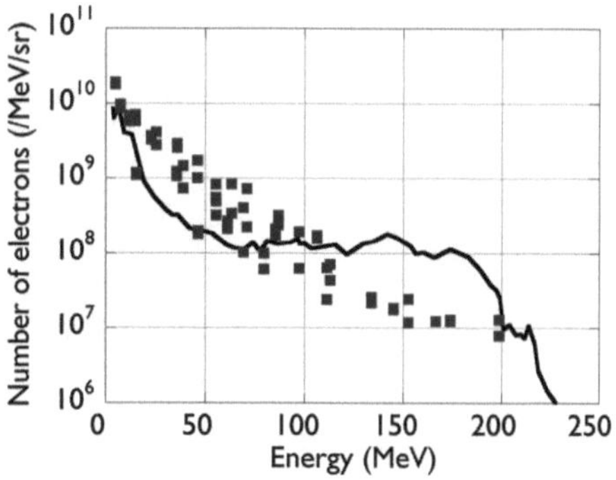

Fig. 11.9 Typical experimental (squares) and calculated (curve) electron spectrum obtained at $n_e = 2.5 \times 10^{19}$ cm^{-3} with a 1 J-30 fs laser pulse focused down to a waist of $w_0 = 18\,\mu$m

generation. Since the very front of the pulse is not self-focused, the erosion will be more severe. The wake then is mostly formed by this fast rising edge, and the back of the pulse has little interaction with the relativistic longitudinal oscillation of the plasma wave electrons. Indeed the increase of plasma wavelength due to relativistic effects means that the breaking and accelerating peak of the plasma wave sits behind most, if not all, of the laser pulse. Hence its interaction and that of the accelerated electrons with the laser pulse is minimised, thus reducing emittance growth due to direct laser acceleration. Thanks to short laser pulses, plasma heating in the forced laser wakefield is significantly lower than in the self-modulated wakefield. This allows to reach much higher plasma wave amplitudes and also higher electron energies as can be seen on Fig. 11.9. Thanks to a limited interaction between the laser and the accelerated electrons, the quality of the electron beam is also improved. Indeed the normalised transverse emittance measured using pepper pot technique has given values comparable to those obtained with conventional accelerators with an equivalent energy (normalised rms emittance $\varepsilon_n = 3\pi$ mm.mrad for electrons at 55 ± 2 MeV) [33].

The 3-D simulations realised for this experiment shows that the radial plasma wave oscillations interact coherently with the longitudinal field, so enhancing the peak amplitude of the plasma wave. This coupled with the aforementioned strong self-focusing are ingredients absent from one-dimensional treatments of this interaction. Even in 2-D simulation, it was not possible to observe electrons beyond 200 MeV, as measured in this experiment, since except in 3-D simulations, both the radial plasma wave enhancement and self-focusing effects are underestimated. Hence it is only in 3-D simulations that $E_{max} \sim E_{wb}$ can be reached. That such large electric fields are generated, demonstrates another important difference between FLW and SMWF regimes, since in the latter, plasma heating by instabilities limits the accelerating electric field to an order of magnitude below the cold wave-breaking limit. It should be noted that the peak electric field inferred for these FLW experiments is in excess of 1 TV/m, considerably larger than any other coherent accelerating structure created in the laboratory.

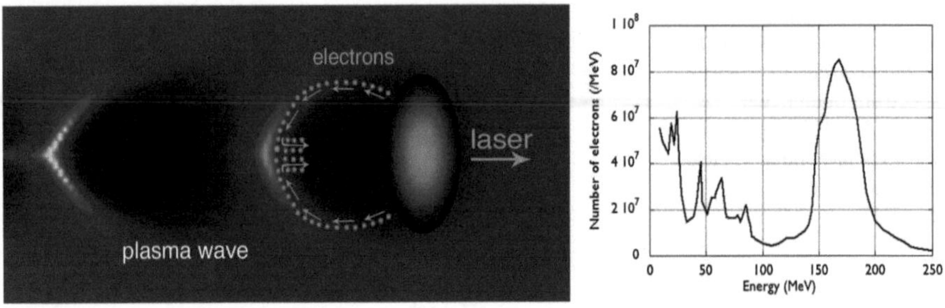

Fig. 11.10 *Left*: Acceleration principle in the bubble regime, *right*: typical quasi monoenergetic electron spectra measured at LOA

11.3.2.3 Bubble Regime

More recently, theoretical work based on 3D PIC simulations have shown the existence of a robust acceleration mechanism called the bubble regime [34]. In this regime, the dimensions of the focused laser are shorter than the plasma wavelength in longitudinal and also transverse directions. Thus, the laser pulse looks like a ball of light with a radius smaller than 10 μm. If the laser energy contained in this volume is large enough, the ponderomotive force of the laser expels radially efficiently electrons from the plasma, which forms a cavity free from electrons behind the laser, surrounded by a dense region of electrons. Behind the bubble, electronic trajectories intersect each other. Electrons are injected in the cavity and accelerated along the laser axis, thus creating an electron beam with radial and longitudinal dimensions smaller than those of the laser (see Fig. 11.10).

The signature of this regime is a quasi monoenergetic electron distribution. This contrasts with previous results reported on electron acceleration using laser-plasma interaction. This properties comes from the combination of several factors:

- The electron injection is different from that in the self-modulated. Injection doesn't occur because of the breaking of the accelerating structure. It is localised at the back of the cavity, which gives similar initial properties in the phase space to injected electrons.
- The acceleration takes place in a stable structure during propagation, as long as the laser intensity is strong enough.
- Electrons are trapped behind the laser, which reduces or suppresses interaction with the electric field of the laser.
- Trapping stops automatically when the charge contained in the cavity compensates the ionic charge.
- The rotation in the phase-space also leads to a shortening of the spectral width of the electron beam [35].

Several laboratories have obtained quasi monoenergetic spectra: in France [5] with a laser pulse shorter than the plasma period, but also with pulses slightly longer than the plasma period in England [3], in the United States [4], then in Japan [36]

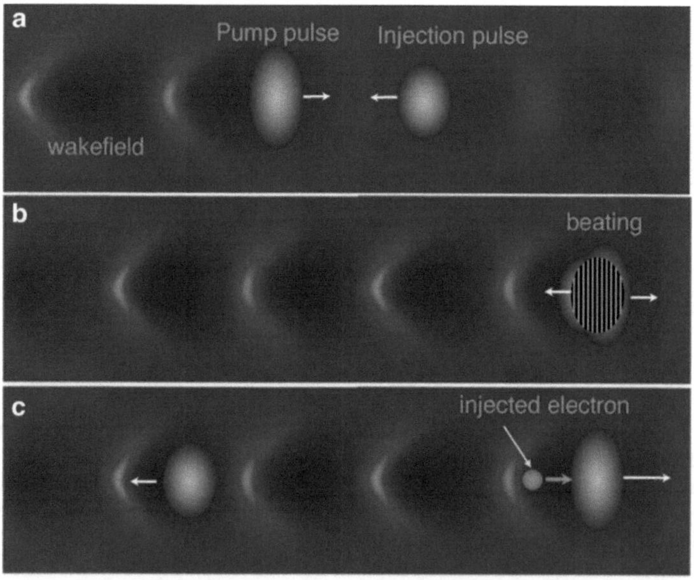

Fig. 11.11 Principle of injection in the counter propagating colliding pulse scheme. (**a**) The two laser pulses have not collided yet; the pump pulse drives a plasma wake. (**b**) The pulses collide and their interference sets up a beat wave that pre-accelerates electrons. (**c**) Preaccelerated electrons are trapped and further accelerated in the wake

and in Germany [37]. The interest of such a beam in important for applications: it is now possible to transport and to refocus this beam by magnetic fields. With a Maxwellian-like spectrum, it would have been necessary to select an energy range for the transport, which would have decreased significantly the electron flux. Electrons in the GeV level were also observed in this regime using in a uniform plasma [38] or in plasma discharge, i.e., a plasma with a parabolic density profile [39] with a more powerful laser which propagates at high intensity over a longer distance.

11.3.2.4 Colliding Laser Pulses Scheme

The control of the electron beam parameters (such as the charge, energy, and relative energy spread) is a crucial issue for many applications. In the colliding scheme successfully demonstrated at LOA, it has been shown that not only these issues were addressed but also that a high improvement of the stability was achieved. In this scheme, one laser beam is used to create the relativistic plasma wave, and a second laser pulse which, when it collides with the main pulse, creates a standing wave which heats locally electrons of the plasma. The scheme of principle of the colliding laser pulses is shown in Fig. 11.11. The control of the heating level gives not only the number of electrons which will be trapped and accelerated but also the volume

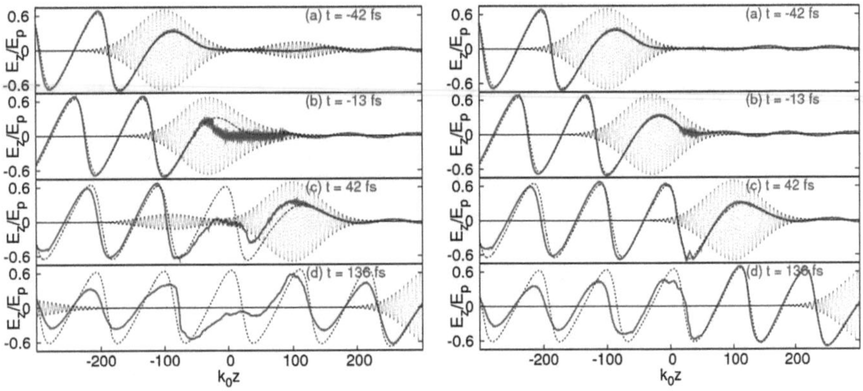

Fig. 11.12 Longitudinal electric field computed at different times in 1D PIC simulation (solid line), and in fluid simulations (dotted line). The transverse electric field is also represented (thin dotted line). Parameters are $a_0 = 2$ and $a_1 = 0.4$, 30 fs duration at FWHM and wavelength $\lambda_0 = 0.8\mu\text{m}$, electron plasma density 7×10^{18} cm^{-3}. *Left*: parallel polarisation, *right*: crossed polarisation

of phase space, or in other words, the energy spread of the injected electrons bunch [40]. In the pioneer work of Esarey et al. [41], a fluid model was used to describe the evolution of the plasma wave whereas electrons were described as test particles. Electron trajectories in the beat wave as well as their energy gain were derived analytically from theory in the case of laser pulses with circular polarisation. It has been shown that this approach fails to describe qualitatively and quantitatively many of the physical mechanisms that occur during and after the laser beams collision [42]. In the fluid approach, the electron beam charge has been found to be one order of magnitude greater than the one obtained in PIC simulations.

For a correct description of injection, one has to describe properly (i) the heating process, e.g. kinetic effects and their consequences on the dynamics of the plasma wave during the beating of the two laser pulses, (ii) the laser pulse evolution which governs the dynamics of the relativistic plasma waves [43]. New unexpected feature have shown that stochastic heating can be also achieved when the two laser pulses are crossed polarised. The stochastic heating can be explained by the fact that for high laser intensities, the electron motion becomes relativistic which introduces a longitudinal component through the $\mathbf{v} \times \mathbf{B}$ force. This relativistic coupling makes it possible to heat electrons. Thus, the two perpendicular laser fields couple through the relativistic longitudinal motion of electrons. The heating level is modified by tuning the intensity of the injection laser beam or by changing the relative polarisation of the two laser pulses [44]. This consequently changes the volume in the phase space and therefore the charge and the energy spread of the electron beam.

Figure 11.12 shows at different times the longitudinal electric field, during and after collision for parallel and crossed polarisation. The solid line corresponds to the PIC simulation results whereas the dotted line corresponds to the fluid calculation.

The laser fields are also represented by the thin dotted line. When the pulses have the same polarisation, electrons are trapped spatially in the beat wave and cannot sustain the collective plasma oscillation inducing a strong inhibition of the plasma wave which persists after the collision. When the polarisations are crossed, the motion of electrons is only slightly disturbed compared to their fluid motion, and the plasma wave is almost unaffected during the collision, which tends to facilitate trapping. Importantly it has been shown that this approach allows a control of the electron beam energy which is done simply by changing the delay between the two laser pulses [6]. The robustness of this scheme has also allowed to carry out very accurate studies of the dynamic of electric field in presence of high current electron beam. This beam loading effect has been used to reduce the relative energy spread of the electron beam.

It has been shown that there is an optimal load which flattened the electric field, accelerating all the electrons with the same efficiency, and producing consequently an electron beam with a very small, 1 %, relative energy spread [45]. In this case, the more energetic electrons are slightly slow down and accelerated at the same energy that the slower one. In case of lower charge, this effect doesn't play any role and the energy spread depends mainly of the heating volume. For higher current, the load is too high and the most energetic electrons slow down too much and get energies even smaller that the slower one [45], increasing the relative energy spread. The optimal load was observed experimentally and supported by full 3D PIC simulations, its corresponds to a current in the 20–40 kA. The decelerating electric field due to the electron beam was found to be in the GV/m/pC range.

11.4 Future of the Laser Plasma Accelerators

Conventional accelerator technology has progressed through a long road paved by scientific challenges. A recent example is the development of superconductivity for high current acceleration in RF cavity, which has required tens of years of theoretical investigations and experiments to understand the physical processes and finally to control the technology which has been successfully used in accelerators such as LEP/LHC (CERN), or HERA (DESY-Hamburg). Laser plasma accelerator researches follow the same road paved with many successful (and unsuccessful) experiments. Thanks to this pioneering works and judging from the incredible results achieved over the last three years, the time has come where a technological approach has to be considered. Two stages laser plasma accelerators schemes should allow the development of few GeV electron beam with a small relative energy spread and a good emittance [46]. In parallel, theoretical and experimental researches should of course be pursued to explore new regimes and to validate theories and numerical codes. The improvement of the laser plasma interaction with the evolution of short-pulse laser technology, a field in rapid progress, will still improve this new and very promising approach which potential societal applications in material science, medicine, chemistry and radio-biology [47]. The ultra short duration (few fs) of the electron beam [48], and consequently his very high current (few kA) comparable to the one delivers at SLAC for LCLS experiment, where very bright x-ray beams were produced by saturating the gain of their free electron laser, indicate that laser plasma accelerators should play a significant role in the compactness of free electron laser design and achievement.

Acknowledgements I acknowledge warmly X. Davoine, J. Faure, S. Fritzler, Y. Glinec, E. Lefebvre, A. Lifschitz, and C. Rechatin who have largely contributed during this last decade to the work I present in this book chapter. I also acknowledge the support of the European Research Council for funding the PARIS ERC project (contract number 226424), of the European Community-Research Infrastructure Activity under the FP6 and of European Community 'Structuring the European Research Area' programs (CARE, contract number RII3-CT-2003-506395), of the EC FP7 LASERLABEUROPE/ LAPTECH Contract No. 228334, of the European Community-New and Emerging Science and Technology Activity under the FP6 'Structuring the European Research Area' program (project EuroLEAP, contract number 028514), and of the French national agency ANR-05-NT05-2-41699 'ACCEL1'.

References

1. T. Tajima, J.M. Dawson, Phys. Rev. Lett. **43**(4), 267 (1979)
2. A. Modena et al., Nature **377**, 606 (1995)
3. S. Mangles et al., Nature **431**, 535 (2004)
4. C.G.R. Geddes et al., Nature **431**, 538 (2004)
5. J. Faure et al., Nature **431**, 541 (2004)
6. J. Faure et al., Nature **444**, 737 (2006)
7. P. Mora, Phys. Fluids B **4**(6), 1630 (1992)
8. L.M. Gorbunov et al., Sov. Phys. JETP **66**, 290 (1987)
9. F. Amiranoff et al., Phys. Rev. Lett. **81**, 995 (1998)
10. J.R. Marquès et al., Phys. Rev. Lett. **76**, 3566 (1996)
11. J.R. Marquès et al., Phys. Rev. Lett. **78**(18), 3463 (1997)
12. C.W. Siders et al., Phys. Rev. Lett. **76**(19), 3570 (1996)
13. N. Matlis et al., Nat. Phys. **2**, 749 (2006)
14. C.E. Clayton et al., Phys. Rev. Lett. **54**, 2343 (1985)
15. C.E. Clayton et al., Phys. Plasmas **14**(1), 1753 (1994)
16. M. Everett et al., Nature **368**, 527 (1994)
17. F. Amiranoff et al., Phys. Rev. Lett. **74**, 5220 (1995)
18. F. Amiranoff et al., Phys. Rev. Lett. **68**, 3710 (1992)
19. B. Walton et al., Opt. lett. **27**(24), 2203 (2002)
20. P. Sprangle et al., Phys. Rev. Lett. **69**, 2200 (1992)
21. T.M. Antonsen, Jr., P. Mora, Phys. Rev. Lett. **69**(15), 2204 (1992)
22. N.E. Andreev et al., JETP Lett. **55**, 571 (1992)
23. C. Joshi et al., Phys. Rev. Lett. **47**, 1285 (1981)
24. D. Umstadter et al., Science **273**, 472 (1996)
25. C.I. Moore et al., Phys. Rev. Lett. **79**, 3909 (1997)
26. C.E. Clayton et al., Phys. Rev. Lett. **81**(1), 100 (1998)
27. V. Malka et al., Phys. Plasmas **8**, 2605 (2001)
28. E. Esarey et al., IEEE Trans. Plasma Sci. **24**(2), 252 (1996)
29. W.P. Leemans et al., Phys. Plasmas **11**(5), 2899 (2004)
30. T. Hosokai et al., Phys. Rev. E **67**, 036407 (2003)

31. C. Gahn et al., Phys. Rev. Lett. **83**(23), 4772 (1999)
32. V. Malka et al., Science **298**, 1596 (2002)
33. S. Fritzler et al., Phys. Rev. Lett. **92**(16), 165006 (2004)
34. A. Pukhov, J. Meyer-ter-Vehn, Appl. Phys. B **74**, 355 (2002)
35. F.S. Tsung et al., Phys. Rev. Lett. **93**(18), 185002 (2004)
36. E. Miura et al., Appl. Phys. Lett. **86**, 251501 (2005)
37. B. Hidding et al., Phys. Rev. Lett. **96**(10), 105004 (2006)
38. N. Hafz et al., Nat. Photon. **2**, 571 (2008)
39. W.P. Leemans et al., Nat. Phys. **2**, 696 (2006)
40. C. Rechatin et al., Phys. Rev. Lett. **102**(16), 164801 (2009)
41. E. Esarey et al., Phys. Rev. Lett. **79**, 2682 (1997)
42. C. Rechatin et al., Phys. Plasmas **14**(6), 060702 (2007)
43. X. Davoine et al., Phys. Plasmas **15**, 113102 (2008)
44. C. Rechatin et al., New J. Phys. **11**, 013011 (2009)
45. C. Rechatin et al., Phys. Rev. Lett. **103**(19), 194804 (2009)
46. V. Malka et al., Phys. Rev. ST Accel. Beams **9**(9), 091301 (2006)
47. V. Malka et al., Nat. Phys. **4**(6), 447 (2008)
48. O. Lundh et al., Nat. Phys. **7**, 219 (2011)

Chapter 12
Ion Acceleration: TNSA

Markus Roth and Marius Schollmeier

Abstract Energetic ions have been observed since the very first laser plasma experiments. The origin was found to be the charge separation of electrons heated by the laser which transfers the energy to the ions accelerated in the field. The advent of ultra intense lasers having pulse length in the femtosecond regime resulted in the discovery of very energetic ions with characteristics quite different from the ones driven by long pulse lasers. Discovered in the late 90s those ion beams have become focus of intense research world wide, because of their unique properties and high particle numbers. Based on their non-isotropic, beam-like behaviour always perpendicular to the emitting surface the acceleration mechanism was called Target Normal Sheath Acceleration (TNSA). In this chapter we will address the physics basics of the mechanism and their dependence on laser and Target parameters. Techniques to explore and diagnose the beams in order to make the useful for applications will be addressed at the end of the chapter.

12.1 Introduction

Since the first irradiation of a target by a laser the generation of energetic ions has been known. The origin of those ions, as we have seen in previous chapters, are the electric fields generated by the charge separation due to the energy transferred from the long pulse laser to the electrons and their respective temperature [1]. The ions are then accelerated in the double layer potential and can reach significant particle

M. Roth (✉)
Institute for Nuclear Physics, TU-Darmstadt, Darmstadt, Germany
e-mail: markus.roth@physik.tu-darmstadt.de

M. Schollmeier
Sandia National Laboratories, Albuquerque, NM 87185, USA

P. McKenna et al. (eds.), *Laser-Plasma Interactions and Applications*,
Scottish Graduate Series, DOI 10.1007/978-3-319-00038-1_12,

energies expanding isotropically in all direction from the target front surface. Since the advent of ultra short pulse lasers having pulse length of less than picoseconds, one of the most exciting results obtained in experiments using solid targets is the discovery of very energetic, very intense bursts of ions coming off the rear, non-irradiated surface in a very high quality, beam like fashion. At the beginning of this century, a number of experiments have resulted in proton beams with energies of up to several tens of MeV behind thin foils irradiated by lasers exceeding hundreds of terawatts of power [2–4]. Since the first observations, an extraordinary amount of experimental and theoretical work has been devoted to the study of these beam characteristics and production mechanisms. Particular attention has been devoted to the exceptional accelerator-like spatial quality of the beams, and current research focuses on their optimisation for use in a number of groundbreaking applications as will be addressed in the second part of this chapter. But first we will focus on the best understood of all the possible acceleration mechanisms, the so called TNSA.

The most part of this chapter has been drawn from Schollmeier [5]. More details and the entire thesis can be downloaded at http://www.gsi.de/forschung/pp/pub/thesis/index.thml. Review articles about TNSA, the diagnostics of short pulse laser plasmas and the application in fast ignition can also be found in [6–10]:

12.2 TNSA: The Mechanism

12.2.1 Initial Conditions

The primary interaction of a high intensity, short laser pulse with a solid target strongly depends on the contrast of the laser pulse, that is the ratio of unwanted, preceding laser light to the main pulse. At peak intensities exceeding $10^{20}\,\mathrm{W/cm^2}$ even a contrast of 10^6 is insufficient not to excite a plasma that is expanding towards the incoming main pulse. As a common source of this unwanted laser light Amplified Spontaneous Emission (ASE) or pre-pulses, caused by a limited polarisation separation in regenerative amplifiers have been identified. The ablative plasma sets the stage for a wealth of uncontrolled phenomena at the interaction of the main pulse with the target. The laser beam can undergo self focusing due to ponderomotive force or relativistic effects, thereby strongly increasing the resulting intensity, the beam can break up into multiple filaments, and finally the beam can excite instabilities that ultimately lead to the production of energetic electrons. Moreover the ablative pressure of blow-off plasma caused by the incident laser energy prior to the main pulse launches a shock wave into the target, which can ultimately destroy the target before the arrival of the main pulse. We address this issue, even not directly related to the TNSA mechanism, because of its influence on the electron spectrum and the thickness of targets that can be used in practise.

12.2.2 General Description

Before we get into the details it is worth to take a step back and qualitatively look at the general picture of the TNSA ion acceleration mechanism. Let us interpret the process generating a proton beam by TNSA as a new variation on a familiar theme – acceleration by a sheath electrostatic field generated by the hot electron component. We assume the interaction of an intense laser pulse well exceeding $10^{18}\,\mathrm{W/cm^2}$ with a solid, thin foil target as the standard case for TNSA. The interaction of the intense laser pulse with the preformed plasma and the underlying solid target constitutes a source of hot electrons with an energy spectrum related to the laser intensity. This cloud of hot electrons penetrates the foil at, as we will see, an opening angle of about 30° and escapes into the vacuum behind the target. The targets capacitance however allows only a small fraction of the electrons to escape before the target is sufficiently charged that escape is impossible for even MeV electrons. Those electrons then are electrolytically confined to the target and circulate back and forth through the target, laterally expanding and forming a charge-separation field on both sides over a Debye length. At the rear surface there is no screening plasma due to the short time scales involved and the induced electric fields are of the order of several TV/m. Such fields can ionise atoms and rapidly accelerates ions normal to the initially unperturbed surface. The resulting ion trajectories thus depend on the local orientation of the rear surface and the electric field lines driven by the time dependent electron density distribution. As the ions start from a cold, solid surface just driven by quasistatic electric fields the resulting beam quality is extremely high as we will see. The whole general description is visualised in Fig. 12.1.

12.2.3 Electron Driver

Current laser systems are not capable of directly accelerating ions yet. Therefore all existing laser ion mechanisms rely on the driving electron component and the resulting strong electric fields cause by charge separation. So the electron driver is extremely important and will be discussed here in detail. As a rule of thumb, particle-in-cell (PIC) calculations [11] have indicated that the so called hot electron electron component has a logarithmic slope temperature that is roughly equal to the ponderomotive potential of the laser beam. This is represented by the cycle-averaged kinetic energy of an electron oscillating in the laser electromagnetic field, $T_{hot} \approx U_{pond} \approx 1\,\mathrm{MeV} \times (I\lambda^2/10^{19}\,\mathrm{W\mu m^2/cm^2})^{1/2}$ in the relativistic regime [12]. The relativistic electrons are directed mainly in forward direction [13], hence the particle distribution function can be simplified by a one-dimensional Maxwell-Jüttner distribution, that is close to an ordinary Boltzmann distribution.

The conversion efficiency from laser energy to hot electrons is not perfect, but only a fraction η is converted. The total number of electrons is

$$n_0 = \frac{\eta E_L}{c\tau_L \pi r_0^2 k_B T_{hot}} \tag{12.1}$$

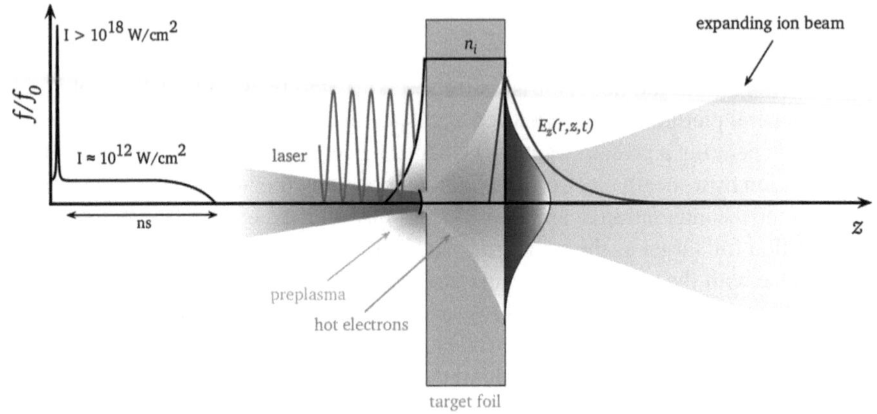

Fig. 12.1 Target Normal Sheath Acceleration (TNSA). A thin target foil with thickness $d = 5 - 50\mu\text{m}$ is irradiated by an intense laser pulse. The laser pre-pulse creates a pre-plasma on the target front side. The main pulse interacts with the plasma and accelerates MeV-energy electrons mainly in forward direction. The electrons propagate through the target, where collisions with the background material can increase the divergence of the electron current. The electrons leave the rear side, resulting in a dense sheath. An electric field due to charge-separation is created. The field is on the order of the laser electric field (TV/m), which ionises atoms at the surface. The ions are then accelerated in this sheath field, pointing in the target normal direction

following a scaling with intensity like

$$\eta = 1.2 \times 10^{-15} I^{0.74} \tag{12.2}$$

with the intensity in $\mathrm{W/cm^2}$ reaching up to 50 % [14]. For ultra-high intensities η can reach up to 60 % for near-normal incidence and up to 90 % for irradiation under 45° [15]. A discussion on which distribution function best fits the experimental data is given in Ref. [16] and in more detail in Ref. [17], leading to the conclusion that it is still not clear from neither theoretical nor experimental data to give a clear answer on the question about the shape of the distribution function.

Given intensities of modern short pulse lasers therefore copious amounts of energetic electrons are generated and, in contrast to thermal electrons in long pulse laser plasmas, are pushed into the target. A fair estimate is a fraction of $N = \eta E_L / k_B T_{hot}$ electrons in the MeV range are created, where E_L is the laser energy. Those electrons have typical energies, that their mean free path is much longer than the thickness of the targets typically used in experiments. While the electrons propagate through the target they constitute a current, which exceed the Alfvén-limit by orders of magnitude. Alfvén found that the main limiting factor on the propagation of an electron beam in a conductor is the self-generated magnetic field, which bends the electrons back towards the source [18]. For parameters interesting for inertial confinement fusion a good review is given in [19]. In order not to exceed the limit of $j_A = \frac{m_e c^3}{e} \beta\gamma = 17\beta\gamma[\mathrm{kA}]$ the net current must be largely

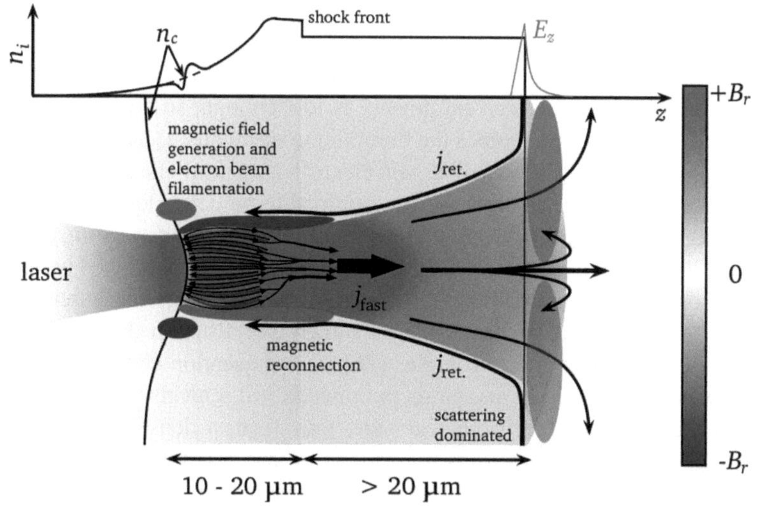

Fig. 12.2 Schematic of laser-generated fast-electron transport. The laser impinges on a pre-plasma with exponential density profile from the *left side*. The light pressure leads to profile steepening, depicted in the one-dimensional scheme on *top*. An ablation plasma creates an inward travelling shockwave that heats, ionises, and compresses the target. Fast electrons are created by the laser, propagating into the dense plasma towards the target rear side. The high electron current j_{fast} can lead to filamentation and magnetic field generation, as well as it drives a return current j_{ret}. The global magnetic field tends to pinch the fast-electron current. Electrons propagating in the dense, solid matter interact by binary collisions with the background material. This leads to a spatial broadening of the electron distribution, that becomes the major effect for longer distances. At the rear side, the electrons form a sheath and build up an electrostatic field E_z (curve in 1D-plot). This can lead to refluxing (recirculation) of the electrons, heating the target even more

compensated by return currents in order to minimise the resulting magnetic field. The return currents will be driven by the charge separation in the laser plasma interaction region and strongly depends on the electrical conductivity of the target as the those currents are lower in energy and thereby affected by the material properties. The large, counter-streaming currents also give rise to instabilities which affects the forward motion of the electrons. The influence of a limited electrical conductivity on the inhibition of fast electron propagation has been addressed in [20] also with respect to space charge separation. Without the return currents the electric field would stop the electrons in a distance of less than 1 nm [21]. The electric field driving the return current in turn, can be strong enough to stop the fast electrons. This effect, known as transport inhibition, is prominent in insulators, but almost negligible in conductors [22].

The propagation of electrons through the target is still an active field of research. As depicted in Fig. 12.2 the laser pushes the critical surface n_c, leading to a steepening of the electron density profile. The motion of the ablated plasma causes a shock wave to be launched into the target, leading to ionisation and therefore a modification of the initial electrical conductivity. As soon as the electrons penetrate the cold solid region, binary collisions (multiple small-angle scattering) with the background material are no longer negligible. These tend to broaden the electron distribution, counteracting the magnetic field effect [23]. For long propagation distances ($z \geq 15\,\mu m$), the current density is low enough, so that broadening due to small-angle scattering becomes the dominating mechanism [24].

The majority of data shows a divergent electron transport. The transport full-cone angle of the electron distribution was determined to be dependent on laser energy, intensity as well as target thickness. For rather thick targets ($d > 40\,\mu m$) this value is around 30° FWHM, whereas for thin targets ($d \leq 10\,\mu m$) published values are in the range of 16° (indirectly obtained by a fit to proton energy measurements) and $\approx 150°$ at most [25]. Just recently it was shown that different diagnostics lead to different electron transport cone angles [26], so the question about the 'true' cone angle dependence with laser and target parameters still remains unclear.

When the electrons reach the rear side, they form a dense charge-separation sheath. The out-flowing electrons lead to a toroidal magnetic field B_θ, that can spread the electrons over large transverse distances by a purely kinematic $E \times B_\theta$ -force [27], sometimes called fountain effect [28]. The electric field created by the electron sheath is sufficiently strong to deflect lower-energy electrons back into the target, which then re-circulate. Experimental evidence for recirculating electrons was found in Refs. [15, 29, 30]. Its relevance to proton acceleration was first demonstrated by Mackinnon et al. [31], who measured a strong enhancement of the maximum proton energy for thin foils below $10\,\mu m$, compared to thicker ones. With the help of computer simulations this energy-enhancement was attributed to an enhanced sheath density due to refluxing electrons. Further evidence of refluxing electrons was also found in an experiment discussed in [32].

Neglecting the complicated interaction for thicknesses below $d \approx 15\,\mu m$, a reasonable estimate for the electron beam divergence is the assumption, that the electrons are generated in a region of the size of the laser focus and are purely collisionally transported to the rear side. This is in agreement with most published data. The broadening of the distribution is then due to multiple Coulomb small-angle scattering, given analytically e.g. by Molie're's theory in Bethe's description [33]. To lowest order the angular broadening $f(\Theta)$ follows a Gaussian (see Ref. [33], Eq. 27)

$$f(\theta) = \frac{2e^{-\vartheta^2}}{\chi_c^2 B}\sqrt{\theta/\sin\theta}, \tag{12.3}$$

where the second term on the right-hand side is a correction for larger angles (from [33], Eq. 58). The angle ϑ can be related to θ by $\vartheta = \theta/\chi_c B^{1/2}$. The transcendental equation $B - lnB = ln\left(\chi_c^2/\chi_{a'}^2\right)$ determines B. The screening angle $\chi_{a'}^2$ is given by $\chi_{a'}^2 = 1.167(1.13 + 3.76\alpha^2)\lambda^2/a^2$, where $\lambda = \hbar/p$ is the deBroglie wavelength of the electron and $a = 0.885 a_B Z^{-1/3}$, with the Bohr radius a_B. α is determined by $\alpha = Ze^2/(4\pi\varepsilon_0\hbar\beta c)$ with the nuclear charge Z, electron charge e, $\beta = v/c$ and $\varepsilon_0, \hbar, c$ denote the usual constants. The variable χ_c is given by

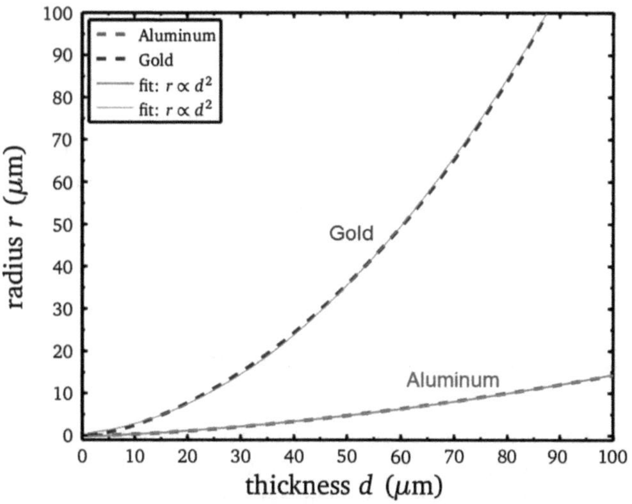

Fig. 12.3 Increase of the radius r of an electron distribution with target thickness d. The calculation was done with Eq. (12.3), taking an energy of $k_B T_{hot} \approx 1\,\text{MeV}$, corresponding to a laser intensity $I = 10^{19}\,\text{W/cm}^2$. (- - -) shows the calculation for gold, (- - -) corresponds to aluminium. Both curves resemble a quadratic increase with thickness d (—) (Color figure online)

$$\chi_c^2 = \frac{e^4}{4\pi\varepsilon_0^2 c^2} \frac{Z(Z+1)Nd}{\beta^2 p^2}, \tag{12.4}$$

with the electron momentum p and $N = N_A\rho/A$ being the number of scattering atoms, determined by Avogadro's number N_A, material density ρ and mass number A. χ_c is proportional to the material thickness d and density ρ as $\chi_c \propto (\rho d)^{1/2}$. Since χ_c determines the width of $f(\theta)$, the angular spread of the electron distribution propagating through matter is proportional to its thickness as well as its density. The analytical formula allows to estimate the broadening of the laser-accelerated electron distribution during the transport through the cold solid target. For a laser intensity $I_L = 10^{19}\,\text{W/cm}^2$ the mean energy (temperature) is $k_B T_{hot} \approx 1\,\text{MeV}$. The increase of the radius r with target thickness d is shown in Fig. 12.3. The electrons were chosen to propagate in aluminium (- - -) and gold (- - -). Al does not lead to a strong broadening due to its low density and Z, compared to the broadening in gold. The graph shows that in both cases the radius at the rear side scales as $r \propto d^2$ (—).

The estimate based on an electron distribution broadening determined by small-angle scattering can be used for an explanation of the measured proton beam profiles. It should be noted, that even though the model seems to be able to calculate the broadening of the forward-propagating fast electron distribution generated by intense laser-matter interaction, it could fail to determine the real number of electrons arriving at the rear side. According to Davies [17] the generation of electromagnetic fields as well as recirculation of the electrons have to be taken into account, both making an estimate and even calculation very difficult. Recent experiments by Akli et al. [34] have shown that this is true at least for thin targets below 20μm, but for thicker foils the assumption of strong recirculation overestimates the number of electrons. Therefore the question if electromagnetic fields and recirculation are essential to determine the fast electron transport from the front to the rear side can still not be satisfactorily answered, though making the assumption of simple collisional broadening a relatively good estimate.

12.2.4 Target Normal Sheath Acceleration: TNSA

After the transport through the target, the electrons end up at the rear side. The laser creates about 10^{13} electrons that are potentially all propagating through the target. The broadening results in transverse extension, that can be estimated by

$$r_{sheath} = r_0 + d\tan(\theta/2), \tag{12.5}$$

where r_0 denotes the laser spot radius, d the target thickness and θ the broadening angle of the distribution, e.g. calculated by Eq. (12.3). The electrons exhibit an exponential energy distribution

$$n_{hot}(E) = n_0 \exp\left(-\frac{E}{k_B T_{hot}}\right) \tag{12.6}$$

with temperature $k_B T$ and overall density n_0 given by Eq. (12.1). The electron density at the rear side (neglecting recirculation) therefore can be estimated to

$$n_{e,0} = \frac{\eta E_L}{c\tau_L \pi (r_0 + d\tan\theta/2)^2 k_B T_{hot}} \tag{12.7}$$

$$\approx 1.5\times 10^{19} \frac{r_0^2}{(r_0 + d\tan\theta/2)^2} \frac{I_{18}^{7/4}}{\sqrt{1+0.73 I_{18}\lambda_{\mu m}^2} - 1} cm^{-3}. \tag{12.8}$$

The last equation was obtained by inserting $E_0 = \sqrt{2I_0/\varepsilon_0 c} \approx 2.7\times 10^{12}\,\mathrm{V/m}$ and Eqs. (12.1) and (12.2) and a practical denotion for the electron temperature based on ponderomotive scaling

$$k_B T_{hot} = m_0 c^2 \left(\sqrt{1 + \frac{I_0[\mathrm{W/cm^2}]\lambda_L^2[\mu\mathrm{m}^2]}{1.37\times 10^{18}}} - 1\right) \tag{12.9}$$

in the first one. I^{18} means that the intensity has to be taken in units of $10^{18}\,\mathrm{W/cm^2}$. The estimate shows that the electron density at the rear side strongly scales with the laser intensity and is inversely proportional to the square of the target thickness. Taking the standard example of a laser pulse with $I = 10^{19}\,\mathrm{W/cm^2}$, focused to a spot of $r_0 = 10\,\mu\mathrm{m}$ and assuming a target thickness $d = 20\,\mu\mathrm{m}$, the angular broadening according to Eq. (12.3) is $\theta = 42°$ (FWHM) for electrons with mean energy k_BT, determined by Eq. (12.9). Hence the electron density at the target's rear side is $n_{e,0} = 1.4 \times 10^{20}\,\mathrm{cm^{-3}}$. This is orders of magnitude below solid density and justifies the assumption of a shielded transport through the target.

The electrons arrive at the rear side and escape into vacuum. The charge separation leads to an electric potential Φ in the vacuum region, according to the Poisson equation. In a one-dimensional consideration it is given as

$$\varepsilon_0 \frac{\partial^2 \Phi}{\partial z^2} = e n_e. \tag{12.10}$$

For a solution it is assumed that the solid matter in one half-space ($z \leq 0$) perfectly compensates the electric potential, whereas for $z \to \infty$ the potential goes to infinity. Its derivative $\partial\Phi/\partial z$ vanishes for $z \to \pm\infty$. In the vacuum region ($z > 0$), the field can be obtained analytically [35]. The electron density is taken as

$$n_e = n_{e,0} \exp\left(\frac{e\Phi}{k_B T_{hot}}\right), \tag{12.11}$$

where the electron kinetic energy is replaced by the potential energy $-e\Phi$. The initial electron density $n_{e,0}$ is taken from Eq. (12.8). The solution of the Poisson equation is found with the Ansatz $e\Phi/k_BT_{hot} = -2\ln(\lambda z + 1)$, where λ is a constant defined by the solution and the $+1$ is necessary to fulfil a continuous solution with the condition $\Phi(0) = 0$ at the boundary to the solid matter. The resulting potential is

$$\Phi(z) = -\frac{2k_BT_{hot}}{e} \ln\left(1 + \frac{z}{\sqrt{2}\lambda_D}\right) \tag{12.12}$$

and the corresponding electric field reads

$$E(z) = \frac{2k_BT_{hot}}{e} \frac{1}{z + \sqrt{2}\lambda_D}. \tag{12.13}$$

In this solution the electron Debye length

$$\lambda_D = \left(\frac{\varepsilon_0 k_B T_{hot}}{e^2 n_{e,0}}\right)^{1/2} \tag{12.14}$$

appears, that is defined as the distance over which significant charge separation occurs [36]. Replacing $k_B T_{hot}$ with Eq. (12.9) and $n_{e,0}$ with Eq. (12.8) leads to

$$\lambda_D \approx 1.37 \mu m \frac{r_0 + d \tan\theta/2}{r_0} \frac{\sqrt{1+0.73 I_{18}\lambda^2} - 1}{I_{18}^{7/8}}. \quad (12.15)$$

The Debye length, or longitudinal sheath extension, on the rear side is on the order of a micrometer. It scales quadratically with target thickness (since $d\tan(\theta/2) \propto d^2$ and is inversely proportional to the laser intensity. Thus, a higher laser intensity on the front side leads to a shorter Debye length at the rear side and results in a stronger electric field. The standard example from above leads to $\lambda_D = 0.6\,\mu\text{m}$.

The maximum electric field is obtained at $z = 0$ to

$$E_{max}(z=0) = \frac{\sqrt{2} k_B T_{hot}}{e \lambda_D} \quad (12.16)$$

$$\approx 5.2 \times 10^{11} V/m \frac{r_0}{r_0 + d\tan\theta/2} I_{18}^{7/8} \quad (12.17)$$

$$= 9 \times 10^{10}\,\text{V/m} \frac{r_0}{r_0 + d\tan\theta/2} E_{12} E_{12}^{3/4}. \quad (12.18)$$

Hence the initial field at $z = 0$ is proportional to the laser intensity and it depends nearly quadratically on the laser's electric field strength. In the last equation the laser's electric field strength is inserted in normalised units of $10^{12}\,\text{V/m}$. By inserting the dependence of the broadening with target thickness from Fig. 12.3, the scaling with the target thickness is obtained as $E_{max}(z=0) \propto d^{-2}$. The standard example leads to a maximum field strength of $E_{max} \approx 2 \times 10^{12}\,\text{V/m}$ just at the surface, that is on the order of TV/m or $MV/\mu m$. It is only slightly less than the laser electric field strength of $E_0 = 8.7 \times 10^{12}\,\text{V/m}$. However, for later times than $t = 0$ the field strength is dictated by the dynamics at the rear side, e.g. ionisation and ion acceleration.

As just mentioned, the electric field strength instantly leads to ionisation of the atoms at the target rear surface, since it is orders of magnitude above the ionisation threshold of the atoms. A simple model to estimate the electric field strength necessary for ionisation is the Field Ionisation by Barrier Suppression (FIBS) model [37]. The external electric field of the laser overlaps with the Coulomb potential of the atom and deforms it. As soon as deformation is below the binding energy of the electron, it is instantly freed, hence the atom is ionised. The threshold electric field strength E_{ion} can be obtained with the binding energy U_{bind} as

$$E_{\text{ion}} = \frac{\pi \varepsilon_0 U_{\text{bind}}^2}{e^3 Z} \quad (12.19)$$

As the electron sheath at the rear side is relatively dense, the atoms could also be ionised by collisional ionisation. However, as discussed by Hegelich [38] the cross section for field ionisation is much higher than the cross section for collisional ionisation for the electron densities and electric fields appearing at the target surface. Taking the ionisation energy of an hydrogen atom with $U_{\text{bind}} = 13.6\,\text{eV}$, the field strength necessary for FIBS is $E_{\text{ion}} = 10^{10}\,\text{V/m}$. It is two orders of magnitude below the field strength developed by the electron sheath in vacuum as shown above. Hence nearly all atoms (protons, carbons, heavier particles) at the rear side are instantly ionised and, since they are no longer neutral particles, they are then subject to the electric field and are accelerated. The maximum charge state of ions found in an experiment is an estimate of the maximum field strength that appeared. This has been used to estimate the sheath peak electric field value [38] as well as the field extension in transverse direction [39,40].

The strong field ionises the target and accelerates ions to MeV-energies, if it is applied for long enough time. The time can be easily calculated by the assumption of a test-particle moving in a static field, generated by the electrons. Free protons were chosen as test-particles. The non-linear equation of motion is obtained from Eq. (12.13). The solution was obtained numerically with MATLAB [41]. It shows that for a proton to obtain a kinetic energy of 5 MeV, the field has to stay for 500 fs in the shape given by Eq. (12.13). During this time the proton has travelled 11.3 µm. The electric field will be created as soon as electrons leave the rear side.

Some electrons can escape this field, whereas others with lower energy will be stopped and will be re-accelerated back into the target. Since the electron velocity is close to the speed of light and the distances are on the order of a micrometer, this happens on a few-fs time scale, leading to a situation where electrons are always present outside the rear side. The electric field being created does not oscillate but is quasi-static on the order of the ion-acceleration time. Therefore ultra-short laser pulses, although providing highest intensities, are not the optimum laser pulses for ion acceleration. The electric field is directed normal to the target rear surface, hence the direction of the ion acceleration follows the target normal, giving the process its name Target Normal Sheath Acceleration – TNSA.

12.2.5 Expansion Models

The laser-acceleration of ions from solid targets is a complicated, multi-dimensional mechanism including relativistic effects, non-linearities, collective as well as kinetic effects. Theoretical methods for the various physical mechanisms involved in TNSA range from analytical approaches for simplified scenarios over fluid models up to fully relativistic, collisional three-dimensional computer simulations.

Most of the approaches that describe TNSA neglect the complex laser-matter interaction at the front-side as well as the electron transport through the foil. These plasma expansion models start with a hot electron distribution that drives the expansion of an initially given ion distribution [14,35,42–48]. Crucial features like the maximum ion energy as well as the particle spectrum can be obtained analytically, whereas the dynamics have to be obtained numerically. The plasma expansion description dates back to 1954 [49]. Since then various refinements of the models were obtained, with an increasing activity after the first discovery of TNSA. These calculations resemble the general features of TNSA. Nevertheless, they rely on somewhat idealised initial conditions from simple estimates. In addition to that, the plasma expansion models are one-dimensional, whereas the experiments have clearly shown that TNSA is at least two-dimensional. Hence these models can only reproduce one-dimensional features, e.g., the particle spectrum of the TNSA process.

Sophisticated three-dimensional computer simulation techniques have been developed for a better understanding of the whole process of short-pulse high-intensity laser-matter interaction, electron transport and subsequent ion acceleration. The simulation methods can be classified as (i) Particle-In-Cell (PIC), (ii) Vlasov, (iii) Vlasov-Fokker-Planck, (iv) hybrid fluid/particle and (v) gridless particle codes; see the short review in Ref. [21] for a description of each method.

The PIC method is the most widely used simulation technique. In PIC the Maxwell equations are solved, together with a description of the particle distribution functions. The method resembles more or less a 'numerical experiment' with only little approximations, hence a detailed insight into the dynamics can be obtained. The disadvantage is that no specific theory serves as an input parameter and the results have to be analysed like experimental results, i.e., they need to be interpreted and compared to analytical estimates.

12.2.5.1 Plasma Expansion Model

As we have seen in previous chapters plasma expansion often is described as an isothermal rarefaction wave into free space. There is quite a large similarity to the expansion models used to describe TNSA. The isothermal expansion model assumes quasi-neutrality $n_e = Zn_i$ and a constant temperature T_e. Using the two-fluid hydrodynamic model for electrons and ions, the continuity, momentum and energy conservation equations are used, usually with the assumption of an isothermal expansion (no temperature change in time), no further source term (no laser), no heat conduction, collisions or external forces and a pure electrostatic acceleration (no magnetic fields). One can find a self-similar solution [50]

$$v(z,t) = c_s + \frac{z}{t} \tag{12.20}$$

$$n_e(z,t) = Zn_i(z,t) = n_{e,0} \exp\left(-\frac{z}{c_s t} - 1\right) \tag{12.21}$$

where v denotes the bulk velocity and $n_i(n_e)$ the evolution of the ion (electron) density. The rarefaction wave expands with the sound velocity $c_s^2 = Zk_B T_e / m_i$. By combining these two equations, replacing the velocity with the kinetic energy $v^2 = 2E_{kin}/m$ and taking the derivative with respect to E_{kin}, the ion energy spectrum

dN/dE_{kin} from the quasi-neutral solution per unit surface and per unit energy in dependence of the expansion time t is obtained as [44]

$$\frac{dN}{dE_{kin}} = \frac{n_{e,0}c_s t}{\sqrt{2Zk_B T_{hot} E_{kin}}} \exp\left(-\sqrt{\frac{2E_kin}{Zk_B T_{hot}}}\right). \tag{12.22}$$

The ion number N is obtained from the ion density as $N = n_{e,0}c_s t$. Additionally, the electric field in the plasma is obtained from the electron momentum equation $n_e eE = -k_B T_e \nabla n_e$ as

$$E = \frac{k_B T_e}{ec_s t} = \frac{E_0}{\omega_{pi} t}, \tag{12.23}$$

with $E_0 = (n_{e,0} k_B T_e/\varepsilon_0)^{1/2}$ and $\omega_{pi} = (n_{e,0} Ze^2/m_i \varepsilon_0)^{1/2}$ denoting the ion plasma frequency. The electric field is uniform in space (i.e. constant) and decays with time as t^{-1}. The temporal scaling of the velocity is obtained by solving the equation of motion $\dot{v} = Zq/mE$ with the electric field from above. This yields

$$v(t) = c_s \ln(\omega_{pi} t) + c_s \tag{12.24}$$

$$z(t) = c_s t \left(\ln(\omega_{pi} t) - 1\right) + c_s t. \tag{12.25}$$

Both equations satisfy the self similar solution. The scaling of the ion density is found as $n(t) = n_0/\omega_{pi} t$.

However, at $t = 0$, the self-similar solution is not defined and has a singularity. Hence the model of a self-similar expansion is not valid for a description of TNSA at early times and has to be modified. Additionally, in TNSA there are more differences: firstly, the expansion is not driven by an electron distribution being in equilibrium with the ion distribution, but by the relativistic hot electrons that are able to extend in the vacuum region in front of the ions. There quasi-neutrality is strongly violated and a strong electric field will built up, modifying the self-similar expansion solution. Secondly, the initial condition of equal ion and electron densities must be questioned, since the hot electron density with $n_e \approx 10^{20}\,\mathrm{cm}^{-3}$ is about three orders of magnitude below the solid density of the rear side contamination layers. This argument can only be overcome by the assumption of a global quasi-neutrality condition $Zn_i = n_e$. Thirdly, it might not be reasonable to assume a model of an isothermal plasma expansion. It can be assumed, however, that the expansion is isothermal for the laser pulse provides ‘fresh’ electrons from the front side, i.e., the assumption is valid as long as the laser pulse duration τ_L. As will be shown below, the main acceleration time period is on the order of the laser pulse duration. This justifies the assumption of an isothermal expansion.

The plasma expansion including charge separation was quantitatively described by Mora et al. [44–46] with high accuracy. The main point of this model is a plasma expansion with charge-separation at the ion front, in contrast to a conventional, self-similar plasma expansion. The plasma consists of electrons and protons, with a step-like initial ion distribution and an electron ensemble being in thermal equilibrium with its potential. The MeV electron temperature results in a charge separation being present for long times. It leads to enhanced ion-acceleration at the front, compared to the case of a normal plasma expansion. This difference is sometimes named the TNSA-effect.

Although being only one-dimensional, the model has been successfully applied to experimental data at more than ten high-intensity short-pulse laser systems worldwide in a recent study [14]. It was separately used to explain measurements taken at the ATLAS-10 at the Max-Planck-Institute in Garching, Germany [51] as well as to explain results obtained at the VULCAN PW [52] (with little modifications). Therefore, it can be seen as a reference model that is currently used worldwide for an explanation of TNSA. Because of its success in the description of TNSA it will be explained in more detail now.

After the laser- acceleration at the foil's front side the electrons arrive at the rear side and escape into vacuum. The atoms are assumed to be instantly field-ionised, leading to $n_i = n_e/Z$. Charge separation occurs and leads to an electric potential ϕ, according to Poisson's equation:

$$\varepsilon_0 \frac{\partial^2 \phi}{\partial z^2} = e\left(n_e(z) - n_i(z)\right). \tag{12.26}$$

The electron density distribution is always assumed to be in local thermal equilibrium with its potential:

$$n_e = n_{e,0} \exp\left(\frac{e\phi}{k_B T_{hot}}\right), \tag{12.27}$$

where the electron kinetic energy is replaced by the potential energy $e\phi$. The initial electron density $n_{e,0}$ is taken from Eq. (12.8). The ions are assumed be of initial constant density $n_i = n_{e,0}$, with a sudden drop to zero at the vacuum interface. The boundary conditions are chosen, so that the solid matter in one half-space ($z \leq 0$) perfectly compensates the electric potential for $z \to -\infty$, whereas for $z \to \infty$ the potential goes to infinity. Its derivative $E = -\partial\phi/\partial z$ vanishes for $z \to \pm\infty$. In the vacuum region (initially $z > 0$), the field can be obtained analytically [35]. The resulting potential is

$$\phi(z) = -\frac{2k_B T_{hot}}{e} \ln\left(1 + \frac{z}{\sqrt{2\exp(1)}\lambda_{D,0}}\right) - \frac{k_B T_{hot}}{e} \tag{12.28}$$

and the corresponding electric field reads

$$E(z) = \frac{2k_B T_{hot}}{e} \frac{1}{z + \sqrt{2\exp(1)}\lambda_{D,0}}. \tag{12.29}$$

The initial electron Debye length is $\lambda_{D,0}^2 = \varepsilon_0 k_B T_{hot}/e^2 n_{e,0}$. The full boundary value problem including the ion distribution can only be solved numerically. The

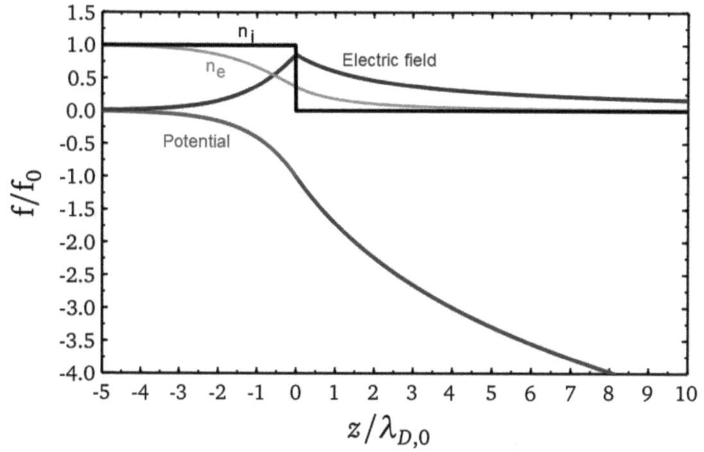

Fig. 12.4 Solution of Eq. (12.26). The potential ϕ (—) was obtained numerically. The analytical solution Eq. (12.28) (—) is in perfect agreement. Both are given in units of $k_B T_{hot}/e$. The electron density n_e (—), normalised to $n_{e,0}$, follows from Eq. (12.27). The normalised ion density n_i (—) is a step-function with $n_i(z<0)/n_{e,0}=1$ and zero for $z>0$. The electric field E (—) is given in units of $k_B T_{hot}/e\lambda_{D,0}$. The coordinate z is given in units of $\lambda_{D,0}$ (Color figure online)

result obtained with MATLAB [41] is shown in Fig. 12.4. The potential ϕ (—) is a smooth function and is in perfect agreement with the analytical solution Eq. (12.28) (—) in the vacuum region. Both are given in units of $k_B T_{hot}/e$. The electron density n_e (—), normalised to $n_{e,0}$, follows from Eq. (12.27). The normalised ion density n_i (—) is a step-function with $n_i(z<0)/n_{e,0}=1$ and zero for $z>0$. The electric field E (—) has a strong peak at the ion front, with $E_{max}=\sqrt{2/\exp(1)}E_0=0.86E_0$. The normalisation field E_0 is given by $E_0=k_B T_{hot}/e\lambda_{D,0}$. The coordinate z was normalised with $\lambda_{D,0}$. The subsequent plasma expansion into vacuum is described in the framework of a fluid model, governed by the equation of continuity (left) and the momentum balance (right):

$$\frac{\partial n_i}{\partial t}+\frac{\partial(v_i n_i)}{\partial z}=0 \qquad \frac{\partial v_i}{\partial t}+v_i\frac{\partial v_i}{\partial t}=-\frac{e}{m_p}\frac{\partial \phi}{\partial z}. \tag{12.30}$$

The full expansion dynamics can only be obtained numerically. Of particular interest is the temporal evolution of the ion distribution and the evolution of the electric field driving the expansion of the bulk. In [5] a Lagrangian code in MATLAB was developed, that solves Eqs. (12.27), (12.28) and (12.30), similar to Ref. [44]. The numerical method is similar to the method described in Ref. [51], however the developed code uses MATLAB's built-in bvp4c-function for a numerical solution of the boundary value problem (BVP) in the ion fluid. The initially constant ion distribution is divided into a grid, choosing the left boundary to be $L\gg_s t$. The boundary value for the potential is $\phi(-L)=0$. At the right boundary (initially at z = 0) the electric field $-\partial\phi_{front/\partial z}=\sqrt{2/e}k_B T_{hot}/e\lambda_{D,front}$ has to coincide with the analytical solution of Eq. (12.29), where the local Debye length has to be determined by the potential at the front:

$$\lambda_{D,front} = \lambda_{D,0} \exp\left(\frac{e\phi_{front}}{k_b T_{hot}}\right)^{-1/2}. \quad (12.31)$$

Initially, the Debye length at the ion front is obtained by inserting Eq. (12.28) in Eq. (12.27) to $\lambda_{D,0,front} = e^{-1}\lambda_{D,0}$. The code divides the fluid region into a regular grid. Each grid element (cell) has a position z_j and an ion density n_j, as well as a velocity v_j. For each time step Δt, the individual grid elements are moved according to the following scheme [51]:

$$z_{j'} = z_j + v_j \Delta t + \frac{e}{2m_p} E \Delta t^2 \quad (12.32)$$

$$v_{j'} = v_j + \frac{e}{m_p} E \Delta t. \quad (12.33)$$

After that, the density of the cell is changed according to the broadening of the cell due to the movement:

$$n_{j'} = n_j \frac{\Delta x_j}{\Delta x_{j'}}. \quad (12.34)$$

At the front, the individual cells quickly move forward, resulting in a 'blow-up' of the cells, that dramatically diminishes the resolution. Thus, after each time-step the calculation grid is mapped onto a new grid ranging from z_{min} to the ion front position z_{front} with an adapted cell spacing. This method is called rezoning. The new values of v_j and n_j are obtained by third-order spline interpolation, providing very good accuracy.

12.2.5.2 Temporal Evolution and Scaling

A crucial point in the ion expansion is the evolution of the electric field strength E_{front}, the ion velocity v_{front} and the position z_{front} of the ion front. Expressions given by Mora are

$$E_{front} \simeq \left(\frac{2n_{e,0}k_B T_{hot}}{e\varepsilon_0} \frac{1}{1+\tau^2}\right)^{1/2}, \quad (12.35)$$

$$v_{front} \simeq 2c_s \ln\left(\tau + \sqrt{1+\tau^2}\right), \quad (12.36)$$

$$z_{front} \simeq 2\sqrt{2e}\lambda_{D,0}\left[\tau \ln\left(\tau + \sqrt{1+\tau^2}\right) - \sqrt{1+\tau^2} + 1\right], \quad (12.37)$$

where $e = \exp(1)$ and $\tau = \omega_{pi}t/\sqrt{2e}$. The other variables in these equations are the initial ion density $n_{i,0}$, the ion-acoustic (or sound) velocity $c_s = (Zk_BT_{hot}/m_i)^{1/2}$, T_{hot} is the hot electron temperature and $\omega_{pi} = (n_{e,0}Ze^2/m_i\varepsilon_0)^{1/2}$ denotes the ion plasma frequency. Due to the charge separation, the ion front expands more than twice as fast as the quasi-neutral solution in Eqs. (12.24) and (12.25). From Eq. (12.36) the maximum ion energy is given as

$$E_{max} = 2k_BT_{hot}\ln^2\left(\tau + \sqrt{1+\tau^2}\right). \tag{12.38}$$

The particle spectrum from Mora's model cannot be given in an analytic form, but it is very close to the spectrum of Eq. (12.22), obtained by the self-similar motion of a fully quasi-neutral plasma expanding into vacuum. The phrase fully quasi-neutral should point out, that in this solution there is no charge-separation at the ion front, hence there is no peak electric field. A drawback of the model is the infinitely increasing energy and velocity of the ions with time, which is due to the assumption of an isothermal expansion. Hence a stopping condition has to be defined. An obvious time duration for the stopping condition is the laser pulse duration τ_L. However, as found by Fuchs et al. [14, 43], the model can be successfully applied to measured maximum energies and spectra, as well as to PIC simulations, if the calculation is stopped at $\tau_{acc} = \alpha(\tau_L + t_{min})$. It was found, that for very short pulse durations the acceleration time τ_{acc} tends towards a constant value $t_{min} = 60\,\text{fs}$, which is the minimum time the energy transfer from the electrons to the ions needs. The variable α takes into account that for lower laser intensities the expansion is slower and the acceleration time has to be increased. It varies linearly from 3 at an intensity of $I_L = 2\times10^{18}\,\text{W/cm}^2$ to 1.3 at $I_L = 3\times10^{19}\,\text{W/cm}^2$. For higher intensities α stays constant at 1.3. Hence the acceleration time is

$$\tau_{acc} = \left(-6.07\times10^{-20}\times(I_L - 2\times10^{18}) + 3\right)\times(\tau_L + t_{min}) \tag{12.39}$$

for $I_L \in [2\times10^{18} - 3\times10^{19}]\,\text{W/cm}^2$,

$$\tau_{acc} = 1.3\times(\tau_L + t_{min}) \tag{12.40}$$

for $I_L \geq 3\times10^{19}\,\text{W/cm}^2$.

Figure 12.5 shows the temporal evolution of the electric field and the ion velocity at the ion front, respectively. The electric field was normalised to E_0, the ion velocity is divided by the sound velocity. There is a very good agreement between the simulated values (○) and the expressions by Mora from Eqs. (12.35) and (12.36) (–). The maximum deviation from the scaling expressions is 1.6 % for the electric field and 0.4 % for the velocity. The electric field evolution, as well as the development of the electron and ion density profiles, are shown in Fig. 12.6. The electric field (—) sharply peaks at the ion front for all times. Initially, the ion density(—) is $n_i = n_0$ for $z \leq 0$ and zero for $z > 0$. The electron density (—) is infinite and decays proportional to z^{-2}. Note the different axes scalings for the electric field and the densities, the

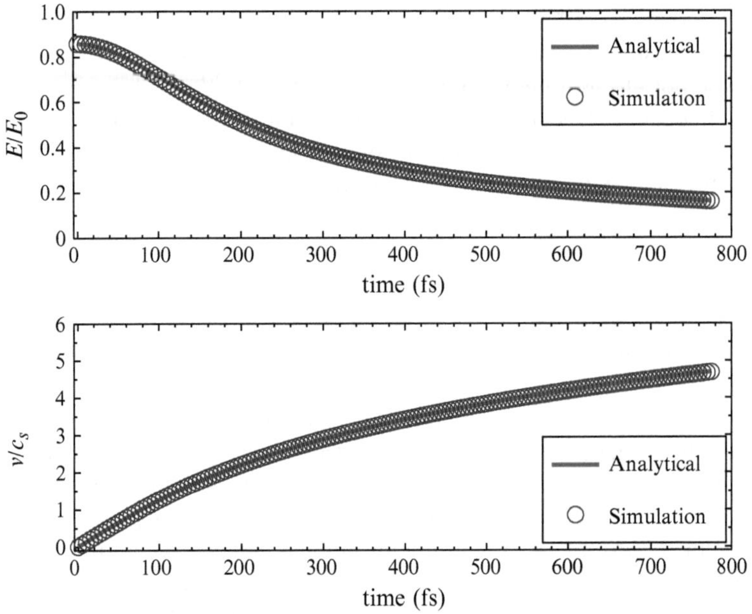

Fig. 12.5 Temporal evolution of the electric field and the ion velocity at the ion front. There is a very good agreement between the simulated values (∘) and Eqs. (12.35) and (12.36) (–)

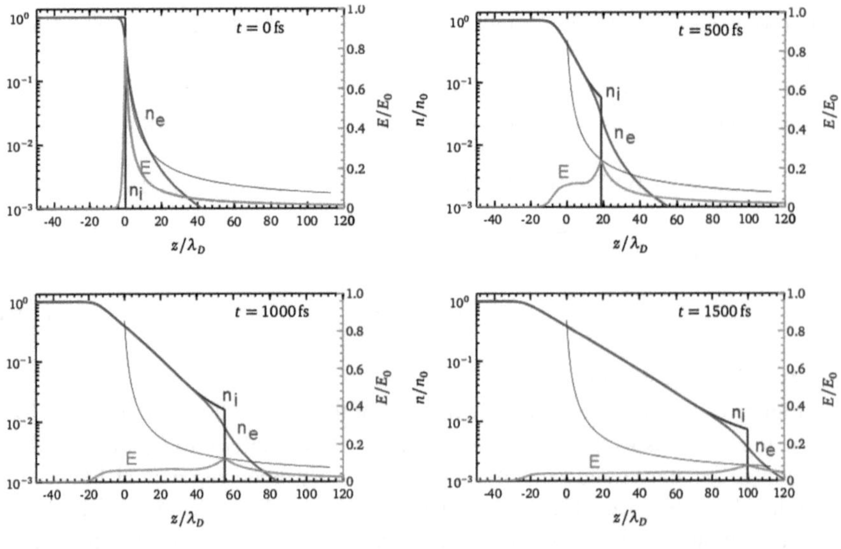

Fig. 12.6 Temporal evolution of the electric field and the ion and electron density, respectively. The electric field(—) sharply peaks at the ion front. The ion density (—) is $n_i = n_0$ for $z \leq 0$ and zero for $z > 0$ for t = 0. The electron density (—) decays proportional to z^{-2}. For later times, at t =(500, 1,000, 1,500) fs, the ions are expanded, forming an exponentially decaying profile (Color figure online)

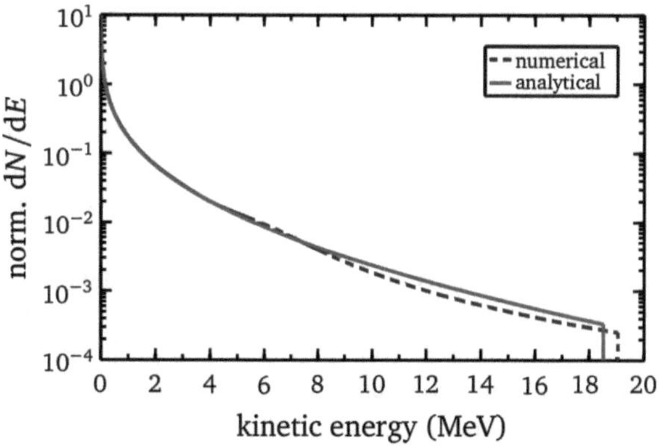

Fig. 12.7 Energy spectrum dN/dE from the simulation compared to the spectrum of a quasi-neutral plasma expansion

latter ones are plotted on a logarithmic scale. For later times, at $t = (500, 1{,}000, 1{,}500)$ fs, the ions are expanded, forming an exponentially decaying profile.

A large part of the expanding plasma is quasi-neutral and can be identified by the constant electric field as derived in Eq. (12.23). At the ion front, the charge-separation is still present, leading to an enhanced electric field that is a factor of two higher than the electric field in the bulk, in agreement to Ref. [44]. This scaling is maintained for the whole expansion. The scaling of the peak electric field value at the ion front with position z, as given by the analytical expressions in Eqs. (12.35) and (12.37), is in perfect agreement with the simulation (–).

The final proton spectrum is shown in Fig. 12.7. The numerical solution (–) is close to the analytical one from the quasi-neutral model by Eq. 12.22 (—). The analytical spectrum is assumed to reach up to a maximum energy, taken from Eq. (12.38). The maximum energy in the simulation is $E_{\text{max,num.}} = 19\,\text{MeV}$, that is in close agreement to the analytical value of $E_{\text{max,analyt.}} = 18.5\,\text{MeV}$. As expected, there is an excellent agreement in the spectra for low energies, since in both cases the expansion is quasi-neutral. For high energies, the numerical spectrum deviates from the self-similar model. The numerical spectrum is lower than the self-similar one even though the ion density of the numerical solution increases close to the ion front, as can be seen in Fig. 12.6 in the deviation of the electron and ion densities close to the front. However, the velocity increase at the front in the simulation is much faster than the self-similar solution, due to the enhanced electric field. Thus, the kinetic energy of the fluid elements close to the ion front is higher than the kinetic energy of fluid elements in a self-similar expansion. The spectrum is obtained by taking the derivative of the ion density with respect to the kinetic energy. In turns out, that the kinetic energy increases stronger than the ion density, hence dN/dE is a little less than the self-similar expansion.

In conclusion, the Lagrangian code and the model developed by Mora show, that TNSA-accelerated ions are emitted mainly in form of a quasi-neutral plasma, with a charge-separation at the ion front that leads to an enhanced acceleration compared to the expansion of a completely quasi-neutral plasma. For later times, if $\omega_{pi}t \gg 1$, the analytical expression of the maximum ion energy in Eq. (12.38) can be used to accurately determine the cut-off energy of TNSA-accelerated ions. The spectral shape of the ions is close to the spectrum of a quasi-neutral, self-similar expansion.

The equations show, that the maximum energy, as well as the spectral shape, strongly scale with the hot electron temperature. The expression for the initial electric field scales as $E \propto k_B T_{hot} n_e$, hence a simplistic estimate would assume that both are equally important for the maximum ion energy. In contradiction to that, the investigation has shown that the maximum ion energy only weakly depends on the hot electron density and is directly proportional to the hot electron temperature. It is worth noting that this finding is in agreement with the results obtained earlier with an electrostatic PIC code by Brambrink [53]. The hot electron density – due to the quasi-neutrality boundary condition – determines the number of the generated ions. Both the number of ions as well as the energy are increasing with time, that again shows that not the shortest and most intense laser pulses are favourable for TNSA, but somewhat longer pulses on the order of a picosecond. This requires a high laser energy to keep the intensity sufficiently high.

Nevertheless, the model is still very idealised, since it is one-dimensional and isothermal, with the electrons ranging into infinity and it neglects the laser interaction and electron transport. An approach with electrons in a Maxwellian distribution always leads to the same asymptotic behaviour of the ion density [54], hence two- temperature [55] or even tailored [56] electron distributions will lead to different ion distributions. There are many alternative approaches to the one described here, including e.g. an adiabatic expansion [46], multi- temperature effects [46, 55], an approach where an upper integration range is introduced to satisfy the energy conservation for the range of a test electron in the potential [48], the expansion of an initially Gaussian shaped plasma [45] or the expansion of a plasma with an initial density gradient [57]. Most of these approaches assume an underlying fluid model, where particle collisions are neglected and the fluid elements are not allowed to overtake each other. Hence a possible wave-breaking or accumulation of particles is not included in the models but requires a kinetic description, e.g. [58, 59]. Furthermore, the transverse distribution of the accelerated ions cannot be determined from a one-dimensional model and requires further modelling. This can be done in the framework of two-dimensional particle-in-cell (PIC) simulations. PIC simulations allow a much more sophisticated description, including relativistic laser-plasma interaction, a kinetic treatment of the particles, as well as a fully three-dimensional approach.

12.3 TNSA: Ion Beam Characteristics

Part of the motivation of the extensive research on laser accelerated ion beams is based on their exceptional properties (high brightness and high spectral cut-off, high directionality and laminarity, short pulse duration), which distinguishes them from those of the lower energy ions accelerated in earlier experiments at moderate laser intensities. In view of this properties, laser-driven ion beams can be employed in a number of groundbreaking applications in the scientific, technological and medical areas. This chapter reviews the main beam parameter and then focuses on established and proposed applications using those unique beam properties.

12.3.1 Beam Parameter

Particle Numbers: One of the striking features of TNSA accelerated ion beams was the fact that the particle number in a forward-directed beam was very high. At present, particle numbers of up to 6×10^{13} protons with energies above 4 MeV have been detected in experiments. This typically leads to a conversion efficiency of laser to ion beam energy that can go up to 9 %. At these high particle numbers, drawn from a very limited source size, for high-energy short-pulse laser systems the depletion of the proton contamination layer at the rear surface becomes an issue. This has been addressed by Allen et al. [60], who determined that 2.24×10^{23} atoms/ cm^3 are at the rear surface of a gold foil, in a layer of 12 thickness. Assuming an area of about 200 μm diameter, the accelerated volume is about $V = 3.8 \times 10^{-11}\,cm^{-3}$. Hence the total number of protons in this area is about $N_{total} = 8.4 \times 10^{12}$, that is close to the integrated number determined in the experiments. Experiments have shown [8] that a rear surface coating of a metal target can provide enough protons up to a thickness of $\approx$ 100 nm, where the layer thickness causes the onset of instabilities in the electron propagation due to its limited electrical conductivity.

Energy spectrum: Based on the acceleration mechanism and the expansion model described earlier the usual ion energy distribution is an exponential one with a cut-off energy that is dependent on the driving electron temperature. Without special target treatment and independently from the target material protons are always accelerated first as they have the highest charge-to-mass ratio. These protons stem from water vapour and hydrocarbon contamination which are always present on the target surface due to the limited achievable vacuum conditions. Protons from the top- most contamination layer on the target surface are exposed to the highest field gradients and screen the electric field for protons and ions coming from the successive layers. The acceleration of particles from different target depths results in a broad energy distribution which becomes broader with the contamination layer thickness. The inhomogeneous electron distribution in the sheath additionally leads to an inhomogeneous accelerating field in transverse direction. The resulting exponential ion energy spectrum constitutes the main disadvantage in laser- ion acceleration.

There are only three groups that have produced a quasi-monoenergetic ion beam with lasers and an energy spread of 20 % or less [61–63] so far. Hegelich et al. [61] have used 20 μm thick palladium foils that were resistively heated before the acceleration. At temperatures of about 600 K the targets were completely de-hydrogenised, but carbon atoms were still remaining on the surface. By increasing the target

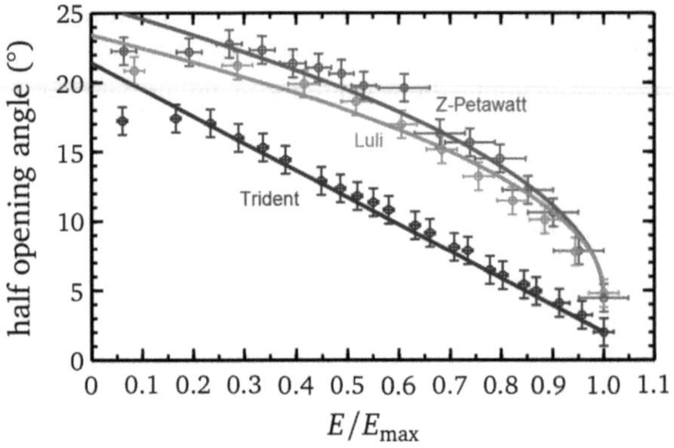

Fig. 12.8 Energy-dependence of the (half) opening angle. The data have been obtained at Trident, at Luli-100TW and at Z-Petawatt. The plots have been normalised to the respective maximum energy of each beam. The opening angle decreases with increasing energy. A parabolic dependency could be fit to the Luli and Z-Petawatt results, the data for Trident has a linear slope

temperature ($T > 1,100\,K$), the carbon underwent a phase transition and formed a mono-layer graphite (graphene) on the Pd surface from which C^{5+} -ions were accelerated to 3 MeV/u with an energy spread of 17 %. An advantage of resistive heating is the complete removal of all hydrogen at once but there are also several disadvantages. The formation of graphene cannot be controlled and the setup requires a very precise temperature measurement.

The second group, Schwoerer et al. [62], has used 5 µm thick titanium foils coated with 0.5 µm thick hydrogen-rich PMMA dots of $(20 \times 20)\,\mu m^2$ on the target back surface. This configuration aimed on limiting the transverse extension of protons such that the proton-rich dot has a smaller diameter than the scale of inhomogeneity of the electron sheath. In this case, all protons experience the same potential. The parasitic proton contamination layer could be reduced by ns-laser ablation and the accelerated protons showed a quasi-monoenergetic energy spectrum peaked at an energy of 1.2 MeV.

The third group, Ter-Avetisyan et al. [63], produced quasi-monoenergetic deuteron bursts by the interaction of a high intensity, high contrast ($> 10^{-8}$) laser with limited mass water droplets. The peak position in the spectrum at 2 MeV had an energy spread of 20 %. This experiment, however, suffered from the bad laser-droplet interaction probability.

Opening angle: Figure 12.8 shows the energy-resolved opening angles for data obtained at Trident (○), at Luli-100 TW (○) and at Z-Petawatt (○), respectively. The plots have been normalised to the respective maximum energy of each beam. These are 19 MeV for TRIDENT, 16.3 MeV for Luli-100 TW and 20.3 MeV for Z-Petawatt, respectively.

Protons with the highest energy are emitted with the smallest opening angle from the source, up to less than 5° half angle. Protons with less energy subsequently

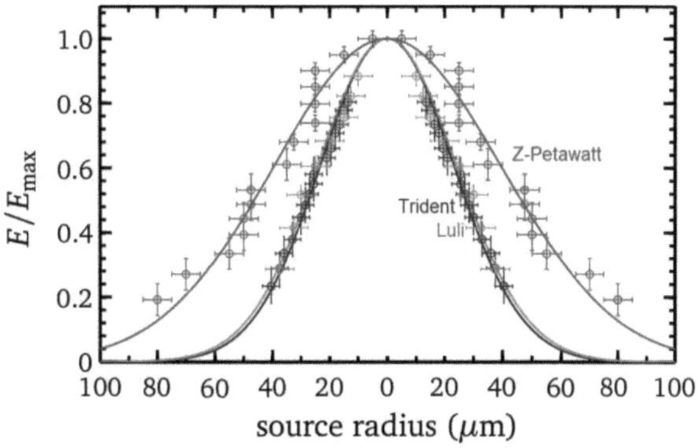

Fig. 12.9 Energy-resolved source sizes for data from TRIDENT, Luli-100TW and Z-Petawatt, respectively. The energy-source size distribution could be fit to a Lorentzian (Eq. 4.1) with FWHM $\sigma = 54.8\,\mu\text{m}$ for TRIDENT, of $\sigma = 56.5\,\mu\text{m}$ for Luli-100 TW and of $\sigma = 92.8\,\mu\text{m}$ for Z -Petawatt, respectively

are emitted in larger opening angles. Below about 30 % maximum energy, the opening angle reaches a maximum and stays constant for lower energies. In most cases, the opening angles decrease parabolically with increasing energy, indicated by the parabolic fits to guide the eye. In some shots, however, the decrease of the opening angle with increasing energy is close to linear as in the example obtained at TRIDENT. The slope of the opening angle with energy is a result of the initial hot electron sheath shape at the target surface, as pointed out by Carroll et al. [64]. According to the reference, a sheath with Gaussian dependence in transverse direction results in a strongly curved opening angle-energy distribution, whereas a parabolic hot electron sheath results in a linear dependency. However, only crude details about the exact modelling of the acceleration process are given in the reference. In [5], a more detailed expansion model has been developed, that is able to explain the experimental results in more detail. It should be noted, that the term 'opening angle' is not equivalent to the beam 'divergence'. The divergence of the protons slightly increases with increasing energy, whereas the emitting area (source size) decreases with proton energy [6, 65]. This results in a total decrease of the opening angle measured experimentally.

Source size: Figure 12.9 shows energy-resolved source sizes for the three laser systems TRIDENT (○), Luli-100 TW (○) and Z-Petawatt (○), respectively. As in the section before, the energy axis has been normalised to the individual maximum energy of the shot, with the maximum energies given in the section before. The source size decreases with increasing energy. Protons with the highest energies are emitted from sources of about 10 µm diameter and less. For lower energies, the source sizes progressively increase, up to about 200 µm diameter for the lowest energies measurable with Radiochromic Film Imaging Spectroscopy (RIS), that are about 1.5 MeV. For even lower energies, the source sizes might be much larger and could reach more than 0.5 mm in diameter [40]. The energy-dependence of the source size well fits a Gaussian, indicated by the lines in Fig. 12.9. The data could be fit by

$$E = \exp\left(\frac{-(4\ln(2)\text{source size})^2}{\sigma^2}\right) \tag{12.41}$$

where 2σ denotes the full width at half maximum (FWHM). This fit allows to characterise the complete energy-dependent source size with one parameter only. The FWHM for Trident with a 10 µm thin gold target is $\sigma = 54.8\,\mu$m. For Luli-100 TW the source size is $\sigma = 56.5\,\mu$m for a 15 µm thin gold foil. A larger source size has been measured at Z-Petawatt with $\sigma = 92.8\,\mu$m and a 25 µm thick gold target.

The energy-dependence of the source size is directly related to the electric field strength distribution of the accelerating hot electron sheath at the source. Protons with high energies have been accelerated by a high electric field. Cowan et al. [65] relate an increasing source size with decreasing energy to the shape of the hot electron sheath, under the assumption of an isothermal, quasi-neutral plasma expansion where the electric field is $E = -(k_B T_e/e)(\delta n_e/n_e)$. A transverse Gaussian electric field distribution would result in a non-analytic expression for the electron density n_e. On the other hand, the realistic assumption of a Gaussian hot electron distribution would result in a radially linearly increasing electric field, in contradiction to the measured data. Hence it is concluded that the quasi-neutral plasma expansion, even though being the driving acceleration mechanism for late times, is not the physical mechanism explaining the observed source sizes. In fact, the source size must develop earlier in the acceleration process, e.g. at very early times when the electric field is governed by the Poisson equation (Eq. 12.10), with $E(z) \propto k_B T_e/\lambda_D \propto \sqrt{k_B T_e n_e}$. With the data from Fig. 12.9 it is concluded, that there must be a radial dependency of $E(z)$, hence a Gaussian decay of either the hot electron temperature or density or both.

Emittance: As we have seen the acceleration of the ions by the TNSA mechanism basically constitutes a quasi electrostatic field acceleration of initially cold (room-temperature) atoms at rest, field ionised and then pulled by the charge separation field. As the electrons are scattered while being pushed through the target, at least for materials with enough electrical conductivity to provide the return currents the transport smoothens the distribution at the rear surface to result in a Gaussian like field shape that expands laterally in time as the electrons start to recirculate. So the initial random ion movement in the beam, represented in an extended phase space and often regarded as an effective beam temperature is very low.

An important parameter in accelerator physics is the transverse emittance of an ion beam. In view of the nature of the ion sources used in conventional accelerators, there is always a spread in kinetic energy and velocity in a particle beam. Each point on the surface of the source emits protons with different initial magnitude and direction of the velocity vector. The emittance ε provides a figure of merit for describing the quality of the beam, i.e., its laminarity [66, 67]. Assuming the beam propagates in the z-direction. Each proton represents a point in the position-momentum space (x, p_x and y, p_y), the phase space. The transverse phase space (e.g. in x-direction) of the TNSA-protons is obtained by mapping the source position (indicated in example by imprinted surface grooves in the RCF) versus the angle of emission p_x/p_z, obtained by the position x of the imprinted line in the RCF and the distance d by $p_x/p_z = x' = \arctan(x/d)$.

In general, the quality of charged-particle beams is characterised by their emittance, which is proportional to the volume of the bounding ellipsoid of the distribution of particles in phase space. By Liouville's theorem, the phase space volume of a particle ensemble is conserved during non-dissipative acceleration and focusing. For the transverse phase-space dimensions, the area of the bounding phase-space ellipse equals $\pi\varepsilon$, where the emittance ε, at a specific beam energy (or momentum p), is expressed in normalised root-mean-square (rms) units as

$$\varepsilon_{N,rms} = (p/mc)[< x^2 >< x'^2 > - < xx' >^2]^{1/2} \tag{12.42}$$

In Eq. (12.42), m is the ion mass, c is the velocity of light, x is the particle position within the beam envelope and $x' = p_x/p_z$ is the particle's divergence in the x-direction. At a beam waist, Eq. (12.42) reduces to $\varepsilon_{N,rms} = \beta\gamma\sigma_x\sigma_{x'}$, where σ_x and $\sigma_{x'}$ are the rms values of the beam width and divergence angle. Several effects contribute to the overall emittance of a beam: its intrinsic transverse 'thermal' spread; intra-beam space charge forces; and non-ideal accelerating fields, for example at apertures in the source or accelerator. For typical proton accelerators (e.g., the CERN SPS or FNAL-Tevatron), the emittance at the proton source is ≈ 0.5 mm-mrad (normalised rms) and up to 20–80 mm-mrad within the synchrotron. The longitudinal phase-space $(z - p_z)$ is characterised by the equivalent, energy-time product of the beam envelope and a typical value, for the CERN SpS, is $\approx 0.1 eV - s$.

The highest quality ion beams have the lowest values of transverse and longitudinal emittance, indicating a low effective ion temperature and a high degree of angle-space and time-energy correlation. In typical TNSA experiments one may estimate an upper limit of the transverse emittance of the proton beam from Eq. (12.42), by assuming that initial beamlets were initially focused to a size $\ll$100 nm, and that the observed width is entirely due to the initial width of the x' distribution. The upper limit of the emittance for 12 MeV protons is $<0.002\,\mathrm{mm-mrad}$. This is a factor of $>$100 smaller than typical proton beam sources, which we attribute to the fact that during much of the acceleration the proton space charge is neutralised by the co-moving hot electrons, and that the sheath electric field self-consistently evolves with the ions to produce an effectively 'ideal' accelerating structure. The remaining irreducible 'thermal' emittance would imply a proton source temperature of $<$100 eV. The energy spread of the laser-accelerated proton beam is large, ranging from zero to tens of MeV, however due to the extremely short duration of the accelerating field ($<10ps$), the longitudinal phase-space energy-time product must be less than 10^{-4} eVs. More details can be found in [65] from which part of this section was extracted.

Ion species: Since the protons are the lightest ions, and having the highest charge-to-mass ratio, they are favoured by the acceleration processes. The protons are present in the target as surface contaminants or as compounds of the target itself or of the target coating. The cloud of accelerated protons then screens the electric field generated by the electrons for all the other ion species. The key for the efficient

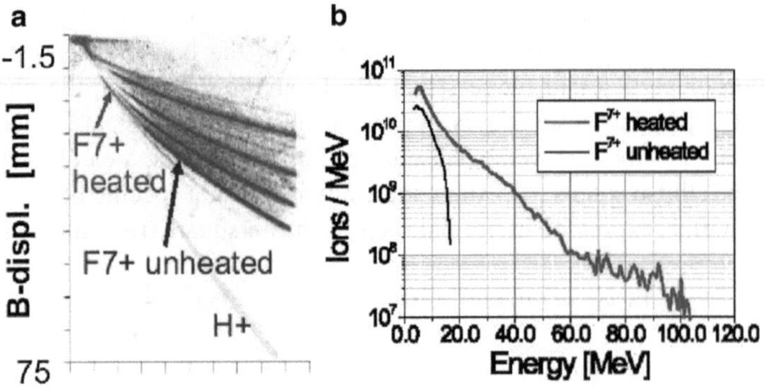

Fig. 12.10 (**a**) overlayed signals of heated and non-heated (*blue*) W/CaF_2 targets: the proton signals vanishes for heated targets, the fluorine signals (especially $F7^+$) go up to much higher energies. (**b**) Corresponding $F7^+$ -spectra: more than 5 MeV/nucleon are achieved for $F7^+$ /ions (From Hegelich et al. [38]) (Color figure online)

acceleration of heavy ions is the removal of any proton or light ion contaminants. In few cases, heating of the target was performed prior to the experiments in order to eliminate the hydrogen contaminants as much as possible, and to obtain a better, controllable ion acceleration. In particular, removing the proton from the targets, or choosing H-free targets, the acceleration of heavier ions was favoured. For the latter, recent experiments have been demonstrated heavy ion acceleration up to more than 5 MeV/u, which corresponds to ion energies usually available at the end of conventional accelerator of hundred of meters in length [38].

First attempts to remove the hydrogen contaminants used resistively heated Al targets up to temperatures of a few hundred degrees. Already the partial removal of the hydrogenous contaminants strongly enhanced the acceleration of heavier ions (i.e. carbon) [8]. A reduction by a factor of ten increased the energy of the carbon ions by a factor of 2.5 and their number by two orders of magnitude. Using tungsten as a thermally stable target resist and coating the rear surface with the material of interest the target could be heated to more than 1,200°. Such targets show no accelerated protons at all, but instead a strongly increased contribution of heavy ions. The maximum energy could be enhanced by a factor of 5 compared to Al targets and the conversion efficiency again by a factor of ten [38] (see Fig. 12.10). In cases, where ohmic heating of target materials of interest is prohibited due to a low evaporation point laser heating has demonstrated to be an appropriate option to remove the contamination layers. In that case the intensity of the laser heating the rear surface and evaporating the proton layers and the timing, with respect to the short pulse, have to be matched carefully. The energy spectrum of the heavy ions together with their charge state distribution also provides detailed information about the accelerating electric field at the rear surface. It was shown, that in a typical experiment collisional ionisation and recombination in flight is negligible, and so the detected charge states directly image the electric field strength because of the field ionisation process. The results that have been obtained match very well the estimated field strength, also predicted by theory and range from $10^{11}\,\mathrm{V/m}$ up to a few $10^{12}\,\mathrm{V/m}$.

The accelerating field deduced from the ion acceleration is also consistent with the observed proton energies with non-heated targets. For example in typical experiments fluorine ions were accelerated up to 100 MeV, i.e. more than 5 MeV/nucleon at a maximum charge state of 7^{+}. This corresponds to an electric field of 2 TV/m, which would have accelerated protons, if present, to energies up to 25 MeV. Those were exactly the maximum energies found in experiments with non-heated targets under similar experimental conditions. Furthermore the conversion efficiency is very high. Similar to the results obtained for proton beams conversion efficiencies of up to 4 % from laser to ion beam energy have been measured. Because of the same accelerating mechanism for protons and heavy ions an excellent beam quality was expected. This part was extracted from [6].

12.3.2 Target Dependence

In the previous section we extensively have looked at the influence of target thickness and conductivity on the driving electron sheath distribution. To summarise, a high conducting ultra-thin target is the most favourable to be used for efficient ion acceleration. Moreover, as the electrons can distribute a part of the energy provided by the laser into Bremsstrahlung, a low Z material is preferable. The ultimate thickness of the target is determined by the limited laser contrast as TNSA requires a sharp density gradient at the rear surface. For the effective acceleration of the ions, an undisturbed back surface of the target is crucial to provide a sharp ion density gradient as the accelerating field strength is proportional to T_{hot}/el_0, where T_{hot} is the temperature of the hot electrons and l_0 is the larger of either the hot-electron Debye length or the ion scale length of the plasma on the rear surface. The limited contrast of the laser causes a shock wave launched by the prepulse that penetrates into the target and causes a rarefaction wave that diminishes the density gradient on the back and therefore drastically reduces the accelerating field. The inward moving shock wave also alters the initial conditions of the target material due to shock wave heating and therefore changes, e.g., the target density and conductivity. As a trade-off one has, however, to note that on the other hand a certain pre-plasma at the front side is beneficial to the production of hot electrons, somehow contradicting the need for high contrast lasers. Also, as the lateral expansion of the electron sheath affects the evolution of the ion acceleration it has been found that to confine the electron by reducing the transverse dimensions of the targets also enhances the ion particle energy. So the ideal target would resemble an ultra-thin, low-Z, highly conducting target with small lateral dimensions and a large pre-plasma at the front side.

Meanwhile, high contrast laser systems are able to irradiate targets as thin as only a few nanometers and we begin to see the change in the accelerating mechanism to RPA [68] or BOA [69] type acceleration, which is beyond the scope of this chapter.

Apart from maximising the ion beam particle energy targets can be used to shape the beam for applications listed below. So ballistic focusing [8, 70–73] and defocusing [8] has been demonstrated by numerous groups, tailoring on a nano-scale using micro-structured targets [74] and layered targets to modify the shape of the energy spectrum [62].

12.3.3 Beam Control

Ballistic focusing of laser accelerated proton beams is known since [70] and has been investigated in detail because of the large importance for proton driven fast ignition [75] and the generation of warm dense matter [76]. Experiments showed that the real source size of a few hundred micrometers could be collimated down to 30 μm using the ballistic focusing off the rear surface of hemispherical targets. However one has to take into account the real sheath geometry of the driving electrons to accurately interpret the proton beam profile. The driving sheath consists of a superposition of the sheath field normal to the inner surface of the hemisphere and the Gaussian electron density distribution caused by the limited source size of the driving laser focus. Therefore the experiments in [70] have indicated that a larger laser focal spot should minimise the second effect and thus result in a better focus quality.

For almost all of the applications a precise control of the beam parameter and a possibility of tailoring the beams is of great importance. As we have seen using the guidance by the shape of the target material we have large control over the spatial ion beam distribution. However it might also be instructive and preferable to look for manipulating the ion beam just using optical methods. The results so far were obtained with a round laser spot, focused as good as possible to obtain the highest intensities. But, as found by Fuchs et al. [77], the laser focal spot shape eventually imprints in the accelerated proton beam. The authors assumed that the bulk of the hot electrons follows the laser focal spot topology and creates a sheath with the same topology at the rear side. The proton beam spatial profile as detected by a film detector was simulated with a simple electrostatic model. The authors took the laser beam profile as input parameter and assumed the electron transport to be homogeneous, with a characteristic opening angle that needed to be fit to match the measured data. The unknown source size of the protons was fit to best match the experimental results. It was shown, that for their specific target thickness and laser parameters, the fitted broadening angle of the electron sheath at the rear side closely matches the broadening angle expected by multiple Coulomb small-angle scattering. However, they could only fit the most intense part of the measured beam and have neglected the lower intense part that originates from rear-side accelerated protons as well. Additionally, there is no information on the dependence of these findings on target thickness.

Using micro-grooved targets a more detailed understanding was achieved. The asymmetric laser beam results in asymmetric proton beam profiles. The energy resolved source size of the protons was deduced by imaging the beam perturbations from the micro-grooved surface into a RCF detector stack. It was shown that the protons with the highest energies were emitted from the smallest source. When the laser focus size was increased, the proton source size increased as well. For symmetric as well as asymmetric laser beam profiles, the source-size dependent energy distribution in both cases could be fit to a Gaussian. This lead to the conclusion that the laser beam profile has no significant contribution to the general expansion characteristics of laser-accelerated protons, but it can strongly modify the transverse beam profile without changing the angle of the beam spread. For a more detailed analysis of the experimental results a code for the Sheath-Accelerated Beam Ray-tracing for IoN Analysis (SABRINA) was developed, which takes the laser beam parameters as input and calculates the shape of the proton distribution in the detector. The electron transport was modelled to closely follow the laser beam profile topology and a broadening due to small-angle collisions was assumed. It was shown that broadening due to small-angle collisions is the major effect to describe the source size of protons for thick target foils (50 μm). In contrast to that, thin target foils (13 μm) show much larger sources than expected due to small-angle collisions. The physical reason behind this observation stays unclear and is most likely the result of electron refluxing. So the shape of the sheath at the rear side of the thick targets can be estimated by a simple model of broadening due to multiple small-angle scattering, but it fails for the description of the sheath broadening in thin targets.

The imprint of the laser beam profile affects the intense part of the proton beam profile. This effect must be present in cases with a round focal spot, too. Therefore a focal spot with a sharp peaked laser beam profile will result in a strongly divergent proton beam as observed in the experiments. The findings also explain that in cases where a collimation of the proton beam is required, e.g., Proton Fast Ignition (PFI) or the injection of the beam into a post-accelerator, not only a curved target surface is necessary, but a large, flattop laser focal spot is indispensable to produce a flat proton- accelerating sheath.

12.4 Applications

Summarising the beam parameters achievable using the TNSA mechanism one concludes:

The measured particle energies so far extends up to tens of MeV (78 MeV protons, 5 MeV/u palladium) and they showed complete space charge- and current neutralisation due to accompanying electrons. In the experiments particle numbers of more than 10^{13} ions per pulse and beam currents in the MA regime were observed. Another outstanding beam parameter is its excellent beam quality with a transverse emittance of less than 0.004π mm mrad and a longitudinal emittance of less than 10^{-4} eVs. Because of these unmatched beam characteristics a wealth of applications were foreseen immediately. Those applications range from:

- injector of high power ion beams for large scale basic research facilities
- new diagnostic techniques for short pulse phenomena, since the short pulse duration allows for the imaging of transient phenomena,
- the modification of material parameters (starting from applications in material science up to warm dense matter research and laboratory astrophysics),
- applications in energy research ('Fast Ignitor' in the inertial fusion energy context),
- medical applications, and as a new laser driven pathway to compact, bright
- neutron sources

12.4.1 Ion Source

An important application for TNSA ion beams is the use as next generation ion sources/accelerators, where the excellent beam quality and the strong field gradients can replace more conventional systems. There are several collaborations actively working on that task, spearheaded by the LIBRA collaboration in the UK, the LIGHT collaboration in Germany and a group at JAEA in Japan.

We briefly address a few aspects of current research in order to apply laser ion beams as a new source:

Collimation and Bunch Rotation of the Accelerated Protons: One of the main drawbacks of laser- accelerated ions and in particular, protons are the exponential energy spectrum and the large envelope divergence of the beam. Different techniques have been developed to modify the energy distribution to produce a more monoenergetic beam as we have seen in the previous chapters. Therefore, special targets were created with thin proton or carbon layers on the rear side, as well as deuterated droplets. Besides, there have been attempts to reduce the initial divergence of the beams by ballistic focusing with the help of curved targets in a hemispherical shape, resulting in a beam focus in a distance of the diameter of the sphere from the laser focus. In a different experiment a laser-triggered microlens was used to select a small energy interval and to focus down the protons with these specific energies to a millimeter spot 70 cm from the target [78].

The total proton yield of typically 10^{13} particles and the observed extremely high phase space density immediately behind the source and prior to any collimator are highly encouraging. As in all cases of sources of secondary particles (antiprotons, muons, rare isotopes and others) transmission efficiency and phase space degradation due to the first collimator need to be carefully examined. In particular, higher than first order focusing properties of the collimator are a serious limitation to the realistically 'usable' fraction of the production energy spectrum as well as of the production cone divergence. As these same limitations may cause a serious degradation of the transverse emittance of the 'usable' protons, the very small production emittance becomes a relatively irrelevant quantity. Instead, an

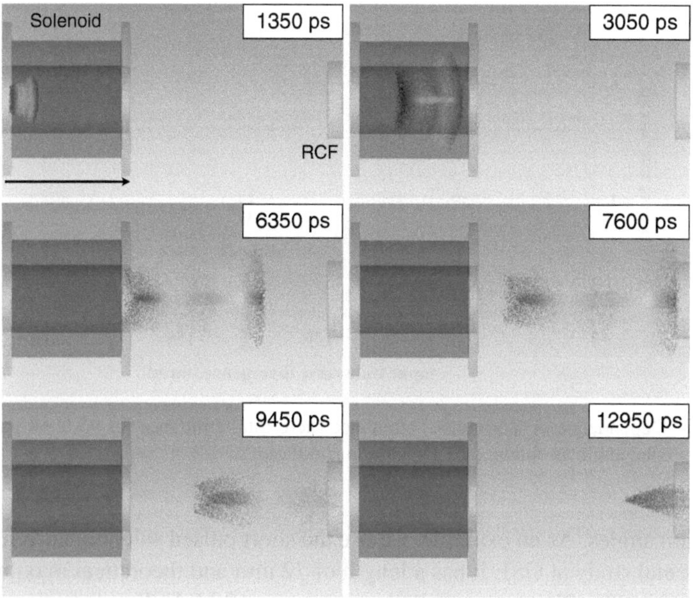

Fig. 12.11 Simulation of the propagation of TNSA protons through a solenoid magnet up to a detector at six points in time. For a better view the accompanying electrons are not shown here. A clear aggregation of protons at the axis due to space charge effects is visible. The proton energy in this case ranges from one to five MeV (Reprinted from Harres et al. [79])

'effective' emittance taking into account transmission loss and blow-up caused by the collimator should be used. In this context space charge (nonlinear) effects are a further source of emittance degradation – probably not the dominant one – to be carefully examined. Recently, we have shown that the collimation of a laser-accelerated proton beam by a pulsed high field solenoid is possible and leads to good results in terms of collimation efficiencies. More than 10^{12} particles were caught and transported by the solenoid. The steadiness of the proton beam after collimation could be proven up to a distance of 324 mm from the target position. Inside the solenoid strong space charge effects occurred due to the co-moving electrons that are forced to circulate around the solenoid's axis at their gyroradius by the strong magnetic field, leading to a proton beam aggregation around the axis (see Fig. 12.11). Details can be found in [79].

More detailed calculations of the injection into ion optical structures have been done by Ingo Hofmann [80, 81]:

Chromatic error of solenoid collimation

In general, pulsed solenoids are a good match to the 'round' production cone of laser accelerated particles; a quadrupole based focusing system appears to be disadvantageous in the defocusing plane of the first lens due to the relatively large

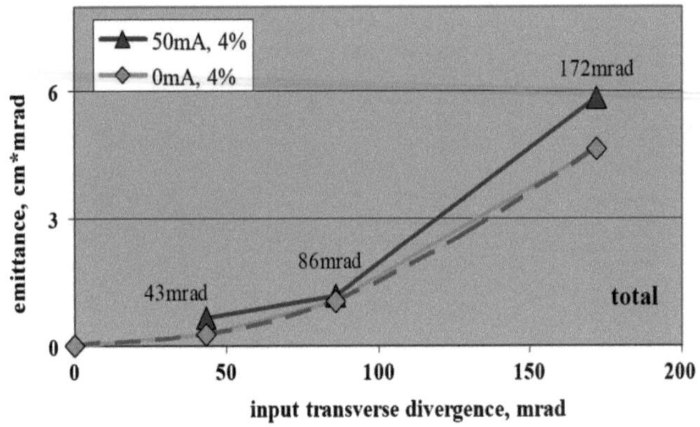

Fig. 12.12 Dependence of 'chromatic' emittance (here total emittance for 95 % of particles) on production cone angle as obtained by Dynamion simulations

production angles. As an example, we use the short pulsed solenoid currently under experimental study at GSI. It has a length of 72 mm and theoretical maximum field strength of 16 T sufficient to parallelise protons at 10 MeV. The distance target spot to solenoid edge is assumed to be 17 mm.

The prevailing higher order effect of a solenoid is the increase of the focal length with particle energy. Due to the de-bunching process different sections along the bunch have different energy and thus focus at different distances. This results in an effective increase of the bunch-averaged emittance to the effect that the tiny initial production emittance should be replaced by a chromatic emittance. In order to examine the expected behaviour in detailed simulation we have employed the Dynamion code [82], which includes higher order effects in amplitudes and energy dependence as well as space charge effects. The latter are based on particle-particle interaction, which limits the space charge resolution.

The solenoid 3D magnetic field has been obtained by direct integration using the coil geometry of the experimental solenoid. In order to quantify the chromatic effect we consider an ensemble of protons with constant energy spread $\Delta E/E = \pm 0.04$ around a reference energy of 10 MeV. Results for final emittance (ignoring space charge) are found in Fig. 12.12 to depend exactly quadratically on the considered production cone opening angle $\delta x'$, which was varied up to $\pm 172\,\text{mrad}(\pm 10^\circ)$. To test the influence of space charge we also simulated a case with the number of N_b protons in the bunch equal to 3×10^9, which is equivalent to a linac current of $I = 50$ mA (using $I = eN_b f_{RF}$ and assuming that each bucket of a $f_{RF} = 108\,\text{MHz}$ sequence is filled identically). For simplicity the bunch intensity was chosen independent of the opening angle. It is noted that the quadratic law still roughly applies. Since for given $\delta x'$ the dependence on $\Delta E/E$ is found practically linear, we can justify the following scaling of the chromatic emittance in the absence of space charge:

$$\varepsilon_x = \alpha_c(\delta x')^2 \frac{\Delta E}{E} \tag{12.43}$$

with $\alpha_c \approx 0.04\,\mathrm{m/rad}$ for the particular solenoid described here. The law is still roughly conserved if space charge is included for the assumed bunch intensity. Note that the chromatic emittance is found practically independent of the initial spot radius r_{spot}– contrary to the production emittance given by the product $r_{spot}\delta x'$.

Transmission through beam pipe: For planned experiments it is important to note that the increase of emittance with energy spread will inevitably lead to transmission loss in the finite acceptance of the following beam pipe. To this end we have assumed a beam pipe of 3 cm radius up to 250 cm distance from the source. We have also assumed a linac current $N_b \approx \Delta E/E$, with $N_b = 2\times10^9$ for the lowest value $\Delta E/E = \pm 0.04$. The increasing transmission loss with distance is mostly due to the large spread of focusing angles as function of the energy spread, and to a lesser extent due to space charge. The surviving energy distribution evaluated at different distances from the source goes down to 35 % for the largest initial energy spread case in the previous example of $\Delta E/E = \pm 0.64$ and correspondingly high current of 560 mA. Obviously an extended beam pipe serves as energy filter.

RF bunch rotation: For most applications of laser accelerated particles, in particular for ion beam therapy, it is desirable to reduce the final energy spread on target to a fraction of a per cent in order to enable focusing on a small target spot. This is achieved by means of a 'bunch rotation' RF cavity applied to the beam after de-bunching to a length suitable for the RF wavelength. The initial short bunch length increases with de-bunching proportional to the distance from the source and the considered energy spread. Capture into the RF bucket of a fraction of beam within a given transverse emittance defines the ultimate 6D extraction efficiency and the 'usable' part of the total production of protons. As reference value we take an energy spread of $\Delta E/E = \pm 0.04$, which can be reduced to a reasonably small value by using a 500 kV/108 MHz bunch rotation RF cavity approximately 250 cm away from the solenoid. This means that only the central part of the totally transmitted energy distribution – about 20 % of it for the 3 cm aperture limitation – can be captured by the RF bucket. Diagnosing the intensity and 6D emittance of this 'usable' fraction in the presence of the background of the fully transmitted spectrum is a challenge to the diagnostics.

So, at the current status a careful study of the transfer efficiency of these beams into conventional transport and focusing structures is crucial and timely, which will be carried out within the next few years given the unique prerequisites present among all the international collaborations. The foremost goal of the proposed effort is to find out the properties of the hereby generated proton/ion beams with the prospect of later applications and to examine the possibilities of collimation, transport, de-bunching and possibly post-acceleration in conventional accelerator structures both theoretically and experimentally.

12.4.2 Diagnostics

A highly, energetic, laminar beam of charge particles with a pulse duration of only a few picoseconds constitutes an ideal tool to diagnose transient phenomena. Like a short burst of x-rays a pulse of laser driven protons can penetrate a target and reveal important information about its parameter. Laser driven protons are complementary to x-rays as the interaction of charged particles are fundamentally different from that of electromagnetic radiation. In the past, ion beams produced by conventional accelerators have already been used to radiograph static and transient samples [83] as well as for the investigation of electric fields in laser-produced plasmas [84–86]. Several experiments with laser-accelerated beams as probe were performed to investigate the evolution of highly transient electric fields evolving from charging of laser-irradiated targets [90, 91]. These fields change on a picosecond time scale. Proton beams from ultra-intense laser–matter interaction are accelerated in a few picoseconds depending on the laser pulse length. Combined with the very low emittance [65] these beams allow for a two-dimensional mapping of the primary target with unprecedented spatial and temporal resolution. Using this technique remnants of relativistic solitons that were generated in a laser– plasma were detected [87], and the accelerating Debye sheath in a TNSA process as well as the ion expansion from the rear side of the target foil could be pictured [88].

Because of the energy spectrum and due to the dispersion of the proton pulse at the point of interaction with the target to be probed different proton energies probe the target at different times. Using the RCF-Stack technique the ions can be separated in energy, where the high energetic particles deposit their energy in the most rear layer while the lower energies are being stopped in the front layers. So, by unfolding the layers one can separate the ion energies and therefore the time of interaction down to picosecond temporal resolution.

12.4.2.1 Energy Loss

The fundamental contrast mechanism for generating image information is energy loss in the sample, and the consequent shift of the energy distribution of transmitted protons. As one proceeds from the shallow layers to the deepest RCF layers, protons having progressively higher incident energies are preferentially recorded. By examining a portion of the image where the sample contained no intervening material, we can deduce the incident proton energy distribution from the depth dependence of the deposited energy, based on the respective response function.

If the incident proton beam passes through a thin sample of thickness δx, they loose energy according to their energy loss rate, and the transmitted proton energy, which is incident on the film detector, is

$$E_f \approx E_i - (dE/dx) \cdot \delta x \approx E_i - \Delta E \tag{12.44}$$

If the sample is thick, so that (dE/dx) is not constant over the energy range ΔE, then the energy loss is given by the integral along the trajectory,

$$\Delta E = \int_0^{\delta x} dE(dx)dx \tag{12.45}$$

A limitation of this technique is that energy-loss, and therefore thickness information, is encoded as a spectral shift of the proton energy distribution due to energy loss. If the object has a large range of thickness, and hence values of the energy loss, early-time (high incident proton energy) information from thick portions of the sample is recorded in the same final proton energy interval as late-time (low incident proton energy) information from thinner portions of the sample. Deconvolution of the spatial, temporal and thickness information is complicated, and even self-consistent solutions may not be unique.

In the ideal limit of no transverse scattering, the resolvable thickness variation over a sample is related only to the energy loss and the exponential fall off of the proton spectrum. This is a strong function of initial proton energy via the energy dependence of (dE/dx). For example, for out test case in which we observe a 2 MeV exponential distribution, a resolvable intensity variation of 1 % implies a resolvable energy loss of 20 keV. At a proton energy of 10 MeV, this would correspond to a CH thickness of <5 μm, and at 4 MeV, a thickness of 2 μm.

12.4.2.2 Transverse Scattering

In addition to continuously slowing down, the protons also undergo multiple small angle Coulomb scattering from the nuclei in the sample. In the energy range of interest, proton multiple scattering can be described by a Gaussian fit to the Moliere distribution, very similar to the case of the electrons we have seen earlier. That is, protons are scattered according to a near Gaussian polar angle distribution,

$$dN/d\Omega \approx \frac{1}{\sqrt{2\pi}\Theta_0} \exp(-\Theta^2/2\Theta_0^2) \tag{12.46}$$

where Θ_0 is given by,

$$\Theta_0 = \frac{13.6\,\text{MeV}}{\beta pc} \sqrt{x/X_0}[1 + 0,038\ln(x/X_0)] \tag{12.47}$$

with X_0 defined as,

$$X_0 = \frac{716.4\,\text{gcm}^{-2} A}{(Z/Z+1)\ln(287/\sqrt{Z})} \tag{12.48}$$

Multiple scattering of the protons as they traverse the sample degrades the spatial resolution possible from ideal energy-loss imaging, but it also can increase the contrast and hence thickness resolution. This is because those protons scattered away from the direct line-of-sight from source to film detector, are moved from the umbral to penumbral region on the film, thus reducing the flux of protons in the direct shadow. The decrease in proton flux associated with a given film plane being sensitive to higher initial energy protons due to the energy loss in the sample, is augmented by a flux reduction from scattering. The very small angle scattered protons, however, increase the net flux of protons in the penumbral region, which can lead to 'limb brightening' effects, which are usual for image techniques based on scattering (rather than absorption).

12.4.2.3 Field Deflection

Probably the most important applications to date of proton probing are related to the unique capability of this technique to detect electrostatic fields in plasmas [89, 90]. This has allowed to obtain for the first time direct information on electric fields arising through a number of laser-plasma interaction processes [87, 91, 92]. The high temporal resolution is here fundamental in allowing the detection of highly transient fields following short pulse interaction. When the protons cross a region with a non-zero electric field they are deflected by the transverse component $E_\perp$ of the field. The proton transverse deflection at the proton detector plane is equal to

$$\Delta r_\perp \approx eL \int_0^b (E_\perp / m_p v_p^2) dl \tag{12.49}$$

where $m_p v_p^2/2$ is the proton kinetic energy, e its charge, b the distance over which the field is present, and L the distance from the object to the detector. As a consequence of the deflections the proton beam cross-section profile undergoes variations showing local modulations in the proton density. Assuming the proton density modulation to be small $\delta n/n_0 \ll 1$, where n_0 and δn are respectively the unperturbed proton density and proton density modulation at the detector plane, we obtain $\delta n/n_0 \approx -div(\Delta r_\perp)/M$ where M is the geometrical magnification. The value of the electric field amplitude and spatial scale can then be determined if a given functional dependence of E can be inferred a priori, e.g. from theoretical or geometrical considerations. More details can be found in the references here in this chapter and in [6], where a part of this paragraph was drawn from.

12.4.3 *Warm Dense Matter*

The creation of extreme states of matter is important for the understanding of the physics covered in various research fields as high-pressure physics, applied material studies, planetary science, inertial fusion energy and all forms of plasma generation generated from solids. The primary difficulties in the study of these states of matter are, that the time scales or the changes are rapid ($\approx 1\,\text{ps}$) while the matter is very dense and the temperatures are relatively low, on the order of a few eV/k_B. With these parameters, the plasma exhibits long- and short-range orders, that are due to the correlating effects of the ions and electrons. The state of matter is too dense and/or too cold to admit standard solutions used in plasma physics. Perturbative approaches using expansions in small parameters for the description of the plasma are no longer valid, providing a tremendous challenge for theoretical models. This region where condensed matter physics and plasma physics converge is the so-called Warm Dense Matter (WDM) regime [93].

WDM conditions can be generated in a number of ways, such as laser-generated shocks [94] or laser-generated x-rays [95, 96], intense ion beams from conventional accelerators [97] or laser-accelerated protons [70–72], just to name a few. Whereas lasers only interact with the surface of a sample, ions can penetrate deep into the material of interest thereby generating large samples of homogeneously heated matter. The short pulse duration of intense ion beams furthermore allows for the investigation of equation of states close to the solid state density, because of the material's inertia preventing the expansion of the sample within the interaction. In addition to that, the interaction of ions with matter dominantly is due to collisions and does not include a high temperature plasma corona as it is present in laser matter interaction.

The generation of large, homogeneous samples of WDM is accompanied by the challenging task to diagnose this state of matter, as usual diagnostic techniques fail under these conditions. The material density results in a huge opacity and the relatively low temperature does not allow traditional spectroscopic methods to be applied. Moreover the sample size, deposited energy and lifetime of the matter state are strongly interrelated and dominated by the stagnation time of the atoms in the probe. Thus high spatial and temporal resolution is required to gain quantitative data in those experiments. Due to the high density of the sample, laser diagnostics cannot be used. The properties of matter could be determined by measuring the expansion after the heating [71] or by measuring the thermal radiation emitted by the sample [70]. However, even more interesting are the plasma parameters deep inside the sample, where the ion heating is most effective. An ideally suited diagnostic technique recently developed is x-ray Thomson Scattering [95, 98, 99].

Figure 12.13 shows a typical experimental scheme to investigate the transformation of solid, low-Z material into the WDM state. The experimental scheme requires a high-energy short-pulse laser and one or more long- pulse laser beams in the same experimental vacuum chamber. In recent years, more and more laser facilities have upgraded their laser systems for such kind of pump-probe experiments. A CPA laser beam above 100 TW power generates an intense proton beam from a thin target foil. The protons hit a solid density sample and heat it isochoric up to several eV/k_B temperature. The long-pulse beam(s) is (are) used to drive an intense x-ray source from a Ti or Saran (contains Cl) foil. The sample is probed by narrowband line-radiation from the Cl- or Ti-plasma. The scattered radiation is first spectrally dispersed by a highly efficient, highly-oriented pyrolytic graphite (HOPG) crystal

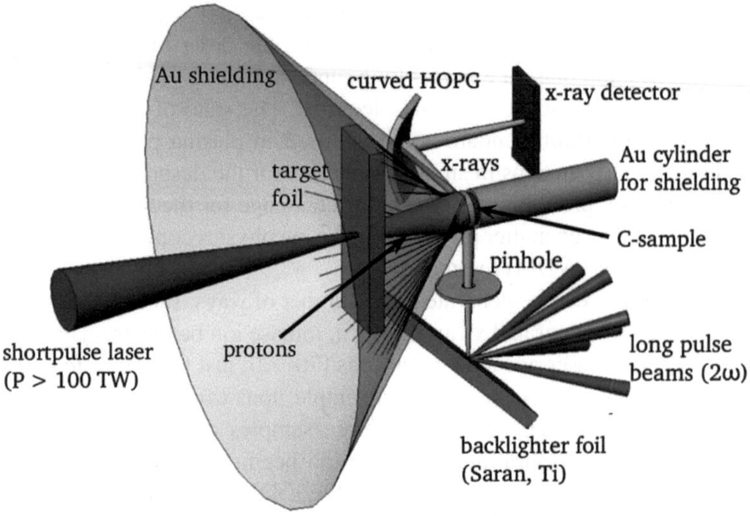

Fig. 12.13 Experimental scheme to investigate the properties of laser-accelerated proton-heated matter by spectrally resolved x-ray Thomson scattering

spectrometer in von Hamos geometry before it is detected. Extensive gold shielding (partially shown in Fig. 12.13) is required to prevent parasitic signals in the scatter spectrometer. From the measured Doppler-broadened, Compton-downshifted signal the temperature and density can be inferred.

Whereas lasers only interact with the surface of a sample, ions can penetrate deep into the material of interest thereby generating large samples of homogeneously heated matter. The short pulse duration of laser produced ion beams furthermore allows for the investigation of equation of states close to the solid state density, because of the materials inertia preventing the expansion of the sample during the interaction. In addition to these unique characteristics, the interaction of ions with matter dominantly is due to collisions and does not include a high temperature plasma corona as it is present in laser matter interaction. The absence of a large radiation background is of importance to the experiment. Large conversion efficiencies have been observed and significant energy can be transferred from the ultra intense laser via the ion beam into the sample of interest. Because of the high beam quality, ballistic focusing has been demonstrated, allowing for an increase in local energy deposition and thus to higher temperatures. The use of hemispherical targets, including cone guided targets to enhance the local proton flux on the material of interest can even enhance the locally deposited energy.

Using laser pulses in excess of 100 J the intense proton beams can heat large targets up to several times the melting temperature. In a milestone experiment at Vulcan TAW last year, the molten fraction in carbon samples heated by intense proton beams was measured [76].

Figure 12.14 shows some results compared with a radiation hydrodynamics simulation, which uses SESAME as equation of state model. We notice that

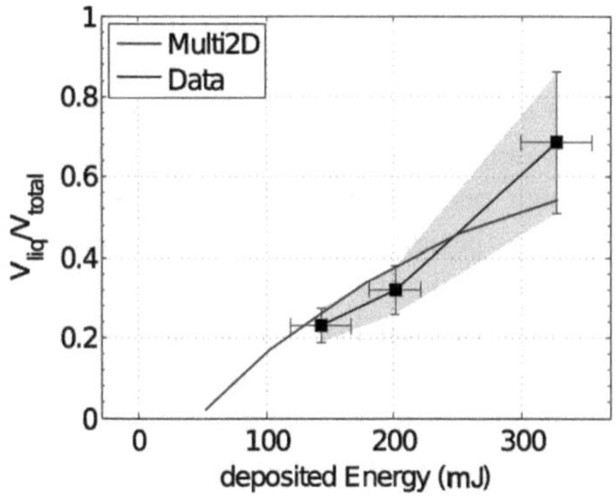

Fig. 12.14 The molten fraction of a carbon sample vs the deposited energy by the proton beam, calculated from the experimental data and simulated with Multi2D [76]

agreement is better at lower temperatures where ionisation is not important, but at higher temperatures the presence of an ionic component may be important.

To summarise, laser accelerated proton beams are very well suited to produce macroscopic samples of warm dense matter. Their unique feature, having pulse duration of only a few picoseconds, while containing more than 10^{12} protons cannot be obtained by conventional ion accelerators.

12.4.4 Fast Ignition

In conventional inertial fusion, ignition and propagating burn occurs when there is a sufficient temperature (5–10 keV) reached within a sufficient mass of DT fuel characterised by a density- radius product greater than an alpha particle range $(\rho r)_\alpha > 0.3$ gcm^{-2}. The necessary conditions for propagating DT burn are achieved by an appropriate balance between the energy gain mechanisms and the energy loss mechanisms. Mechanical work (PdV), alpha particle energy deposition and, to a smaller extent, neutron energy deposition are the principal energy gain mechanisms in deuterium-tritium fuel. Electron heat conduction and radiation are the principal energy loss mechanisms. When the rate of energy gain exceeds the rate of energy loss for a sufficient period of time, ignition occurs. Because of the short burn time and the inertia of the fuel the contribution of expansion losses is negligible. Fast ignition (FI) [100, 101] was proposed as a means to increase the gain, reduce the driver energy and relax the symmetry requirements for compression, primarily in direct drive inertial confinement fusion (ICF). The concept is to pre-compress the cold fuel and subsequently to ignite it with a separate short pulse high-intensity laser or particle (electron or ion) pulse. FI is being extensively studied by many groups worldwide, using short pulse lasers or temporally compressed heavy-ion beams. There are several technical challenges for the success of laser- driven FI. Absorption of the ignitor pulse generates copious relativistic electrons, but it is not yet known whether these electrons will propagate as a stable beam into the compressed fuel to deposit their energy in a small volume. In principle, heavy-ion beams can have advantages for FI. A focused ion beam may maintain an almost straight trajectory while traversing the coronal plasma and compressed target, and ions have an excellent coupling efficiency to the fuel and deliver their energy in a well-defined volume due to the higher energy deposition at the end of their range (Bragg peak) [102]. With the discovery of the TNSA ions with excellent beam quality the idea of using those beams for FI was introduced [75, 103]. Protons do have several advantages compared with other ion species [104] and electrons. First, because of their highest charge-to-mass ratio, they are accelerated most efficiently up to the highest energies. They can penetrate deeper into a target to reach the high density region, where the hot spot is to be formed, because of the quadratic dependence of the stopping power on the charge state. And finally they do, like all ions, exhibit a characteristic maximum of the energy deposition at the end of their range, which is desirable in order to heat a localised volume efficiently.

The basic idea is to use multiple, short pulse lasers irradiating a thin foil. The protons were accelerated off the rear surface of the foils and, because of the parabolic geometry, are focused into the compressed fuel. One of the requirements for proton fast ignition (PFI) is the possibility of focusing the proton beams into a small volume. It has recently been demonstrated that proton beam focusing is indeed possible and spot sizes of about 40 μm have been achieved. This is comparable to what is required by PFI. Larger irradiated areas on the target front surface as required for PFI would flatten the electron distribution at the rear surface. This not only might result in a single divergence angle for different energies but also in a much smaller initial divergence angle that could be compensated in order to reach the required focal spot diameters.

The pulse length at the source is in the right order of magnitude for PFI, which was already indicated in first experiments on ion acceleration. The protons are not monochromatic but rather have an exponential energy distribution. This seemed to be a concern at the beginning because of dispersion and pulse lengthening. A close distance to the pellet on the other hand raises the concern if the thin metallic foil, that is to be the source of the protons, can be kept cold enough not to develop a density gradient at the rear surface which would diminish the accelerating field. A second concern was related to the stopping power. Because of the difference in initial velocity, the energy deposition of protons with different kinetic energy is spread over a larger volume. Slower protons are stopped at a shorter distance and do not contribute to the creation of the ignitor spark. Fortunately, further work has relieved those concerns. Simulations by Basko et al. (presented at the FI-Workshop 2002, Tampa, Florida) have shown that the protective shield placed in front of the source can withstand the x-ray flux of the pellet compression and keep the rear surface of the source foil cold enough for the acceleration via TNSA. A thickness of a few tens of microns on the other hand provides thermal shielding as well as sufficient mechanical stability. The distance between the source and the ignition spot can be a few millimetres. If this distance is too short for the compression geometry (e.g. not using a closely coupled hohlraum) the distance can be adjusted using a similar cone target, as for conventional FI. For the concern on the hydrodynamic stability of the proton source foil the proposed usage of a cone target similar to the one proposed

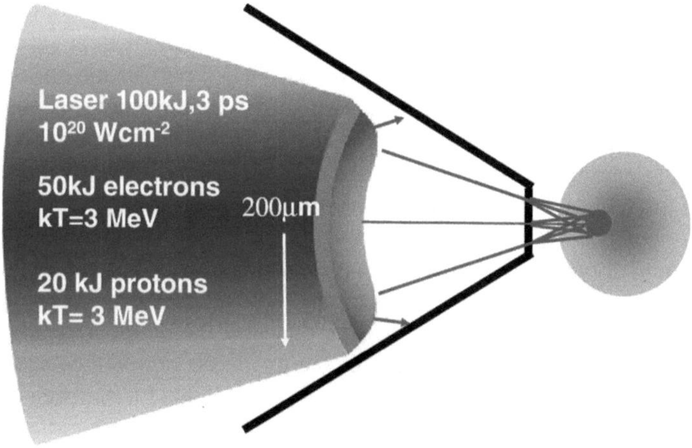

Fig. 12.15 Proposed concept of using cone-guided PFI (Courtesy of M. Key, LLNL)

for electron FI [105, 106] has solved most of the problems by shielding the foil from primary soft x-rays generated in the compression of the capsule (see Fig. 12.15). Furthermore it was demonstrated that small-scale plasma density gradients at the rear side of the proton source target caused by target preheat have no significant (less than 10 %) impact on the TNSA mechanism [43].

A big surprise was the fact that a monochromatic proton beam is actually not the optimum to heat a hot spot in a fusion target. Numerical simulations have shown that one has to take into account the decrease of the stopping power of the nuclear fuel with increasing plasma temperature. So an exponential energy spectrum, like the one generated by this mechanism is the most favourable one. The first protons with the highest energies penetrate deep into the fuel. By the time the proton number increases and the target temperature rises, the stopping power is reduced, thereby compensating for the lower initial energy of the incoming protons. Thus the majority of the protons deposit their energy within the same volume.

Existing short pulse lasers have already demonstrated intensities, which are sufficient for generating proton energy spectra required for PFI. Regardless of the nature of the ignitor beam, calculations show a minimum deposited energy required for FI of the order of a few tens of kilojoules. There have been many experiments at different laser systems accelerating proton beams. Interestingly, the laser to ion beam conversion efficiency seems to strongly increase with total laser energy from a fraction of a percent up to more than 10 %. Carefully extrapolating the conversion efficiencies to multi-kilojoule laser systems conversion efficiencies of more than 10 % can be expected, which would result in the need for a few hundred kilojoules of short pulse laser energy for PFI.

The most detailed theoretical analysis so far has been published by Temporal et al. [107] for a proton beam with an exponential energy spectrum. Following their assumptions a total proton energy of about 26 kJ at an effective temperature of 3

MeV is required. The moderate temperature was found to be an optimum between the need for high temperatures to minimise the pulse lengthening caused by the velocity spread and the stopping range for the major part of the spectrum. It is interesting to note that the protons, which effectively heat the hot spot contain only 10 kJ of the total energy and range from 19 to 10 MeV. If it could be possible to shape the energy spectrum of the laser accelerated protons it would strongly influence the required laser beam energy. The total number of protons needed for ignition is close to 10^{16}. Is it conceivable to get a consistent scenario for those requirements? A typical proton beam temperature of 3 MeV is commonly obtained in experiments at $5 \times 10^{19}\,\mathrm{Wcm}^{-2}$. Assuming a pulse length of 4 ps (which would increase the damage threshold of modern dielectric compressor gratings) and a conversion efficiency of 10 % a total laser energy of 260 kJ would be needed.

The use of a cone-guided geometry, like in conventional FI, has been considered to be of great advantage. The source foil can be shielded from the radiation caused by the primary drivers, the source-to-hot spot distance can be tailored precisely and the pellet can be protected from heat during the injection into the target chamber. A recent experimental campaign to study the influence of the cone walls on the propagation and the transport of TNSA protons has shown that despite of the influence of self-generated electric fields in the cone walls by the recirculating electrons good focusing may be achievable.

After the initial introduction of laser accelerated proton beams for FI theoretical studies have not only quantified the required beam parameters [102, 107, 108], but also recently introduced sophisticated scenarios that have greatly relaxed those parameters. A recent study proposed a combination of two spatially shaped proton beam pulses with a total beam energy that match laser systems which might be available in the not too distant future [109, 110]. The most recent scenario looks for a ring shaped proton beam to impact into the dense fuel and further compress the hot spot and a subsequent pulse of protons in the centre to ignite the double-compressed core. This would cause an energy, which is further reduced by a factor of two compared to the model above. Recently (2010) such laser proton beams have been generated using advanced cone geometries [111].

12.4.5 Medical Applications

Right after the discovery of TNSA ion beams also the prospects for medical applications have been in the focus of research. Besides the possibility of transmuting short-lived isotopes for Positron Emission Tomography (PET) the main interest was in the use as a driver for hadron therapy. Hadron therapy [112–116] is the radiotherapy technique that uses protons, neutrons or carbon ions to irradiate cancer tumours. The use of protons and C ions in radiotherapy has several advantages to the more widely used x-ray radiotherapy. First of all, the proton beam scattering on the atomic electrons is weak and thus there is less irradiation of healthy tissues in the vicinity of the tumour. Second, the slowing down length for a proton with given energy is fixed, which avoids undesirable irradiation of healthy tissues at the rear side of the tumour. Third, the well localised maximum of the proton energy losses in matter (the Bragg peak) leads to a substantial increase of the irradiation dose in the vicinity of the proton stopping point.

The proton energy window of therapeutic interest ranges between 60 and 250 MeV, depending on the location of the tumour. Proton beams with the required parameters are currently obtained using conventional charged particle accelerators such as synchrotrons, cyclotrons, linacs [117]. The use of laser based accelerators has been proposed as an alternative [118–122], which could lead to advantages in term of device compactness and costs.

A laser accelerator could be used simply as a high efficiency ion injector for the proton accelerator, or could replace altogether a conventional proton accelerator. Because of the broad energy and angular spectra of the protons, a energy selection and beam collimation system will be needed [123, 124]. Typically, $\Delta E/E \approx 10^{-2}$ are required for optimal dose delivery over the tumour region. All-optical systems have also been proposed, in which the ion beam acceleration takes place in the treatment room itself and ion beam transport and delivery issues are minimised. In this case the beam energy spread and divergence would have to be minimised by controlling the beam and target parameters. The required energies for deep-seated tumours (>200 MeV) are still in the future, but appear within reach considering the ongoing developments in the field. A demanding requirement to be satisfied is also the system duty factor, i.e., the fraction of time during which the proton beam is available for use that must not be smaller than 0.3.

With the recent experimental results of ion beams in the range up to 80 MeV the lower threshold for medical applications has been achieved. However for deep seated tumours it is questionable if the TNSA mechanism still is the best option or if new mechanism should be explored that not only lead to higher particle energies, but also offer a much smaller energy dispersion to begin with.

12.4.6 Neutron Source

Since the first experiments with ultra-intense lasers nuclear reaction have been observed and also used to diagnose the hot core part of the laser plasma interaction [125]. In addition to the generation and detection of radio-isotopes and transmuted nuclei, neutrons are released either as a result of intense Bremsstrahlung or by electron or ion impact. Because of the large conversion efficiency of laser to ion beam energy and large cross section for subsequent proton neutron reactions, laser driven neutron sources based on the TNSA mechanism have become a focus of modern nuclear research.

One has to distinguish between the different neutron generation mechanisms. At proton beam particle energies in the MeV range the interaction and neutron generation relies on the excitation of giant resonances that result in single (p, n) or multiple (p, Xn) neutron emission. The cross section can be quite large and is energy dependent peaking at characteristic proton impact energies. In the case of two particles in the exit channel, the neutron spectrum is monoenergetic for a given projectile energy and neutron emission angle. However, since the angle and energy spread of laser-emitted particles is large, only strongly exothermal reactions ($Q \gg E_{proj}$) will yield roughly monoenergetic neutrons. Which process takes place in a particular case depends on the combination of target, projectile and momentum transfer. The cross-sections for these processes are in the range of 100 mbarn up to one barn and therefore quite large.

As the driving ion beam is ultra-short and the release mechanism is prompt the neutron pulses are very short and originate from a very small region maximising the net flux on secondary samples. Such a probe exists in the form of fusion neutrons. They are generated by the d(d, n)3He fusion reaction in deuterated targets, and their use as a laser-plasma diagnostic is not fundamentally new. When neutrons are produced from laser accelerated ions in the bulk of an irradiated $(CD_2)_n$ target, they are emitted within a few ps from a volume of the order of a few $(10\,\mu m)^3$. During the neutron pulse, in a distance of several millimeters from the target, fast neutron fluxes of $10^{19}/(cm^2 s)$ can be achieved, which is four orders of magnitude higher than current continuous research reactors can deliver.

In the past the neutron emission caused by (γ, n) and (p, n) reactions from the target have been measured at moderate laser intensities. A typical detector setup is a silver activation detector attached to a photo multiplier tube (PMT). On typical shots, the neutrons are generated by (γ, n) reactions within the target (caused by the bremsstrahlung photons from the relativistic electrons) and by (p, n) reactions of our proton beam impacting on the RCF screen or a dedicated secondary production target. This can be e.g. a target of deuterised plastic (CD2), which was irradiated by a beam of TNSA accelerated deuterons. One can observe the yield of neutrons from (d, d) fusion reactions.

To optimise laser driven neutron sources one can perform simulation studies using the consolidated findings about the particle beam characteristics obtained from laser experiments [126]. The optimisation will be according to the absolute neutron yield, the angle as well as the spectral distribution. For neutron generation we attempt a two-stage target design where the TNSA ion beam irradiates a secondary sample. The advantage of this design is that we can optimise the proton or deuteron generation using different targets in the first stage. According to the beam properties obtained from the first stage it will be possible to optimise the neutron production via proton- and deuteron-induced neutron disintegration reactions, respectively, in the second stage. The neutron production target design (second stage) allows the adaptation to the desired application.

In earlier experimental campaigns at the PHELIX laser facility at GSI Helmholtzzentrum für Schwerionenforschung $> 10^9$ neutrons per shot from proton-induced reactions in copper have been produced. The integrated number of protons was 10^{12} to 10^{13}. Each neutron yield in these experiments already exceeded that from the accelerator driven neutron source FRANZ [127] in Frankfurt (Germany) by

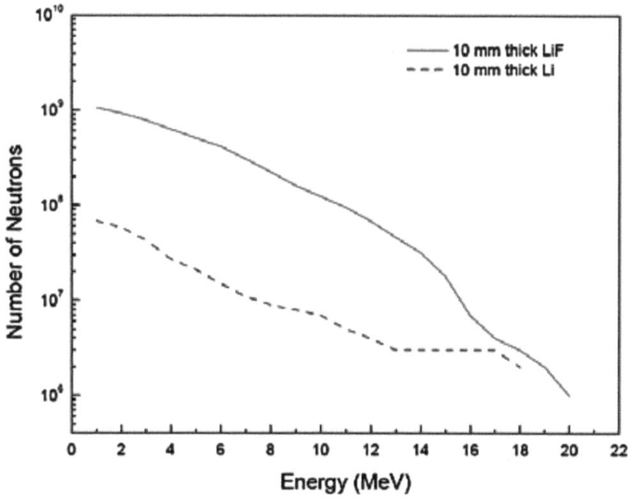

Fig. 12.16 Experimental scheme to investigate the properties of laser-accelerated proton-heated matter by spectrally resolved x-ray Thomson scattering

five orders of magnitude. With the help of the GEANT4 code [128] we can simulate the expected neutron yield for TNSA protons using experimental input spectra. We have simulated the neutron spectra and the production efficiencies using several isotopes within the second-stage target. The thickness of these different targets was 10 mm. The highest production efficiencies were obtained from proton-induced reactions at lithium, beryllium, boron and vanadium in their natural abundance. As a benchmark one can compare the simulation results for copper with experimental data where one finds a good agreement.

The neutron spectra from proton-induced reactions in beryllium and lithium show a high particle number in the lower energy range and in the range around several MeV, respectively. This is of interest in transmutation studies and nuclear material science. Figure 12.16 demonstrates the difference of the simulated neutron spectra using lithium (Li) in the natural abundance and the compound lithium-fluoride (LiF). The properties of the initial proton spectrum which was used in the simulation were obtained from experimental results at the Phelix facility. The initial particle number was 10^{13} and the maximum proton energy was 25 MeV.

In addition, the neutron yield using LiF is much higher than the neutron yield from proton-induced reactions in Li. The explanation is that the mass density of LiF is higher due to the inter-atomic compounds. LiF has a mass density of $2.64\,\mathrm{g/cm^3}$ and the mass density of Li is only $0.53\,\mathrm{g/cm^3}$. This demonstrates the attraction to use composite targets in future studies of laser driven neutron sources. In future developments of the optimisation of laser driven neutron sources we will use more sophisticated composite target designs for the adaptation to the desired application.

References

1. S.J. Gitomer, R.D. Jones, F. Begay et al., Phys. Fluids **29**, 2679 (1986)
2. E.L. Clark, K. Krushelnick, J.R. Davies et al., Phys. Rev. Lett. **84**, 670 (2000)
3. R.A. Snavely et al., Phys. Rev. Lett. **85**, 2945 (2000)
4. A. Maksimchuk et al., Phys. Rev. Lett. **84**, 4108 (2000)
5. M. Schollmeier, *Optimization and control of laser-accelerated proton beams*, Dissertation - Technische Unitversität Darmstadt, 2008
6. M. Borghesi et al., Fusion Sci. Technol. **49**(3), 412 (2006)
7. M. Roth, J. Instrum. **6**, R09001 (2011)
8. M. Roth et al., Phys. Rev. ST-AB **5**, 061301 (2002)
9. B.G. Logan et al., Fusion Sci. Technol. **49**(3), 399 (2006)
10. M. Roth, Plasma Phys. Contr. F. **51**, 014004 (2009)
11. S.C. Wilks et al., Phys. Rev. Lett. **69**, 1383 (1992); S.C. Wilks, Phys. Fluids B **5**, 2603 (1993); B.F. Lasinski et al., Phys. Plasmas **6**, 2041 (1999)
12. S.P. Hatchett et al., Phys. Plasmas **7**, 2076 (2000)
13. P.A. Norreys et al., Phys. Plasmas **6**, 2150 (1999)
14. J. Fuchs et al., Nature Phys. **2**, 48 (2006)
15. Y. Ping et al., Phys. Rev. Lett. **100**, 085004 (2008)
16. M.H. Key et al., Phys. Plasmas **5**, 1966 (1998)
17. J.R. Davies, Phys. Rev. E **65**, 026407 (2002)
18. H. Alfven, Phys. Rev. **55**, 425 (1939)
19. J.R. Davis, Phys. Rev. E **69**, 065401 (R) (2004)
20. D. Batani et al., Phys. Rev. E **65**, 066409 (2002)
21. A.R. Bell et al., Plasma Phys. Contr. F. **48**, 37 (2006)
22. F. Pisani et al., Phys. Rev. E **62**, R5927 (2000)
23. J. Honrubia et al., Laser Part. Beams **24**, 217 (2006)
24. J.J. Santos et al., Phys. Plasmas **14**, 103107 (2007)
25. R.B. Stephens et al., Phys. Rev. E **69**, 066414 (2004)
26. K.L. Lancaster et al., Phys. Rev. Lett. **98**, 125002 (2007)
27. Y. Sentoku et al., Phys. Plasma **14**, 122701 (2007)
28. A. Pukhov, Phys. Rev. Lett. **86**, 3562 (2001)
29. J.J. Santos et al., Phys. Rev. Lett. **89**, 025001 (2002)
30. H. Nishimura et al., PLasma Phys. Contr. F. **47**, 823 (2005)
31. A.J. MacKinnon et al., Phys. Rev. Lett. **88**, 215006 (2002)
32. M. Schollmeier et al., Phys. Plasmas **15**, 053101 (2008)
33. H.A. Bethe, Phys. Rev. **89**, 1256 (1953)
34. K.U. Akli et al., Phys. Plasmas **14**, 023102 (2007)
35. J.E. Crow, P.L. Auer, J.E. Allen, J. Plasma Phys. **14**, 65 (1975)
36. N.A. Krall, A.W. Trivelpiece, *Principles of Plasma Physics* (San Francisco Press Inc., San Francisco, 1986)
37. S. Augst et al., Phys. Rev. Lett. **63**, 2212 (1989)
38. M. Hegelich et al., Phys. Rev. lett **89**, 085002 (2002)
39. P. McKenna et al., Phys. Rev. Lett. **98**, 145001 (2007)
40. J. Schreiber et al., Appl. Phys. B **79**, 1041 (2004)
41. The Mathworks Inc. MATLAB. Website (2011), Available online at http://www.mathworks.com/. Accessed Jul 2011
42. J. Schreiber et al., Phys. Rev. Lett. **97**, 045005 (2006)
43. J. Fuchs et al., Phys. Plasmas **14**, 053105 (2007)
44. P. Mora, Phys. Rev. Lett. **90**, 185002 (2003)
45. P. Mora, Phys. Plasmas **12**, 112102 (2005)
46. P. Mora, Phys. Rev. E **72**, 056401 (2005)
47. B.J. Albright et al., Phys. Rev. Lett. **97**, 115002 (2006)

48. M. Passoni, M. Lontano, Laser Part. Beams **22**, 163 (2004)
49. L.D. Landau, E.M. Lifshitz, *Mekhanika sploshnykh sred (Mechanics of Continuous Media)* (Gostekhizdat, Moskva, 1954)
50. S. Eliezer *The Interaction of High Power Lasers with Plasma* (IOP Publishing, Bristol/Philadelphia, 2002)
51. M. Kaluza et al., Phys. Rev. Lett. **93**, 045003 (2004)
52. L. Robson et al., Nature Phys. **3**, 58 (2007)
53. E. Brambrink, *Untersuchung der Eigenschaften lasererzeugter Ionenstrahlen* PhD Thesis, Technische Universität Darmstadt (2004)
54. A.V. Gurevich et al., Sov. Phys. JETP **22**, 449 (1966)
55. M. Passoni et al., Phys. Rev. E **69**, 026411 (2004)
56. N. Kumar, A. Pukhov, Phys. Plasmas **15**, 053103 (2008)
57. T. Grismayer, P. Mora, Phys. Plasmas **13**, 032103 (2006)
58. T. Grismayer et al., Phys. Rev. E **77**, 066407 (2008)
59. V.Yu. Bychenkov et al., Phys Plasmas **11**, 3242 (2004)
60. M. Allen et al., Phys. Rev. Lett. **93**, 265004 (2004)
61. B.M. Hegelich et al., Nature **439**, 441 (2006)
62. H. Schwoerer et al., Nature **439**, 445 (2006)
63. S. Ter-Avetisyan et al., Phys. Rev. Lett. **96**, 145006 (2006)
64. D.C. Caroll et al., Phys. Rev. E **76**, 065401(R) (2007)
65. T.E. Cowan et al., Phys. Rev. Lett. **92**, 204801 (2004)
66. S. Humphries, *Charged Particle Beams* (Wiley, New York, 2008) Available online at http://www.fieldp.com/cpb.html
67. F. Nürnberg et al., Rev. Sci. Instrum. **80**(3), 033301 (2009)
68. A.P.L. Robinson et al., New J. Phys. **10**, 013021 (2008)
69. L. Yin et al., Phys. Plasmas, **14**, 056706 (2007)
70. P.K. Patel et al., Phys. Rev. Lett. **91**, 125004 (2003)
71. G.M. Dyer et al., Phys. Rev. Lett. **101**, 015002 (2008)
72. P. Antici et al., J. Phys. France **133**, 1077 (2006)
73. R.A. Snavely et al., Phys. Plasmas **14**, 092703 (2007)
74. E. Brambrink et al., Phys. Rev. Lett. **96**, 154801 (2006)
75. M. Roth et al., Phys. Rev. Lett. **86**, 436 (2001)
76. A. Pelka et al., Phys. Rev. Lett. **105**, 265701 (2010)
77. J. Fuchs et al., Phys. Rev. Lett. **91**, 255002 (2003)
78. T. Toncian et al., Science **312**, 410 (2006)
79. K. Harres et al., Phys. Plasmas **17**, 023107 (2010)
80. I. Hofmann et al., in *Proceedings of the HIAT09*, Venice, 6–12 June 2009
81. I. Hofmann et al., Phys. Rev.-STAB **14**, 031304 (2011)
82. S. Yaramishev et al., Nucl. Instrum. Meth. A **588**, 1 (2006)
83. D. West, A.C. Sherwood, Nature **239**, 157 (1972)
84. J. Cobble et al., J. Appl. Phys. **92**, 1775 (2002)
85. M. Borghesi et al., Appl. Phys. Lett. **82**, 1529–1531 (2003)
86. A. Mackinnon et al., Rev. Sci. Instrum. **75**, 3531–3536 (2004)
87. M. Borghesi et al., Phys. Rev. Lett. **88**, 135002 (2002)
88. M. Borghesi et al., Laser Part. Beams **25**, 161–167 (2007)
89. M. Borghesi et al., Phys. Plasmas **9**, 2214 (2002)
90. M. Borghesi et al., Rev. Sci. Instrum. **74**, 1688 (2003)
91. A.J. Mackinnon et al., Rev. Sci. Instrum. **75**, 3531 (2004)
92. M. Borghesi et al., Appl. Phys. Lett. **82**, 1529 (2003)
93. R.W. Lee et al., J. Opt. Am. B **20**, 770 (2003)
94. E. Garcia Saiz et al., Nature Phys. **4**, 093306 (2008)
95. S.H. Glenzer et al., Phys. Rev. Lett. **90**, 175002 (2003)
96. S.H. Glenzer et al., Phys. Rev. Lett. **98**, 065002 (2007)
97. N.A. Tahir et al., Phys. Scrip. T123 **2008**, 014023 (2008)

98. O.L. Landen, J. Quant. Spectrosc. Radiat. Trans. **71**, 465 (2001)
99. M. Schollmeier et al., Laser Part. Beams **24**, 335 (2006)
100. M. Tabak et al., Phys. Plasmas **1**, 1626 (1994)
101. M. Temporal, J.J. Honrubia, S. Atzeni, Phys. Plasmas **9**, 3098 (2002)
102. A. Caruso, V.A. Pais, Nucl. Fusion **36**, 745 (1996)
103. H. Ruhl et al., Plasma Phys. Rep. **27**, 263 (2001)
104. V.Yu. Bychenkov, Plasma Phys. Rep. **27**, 1076 (2001)
105. P.A. Norreys et al., Phys. Plasmas **7**, 372 (2000)
106. R. Kodama et al., Nature **412**, 798 (2001)
107. M. Temporal et al., Phys. Plasmas **9**, 3098 (2002)
108. J. Fuchs et al., Phys. Rev.Lett. **99**, 015002 (2007)
109. M. Temporal, Phys. Plasmas **13**, 122704 (2006)
110. M. Temporal, Phys. Plasmas **15**, 052702 (2008)
111. O. Deppert et al., to be published
112. W.T. Chu, B.A. Ludewigt and T.R. Renner, Rev. Sci. Instrum. **64**, 2055 (1993)
113. V.S. Khoroshkov, K.K. Onosovsky, Instrum. Exp. Tech. **38**, 149 (1995)
114. G. Kraft, Nucl. Instrum. Meth. Phys. Res. Sect. A-Accel. Spectrom. Dect. Assoc. Equip. **454**, 1 (2000)
115. R.R. Wilson, Radiology **47**, 487 (1946)
116. U. Amaldi, Nucl. Phys. A **654**, 375C (1999)
117. W. Wieszczycka, W.H. Scharf, in *Proton Radiotherapy Accelerators*, 323 (World Scientific, River Edge, 2001)
118. S.V. Bulanov et al., Phys. Lett. A **299**, 240 (2002)
119. S.V. Bulanov et al., Plasma Phys. Rep. **28**, 453 (2002)
120. E. Fourkal, C. Ma, Med. Phys. **30**, 1448 (2003)
121. E. Fourkal et al., Med. Phys. **31**, 1883 (2004)
122. V. Malke et al., Med. Phys. **31**, 1587 (2004)
123. E. Fourkal et al., Med. Phys. **29**, 1326 (2002)
124. E. Fourkal et al., Med. Phys. **30**, 1660 (2003)
125. M. Guenther et al., Phys. Plasmas **18**, 083102 (2011)
126. M.M. Guenther et al., Fus. Sci. Technol. **61**(1), 231–236 (2012)
127. U. Ratzinger et al., Conf. Proc. IPAC'10, Kyoto, Japan (2010)
128. S. Agostinelli et al., Nucl. Instrum. Meth. A **506**, 250 (2003)

Chapter 13
Coherent Light Sources in the Extreme Ultraviolet, Frequency Combs and Attosecond Pulses

Matt Zepf

Abstract Converting laser radiation into coherent extreme ultraviolet radiation via high-order harmonic processes allows the creation of extremely broadband spectra with well-behaved phase structure. Such spectra will exhibit attosecond temporal structure – either in the form of an attosecond pulse train or an isolated attosecond pulse. The basic principles of achieving such broad, phase controlled spectra and the two prevalent non-linear media (extended gaseous media and step-like plasma-vacuum interfaces) will be discussed.

13.1 Introduction

The history of science and human discovery is also a history of describing and recording nature to develop and communicate scientific theories and concepts. The strong influence of vision on how we perceive the world around us makes images of natural processes extremely powerful in shaping our understanding of the world. Thus it comes as no surprise that using lenses as a means of enhancing our vision (as magnifying glasses) dates back to Greek and Roman times. The impact of using pairs of glass lenses to form telescopes by Galileo challenged and led to a transformation of our worldview. However, the recording of what was observed had to be performed by hand with its obvious limitations in terms of speed and accuracy. Since the development of the first photographic materials in the early part of the nineteenth century by Niépce and Daguerre, photographic recording has undergone an extremely rapid development, culminating in pixelated semi-conductor devices such as CCDs (charge-coupled devices), which allow rapid acquisition and storage of digital images. Digital images are not only a simplification of the processes

M. Zepf (✉)
School of Mathematics and Physics, Queen's University of Belfast, BT7 1NN, UK
e-mail: m.zepf@qub.ac.uk

P. McKenna et al. (eds.), *Laser-Plasma Interactions and Applications*, Scottish Graduate Series, DOI 10.1007/978-3-319-00038-1_13,

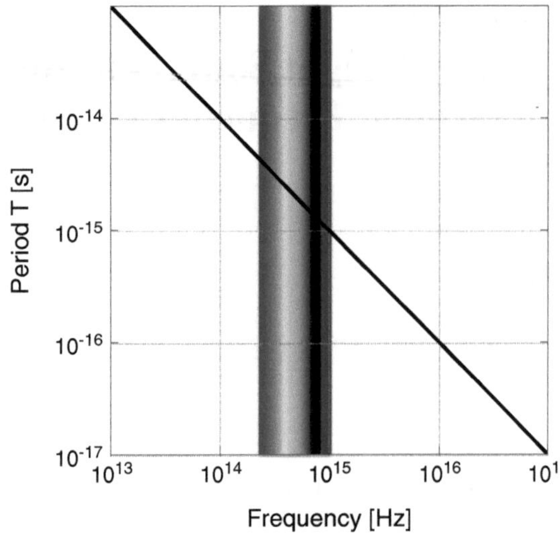

Fig. 13.1 Single cycle limit vs. carrier frequency. Note that achieving pulse durations below 1 fs requires frequencies beyond the optical spectrum

possible until then, they are also the essential ingredient to novel forms of microscopy using high quality X-ray beams for lensless microscopy and hence the reconstruction of objects on a nm to scale [1], which can also be combined with femtosecond (10^{-15} s) temporal resolution. To achieve recording on such extreme temporal and spatial scales requires a photon source, that can provide sufficiently well controlled (coherent, short) pulses of light. Thus, extending the realm of natural phenomena that can be investigated using photons relies as much on the development of light sources as on the development of suitable detection systems. In particular recording dynamic events without motion blurring requires either a bright continuous light source combined with a detector with sufficient temporal resolution (shuttering/gating) or a slow detector with a bright sufficiently short pulse light source. Note that the two approaches differ substantially in terms of what can realistically be achieved. While mechanical shutters are ultimately limited (by inertia) to shutter speeds of around a microsecond and electrically gated shutters are limited to temporal windows of 50–100 ps by the risetime of the electrical pulses driving them, the limit for photon light sources is simply given by the Fourier-transform limit of the spectral width of the pulse. Thus the ultimate limit for a given pulse of a frequency f is a temporal duration of the associated optical cycle $T = 1/f$.[1] Figure 13.1 shows the duration of a single optical cycle vs. the centre frequency.

What is clear from Fig. 13.1 is that the shortest pulse duration that can be achieved using optical pulses is limited to just above 1 fs. While this is an extremely short pulse, it is still long compared to the dynamics of a bound electron. For

[1]Note that the temporal resolution of electrically gated devices such as Pockels cells and gated MCPs is also limited by the effective bandwidth of the electrical signals and their dispersion.

example the timescale associated with the Bohr-orbit in the ground state of hydrogen is $\tau_{Bohr} = v/r_{Bohr} = 2.4 \times 10^{-17}$ s = 24 as (also known as the atomic unit of time). Consequently, freezing the dynamics of electrons under such conditions requires pulses in the attosecond regime and therefore light pulses in the extreme ultraviolet part of the electromagnetic spectrum and beyond [2–4].

13.1.1 Requirements for Attosecond Pulses

The challenge is therefore to generate and control light under such conditions. To understand the requirements that we need to meet, it is useful to first recap how ultrafast pulses close to the single cycle limit are generated using optical lasers. The first condition that needs to be met by any light source that aims to be shorter than a given pulse duration is that the bandwidth must be sufficiently large to support the pulse duration. The Fourier transform limit of a given spectrum is given by $\Delta\nu\Delta t = \beta$, whereby β is constant of the order of unity that depends on the exact spectral shape. For example, for a Gaussian spectral shape $\beta = 0.441$, while a sech^2 has $\beta = 0.315$.

While large spectral bandwidth is a necessary requirement, it is clearly not sufficient (think of a light-bulb!). The key parameter that distinguishes a light-bulb from a femtosecond optical pulse is the spectral phase. The Fourier-transform limited pulse duration is only achieved if the phase of all spectral components is identical. Hence, one needs to find a means of producing a broad spectrum with identical spectral phase. In the optical regime, the characteristics of our light pulse are controlled by designing the optical cavity to ensure that only those photons that match our requirements are allowed to propagate in the oscillator, while the unwanted photons have a net-loss in each round-trip and thus die away exponentially.

Remember that oscillator cavities only support discrete frequencies (oscillator modes) separated by a frequency $\Delta\nu_{comb} = c/2L$ (i.e. the cavity round-trip length $2L$ is a multiple of the wavelength λ). Hence an oscillator produces a comb of equally spaced spectral peaks. In time, such a frequency comb with constant spectral phase corresponds not to a single short pulse but a *pulse train* (Fig. 13.2) [5, 6]. That this should be the case is also easily understood from the basic layout of an oscillator which contains a pulse that is short compared to the cavity length (Fig. 13.3). Every time the pulse reflects from the partial reflector, part of the pulse is also transmitted and the separation between each of the transmitted pulses must just be the cavity round-trip time of the oscillator. The task of achieving a transform limited pulse is therefore equivalent to achieving a fixed relative phase between the individual modes of the oscillator – a mode-locked frequency comb. For a reflective cavity with no dispersive elements this will be fulfilled to very good approximation. However a laser cavity must include a gain medium, which contributes dispersion (i.e. frequency dependent phase) to the cavity and thus results in the phase between two modes varying from one round-trip to the next. To regain the fixed relative phase

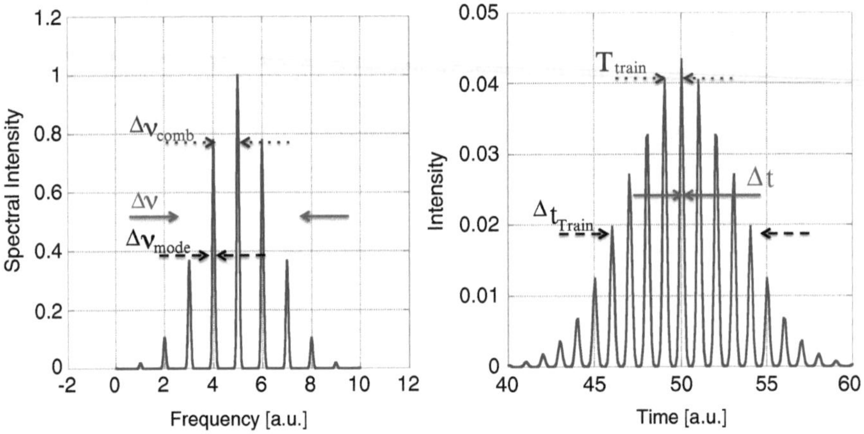

Fig. 13.2 Fourier-transfom of a frequency comb under the assumption of ideal (constant) spectral phase. Note the correspondence between spectrum and time of the various structures, with corresponding features in frequency and time highlighted by corresponding colour and lines. The temporal duration of each attosecond spike $\Delta t \Delta \nu$ is set by the total spectral width $\Delta \nu$ (solid line), the separation of each peak in the frequency comb $\Delta \nu_{comb}$ (i.e. mode-separation or harmonic separation) determines the period of pulse-train T_{train} (dots) while the width of each individual (harmonic or mode) peak $\Delta \nu_{mode}$ determines the temporal width of the train as a whole Δt_{train} (dashed line)

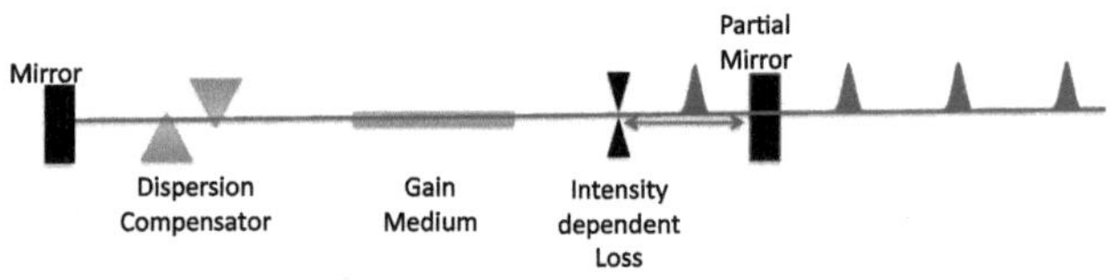

Fig. 13.3 Schematic of a mode-locked oscillator showing the key components required for operation. The dispersion compensator (shown here as a prism pair) ensures that the relative spectral phase of each mode stays constant from one round-trip to the next. The intensity dependent loss mechanism (shown here as an aperture placed at the focus of a Kerr-lens) ensures that only intense pulses can propagate through the cavity with a net gain. The separation of the peaks in the pulse-train transmitted through the partial mirror corresponds to one oscillator round-trip in reality

between these modes (i.e. mode-locked operation), a short pulse cavity must include additional dispersive elements that compensate the dispersion in the gain medium. Common approaches to achieving this goal are a prism-pair or dispersive ('chirped') mirrors (Fig. 13.3). Compensating the dispersion allows the relative spectral phase of the modes to remain fixed. It does not, however, result in the selection of modes with identical spectral phase. Since the shortest pulse for a given average power in the oscillator cavity corresponds to the highest peak intensity, the selection of a short pulse is achieved by introducing an intensity dependent loss mechanism (such as a Kerr-lens with an aperture at its focus or a saturable absorber).

It is easy to see that such an optical layout will produce a pulse-train of ultra-short pulses with a repetition rate corresponding to the cavity round-trip time $T_{train} = 1/\Delta\nu_{comb} = 2L/c$. For time resolved applications using an isolated pulse is of course highly desirable. In the optical regime selecting a single pulse from such an oscillator can be straightforwardly achieved by electro-optical switching using a Pockels-cell, since typical cavity round-trip times (and therefore interpulse spacings) are $\sim 10\,\text{ns}$. The changed temporal structure (by selecting a single pulse) must result in a change of the spectral structure. Selecting a single pulse corresponds to a pulse-train with infinite spacing. Since the separation of the pulses in time is inversely proportional to the comb-spacing in frequency, we obtain $\Delta\nu_{comb} = 0$ and hence a continuous spectrum.

At this point it is worth emphasising some basic corollaries from our discussion (Fig. 13.2).

1. Although the lasing medium produces, and can amplify, emission from a continuous range of frequencies the temporal interference in a cavity results in a spectral structure commonly referred to as a frequency comb.
2. Achieving constant spectral phase across the frequency comb results in the individual pulses having a duration corresponding to the Fourier Transform Limit.
3. Isolating a single pulse from the train changes the spectrum from a frequency comb to a continuous spectrum while retaining the same shape of the envelope.
4. The temporal separation of the pulses is connected to the frequency separation of the individual modes as $T_{train} = 1/\Delta\nu_{comb}$
5. The temporal duration of each pulse in the train is determined by the width of the spectral envelope $\Delta t = \beta/\Delta\nu$
6. The width of the temporal envelope in the pulse-train Δt_{train} is determined by the spectral width $\Delta\nu_{mode}$ of the individual frequency comb spikes

13.2 Producing an Attosecond Pulse – Harmonic Generation

From the previous discussion it is clear that scaling the principle of a femtosecond oscillator to an attosecond pulse(-train) requires the production of a phase-locked frequency distribution (-comb) with an overall spectral width $\Delta\nu$ sufficient to support the desired pulse-duration. Once this has been achieved the remaining challenge is to isolate an individual pulse from the train (or to ensure only one is produced in the first instance). As can be seen from Fig. 13.1 achieving a pulse duration below 100 as requires a central frequency of $> 10^{16}$ Hz or expressed in terms of wavelength of $< 30\,\text{nm}$. The lack of optical components with similar quality to those found in visible/infra-red lasers in terms of reflectivity and transparency and the lack of easily pumped gain media result in the oscillator approach developed at optical wavelengths no longer being directly applicable.

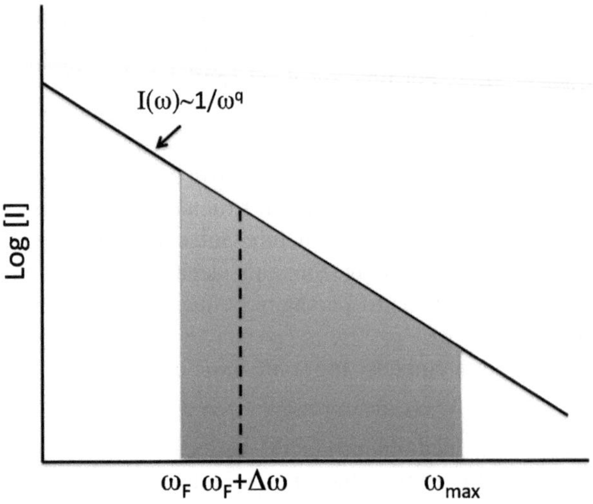

Fig. 13.4 Schematic of a harmonic spectrum with power-law decay ω^{-q} (plotted with both axes logarithmic). The spectral width and hence the Fourier-transform limited pulse duration is determined only by the low-frequency cut-off of the filter ω_F and the parameter q

However the principles derived from mode-locked oscillators still apply. The frequency comb technique is extended to short wavelength by generating harmonics (integer multiples) of the laser frequency at frequencies $\omega_m = m\omega_{Laser}$. Harmonics can be generated in any medium that displays a strong non-linear response to the incident laser light. Consider an electron bound in an atomic potential: as long as the potential shape is parabolic to good approximation, the electron will respond as a simple harmonic oscillator, which oscillates and emits light only at the frequency of the incident light wave. Since all atomic potential wells are parabolic to good approximation for sufficiently small amplitudes (cf. Taylor expansion), the polarisation at low intensities is simply linear with the applied electric field $\mathbf{P}^{(1)} = \varepsilon_0 \chi^{(1)} \mathbf{E}$ and the superposition of the external and re-emitted field results in the well-known linear refractive index n. At higher intensities the potential shape generally deviates from an ideal parabola and the polarisation of the medium contains higher order terms $|P^{(n)}| \sim \chi^{(n)} E^n$. In terms of the electron's motion this simply implies that the electron's displacement x must be described as a series that requires higher frequencies $2\omega, 3\omega, \ldots, n\omega$ to describe the motion and the emitted field accurately, and thus the emission of higher integer multiples of the laser frequency (harmonics). However, the production of harmonics on its own is not sufficient to achieve as substantial reduction of the pulse duration. To lead to a substantial increase in the effective bandwidth harmonic spectrum must decay sufficiently slowly. Assuming a simple power law decay of a given spectrum [7], such that $I(\omega) \sim 1/\omega^q$ and a simple step filter that transmits above some critical frequency ω_F one obtains an effective bandwidth of the spectrum collected by the filter of $\Delta\omega = (2^{1/q} - 1)\omega_F$ (Fig. 13.4). Note that the transform limited pulse duration is purely determined by

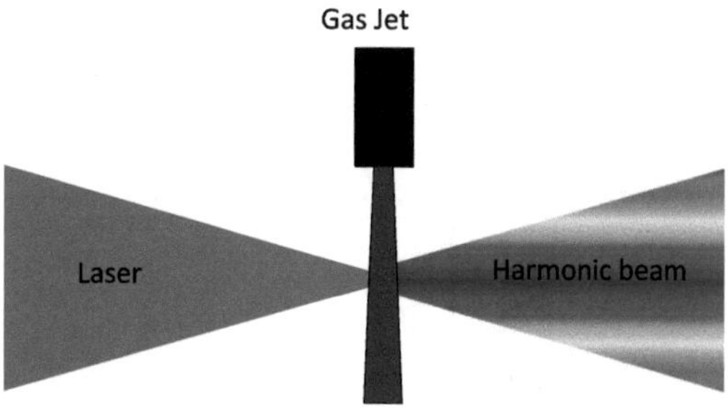

Fig. 13.5 Schematic of a simple HHG experiment with a gaseous non-linear medium. Typical parameters would be lasers with few to 10's of optical cycles with intensities close to the saturation intensity for tunnel-ionisation for the medium of interest (typically 10^{14}–10^{15} W cm^{-2})

the decay of the spectrum and the lowest transmitted frequency ω_F. Any constraint on the maximum frequency ω_{max}in the transmitted spectrum has no bearing on the pulse duration, so long as $\omega_{max} > \omega_F + \Delta\omega$.

While harmonics generated from bound electrons via the perturbative process described above find widespread use in the frequency doubling of lasers in crystals and even some higher order direct processes in gases such as 3rd harmonic generation, their intensity generally decays too rapidly to higher orders to be a source of a broad harmonic frequency comb suitable for producing a higher central frequency with increased spectral bandwidth. Note that bulk solid media can be ruled out as a suitable non-linear medium for attosecond pulse generation, because they strongly absorb the high frequencies required for attosecond pulse production. Therefore we will need to look beyond the perturbative harmonic generation from bound electrons i.e. consider laser fields comparable or large with respect to the binding field strength and consider only gaseous media or surfaces.

13.2.1 Non-Linear Medium 1: Harmonic Generation from Gaseous Targets (HHG)

As the intensity is increased further – e.g. by placing a jet of atoms into the focus of a femtosecond laser (Fig. 13.5) – to an appreciable percentage of the Coulomb field of the atom, the perturbative approximation breaks down, because the electron is no longer trapped in the potential well of the atom, but can instead tunnel ionise. Once ionised the electrons motion can be described to good approximation by the motion of a free electron in the laser field [8]. For linear polarisation there is a high probability that the electron will recollide with the atom and emit a photon with an energy of $h\nu = I_p + W_{kin}$ (where I_p is the ionisation potential, W_{kin} the kinetic energy of the returning electron). By solving the equation of motion for an electron that tunnel ionises at a time t_0 in a field given by $E(t) = E_0 \cos(\omega t)$ one finds that the electrons velocity at a given point in time depends on t_0 as

$$v(t,t_0) = -\frac{eE_0}{m\omega}[\sin(\omega t) - \sin(\omega t_0)] \tag{13.1}$$

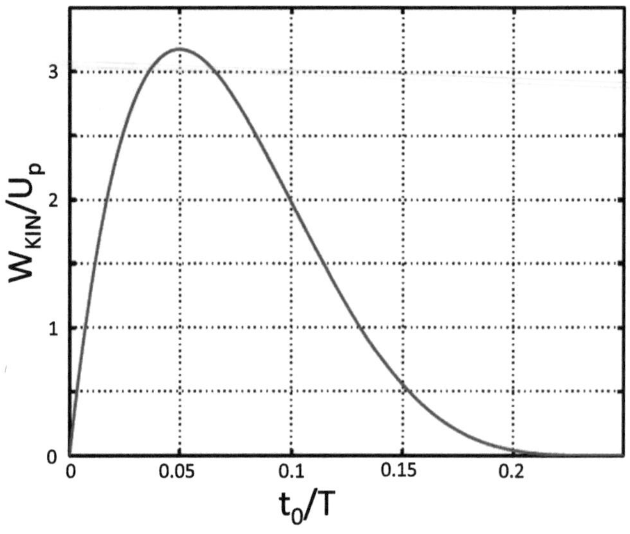

Fig. 13.6 Kinetic energy upon first return versus tunnelling time in units of the laser period T. The maximum possible return energy is $3.17 \times$ the ponderomotive energy U_p. Note that for all return energies apart from $3.17U_p$ two distinct tunnelling times exist that lead to the same return energy, corresponding to the long and short trajectories respectively. The pattern repeats in the second half-cycle $t_0/T > 0.5$. Note that not all tunnelling times t_0 lead to solutions that return to the parent ion within the first optical cycle

The return energy therefore also depends on t_0 and reaches a maximum value of $W_{kin} = 3.17U_p$ for electrons which ionise 17° after the peak of the electric field of the laser (Fig. 13.6, where the ponderomotive energy $U_p = e^2E^2/me\omega^2 = 9.33 \times 10^{-14}I[\mathrm{W\,cm^{-2}}]\lambda^2[\mu\mathrm{m}^2]$ is the kinetic energy of a free electron oscillating in the laser field). Therefore the highest possible photon energy is given by the sum of the kinetic energy and the ionisation potential.

$$h\nu_{max} = I_p + 3.17U_p \tag{13.2}$$

Thus, harmonic generation in the tunnel-ionising regime is generally thought of as a three-step process [8]:

1. Electron tunnel ionises
2. Electron gains energy in the laser field
3. Harmonic photons are emitted upon recombination with the atom

Unlike perturbative harmonic generation discussed earlier, the probability of generating a photon at a given harmonic order m, is only weakly dependent on the harmonic order. We might expect photon energies that correspond to electrons 'born' at the peak of the field, where the ionisation rate is highest (Fig. 13.7), to be somewhat stronger and thus favouring higher harmonic orders. However this effect is off-set by the fact that the recombination probability is higher for electrons with

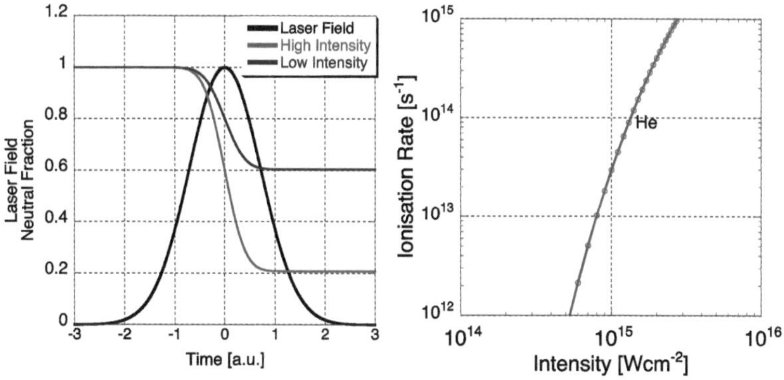

Fig. 13.7 Temporal dependence of ionisation fraction (*left*) and intensity dependence of ionisation rate (*right*). The remaining neutral fraction at the peak of the pulse strongly depends on the pulse intensity

lower return energy, resulting a slow decay of the harmonic spectrum (Fig. 13.9). As discussed earlier, the effective bandwidth of a slowly decaying frequency comb can be very large and would be sufficient for transform limited pulses of 10's of attoseconds. Note that harmonics are generated twice per optical cycle with a separation of $T/2$, since the ionisation dynamics are symmetrical with respect to the sign of the electric field. From our previous discussion on laser cavities we see that harmonics must therefore be separated by $\Delta\nu = 2/T = 2\nu_{Laser}$ and therefore only odd harmonic orders $n = 1,3,5,$ are generated. From the physical picture of the dynamics on a single atom level described above, it may seem surprising that one observes emission of well defined harmonics at all. Since electrons can return with a continuum of energies $W_{kin}(t_0) = 0\ldots3.17U_p$ one might expect to observe the emission of a continuous photon energy spectrum instead of a discrete frequency comb. However we can easily see that even harmonic orders generated in one half-cycle are simply cancelled by destructive interference with the even harmonics generated in the following half-cycle. The interference between waves generated in each half cycle also explains the observation of well-defined harmonics per se. Similar to a Fabry-Perot etalon, the sharpness of a peak increases with increasing number of interfering beams or, in our case, an increasing number of attosecond pulses. Therefore the appearance of odd harmonics can be understood as interference between the individual attosecond pulses in the pulse train. Isolating a single attosecond pulse from this train must therefore result in a continuous spectrum being observed as is indeed the case.

13.2.1.1 Short Wavelength Limit of HHG

From Eq. 13.2 it is clear that to produce harmonics with the shortest wavelength requires the highest possible value of both U_p (and therefore $I\lambda^2$) and I_p. However, these two parameters are not independent of each other, and the maximum possible value of U_p depends on I_p and the pulse duration. Since ionisation is an intrinsic part of the HHG process, the medium (typically a neutral noble gas or noble gas ion) will become depleted during the laser pulse (Fig. 13.7). The maximum intensity I_{SAT} for a given species is given by the point where the remaining amount $N(I)$ of the generating species at the peak of the pulse falls below a given threshold (e.g. $N(I_{SAT})/N_0 < 1/e$). The dominant ionisation mechanism for HHG experiments typically tunnel ionisation, which is well approximated by the ADK formula (Ammosov, Delone, Krainov [9]). Figure 13.7 shows the tunnel ionisation rate as a function of intensity. It is clear that even for very short pulses the ionisation probability will rapidly approach unity during the laser pulse even for moderate intensities.

$$W = \int R(t)dt \sim \frac{R_{Imax}\tau}{2} \tag{13.3}$$

Figure 13.7 shows the temporal dependence of ionisation during a laser pulse. For the higher intensity pulse the cumulative ionisation during the pulse has significantly depleted the neutral species before the peak of the pulse, while for the lower intensity pulse most atoms remain neutral. While HHG from ions [10] is possible, it is much harder to achieve a strong response from the entire medium due to macroscopic phase-matching considerations discussed below. The strong preference for neutral media arising from this makes the neutral atoms with the highest ionisation potentials the preferred choice as HHG medium (He 25.4 eV, Ne 21.6 eV, Ar 15.8, Kr 14 eV, Xe 11 eV). Equation 13.2 suggests that the use of long wavelengths can be used to extend the cut-off for neutral media by using very long laser wavelengths. While this is indeed true and photon energies exceeding 1 keV have been produced with long wavelength lasers, the benefit of this approach is offset to a certain degree by the strong wavelength scaling of harmonic emission probability $P \propto \lambda^{-6}$ [11]. This scaling can be understood in qualitative terms to be due to wave packet spreading after ionisation reducing the amplitude of the electron wave function at the point of recollision and thus substantially reducing the recollision probability. This dependence on the wave packet evolution between ionisation and recollision points to the importance of the 2nd phase of the 3-step process in understanding the behaviour of HHG.

13.2.1.2 HHG Phase

Thus HHG from gaseous targets appears to be a promising route to generating attosecond pulses. However, before we consider our quest for an attosecond pulse complete we must first consider the phase structure of the harmonics, the route to achieving appreciable conversion efficiencies and selection of a single attosecond pulse. For pertubative harmonics generated from bound electrons, the phase of the harmonic is essentially dictated by that of the driving laser. In this case we can view the harmonic generation process as a strongly driven oscillator, where the relative

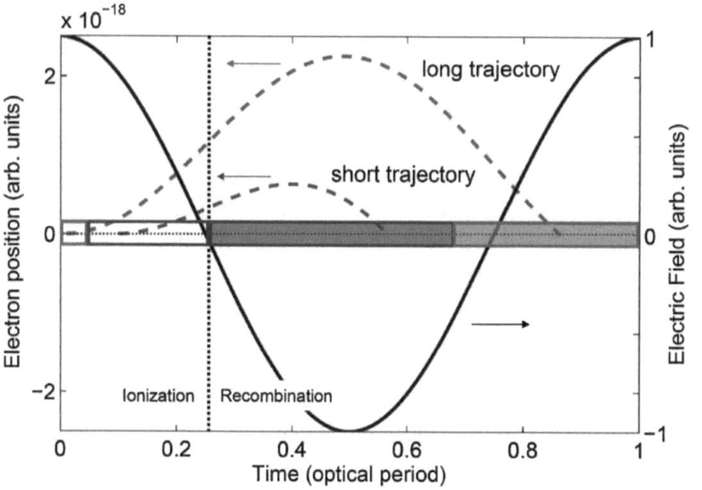

Fig. 13.8 Illustration of the two distinct electron trajectories in HHG. Times of ionisation and recombination for the long (larger area) and short (smaller area) trajectory in a laser field (solid curve). Two example trajectories are highlighted for a short trajectory ionised at $t_0 = 0.1$ and long trajectory ionised at $t_0 = 0.01$ (dashed curves). They are plotted as electron position with respect to the parent ion vs. time

phase between the driving force and the electron oscillation depends only on the ratio of the laser frequency to the resonance frequency of the oscillator. For HHG generated via the 3-step process described above, the situation is significantly more complex. From a purely classical view of the electron trajectory after ionisation, it is clear that the time of recombination depends on the laser phase at the point of ionisation and hence the time elapsed between ionisation and recombination is a function of the emitted photon energy, implying that we would expect a chirp in the spectrum of each individual attosecond burst. Close inspection of the equations of motion shows that during an optical half-cycle all return energies can occur for 2 distinct values of ionisation phase, with the exception of the cut-off energy, which only occurs for a single well defined value of t_0 (Fig. 13.6). These distinct quantum paths are referred to as the long and short trajectory respectively (Fig. 13.8) and for each trajectory the relative phase between the harmonic and the laser must be different since they relative phase clearly depends on the time of return t_R. The full quantum mechanical picture must take into account the phase of the electron wave-packet, which is proportional to the quasi-classical action integral along the path of the free electron $S(t_0, t_R)$ [12].[2] The accumulated phase of the electron is

$$\phi(m) = m\omega t_R - \frac{1}{\hbar} S(t_0, t_R) \tag{13.4}$$

[2]The action is the product of electron energy and time.

This distinct emission phase for short and the long path results in interference between the two quantum paths. In practise, optimising the conversion efficiency and selecting an isolated attosecond pulse requires phase matching (discussed below), which automatically results in the selection of one quantum path over the other. For the further discussion it is always assumed that one quantum path is being selected via phase matching. However, the fact that relative phase between the harmonic and the laser depends on the time and mean energy between ionisation and recollision, implies a different phase for each harmonic order. Therefore the HHG spectrum is intrinsically chirped and thus the attosecond pulses in the train are longer than the FTL. Controlling this chirp requires the use of suitably dispersive filters or chirped (dispersive) XUV multi-layer mirrors.

13.2.1.3 Isolating a Single Attosecond Pulse

For time-resolved experiments having only one pulse greatly simplifies the measurement and hence one would like to find a way to isolate a single attosecond pulse. Applying an optical switch as in the case of a mode-locked oscillator separately from the production of the pulse-train appears impractical because the temporal separation of the individual pulses is extremely short. In practise therefore the selection of an individual attosecond pulse is performed by controlling the properties of the laser generating the harmonics. There are two main approaches to isolating an individual attosecond pulse:

(i) Intensity gating: For very short pulses (~ 5 fs) the neighbouring optical half-cycles relative to the peak cycle already have an appreciable lower intensity. The intensity dependence of the HHG process is mainly determined by the tunnel ionisation rate $R(I)$, which can be approximated as $R \propto I^5 \ldots I^7$ in the regime of interest. This implies a strong suppression of the neighbouring half cycle in terms of conversion efficiency. Secondly, the cut-off harmonic scales linearly with I, implying that there is a spectral region which is only produced by the strongest half-cycle (Fig. 13.9). In this spectral region only a single burst of XUV radiation is produced during each pulse and thus an appropriate filter or multi-layer mirror can be used to select the desired spectral range [13].

(ii) Polarisation gating: For longer pulses, the efficiency from one cycle to the next can be controlled by exploiting the polarisation dependence of the HHG process [14, 15]. The strong reduction in efficiency with ellipticity of the laser light can be understood in terms of the trajectory of the free electron in the laser field. In the case of circular polarisation the electrons will not generally return to the parent ion and hence no XUV photons are emitted. If a laser pulse with time varying ellipticity $\varepsilon(t)$ can be produced it will only produce harmonics efficiently for those cycle which are close to linearly polarised, i.e. $\varepsilon(t) \sim 0$). Such a polarisation state can be produced by superimposing two delayed pulses with left- and right-circular polarisation respectively. For sufficiently short pulses the laser will only be linearly polarised for one half-cycle, resulting in the emission of an isolated attosecond pulse.

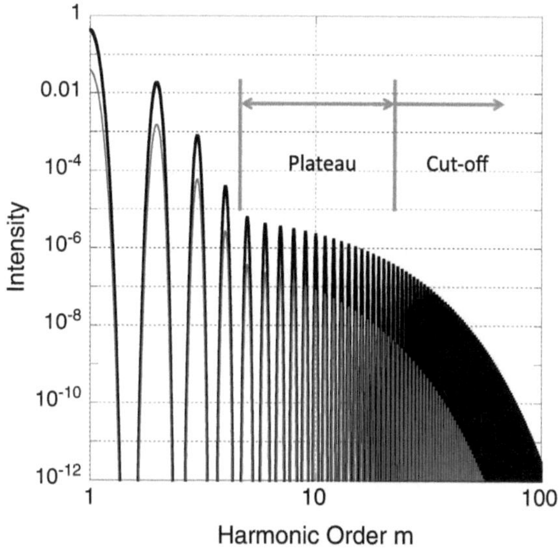

Fig. 13.9 Typical harmonic spectrum exhibiting rapid decay to a plateau region and exponential cut-off. The two spectra are similar except for the strength and position of the cut-off corresponding to a spectrum generated with lower (red) and higher intensity (black) for a few cycle laser. If the cut-off region of the black spectrum is uniquely produced by one cycle it becomes spectrally continuous and with suitable filtering will result in an isolated attosecond pulse (Color figure online)

13.2.1.4 The Role of Phase Matching

The macroscopic response of the medium must depend on the coherent sum of all emitters at the point of the observer [16]. Clearly achieving the maximum harmonic signal therefore requires that all emitters add in phase, or at the very least, not destructively i.e. with a phase difference of $< \pi$. For perfect coherent overlap of the fields emitted from N individual atoms, the electric field will simply be $E = NE_{atom}$ and the measured intensity $I \propto N^2$. The total number of atoms contributing to the harmonic signal is simply $N = n_a LA$, where n_a is the atomic density, L the length and A the effective area of the focal spot. Including the probability of a harmonic photon being emitted for a given atom P_m, the maximum achievable intensity is therefore $I(m) \propto (n_a LAP_m)^2$. For otherwise fixed parameters, the effective length L_{eff} is limited to either

- the maximum length L_I over which a sufficiently high intensity can be maintained (usually due to defocusing or absorption of the laser radiation)
- the absorption length $L_{abs} = (n_a * \sigma)^{-1}$ of the harmonic radiation [16], where σ is the absorption cross-section.
- the coherence length L_c by dispersion between the harmonic and the laser field.

The coherence length $L_c = \pi/\Delta k$ is the length over which a signal can grow without destructive interference in extended non-linear medium, where $\Delta k = k_m - mk_{Laser}$. Here m is the harmonic order k_i is the wave vector of the harmonic or laser respectively. Clearly in vacuum $k_m = mk_{Laser}$ and one has perfect phase matching $\Delta k = 0$. In the presence of a medium the phase matching corresponds to the harmonic and the laser driving the interaction having identical phase velocities and therefore an identical refractive index n. The effect of any mismatch $\Delta k > \pi$ is very substantial and the phase matching form-factor $F(\Delta k)$ is shown in Fig. 13.10.

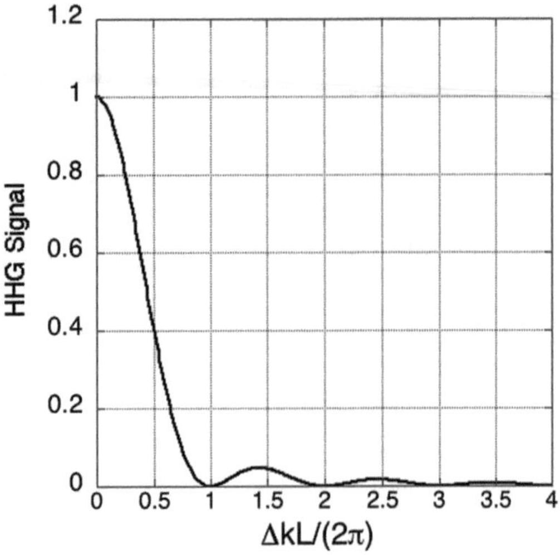

Fig. 13.10 Phase matching factor $F(\Delta k)$ as a function of phase mismatch ΔkL. Note the rapid decay beyond a total mismatch of π radians

Phase matching can only be achieved over a narrow time window, since the continuous ionisation of the medium leads to time varying contributions to the dispersion (Figure 13.7). Optimised harmonic generation can thus be summarised by achieving phase matching over the maximum possible length allowed by absorption L_{abs} at the peak of the laser pulse. The ideal choice of medium (in the presence of phase-matching) is therefore determined by optimising the ratio P/σ.

In practice, all dispersive terms are wavelength dependent and thus phase matching can in principle only be achieved by balancing the different contributions to the dispersion. The wavelength dependence of the dispersive terms implies that one would expect this to be exactly possible for only one wavelength, though achieving $L_c > L_{abs}$ may be possible over a fairly wide range of wavelengths. In practice, the refractive index for the high order harmonic can be assumed to be $n_m = 1$. In the case of phase matching in a capillary waveguide phase matching is dominated by laser propagation effects [17]

$$k_{Laser} \approx \frac{2\pi}{\lambda} + \frac{2\pi p(1-\eta)\delta(\lambda)}{\lambda} - p\eta N_a tmr_e\lambda - \frac{u_{11}^2\lambda}{4\pi a^2} \tag{13.5}$$

where the terms are the vacuum k-vector, the neutral atom dispersion, the plasma dispersion and the waveguide dispersion (with p: pressure in atm η: ionisation fraction, N_{atm}: number density at 1 atmosphere, r_e: classical electron radius, δ: neutral gas dispersion.

In free propagating geometries the waveguide dispersion term would be replaced by the Guoy shift [18] which has the useful property of changing sign in the focus. For all practically relevant circumstances, the dominant term with $(n-1) < 0$ is the refractive index due to free electrons while the leading term with $(n-1) > 0$ is refractive index of the neutral atoms. As a rule of thumb, the free electron dispersion is around 20–50× greater than that neutral dispersion of the gas thus implying that phase matching should be achieved at ionisation levels of a few % [17]. This constrains the highest harmonic that can be achieved due to Eq. 13.2 and thus is applicable only to harmonic orders up to 30 using a Ti:Sapphire laser at 800 nm and that phase matching ions with a charge state $Z > 1$ is impossible since no neutral atoms are available to balance the dispersion of the free electrons. The λ^2 scaling of the cut-off allows phase matching using this approach to be extended to shorter wavelengths, but at the cost of a much weaker ($P \sim \lambda^{-6}$) single atom response and thus limited overall response [19].

13.2.1.5 Quasi-Phase Matching (QPM)

Thus, while true phase matching ($\Delta k = 0$) is desirable, it is only achievable in a narrow parameter space. Therefore optimisation of other parameters such as the laser intensity, ionisation potential and P/σ is significantly constrained by the need to maintain phase matching. So called quasi-phase matching provides an alternative to ensure rapid signal growth of harmonic radiation. The principle of quasi-phase matching is illustrated in Fig. 13.11 [20], and simply relies on suppressing the out-of-phase contributions along the propagation path. This can be done by any means, which varies the harmonic generation efficiency (intensity, medium etc.). Figure 13.11 illustrates the effect different QPM scenarios and compares these with a situation where $\Delta k \neq 0$. In the mismatched case, the signal grows for one coherence length L_c and oscillates between 0 and the maximum value achieved after one coherence length thereafter i.e. there is no advantage to using a medium longer than L_c in this case. The operating principle of QPM can be seen clearly in the ideal QPM case: The harmonic intensity initially increases for a length of L_c, for the subsequent coherence length the phase between the drive laser and harmonic field continues to slip but the overall signal level remains constant since HHG is suppressed. This process continues periodically leading to rapid signal growth. The signal will then grow quadratically with the number of QPM periods N_{QPM} (consisting of a HHG or ON zone and an suppressed or OFF zone). Recent advances have shown that interchanging noble gas with hydrogen jets allows the HHG signal to grow at the theoretical rate of N_{QPM}^2 [21] thus decoupling the challenge of phase matching from other relevant parameters.

13.2.2 Non-Linear Medium 2: Harmonic Generation from Plasma Vacuum Interfaces (SHHG)

From our initial considerations it has become clear that attosecond pulse production requires a medium with a strong non-linear response that is capable of providing a harmonic frequency comb with a well-defined phase behaviour. There are two main areas in which one would like to go beyond the performance currently available with HHG. Firstly, higher pulse energy would be highly desirable for a number of

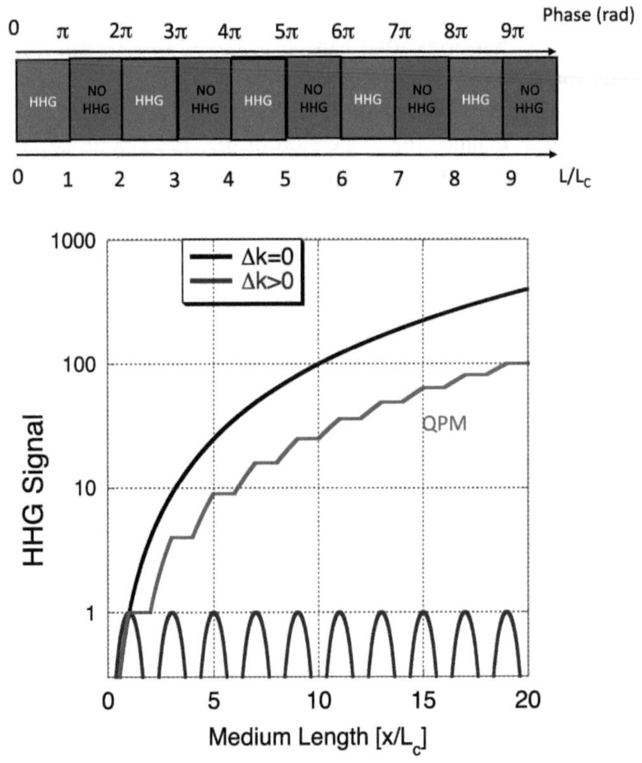

Fig. 13.11 Quasi-phase matching (QPM) allows coherent build-up of signal in the presence of wave vector mismatch ($\Delta k \neq 0$). The harmonic source term must be modulated to suppress the harmonic production over each alternate coherence length L_c (marked 'NO HHG') resulting in constructive interference between the HHG zones marked 'HHG'. The signal growth for a medium with ideal QPM is compared to perfect phase matching and mismatch in the absence of QPM in the lower graph

possible applications. Secondly, the highest harmonic order that can be produced with reasonable efficiency is constrained to below a few hundred eV photon energy. The energy in a given attosecond pulse is determined by the energy of the drive laser pulse and the conversion efficiency. The conversion efficiency is quite low, owing in part to the difficulty of phase matching at shorter wavelengths and limitations on the effective density length product due to absorption and defocusing, while the low intensity required for optimal HHG of $< 10^{15}\ \mathrm{W\,cm^{-2}}$ makes it hard to exploit high peak powers and pulse energy available with current ultra-fast lasers. For example a 20 cm diameter petawatt power laser would require a focal length of 2 km![3] Plasma surfaces driven at relativistic intensities (SHHG) provide an attractive alternative to

[3]This calculation assumes a diffraction limited spot of 1 cm size and therefore a ratio of focal length to beam diameter of $f/D \sim 10^4$. While one could consider going out of focus, this is undesirable

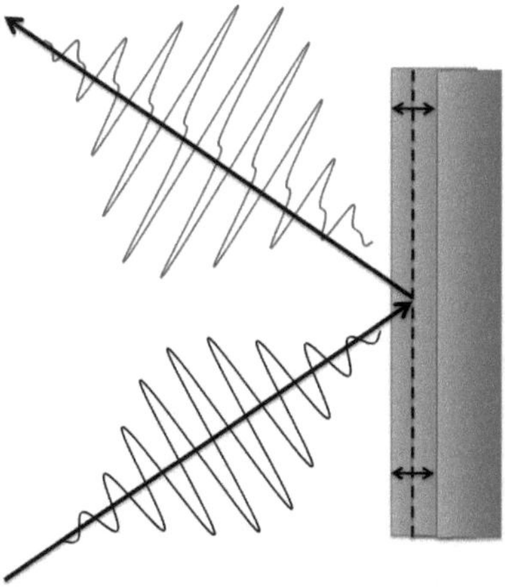

Fig. 13.12 Schematic of the Relativistically Oscillating Mirror (ROM) harmonic generation process. The force of the electric field on the plasma surface at the plasma vacuum interface leads to a periodic oscillation of the point at which the incoming laser is reflected by the plasma. This oscillation of the reflection point (indicated as a dashed line) leads to strong modification of the reflected waveform and the emission of harmonics of the laser frequency (for a multi-cycle pulse)

HHG in gaseous media and in particular a route to intense attosecond pulses. The primary mechanism of interest is the so-called Relativistically Oscillating Mirror (ROM) process, although there are other processes which can convert the optical laser light into higher orders (see [22, 23] for an in-depth reviews of SHHG).

13.2.2.1 The ROM Mechanism

Figure 13.12 shows the basic concept of up shifting via the ROM process. An initially solid target is illuminated by an intense laser with sufficiently high contrast to result in a step-like plasma vacuum interface. The plasma surface experiences the force of the laser and oscillates around its rest position with a mean kinetic energy of the order of the ponderomotive energy U_p. At high intensities, the ponderomotive potential U_p exceeds the rest-mass energy of the electron (511 keV) and the motion of the surface becomes relativistic – i.e. the surface oscillates by an appreciable fraction of a laser wavelength during each optical cycle resulting in a periodic distortion of the reflected waveform and hence harmonic generation. This occurs at $I\lambda^2 = 1.3 \times 10^{18}\,\mathrm{W\,cm^{-2}\mu m^2}$ and for relativistic interactions the laser strength is typically referred to by the normalised vector potential $a_0 = (I\lambda^2/1.3 \times 10^{18}\,\mathrm{W\,cm^{-2}\mu m^2})^{1/2}$. Unlike the case of HHG in gaseous targets, where the electron

from the point of view of the spatial phase which tends to be excellent only in focus due to the inherent spatial filtering of the laser beam in focus.

density is much less than the critical density n_c, solid targets are have $n_e/n_c \gg 1$ and hence reflect the incident laser radiation. The surface oscillations imply an oscillation of the apparent reflection point (ARP) at which the incident laser light is reflected [24, 25]. Note that even for low intensities, where the oscillation of the ARP is sinusoidal the resulting modulation and associated distortion of the phase of the reflected waveform will already give rise to harmonics. For higher intensities, the oscillation of the surface becomes increasingly non-sinusoidal giving rise to stronger harmonics. The underlying process in the case of a relativistically oscillating mirror is in many ways similar to the process of the relativistic Doppler up shift described by Einstein [26]. For a mirror moving with a constant velocity v close to the speed of light c an observer would detect reflected radiation at a frequency of $\omega_r = \omega_0(1+v/c)/(1-v/c) \approx 4\gamma^2$ (where ω_0 is the laser frequency). In the case of ROM instead of a constant value of γ describing the motion of the mirror surface, one now has Lorentz-factor that is a function of time $\gamma(t)$. The initial theoretical approach – a physical picture first proposed by Bulanov et al. [27] – was therefore to describe the harmonics observed in PIC simulations in terms of the reflection of the incident laser off a moving mirror oscillating at the laser frequency ω_0. A detailed semi-analytical moving mirror model was developed by Lichters et al. and was found to be in good agreement with PIC simulations [24]. This demonstrated that the picture of the moving mirror captures the essence of the harmonic generation process. Experiments performed in the mid 1990s observed harmonic spectra [28,29], where the conversion efficiency $\eta(m)$ of a given harmonic order m followed a power-law scaling $\eta(n) \sim m^{-q}$, where q is an intensity dependent exponent that increased from $q = 5.5$ to $q = 3.3$ when the intensity was varied from 5×10^{17} to $10^{19}\,\mathrm{W\,cm^{-2}}$ [29]. A quantitative understanding of ROM spectra was first given by Gordienko et al. and Baeva et al. [25, 30], based on the dynamics of the ARP. By assuming a boundary condition for the incident and reflected electric field at the ARP $E_r + E_i = 0$[4] it was found that the harmonic spectrum assumes an asymptotic spectral shape in the so-called relativistic limit (where $\gamma_{max} \gg 1$). The spectrum retains a power law scaling for the conversion efficiency in the relativistic limit with the efficiency of the m-th harmonic reaching $\eta(m)\ m^{-q_{REL}}$, with $q_{REL} = 8/3$ [25]. This slow decay has been identified as being sufficient to support pulse duration in the zeptosecond regime [30] and extremely high intensity X-ray radiation [31].

13.2.2.2 Short Wavelength Limit of ROM

Naively, one would expect the short wavelength limit of ROM harmonics to be determined by the peak γ of the surface to $\omega_{max} \sim 4\gamma^2$ as predicted by the Doppler-upshift from a mirror moving at constant velocity. However the spectra,

[4]This boundary condition is not always met but provides a useful guide to the typical scaling of ROM spectra [22].

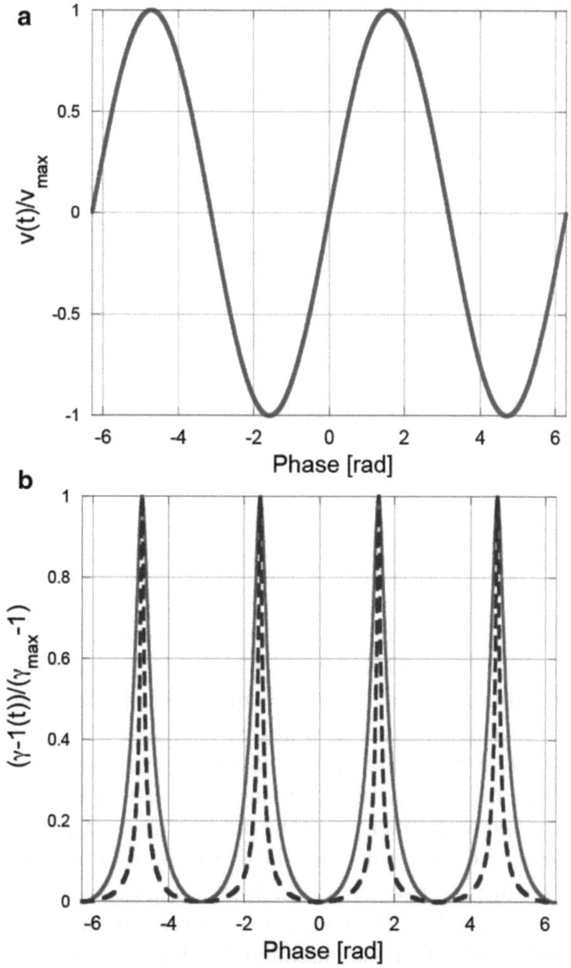

Fig. 13.13 The Lorentz factor γ is sharply spiked even for a very smooth dependence of velocity with time. Here a simple sinusoidal velocity dependence ($v(\phi) = v_{max}\sin(\phi)$) is chosen to illustrate the dependence of $v(t)$ on $\gamma(t)$. The variation of the Lorentz factor for two different peak velocities $v_{max} = 0.995c$ ($\gamma_{max} \sim 6$, solid line) and $v_{max} = 0.985c$ ($\gamma_{max} \sim 22$, dashed line) is shown in the lower plot. The width of the individual γ-spikes is much less than the oscillation period and reduces linearly with increasing γ_{max}

both experimentally [32] and in simulations [25], extend far beyond this limit. The theoretical prediction is that the $q = 8/3$ scaling still applies up to an order $n_{RO} \sim 8^{1/2}\gamma_{max}^3$, beyond which the conversion efficiency decreases exponentially or rolls over. The temporal dynamics of $\gamma(t)$ are essential to understanding the substantially larger frequency up shift and hence the short wavelength limit of ROM [25, 30]. Even assuming a very smooth variation of the actual surface velocity with time (e.g. $v(t) \sim \sin(\omega t)$ as in Fig. 13.13) results in a corresponding variation of $\gamma(t)$ that is sharply peaked. Returning to Einsteins theory of relativistic Doppler up shift one would therefore expect the up shifting process to be restricted to a timescale of the order of the temporal width of each γ-spike – substantially shorter than an optical half cycle – and the maximum up shift to take place when the Lorentz factor reaches its maximum γ_{max}. Since the emission of high harmonic orders only takes place

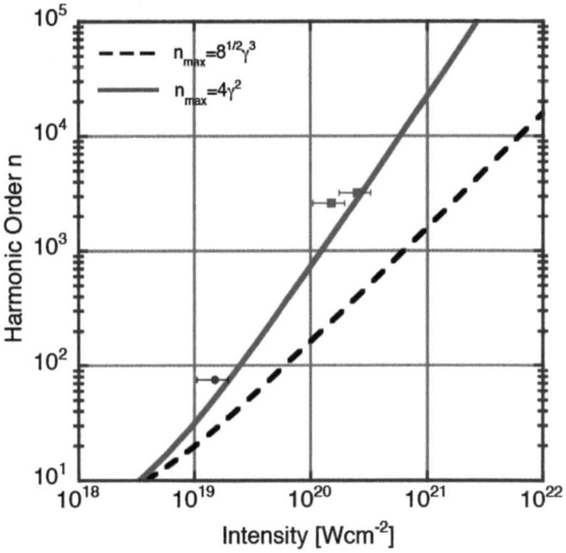

Fig. 13.14 Scaling of the highest harmonic with laser intensity. The data from the Vulcan laser experiments [32] clearly follows the γ^3 law for an oscillating surface

for large values of γ a sharp temporal localisation of the emitted harmonics results and the harmonic radiation is emitted in a burst on the timescale of attoseconds. As illustrated in Fig. 13.13, the temporal duration of the γ-spikes reduces for increasing intensity as $T_{Spike} \sim T_0/\gamma_{max}$ (with $T_0 = 2\pi/\omega_0$) [25]. The pulses of duration T_{Spike} are up-shifted and compressed by the factor of $4\gamma_{max}^2$ – familiar from the continuously moving relativistic mirror. As a result the harmonics are emitted in short temporal bursts with $T_{burst} \sim T_{Spike}/\gamma_{max}^2 \sim T_0/\gamma_{max}^3$ and hence, from Fourier-theory, must contain significant spectral components up to frequencies of $O \sim \omega_0 \gamma_{max}^3$. In effect, the high energy cut-off and the ultimate slope of the spectrum is governed by the temporal compression and truncation of the electromagnetic pulse rather than the maximum up shift expected from a relativistic mirror moving at constant γ. Experimental data (Fig. 13.14) obtained with the Vulcan laser shows that the highest harmonics observed follow the γ_{max}^3 trend – a powerful indication that the theoretical framework of ROM harmonics captures the essential physics correctly.

An estimate for magnitude γ_{max} can be obtained from the motion of a free electron in a laser field where $\gamma_{max} = (1 + 3.6 \times 10^{-19} I\lambda^2)^{1/2}$. Note that this applies only for gradients which are a significant fraction of the laser wavelength λ or greater. In the limit of very steep gradients the laser field at the surface is reduced and the higher plasma density leads to a larger restoring force. The influence of the peak plasma density in the limit of step-like density profiles can be quantified in terms of the similarity parameter $S = n_e/(a_0 n_c)$ [25, 33]. For constant S the surface dynamics of the plasma remain similar – particularly with regards to the velocity and phase of the ARP.

13.2.2.3 ROM Phase

In the previous section it was argued that the ultimate spectral extent of the ROM harmonics arise from temporal truncation of the up shifting process. This implies that the spectral phase of the highest harmonic orders must be constant or very close to constant (at least if we restrict the analysis to a single attosecond burst) and a pulse consisting of in the highest frequency part of the spectrum is therefore near transform limited. That this should be so, can be understood by a simple Gedanken experiment (thought experiment). If one takes a beam of light with a spectral width $\Delta\nu$ and truncates this to a duration Δt such that $\Delta\nu\Delta t \ll 1$ one obtains a beam with $\Delta\nu' \gg \Delta\nu$. The condition $\Delta\nu\Delta t \ll 1$ implies that Δt is much less than the coherence time $t_c \sim 1/\delta\nu$. The carrier oscillation within the time window Δt must therefore have full temporal coherence, i.e. flat spectral phase. This transform limited phase structure for the highest harmonics is predicted to result in pulses in the zeptosecond regime ($1\,\mathrm{zs} = 10^{-21}\mathrm{s}$) [30] under ideal conditions.[5] There are however contributions to both spatial and temporal phase that can lead to a departure from this ideal scenario. The peak plasma density in a step-like plasma gradient effectively changes the resonance frequency of the system. As in a simple harmonic oscillator the ratio of driving frequency to resonance frequency determines the relative phase of driver and oscillator. In the case of a plasma surface this can be parametrised by the S-parameter mentioned above [33] and if the S-parameter varies in time or space (as it certainly will given the dependence on the laser strength a_0) the phase will vary temporally and spatially. Spatially this leads to phase-front curvature while temporally this results in a change in the periodicity of the pulse train. A larger effect is the motion of the critical surface under the immense laser pressure ($P = I/c \approx$ Gbar). This pressure leads to a deformation of the critical density surface and a continuous underlying motion into the target (hole boring). This effect also leads to a departure from perfect periodicity and hence spectral changes [34] as well as a red-shift of the spectrum due to the Doppler-effect [35]. Finally, the radial deformation (denting) determines the observed angular distribution of ROM harmonics [36]. Such effects affect the spectral shape, but do not affect the duration of each individual attosecond burst of radiation.

13.2.2.4 Single Attosecond Pulses

To date single attosecond pulses have not been achieved from SHHG interactions, though trains of attosecond pulses have been observed [37]. However, the principles established for HHG remain the same for SHHG. In particular for the ROM process,

---

[5] A $\tau = 300\,\mathrm{zs}$ pulse corresponds to a spatial extent in propagation direction of $\Delta x = \tau c = 1$. Maintaining the integrity of the pulse front of such a small extent in the propagation direction would be extremely challenging.

the scaling of the highest harmonic is more rapid than for HHG ($\gamma^3 \propto I^{3/2}$) and the pulse to pulse separation in the attosecond pulse-train is greater (T_0 compared to $T_0/2$). Thus the requirements regarding the pulse-duration are more relaxed than for HHG. The much higher laser power required raises a laser-technological challenge of providing few-cycle pulse-duration *and* high intensity concurrently, with only the latest generation of laser based on OPCPA technology capable of such performance [38]. Polarisation dependence for ROM harmonic is somewhat different than for HHG [24]. For oblique incidence circular and linear polarisation have comparable efficiency, thus precluding polarisation gating. However at normal (near-normal) incidence the oscillating component of the laser-forces vanish (are suppressed) and polarisation gating becomes viable – albeit at some cost to overall efficiency of the process [39].

13.3 Conclusion

Converting intense optical laser radiation to high order harmonics of the incident laser light is an excellent means of achieving phase controlled spectra with large spectral width – and hence attosecond pulses. HHG in gaseous targets is a highly effective means of producing phase-locked spectra with a spectral width sufficient to support attosecond pulses and is the work-horse of attosecond science to date [3]. The only significant limitation is the relatively low single shot yields which are the result of challenging, time and space dependent phase matching considerations and, to a certain extent, practical difficulties in using lasers with extreme peak powers in the PW regime effectively for HHG due to geometrical constraints. However the relative ease and versatility of gas targets ensures that the development effort for HHG has not yet reached it's conclusion and schemes such as QPM may yet substantially transform what is possible with this source of attosecond XUV pulses. SHHG in general and the ROM mechanism in particular has the potential to increase the pulse brightness of attosecond pulses by many orders of magnitude. While experimental results to date are very encouraging, SHHG poses substantial additional complications with respect to targetry.

References

1. R. Neutze et al., Nature **406**, 752–757 (2000)
2. P. Corkum, F. Krausz, Nat. Phys. **3**, 381 (2007)
3. F. Krausz, M. Ivanov, Rev. Mod. Phys. **81**, 163 (2009)
4. M. Nisoli, G. Sansone, Prog. Quant. Electron. **33**, 17 (2009)
5. T. Brabec, F. Krausz, Rev. Mod. Phys. **72**, 545 (2000)
6. S. Backus et al., Rev. Sci. Instrum. **69**, 1207 (1998)
7. G. Tsakiris et al., New J. Phys. **8**, 19 (2006)
8. P.B. Corkum, Phys. Rev. Lett. **73**, 1994 (1993)

9. M. Ammosov, N. Delone, V. Krainov, JETP **91**, 2008 (1986)
10. S.G. Preston et al., Phys. Rev. A **53**, R31 (1996)
11. M.V. Frolov et al., Phys. Rev. Lett. **100**, 173001 (2008)
12. H.J. Shin et al., Phys. Rev. A, **63**, 053407 (2001)
13. G. Sansone et al., Science **314**, 443 (2006)
14. P.B. Corkum et al., Opt. Lett. **19**, 1870 (1994); M. Ivanov et al., Phys. Rev. Lett. **74**, 2933 (1995)
15. K.S. Budil et al., Phys. Rev. A **48**, R3437 (1993)
16. E. Constant et al., Phys. Rev. Lett. **82**, 1668 (1999)
17. C.G. Durfee et al., Phys. Rev. Lett. **83**, 2187 (1999)
18. M. Lewenstein et al., Phys. Rev. A **49**, 2117 (1994)
19. M.C. Chen et al., Phys. Rev. Lett. **105**,173901 (2010)
20. A. Paul et al., Nature **421**, 51 (2003)
21. A. Willner et al., Phys. Rev. Lett. **107**, 175002 (2011)
22. C. Thaury, F. Qur, J. Phys. B: At. Mol. Opt. Phys. **43**, 213001 (2010)
23. U. Teubner, P. Gibbon, Rev. Mod. Phys. **81**, 445 (2009)
24. R. Lichters, J. Meyer-ter-Vehn, A. Pukhov, Phys. Plasmas **3**, 3425 (1996)
25. T. Baeva et al., Phys. Rev. E **74**, 046404 (2006)
26. A. Einstein, Ann. Phys. (Leipzig) **17**, 891 (1905)
27. S.V. Bulanov, N.M. Naumova, F. Pegoraro, Phys. Plasmas **1**, 745 (1994)
28. P. Gibbon, Phys. Rev. Lett. **76**, 50 (1995)
29. P. Norreys et al., Phys. Rev. Lett. **76**, 1832 (1996)
30. S. Gordienko et al., Phys. Rev. Lett. **93**, 115002 (2004)
31. S. Gordienko et al., Phys. Rev. Lett. **94**, 103903 (2005)
32. B. Dromey et al., Phys. Rev. Lett. **99**, 085001 (2007)
33. D. an der Brügge, A. Pukhov, Phys. Plasmas **14**, 093104 (2007)
34. M. Behmke et al., Phys. Rev. Lett. **106**, 185002 (2011)
35. M. Zepf et al., Phys. Plasmas **3** 3242 (1996)
36. B. Dromey et al., Nat. Phys. **2**, 456 (2006).
37. Y. Nomura et al., Nat. Phys. **5**, 124 (2009)
38. F. Tavella et al., Opt. Lett. **32**, 2227 (2007)
39. S. Rykovanov et al., New J. Phys. **10**, 025025 (2008)

Part V
Tools and Instrumentation

Chapter 14
Hydrodynamic Simulation

Alex P. L. Robinson

Abstract The main aim of this lecture is to provide a broad overview of the area of hydrodynamic simulation. The provision of introductions to a couple of basic algorithms for solving the equations of gas dynamics is a secondary objective. Hydrodynamic simulation in the context of laser-plasma physics and inertial fusion is now a large and mature field, deserving of an entire book (or books...) for a proper treatment. Individual topics will not be treated in great depth, and mathematical detail is avoided where possible. It is hoped that the reader will understand the key aspects of hydrodynamic simulation and the ability to write a very simple 1D hydro-solver with a view to using this knowledge as a "springboard" for more in-depth study.

14.1 Hydrodynamic Regimes of Laser-Plasma Interactions

There are many problems in the field of laser-plasma interactions where a hydrodynamic or magneto-hydrodynamic model will be an accurate description of the plasma. The ablative implosion of DT shells to produce highly compressed fusion fuel in Inertial Confinement Fusion is perhaps the area most clearly dominated by hydro-code simulation. On the other hand, the study of electron acceleration in laser-driven wakefields cannot be properly studied using a hydro-code. The suitability of any numerical simulation technique for a given problem has to be assessed by considering how valid it is to describe a physical system in a particular way. We shall therefore start our discussion of hydrodynamic simulation by looking at the characteristics of the fluid description.

A.P.L. Robinson (✉)
Central Laser Facility, STFC Rutherford-Appleton Laboratory, Chilton, Oxfordshire, UK
e-mail: alex.robinson@stfc.ac.uk

P. McKenna et al. (eds.), *Laser-Plasma Interactions and Applications*, Scottish Graduate Series, DOI 10.1007/978-3-319-00038-1_14,

At the heart of the fluid description is the notion that the microscopic behaviour of the constituent particles can be ignored and that one only has to consider the macroscopic properties of the mass of particles, which include: the mass density (ρ), the fluid velocity ($\mathbf{v}$), and the pressure of the fluid (P). Generally the fluid description of plasmas will include the possibility of magnetic fields ($\mathbf{B}$) as well (as in MHD, *Magneto-Hydrodynamics*), however we shall ignore this for the time being.

What conditions must be satisfied in order for this macroscopic description to be accurate? This can be answered by examining how we arrive at these quantities from the kinetic description. In the kinetic description, one describes a plasma in terms of a distribution function, $f(\mathbf{r},\mathbf{v})$, for each species. The fluid quantities are *moments* (i.e. integrals over velocity space) of the distribution function. The n-th moment is defined by,

$$M_n(\mathbf{r}) = \int \mathbf{v}^n f(\mathbf{r},\mathbf{v}) d\mathbf{v}^3. \tag{14.1}$$

The particle density is equal to the zeroth moment, the components of the fluid velocity are given by the components of the first moment, so forth and so on. Suppose that the plasma is highly collisional on the time-scale (τ_H) and length-scale (L_H) of hydrodynamic interest. In the presence of strong collisions one will find that locally the distribution function will be very close to a Maxwellian,

$$f(\mathbf{r},\mathbf{v}) = n(\mathbf{r}) \left(\frac{m}{2\pi k_B T(\mathbf{r})} \right)^{3/2} \exp\left[-\frac{m(\mathbf{v}-\mathbf{V}(\mathbf{r}))^2}{2k_B T\mathbf{r}} \right], \tag{14.2}$$

a state which is referred to as *Local Thermodynamic Equilibrium* (LTE). Since the assumption of strong collisionality ensures that one will have a Maxwellian everywhere – varying only in terms of $n(\mathbf{r})$,$\mathbf{v}(\mathbf{r})$, and $T(\mathbf{r}$ – there is no need to solve for the distribution function, since one can take the first three moments (continuity, momentum and energy) and then close the set of equations by making some slight approximations (particularly concerning thermal conduction). By *strong* collisions we mean that a typical particle will have undergone many collisions on the time-scale of interest. So if the collision rate is ν, this means that $\nu \gg 1/\tau_H$. The term "strong" collisions also implies that the mean free path, λ_{mfp}, of a particle is small compared to the length-scale of interest, i.e. $L_H \gg \lambda_{mfp}$. For reference, the electron-ion collision time in a Maxwellian plasma is given by,

$$\tau_{ei} = 2.4 \times 10^{-9} \left[\frac{T_e}{\mathrm{eV}} \right]^{3/2} \left[\frac{n_i}{10^{20}\mathrm{m}^{-3}} \right]^{-1} \frac{1}{Z^2 \log \Lambda_{ei}}. \tag{14.3}$$

If, however, we are dealing with a system where collisions do not have enough time to return the system to a locally Maxwellian distribution then in general one will have to treat the system fully kinetically. Here the fluid description will not be valid.

If it is valid to use a fluid description, then potentially huge savings in computational effort can be made. Kinetic modelling is hugely demanding since the distribution function is six-dimensional. Even with modern computers there are many problems in plasma physics that are simply intractable in terms of kinetic modelling due to the disparity between the kinetic scales and the system scales. Fluid modelling, in contrast, is concerned with a relatively few macroscopic variables which are all 3D. Furthermore one is not bound to resolve small kinetic scales, which saves further effort. Computational Fluid Dynamics (CFD) has thus become quite a mature field, and hydrodynamic modelling of laser-plasmas, particularly in the context of ICF has become highly sophisticated. In the rest of this chapter we will provide an introductory guide to hydrodynamic simulation of laser-plasmas.

Key Points

- The hydrodynamic description is valid if the plasma is close to a state of Local Thermodynamic Equilibrium (LTE), where the distribution function is nearly Maxwellian at all points.
- By estimating collision times and mean free paths, one can check these against the hydrodynamic scales of interest to ensure that $\nu \gg 1/\tau_H$ and $L_H \gg \lambda_{mfp}$.

14.2 Architecture of a Hydro-code

The main computational cycle of a hydro-code can be thought of as a set of simple conceptual steps:

1. **Core hydrodynamics solve**: Solve the core set of advection-compression hydrodynamics equations for one time step.
2. **Evolve mesh**: If the mesh is non-static, then update the mesh.
3. **Energy transport solve**: Inject energy (e.g. laser heating) and calculate all energy transport by non-advective means (e.g. thermal conduction or radiation).

Within this very top-level view of hydro-code architecture, we can identify a set of key elements.

- **Mesh**: The nature of the grid on which the fluid variables are represented.
- **Hydro scheme**: The numerical scheme used for solving the hydrodynamic equations.
- **Heating**: i.e. local energy deposition due to laser beams.
- **Energy transport**: Particularly thermal conduction and radiative transport.
- **Equation of state**: The relation of pressure to internal energy not always being that of an ideal gas.

The first element – the gridding or mesh scheme – demarcates quite different types of hydro-code. In the Lagrangian approach to solving the hydrodynamic equations, the mesh co-moves with the fluid elements and no fluid advects *through* the mesh. In the Eulerian approach, the mesh is completely static and the fluid advects through the mesh. One modern approach to hydro-codes actually blends these methods, and is thus called the Arbitrary-Lagrangian-Eulerian (ALE) method.

In all gridding schemes, one must numerically solve the hydrodynamic equations during each time-step. The algorithms that have been developed for this do not just consider stability and accuracy (although both are important), as whether or not an algorithm is conservative and positivity maintaining (of mass and energy density) is just as important. In laser-plasma simulations, one will often deal with strong shocks, so it is important that the scheme is not too diffusive. On the other hand, one also wants a numerical scheme that is easy-to-code, does not lead to a slow execution and one which might also be easy to implement in parallel computations.

Although energy transport (radiative or by thermal conduction) and energy deposition (e.g. by laser beams) should formally be part of the hydrodynamics equations, energy transport is usually a separate solve (this is re-iterated later on), or a series of separate solves. Apart from making codes easier to write, this also makes these physics elements more modular and one is therefore able to "turn off" (artificially that is) these physical processes in order to examine their role in any given simulation.

At the heart of the hydro-code is the mesh and the scheme for solving the hydrodynamic equations. All other physics is built on this "frame" or "skeleton". These two aspects are therefore the most important aspects to understand.

Key Points

- The solution of the key equations is normally separated into a solve for the core hydrodynamics (or advective transport), and energy transport.
- The grid or mesh used may also evolve during the simulation, and the type of grid used is a major distinction between different types of code, especially Eulerian versus Lagrangian.

14.3 Core Hydrodynamics

14.3.1 Key Equations

In order to understand how to devise suitable algorithms for solving the hydrodynamic equations, we first need to know what these equations are, and determine their mathematical properties. In many treatment of gas hydrodynamics, the following set of hydrodynamic variables is usually chosen: mass density (ρ), fluid velocity ($\mathbf{v}$), pressure (P; which we take to be isotropic), and total energy (ε). In this case the set of hydrodynamic equations will then be written as:

$$\frac{\partial \rho}{\partial t} + \nabla.(\rho \mathbf{v}) = 0, \tag{14.4}$$

$$\rho \frac{\partial \mathbf{v}}{\partial t} + \rho(\mathbf{v}.\nabla)\mathbf{v} = -\nabla P, \tag{14.5}$$

$$\frac{\partial \varepsilon}{\partial t} + \nabla.(\mathbf{v}(\varepsilon + P)) = 0. \tag{14.6}$$

This set of equations must be closed by an Equation of State, $P = f(\varepsilon - \frac{1}{2}\rho \mathbf{v}^2)$. In the case of an ideal gas this would be $P = (\gamma - 1)(\varepsilon - \frac{1}{2}\rho \mathbf{v}^2)$, where γ is the ratio of specific heat capacities. Note that the only processes affecting the energy equation (Eq. 14.6) are advection and compressional work. In laser-plasma simulations, we will have two species (electrons and ions) that can be driven out of temperature equilibrium by processes such as laser heating. However here we will just concentrate on the simple 'gas dynamics' set of equations.

In fact, the processes of thermal conduction, laser heating, etc. are not included in Eqs. 14.4, 14.5, and 14.6. This is a very standard approach in hydrodynamic simulation is to split such processes, which should appear as extra terms in the energy equation, into separate 'solves'. In general such an approach is known as 'operator splitting', a subject that attracts a fair amount of debate! Nontheless the approach has been shown to work well in laser-plasma hydro-codes, so it is the approach adopted here.

Each equation of 14.4, 14.5, and 14.6 expresses the conservation of a particular quantity, namely mass, momentum and energy respectively. Thus one often finds such a set being referred to as a 'set of conservation laws'. We should also note that, if one were to linearize this system then one would reduce it to a linear hyperbolic system, i.e. a system with a full set of real wave speeds. Thus the system is also classified as a nonlinear hyperbolic system.

14.3.2 Eulerian System and Conservative Form

We can now turn our attention to finding methods for solving these equations, and we will begin by considering this in a relatively general form. Firstly let us consider how we might solve Eqs. 14.4, 14.5, and 14.6 on a fixed grid – this is the *Eulerian* approach. These equations could be solved by applying the techniques that are commonly known for the numerically integration of ODEs (Euler method, etc.). The difficulty with this "straightforward" approach is that one can fail to conserve important quantities such as mass, momentum, and energy. There are also problems with ensuring that quantities remain positive and don't exhibit unphysical oscillations. We can see a way to devise appropriate numerical schemes by recasting Eqs. 14.4, 14.5, and 14.6 in a *fully conservative* form as follows:

$$\frac{\partial \rho}{\partial t} + \nabla.\mathbf{m} = 0 \tag{14.7}$$

$$\frac{\partial \mathbf{m}}{\partial t} + \nabla.\left(\frac{\mathbf{mm}}{\rho} + P\mathbf{I}\right) = 0 \tag{14.8}$$

$$\frac{\partial \varepsilon}{\partial t} + \nabla.\left(\frac{\mathbf{m}(\varepsilon + P)}{\rho}\right) = 0 \tag{14.9}$$

Note that this has involved a change of variables from $(\rho, \mathbf{v}, \varepsilon)$ to $(\rho, \mathbf{m}, \varepsilon)$, where $\mathbf{m} = \rho\mathbf{v}$. We can now see that we have re-written the set in the form of

$$\frac{\partial \mathbf{U}}{\partial t} + \nabla.\mathbf{F}(\mathbf{U}) = 0, \tag{14.10}$$

so if we now consider the evolution of these quantities in a small volume, V, around a point, $\mathbf{r}$, we will find that,

$$\frac{\partial}{\partial t}\int_V \mathbf{U}dV = -\int_V \nabla.\mathbf{F}(\mathbf{U})dV, \tag{14.11}$$

and on applying Gauss's theorem we transform this to,

$$\frac{\partial}{\partial t}\int_V \mathbf{U}dV = -\int_S \mathbf{F}(\mathbf{U}).d\mathbf{S}, \tag{14.12}$$

where S now represents the surface of this volume. Therefore $\mathbf{F}$ represents a set of fluxes of mass, momentum, and energy. This shows us a route to constructing conservative schemes on a fixed grid – by exploiting this set of fluxes and this conservative form of the equation set. All of the schemes for solving the hydrodynamic equations in an Eulerian framework exploit the conservative form.

14.3.3 Lagrangian Form

Let us now consider a radically different approach to solving Eqs. 14.4, 14.5, and 14.6 – the *Lagrangian* approach. Once again, the Lagrangian approach is based on transforming the equations. This time we will start by defining a new differential operator, D/Dt,

$$\frac{D}{Dt} = \frac{\partial}{\partial t} + v\frac{\partial}{\partial x}, \tag{14.13}$$

which we only define in 1D. The meaning of this mathematical construct in physical terms is the rate of change of a quantity of a fluid element over time as we track the fluid element along its trajectory. We can now transform the 1D versions of Eqs. 14.4, 14.5, and 14.6 into,

$$\frac{D\rho}{Dt} = -\rho\frac{\partial v}{\partial x} \tag{14.14}$$

$$\frac{Dv}{Dt} = -\frac{1}{\rho}\frac{\partial P}{\partial x} \tag{14.15}$$

$$\frac{D\varepsilon}{Dt} = -\varepsilon\frac{\partial v}{\partial x} - \frac{\partial vP}{\partial x} \tag{14.16}$$

Thus a very different way to deal with the hydrodynamic equations is to track a set of fluid elements via a set of freely moving nodes, i.e. a fully moving grid. The fluid properties can then be evolved purely by computing spatial derivatives. This approach looks highly appealing from a computational point of view, as well as from the point of avoiding unphysical behaviour (such as computing negative densities). In 2D and 3D there are a number of complications. It has been a highly successful method nonetheless.

14.3.4 Shocks

Having shown the equations that are to be solved, we now need to move on and consider what types of solutions the equations permit. A solver that is notionally accurate, may still be of little use if it does not evolve the correct type of solution. It is important to recognize that Eqs. 14.4, 14.5, and 14.6 permit *discontinuous* solutions, or shocks. Shocks can emerge from an initially smooth solution, and are virtually unavoidable in most laser-plasma problems.

The values that the hydrodynamic variables take on either side of the shock are not arbitrary, and are in fact related by the Rankine-Hugoniot equations, which are in turn derived by considering conservation of mass, momentum flux and energy flux across a shock. If one denotes the values taken on either side of the shock by subscripts 1 and 2, then in the limit of a very strong shock where $p_2 \gg p_1$, one finds, for example, that,

$$\frac{\rho_2}{\rho_1} = \frac{\gamma+1}{\gamma-1}, \tag{14.17}$$

where γ is the ratio of specific heat capacities. Thus for an ideal gas $\rho_2/\rho_1 \to 4$ in the case of a strong shock.

From the point of view of numerical solutions, shocks clearly pose a challenge as numerical algorithms are often diffusive in nature. As a result there has been a great amount of development of algorithms, which can handle shocks in a proper way.

14.3.5 Rarefaction Waves

Suppose that we re-consider the problem where initially we have two uniform regions. Clearly if the Rankine-Hugoniot relations aren't satisfied then this cannot describe a steady shock. Another clear possibility is that one region undergoes expansion and fluid is rarefied in the process. The extreme example of this is the expansion of plasma into a vacuum. In order to illustrate this aspect of hydrodynamic behaviour, we will review this particular case. If one takes Eqs. 14.4, 14.5, and 14.6 then these can be written in 1D as,

$$\frac{\partial \rho}{\partial t} + \frac{\partial(\rho u)}{\partial x} = 0, \tag{14.18}$$

$$\rho\frac{\partial u}{\partial t} + \rho u\frac{\partial u}{\partial x} = -c^2\frac{\partial \rho}{\partial x}, \tag{14.19}$$

$$P\rho^{-\gamma} = P_0\rho_0^{-\gamma}, \tag{14.20}$$

where $c = \sqrt{\gamma P/\rho}$ is the local sound speed. Initially we have a uniform stationary fluid in the region $x < 0$ with $\rho = \rho_0$ and $P = P_0$, and vacuum in the region $x > 0$. We look for solutions that will satisfy the physical boundary conditions in terms of $z = x/t$. The equations for density and velocity can then be written as,

$$\begin{bmatrix} u-z & \rho \\ \dfrac{c^2}{\rho} & u-z \end{bmatrix} \begin{bmatrix} \dfrac{\partial \rho}{\partial z} \\ \dfrac{\partial u}{\partial z} \end{bmatrix} = 0.$$

In order to obtain a non-trivial solution to this, we must equate the determinant of the two by two matrix to zero. This yields,

$$u = z + c. \tag{14.21}$$

One can also take the adiabatic equation of state and obtain,

$$\frac{\partial c}{\partial z} = \frac{c(\gamma-1)}{2\rho}\frac{\partial \rho}{\partial z}, \tag{14.22}$$

and from the continuity equation one can also write,

$$(u-z)\frac{\partial c}{\partial z} + \frac{c(\gamma-1)}{2}\frac{\partial u}{\partial z}. \tag{14.23}$$

Since we have just seen that $c = u - z$, we can immediately integrate this last equation (noting that when $u = 0$, $c = c_0$ from the initial conditions) to get,

$$c = c_0 - \frac{\gamma - 1}{2} u. \tag{14.24}$$

So we can now write down the explicit solution for the range $-c_0 \leq x/t \leq \frac{2c_0 t}{\gamma-1}$,

$$u = \frac{2}{\gamma+1} \left(\frac{x}{t} + c_0 \right), \tag{14.25}$$

$$c = c_0 - \frac{\gamma - 1}{\gamma+1} \left(\frac{x}{t} + c_0 \right), \tag{14.26}$$

$$\rho = \rho_0 \left(1 - \frac{\gamma - 1}{\gamma+1} \left(\frac{x}{c_0 t} + 1 \right) \right)^{\frac{2}{\gamma-1}}. \tag{14.27}$$

This example illustrates the key features of rarefaction waves in general – smooth linear variation in velocity, smooth power law variation in density, and a characteristic scale length of $c_0 t$. This specific problem is often used as a model for the disassembly of highly compressed hot fusion fuel, so the rarefaction wave is a regular feature of LP/ICF hydrodynamics, and is not some esoteric obscurity!

14.3.6 The Riemann Problem

The two preceding sub-sections imply that solving hydrodynamic problems is actually rather difficult. This is because the simplest problem, that is two semi-infinite regions in contact with one another, can produce two very different kinds of solution. This problem is known as the *Riemann Problem*, and it is critical in hydrodynamics. Analytic solution of certain problems is possible, but here we will not delve into the mathematical details. Qualitatively there are three fundamental solutions. Two we have just met – the shock and the rarefaction wave. The third important possibility is that there is a region of uniform flow.

In terms of numerical simulation the important point is whether a solver is capable of correctly dealing with a Riemann problem. This means that an algorithm must include a *Riemann Solver* either in an implicit sense or in a direct sense. In Lagrangian codes the Riemann Problem is solved naturally. In Eulerian codes it is not, and this has lead to developing a number of very sophisticated numerical schemes based on Riemann Solvers, the most notable being *Godunov's Method*. Godunov's Method is based on the idea that we can view the discretized hydrodynamic data as being a set of uniformly filled cells with jumps at the interfaces. Therefore the interface between each pair of cells becomes a "miniature" Riemann Problem. One can now use the analytic solution to accurately determine the numerical fluxes between the cells and thus evolve the solution by one time step. Many advanced algorithms for Eulerian hydrodynamics are based on this approach.

14.3.7 Testing Codes Against Analytic Solutions

An examination of the mathematical properties of the hydrodynamic equations not only provides insights into how to solve them numerically, but it also reveals analytic solutions that can be used as test cases for a computational physicist to check his or her code. A commonly used problem is Sod's problem (named after Gary Sod), which is a 1D shock tube problem. Here we will use blast waves as a test problem. If one starts with a fluid that is initially completely stationary, but the central region is at much higher pressure, then this will produce a strong explosion. At a much later time, one will observe an expanding region of disturbed fluid bounded by a shock, beyond which is undisturbed fluid.

One can extract the spatial extent of the disturbed region from dimensional analysis alone. One does this by assuming that the radius, R, can only be a function of the initial energy deposited (E), the density of the undisturbed fluid (ρ_0), and the time (t). In order to obtain something with dimensions of length only from E, ρ_0, and t one is lead to the conclusion that,

$$R = \left(\frac{Et^2}{\rho_0}\right)^{1/5}. \tag{14.28}$$

So one can test one's code by checking that it reproduces a $R \propto t^{2/5}$ scaling. What is somewhat more complex to show is that behind the shock, the disturbed fluid will evolve in a self-similar fashion according to the similarity variable,

$$\varepsilon = \frac{r}{R(t)} = r\left(\frac{\rho_0}{Et^2}\right)^{1/5}. \tag{14.29}$$

So, for example, behind the shock one finds that $\rho = \rho_0 G(\varepsilon)$, where $G(\varepsilon)$ is a function one obtains from the full self-similarity analysis. Even without performing the full analysis, one can also exploit this as another way to check one's code.

There are a number of other test problems with either analytic solutions or 'reference solutions' that one can use to check that one's code is working properly. It is critically important that one validates the core hydrodynamic solver, so this is not a step that can be skipped!

In conclusion we have just outlined two general ways to solve Euler's equations of gas dynamics. The same considerations apply to the sort of equation set one would have in a laser-plasma hydro-code. In the following sections we will actually describe a set of different algorithms – 1 for the Lagrangian method, and 3 for the Eulerian method. Additionally we have also discussed analytic solutions which can be used to validate hydro-solvers.

Key Points

- The key hydro-dynamic equations that one must solve are:
-
$$\frac{\partial \rho}{\partial t} + \nabla.(\rho \mathbf{v}) = 0 \tag{14.30}$$
-
$$\rho \frac{\partial \mathbf{v}}{\partial t} + \rho(\mathbf{v}.\nabla)\mathbf{v} = -\nabla P \tag{14.31}$$
-
$$\frac{\partial \varepsilon}{\partial t} + \nabla.(\mathbf{v}(\varepsilon + P)) = 0 \tag{14.32}$$
- These equations can be written in *conservative* form which show how Eulerian algorithms can be devised.
- These equations can also be written in terms of following a set of fluid elements, which show how Lagrangian algorithms can be devised.
- One must also consider analytic solutions of these equations, so that one can rigorously validate one's code.

14.4 Solution in 1D by Lagrangian Method

The first algorithm that we'll look at is the 1D Lagrangian algorithm. The key equations for this are Eqs. 14.14, 14.15, and 14.16. We shall discretize the fluid into a set of N fluid elements (or cells) that are in contact. We shall define a set of $N+1$ cell walls, where we define a wall position at the nth time step, $x^n_{w,k}$. We also define the velocity at the cell walls too, but at a set of staggered points in time, $v^{n+1/2}_{w,k}$. The density and pressure are defined at the cell centres, which we denote by the positions $\{k+1/2\}$.

The motion of the fluid elements is tracked by following the motion of the walls. The walls positions are updated via,

$$x^{n+1}_{w,k} = x^n_{w,k} + v^{n+1/2}_{w,k} \Delta t, \tag{14.33}$$

where Δt is our time step.

We can define the width of each cell as $\Delta_{k+1/2} = x^n_{w,k+1} - x^n_{w,k}$, and since the mass of each element must be conserved, we can write the continuity equation, or the density update, as follows:

$$\rho^{n+1}_{k+1/2} = \rho^n_{k+1/2} \frac{\Delta^n_{k+1/2}}{\Delta^{n+1}_{k+1/2}}. \tag{14.34}$$

The change in internal energy due to PdV work is handled by noting that Eq. 14.16 will become,

$$P^{n+1}_{k+1/2} = P^{n}_{k+1/2}\left(\frac{\rho^{n+1}_{k+1/2}}{\rho^{n}_{k+1/2}}\right)^{\gamma}. \tag{14.35}$$

if we use pressure instead of internal energy and assume an adiabatic equation of state. In other words we use $P\rho^{-\gamma} = const.$ as our equation of state.

Finally we must accelerate the fluid elements according to the pressures acting on the cell walls. This is done by finite-differencing Eq. 14.15 to obtain,

$$v^{n+1/2}_{w,k} = v^{n-1/2}_{w,k} + \frac{2\Delta t}{\rho^{n}_{k+1/2}+\rho^{n}_{k-1/2}}\frac{P^{n}_{k+1/2}-P^{n}_{k-1/2}}{x^{n}_{c,k+1/2}-x^{n}_{c,k-1/2}}. \tag{14.36}$$

This method is not quite complete, despite the fact that we apparently discretized all of the relevant equations. The reason being that the discretized set is prone to producing non-physical small-scale oscillations in the vicinity of shocks. To remedy this, one needs to introduce *artificial viscosity*. Although the introduction of non-physical terms to fix non-physical behaviour is not desirable, it is the simplest solution to this problem. As it turns out, it will not 'damage' the physical nature of the solution to any great extent. The artificial viscosity is introduced by changing the pressure used in Eq. 14.36 to the effective pressure, $P^{\star} = P + Q$, where Q is the artificial viscosity. Being artificial, there are a number of essentially arbitrary choices that can be made for Q, but here we will choose the following form,

$$Q = -B\rho\Delta^2 c_s\left|\frac{\partial v}{\partial x}\right|, \tag{14.37}$$

where c_s is the local sound speed, and B is an arbitrary coefficient.

Now let us "code up" these equations, and test them against the following initial value problem:

$$\rho = 1, \tag{14.38}$$

$$v = 0, \tag{14.39}$$

$$P = 1 + 10e^{-(x-x_0)^2/8} \tag{14.40}$$

So we have a uniform fluid at rest everywhere, with a small hot spot in the centre. Results from this calculation are shown below in Fig. 14.1. This shows that the Lagrangian method is able to confidently reproduce the classic blast wave, including the shock between the disturbed and undisturbed regions. In this calculation used 4,000 cells and a fluid initially 40 units in length. The artificial viscosity coefficient, B, was set to 0.5.

We can go a step further in looking at these results by tracking the motion of the shock, and comparing this to the predictions of Eq. 14.28. This is done in Fig. 14.2, where the position of the shock is tracked against time, and then plotted in log – log

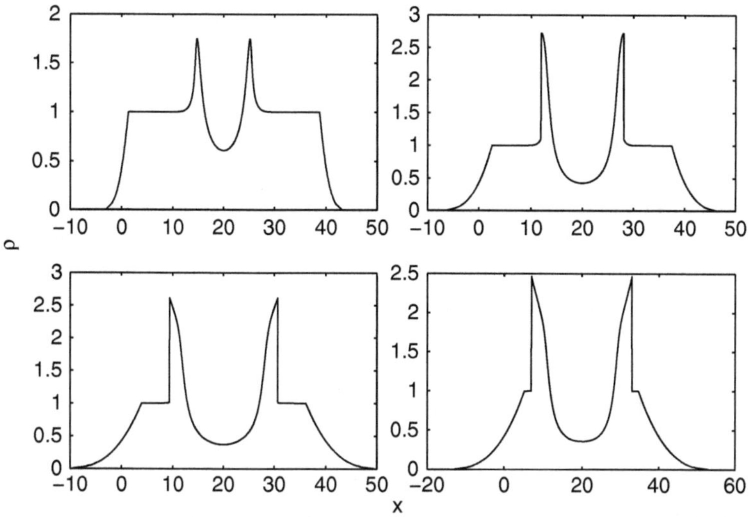

Fig. 14.1 Density profiles at different times in the case of Lagrangian simulation of test problem

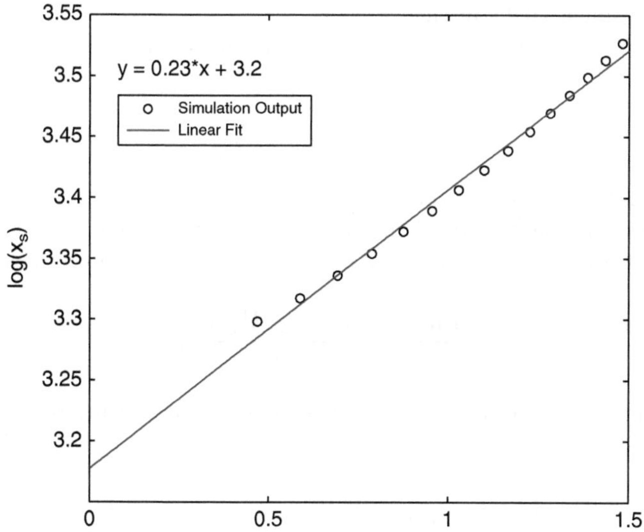

Fig. 14.2 Shock front position versus time in the case of Lagrangian simulation of test problem

form. A linear fit reveals that the power law fit to the simulation data is very close to $\propto x^{1/5}$, i.e. we are very close to the scaling predicted by Eq. 14.28. This indicates that our simple code is able to reproduce physical results.

We therefore see that the Lagrangian approach is easy to code, and can reproduce physical results. We have had to include a clearly non-physical element in the algorithm to ensure stability. Without it we would never be able to reproduce the results shown in Figs. 14.1 and 14.2. It has not obviously impaired our ability to obtain results that in agreement with analytic theory. Although undesirable it is perhaps tolerable. Given these observations, it is clear why the Lagrangian approach has been so popular in laser-plasma and laser fusion research. In one dimension, the Lagrangian approach is probably preferable to an Eulerian approach. One advantage over Eulerian methods is that it is very easy to track material interfaces in Lagrangian codes. In 2D and 3D, the Lagrangian method does faces additional problems, such as the possibility of reversing the order of grid points ('bow-tie' problem), which makes the method less appealing.

14.5 Solution in 1D by Centred-Upwind Method

The second algorithm that we will look at in detail is for 1D Eulerian hydrodynamics. This one, although it is ultimately based on Godunov's method, does not directly utilize a Riemann solver. The method is known as the Kurganov-Noelle-Petrova (KNP) scheme. It is simpler than the advanced Riemann-Solver-based methods, but it is still quite accurate and robust. The method has been employed successfully in multi-dimensional MHD by Ziegler.

The KNP scheme is actually a general method for solving systems of the form,

$$\frac{\partial \mathbf{u}}{\partial t} + \frac{\partial \mathbf{F}(\mathbf{u})}{\partial x} = 0, \tag{14.41}$$

and the multi-dimensional generalization of this. As we have seen, the hydrodynamic equations can be written in precisely this form. In the KNP scheme, one defines the hydrodynamic variables at the cell centres and then integrates as follows,

$$\frac{d\mathbf{u}_j}{dt} = -\frac{\mathbf{H}_{j+1/2} - \mathbf{H}_{j-1/2}}{\Delta x}, \tag{14.42}$$

where $\mathbf{H}$ are numerical fluxes across the cell boundaries. In the KNP scheme the numerical fluxes are given by,

$$\mathbf{H}_{j+1/2} = \frac{a^+_{j+1/2}\mathbf{F}(\mathbf{u}_j) - a^-_{j+1/2}\mathbf{F}(\mathbf{u}_{j+1})}{a^+_{j+1/2} - a^-_{j+1/2}} + \frac{a^+_{j+1/2}a^-_{j+1/2}\left(\mathbf{u}_{j+1} - \mathbf{u}_j\right)}{a^+_{j+1/2} - a^-_{j+1/2}}, \tag{14.43}$$

where we have assumed a zeroth order interpolation of the hydrodynamic variables across each cells. Higher order interpolation across cells is not much more difficult. The $a^{\pm}$s are local speeds at the cell surfaces and are determined by,

$$a^+_{j+1/2} = \mathtt{max}\left\{(v + c_s)_{i+1}, (v + c_s)_i, 0\right\}, \tag{14.44}$$

$$a^-_{j+1/2} = \mathtt{min}\left\{(v - c_s)_{i+1}, (v - c_s)_i, 0\right\}. \tag{14.45}$$

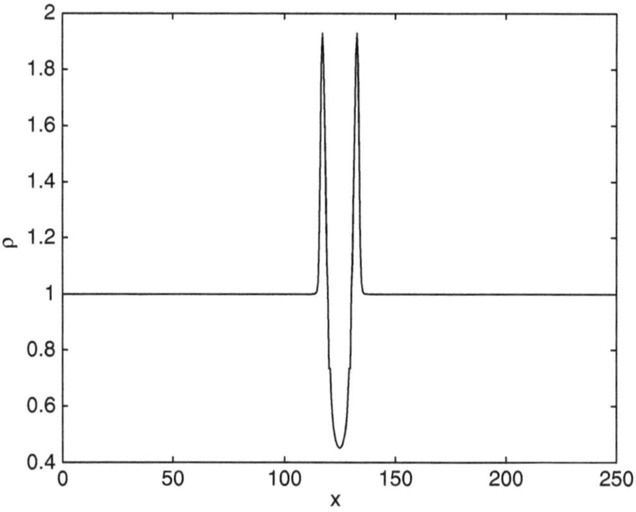

Fig. 14.3 Density profile produced by Central-Upwind method at t =2 in the case of the test problem

If we now apply this to the hydrodynamic equation as cast in conservative form, i.e. Eqs. 14.7, 14.8, and 14.9. We see that, in 1D, $\mathbf{u} = (\rho, m = \rho v, e)$. The flux vector is given by,

$$\mathbf{f} = \begin{bmatrix} m \\ \dfrac{m^2}{\rho} + P \\ \dfrac{m(e+P)}{\rho} \end{bmatrix}$$

With a suitable equation of state, e.g. $P = (\gamma - 1)(e - \rho v^2/2)$, one now has a complete scheme for an Eulerian hydro-solver. Note the absence of any kind of artificial viscosity.

This method can now be coded up, and its performance can be tested against the test problem that we have previously used. Sample output is shown in Fig. 14.3 below.

14.6 Thermal Transport

Energy transport is the most critical part of hydrodynamic codes after the core hydrodynamic solve. Thermal conduction is often important, especially so in laser-plasma simulations, e.g. laser driven ablation. If the scale length associated with a temperature variation is much longer than the electron mean free path then the thermal flux should be given by Fourier's Law,

$$\mathbf{q} = -\kappa \nabla T, \tag{14.46}$$

where $\kappa = \kappa(\rho, T)$ should be given by the Spitzer-Harm thermal conductivity. If this is valid then thermal conduction can be handled by solving the relevant parabolic system, e.g. in 1D we have,

$$\frac{\partial T}{\partial t} = \frac{2}{3nk_B} \frac{\partial}{\partial x} \left(\kappa \frac{\partial T}{\partial x} \right), \tag{14.47}$$

using implicit differencing and a matrix solve. So the basic problem of thermal conduction is well catered for by a large body of reliable numerical methods.

A much bigger challenge arises when one encounters a situation where classical transport might break down. This can readily occur in laser-plasma scenarios where temperature gradients become extremely large and the characteristic scale length of temperature variation shrinks to values that are not far from the electron mean free path. In such situations the Spitzer-Harm conductivity must breakdown. Ultimately any heat flow based on Fourier's law must break down too, as the heat flow shouldn't exceed the *free-streaming* limit where $q_{FS} = n_e k_B T c_s$.

A number of different approaches have been tried to deal with this problem. The simplest approach is to utilize a *flux limiter*. A flux limiter simply caps the heat flow at a specified fraction of the free streaming limit. This is sensible in the sense that it ensures that the heat flux can't go unphysically high, but it is problematic in the sense that proscribing a specified limit might be inaccurate. A more accurate approach is the *convolution* approach, which is based on the linearized Vlasov-Fokker-Planck equation and Fourier transforms. This is still based on certain approximations however. The best solution would be to fully incorporate a Vlasov-Fokker-Planck solver into a hydrodynamics code. This would probably still be far too slow and expensive for the largest simulations. Non-local thermal conduction is therefore still a problem for hydrodynamic simulation.

14.7 Radiation Transport

In high temperature plasma physics, one does not only have to consider thermal conduction, but also the transport of energy due to short wavelength photons. In many hydrodynamic problems relating to laser-plasma studies, energy transport due to soft x-rays is not negligible, and can even be of central importance, e.g. indirect-drive ICF. In general, radiation transport is a very demanding problem, and we will only touch on this important topic here. We can see this in mathematical form by considering a simple form of the radiative transfer equation, which is essentially a kinetic equation for photons,

$$\frac{1}{c}\frac{\partial I_\nu}{\partial t} + \hat{\mathbf{n}}.\nabla I_\nu = j_\nu - k_\nu I_\nu. \tag{14.48}$$

In the above equation, $I_\nu = I_\nu(\mathbf{r}, \hat{\mathbf{n}}, t)$ is *spectral intensity*. By *spectral intensity* we mean that the energy transported in time dt in the direction $\hat{\mathbf{n}}$ through area dA in the frequency interval $d\nu$ is,

$$dE = I_\nu \hat{\mathbf{n}}.\mathbf{dA} d\Omega d\nu dt. \tag{14.49}$$

The term j_ν denotes emission of radiation at this frequency, and k_ν is the absorption coefficient (or opacity). The scattering of photons is not included in this equation. If no approximations can be made then it is clear that we need to solve a fully kinetic problem (up to 6D) for the radiation which will impose significant computational overheads on the simulation code, even if it is valid to neglect the time derivative and obtain a reduced equation.

If the mean free path of the photons is very short compared to the hydrodynamic length scales then one may use the diffusion approximation along with an assumption of equilibrium. Here it is assumed that the time derivative in the transfer equation can be neglected, that the plasma is in Local Thermodynamic Equilibrium, and, crucially that the radiation field is nearly isotropic and Planckian. In this approximation the total radiation flux, $\mathbf{S}$, is reduced to the form of,

$$\mathbf{S} = -\frac{c}{3\kappa}\nabla U, \tag{14.50}$$

where κ is the opacity, U is the radiation energy density which is assumed to be close to $U_p = \pi^2 (k_B T)^4 / 15\hbar^3 c^3$ which is the energy density of the equilibrium (Planck) radiation field. We thus have what is essentially a diffusion equation and can be solved by methods appropriate to that class of equations. Note that in this approximation, radiative transport is very close to thermal conduction. In terms of modifications to the hydrodynamic equations, this essentially becomes a transport term in the energy equation. The opacity is a material property that can be pre-calculated or obtained in tabulated form. There are levels of approximation that improve upon the diffusive case without becoming as demanding as the full radiative transfer equation.

14.8 Other Physics

14.8.1 Equation of State

While many problems that are tackled in the magneto-hydrodynamic behaviour of magnetically confined laboratory plasmas and certain astrophysical problems deal with plasmas that are well described by an ideal gas equation of state (EOS), the majority of laser-plasma problems will involve significant regions where the plasma EOS is far from that of an ideal gas. There are three general reasons why this might be so.

Firstly, a lot of laser-plasma problems involve the irradiation of initially cold solids. From a fluid dynamical viewpoint, cold solids will essentially have zero pressure until they have been heated to a sufficiently high temperature. Energy must be supplied to material to overcome the potential energies, i.e. the physical processes of melting, boiling, ionizing, etc. Throughout this entire transition one must expect the fluid to be far from the ideal EOS.

Secondly, at high densities there is the issue of electron degeneracy. Remember that the Fermi energy is,

$$\varepsilon_F = \frac{\hbar^2}{2m_e}\left(3\pi^2 n_e\right)^{2/3}. \tag{14.51}$$

Therefore at an electron density of $10^{32}\,\mathrm{m}^{-3}$ one finds that the Fermi energy is ε_F =780 eV. So the cold compressed DT fuel will be degenerate.

Thirdly there is the problem of "warm dense matter" where one may be dealing with matter that is both strongly coupled and degenerate, but well outside the parameter space of normal condensed matter. This area is actually a research frontier, where different EOS models may even may markedly different predictions.

Whatever the reason might be, inclusion of an appropriate EOS is an important matter both for hydro-code developers and users. Nowadays both EOS tables are available (e.g. SESAME) and EOS models (e.g. QEOS) in the literature.

14.8.2 Laser Heating and Energy Deposition

In only a very few problems will one be able to specify in the entire drive energy as an initial condition. Usually drive energy is supplied over a time that is a large fraction of the simulation time of interest. An obvious example is an ICF implosion where the total laser irradiation time might be more than 20 ns while the simulation time of interest might be only 25 ns. Hydrodynamic codes must therefore include energy deposition by lasers and other external drivers.

Since the hydrodynamic equations on their own to do not allow for self-consistent laser-plasma interactions, heating by lasers is essentially treated as a simple energy deposition. This means that there will be some prescription which will heat the fluid at a specified rate where it interacts with the laser (particularly at the critical surface). This may be quite a sophisticated prescription involving propagating laser beams by ray-tracing, and accurate accounting for absorption by physical processes such as inverse bremsstralhung.

On the other hand, these prescriptions will always have their limitations, which both developers and users need to be aware of. While inverse bremsstrahlung will be highly efficient under certain circumstances, the absorption efficiency decreases at very high intensities. At these higher intensities other processes come into play (e.g. resonance absorption), and the generation of suprathermal electrons (a.k.a. "hot" electrons) also becomes significant. Hot electrons will not be treated at all by the core set of hydrodynamic equations so codes often require modifications to even adequately handle the effect of hot electrons.

From the perspective of a hydro-code user the most important thing is to know what models are used in laser heating and absorption and to thus understand their limitations and applicability.

14.9 Summary

I hope that this lecture has provided the reader with general overview of the key aspects of hydrodynamic simulation. It is well worth attempting to write a simple 1D hydrocode yourself, as such an activity can often be more instructive than straightforward "book-learning". In terms of further reading, I have two immediate recommendations. For the relevant physics, *The Physics of Inertial Fusion* by Atzeni and Meyer-ter-Vehn is an excellent resource [1]. In terms of numerical methods, *Computational Fluid Dynamics* by Chung [2] is a reasonable starting point. Readers who want an introduction to fluid dynamics in general may wish to read Faber's book [3] alongside Atzeni and Meyer-ter-Vehn. Shock waves and other aspects of high energy density hydrodynamics are well covered by Zel'dovich and Raizer's classic text [4]. Other textbooks that cover computational fluid dynamics are available (e.g. Jardin [5]), and there is an extremely large body of peer-review literature that the reader may wish to consult at a later stage.

References

1. S. Atzeni, J. Meyer-ter-Vehn, *The Physics of Inertial Fusion* (Oxford University Press, New York, 2004)
2. T.J. Chung, *Computational Fluid Dynamics* (Cambridge University Press, New York, 2002)
3. T.E. Faber, *Fluid Dynamics for Physicists* (Cambridge University Press, New York, 2005)
4. Ya.B. Zel'dovich, Yu.P. Raizer, *Physics of Shock and High-Temperature Hydrodynamic Phenomena* (Dover Publications, Mineola , 2002)
5. S. Jardin, *Computational Methods in Plasma Physics* (CRC Press, Boca Raton, 2010)

Chapter 15
Particle-in-Cell and Hybrid Simulation

Alex P. L. Robinson

Abstract The aim of this lecture is to provide a very short 'primer' course on the PIC method. This will cover the basics of PIC algorithms and numerical methods, potential pit-falls of the method, and the important extensions to the method (including so-called 'hybrid' codes). This lecture is intended to be a starting point for further study, however enough details are given for a student to write their own 1D PIC code with some extra work.

15.1 The Kinetic Equation

The mathematical foundation of the kinetic approach is found in the kinetic equation for the distribution function, $f = f(\mathbf{r}, \mathbf{p}, t)$ of a given species,

$$\frac{\partial f}{\partial t} + \frac{\mathbf{p}}{\gamma m}\nabla_{\mathbf{r}} f + q(\mathbf{E} + \frac{\mathbf{p}\times\mathbf{B}}{\gamma m}\nabla_{\mathbf{p}} f = C(f) + S(f). \tag{15.1}$$

If the left hand side is equated to zero then this becomes the well known *Vlasov* equation which models a collisionless plasma. The right hand side contains the collision operator, $C(f)$, and potentially a source term, $S(f)$ if ionisation/recombination processes are present. The single-particle distribution function, f, is valid if there are no strong correlations between individual particles, which is valid in weakly-coupled plasmas.

The collision operator can be specified mathematically via the Fokker-Planck equation, although since we are initially concerned with fully ionised, collisionless plasmas we will not consider either $C(f)$ or $S(f)$ in much detail. Importantly, one

A.P.L. Robinson (✉)
Central Laser Facility, STFC Rutherford-Appleton Laboratory, Chilton, Oxfordshire, UK
e-mail: alex.robinson@stfc.ac.uk

P. McKenna et al. (eds.), *Laser-Plasma Interactions and Applications*, Scottish Graduate Series, DOI 10.1007/978-3-319-00038-1_15,

will need to couple the kinetic equation to Maxwell's equations, noting that only two are actually required,

$$\frac{\partial \mathbf{E}}{\partial t} = -\frac{\mathbf{j}}{\varepsilon_0} + c^2 \nabla \times \mathbf{B}, \tag{15.2}$$

$$\frac{\partial \mathbf{B}}{\partial t} = -\nabla \times \mathbf{E}. \tag{15.3}$$

The source term (current density), $\mathbf{j}$, can be obtained from the first moment of the distribution function. One now has a closed set of equations.

The problem with using this directly, i.e. by gridding f and finite differencing the Vlasov equation, as a basis for numerical simulation is that this rapidly becomes very computationally demanding. The distribution function at any given time can be up to six dimensional in nature. If N grid points are used for each dimension, then the storage requirement will scale as N^6. For storing each grid point as an eight byte double precision number this means that if $N = 100$ one would require over 7,000 GB to store this data structure. For lower dimensional systems (e.g. $f(x, p_x)$) the requirements are feasible, and indeed finite difference Vlasov simulation for those systems are possible. Nonetheless a reduction of the computational overheads is required.

15.2 The Particle Approach

One way to achieve 'information reduction' is to note that any distribution function can be approximated by a set of 'clouds' or 'macroparticles' that have some finite volume in phase space. Suppose we were to track a finite set of such macroparticles that each had a fixed form. What equations would have to be solved? Let,

$$f = \sum_i g_i(\mathbf{r} - \mathbf{r}_i(t), \mathbf{p} - \mathbf{p}_i(t)). \tag{15.4}$$

If we substitute this into the Vlasov equation and resolve the time derivative by using the chain rule then we quickly find that,

$$\frac{d\mathbf{r_i}}{dt} = \frac{\mathbf{p_i}}{\gamma_i m_i}, \tag{15.5}$$

and

$$\frac{d\mathbf{p_i}}{dt} = \frac{q_i\,(\mathbf{E} + \mathbf{v} \times \mathbf{B})}{\gamma_i m_i}. \tag{15.6}$$

In other words, we find that the 'macroparticles' can be evolved as a set of single particles. Although this is intuitively obvious, it shows that particle simulation is a fully consistent approximation to the Vlasov-Maxwell system. This approach has other advantages as well. A finite difference approach allows the possibility that a poor algorithm will drive f negative in some region of phase space, something which is impossible in this approach. One also notes that although a Vlasov code may have to handle regions of phase space that are empty, a PIC code does not, and it is therefore more efficient in this sense. One therefore expects this to produce highly 'robust' codes. On the other hand, this then raises questions about how accurate this method will be for a given number of macroparticles, and whether it leads to unphysical behaviour or phenomena. Since we will still solve Maxwell's equations on a grid that the macroparticles move through, we will have to interpolate onto the grid to obtain current densities and interpolate from the grid to obtain the electromagnetic fields at each macroparticle. This interpolation has the potential to cause a number of non-physical effects. These matters will be considered in more detail later on.

The central PIC algorithm will therefore proceed as follows:

1. Move the macroparticles.
2. Interpolate onto grid to obtain current densities.
3. Update the electromagnetic fields.
4. Interpolate from the grid to obtain EM fields at each macroparticle.
5. Update the momentum of the macroparticles.

Clearly choices must be made about the overall integration scheme. One common choice is to store the particle positions at $t_n = n\delta t$, and the particle momenta at $t_{n+1/2} = (n+1/2)\delta t$,i.e. the 'Leapfrog' scheme. Once these details are determined we can deal with each of these key steps in turn. In what follows we will describe things primarily in terms of a 1D3P, relativistic and electromagnetic code, but we will add important details that are relevant to 2D and 3D algorithms where appropriate.

15.3 Particle Pusher

The position and momentum updates constitute what might be termed the "particle pusher" part of the algorithm. The simplest step is the update of the macroparticle positions. Assuming that we have adopted a Leapfrog scheme, the step that must be performed in the 1D3P algorithm on each macroparticle at the $(n+1)$-th time-step is,

$$\frac{x^{n+1} - x^n}{\Delta t} = \frac{p_x^{n+1/2}}{\gamma^{n+1/2} m}. \tag{15.7}$$

The momentum update is only slightly more complex, and the complicating factor is the $\mathbf{v} \times \mathbf{B}$ term in the Lorentz force. One algorithm that does this effectively is the *Boris* algorithm, which separates the electric field and the $\mathbf{v} \times \mathbf{B}$ rotation. There are three main stages to this. Firstly one accelerates the macroparticle in the electric field for half a time step,

$$\mathbf{p_i^-} = \mathbf{p_i^{n-1/2}} + \frac{1}{2}q\mathbf{E^n}\delta t. \tag{15.8}$$

Next the $\mathbf{v}\times\mathbf{B}$ rotation is performed. Since this does no work on the macroparticle, γ is constant throughout this step, and we denote this as γ^*. The difference equation that is solved, for $\mathbf{p}_i^+$, is,

$$\frac{\mathbf{p}^+ - \mathbf{p}^-}{\Delta t} = \frac{q}{2\gamma^* m}\left(\mathbf{p}^+ + \mathbf{p}^-\right)\times\mathbf{B^n}. \tag{15.9}$$

This can be re-arranged into a set of three coupled equations (assuming 3P) for $\mathbf{p}^+$, which can be cast in matrix form and solved by inverting this matrix.

Finally, the macroparticle is again accelerated by the electric field for half a time step.

$$\mathbf{p^{n+1/2}} = \mathbf{p}^+ + \frac{1}{2}q\mathbf{E^n}\delta t. \tag{15.10}$$

15.4 Interpolating to and from the Grid

The interpolation between particles and the grid and vice versa (Particle-Grid Interpolation; PGI), depends on what choice is made for the spatial form of the macroparticles, or what is also called the *shape factor*, $S(\mathbf{r})$. In 1D, the i-th macroparticle will have an associated number of (real) particles per unit area, $\mathcal{N}_i$, and charge per unit area σ_i. The particle number and charge density in the j-th spatial cell is then given by,

$$n_j = \sum_i \mathcal{N}_i S(x_j - x_i)/\Delta x, \tag{15.11}$$

and,

$$\rho_j = \sum_i \sigma_i S(x_j - x_i)/\Delta x. \tag{15.12}$$

Other quantities such as current density can be obtained in the same way. In order to interpolate, for example, the E_x component of the electric field from the grid onto the particle in 1D, one has,

$$E_{x,i} = \sum_j E_j S(x_j - x_i). \tag{15.13}$$

One cannot proceed without making a definite choice for $S(x)$. Choosing a very simple top hat function of width Δx,i.e. *Nearest Grid Point* weighting, will result in simple coding whereas using linear interpolation or quadratic splines will require somewhat more coding. On the other hand higher order interpolation improves accuracy, smoothness, and other aspects (including energy conservation). To illustrate this, consider the case of linear interpolation in 1D, each macroparticle will contribute to only two grid points and vice versa, i.e.

$$\rho_j = \frac{\sigma_i}{\Delta x}\frac{x_{j+1} - x_i}{\Delta x}, \tag{15.14}$$

and,

$$\rho_{j+1} = \frac{\sigma_i}{\Delta x}\frac{x_i - x_j}{\Delta x}. \tag{15.15}$$

For the electric field we would then have contributions from only two grid points,

$$E_{x,i} = E_{x,j}\frac{x_{j+1} - x_i}{\Delta x} + E_{x,j+1}\frac{x_i - x_j}{\Delta x}. \tag{15.16}$$

PGI in 2D and 3D obviously involves more work, but is not conceptually different. PGI is a very critical part of the PIC algorithm in the sense that it is the point where the majority of non-physical behaviour can emerge. Firstly, PGI opens the possibility of a macroparticle exerting a force on itself. If, however, one ensures that the same $S(x)$ is used in both interpolation steps then this self-force can be eliminated. Secondly, using the same $S(x)$ throughout will ensure conservation of momentum (at least with periodic boundary conditions). It is possible to devise energy conserving algorithms, but these will then not conserve momentum exactly.

15.5 Solving Maxwell's Equations

Once PGI has taken place, one now has the source terms required for advancing the electromagnetic fields by one time step. The discussion here starts with the general approach used in 2D and 3D. The standard approach is to define the fields on the grid not at cell centres but with the electric fields along the cell edges and the magnetic fields normal to the cell faces as was proposed by Yee. Once this is done, one can difference Maxwell's equations in a way that is second order accurate in time and space. Here this is illustrated for the case of (E_x, E_y, B_z) in 2D. The scheme is a leapfrog method in essence, with the electric field stored at $n\Delta t$ and the magnetic flux densities and current densities at $(n+1/2)\Delta t$. The induction equation becomes,

$$\frac{B^{n+1/2}_{z,i,j} - B^{n-1/2}_{z,i,j}}{\Delta t} = -\frac{E^n_{y,i+1,j+1/2} - E^n_{y,i,j+1/2}}{\Delta x} + \frac{E^n_{x,i+1/2,j+1} - E^n_{x,i+1/2,j}}{\Delta y}, \tag{15.17}$$

and the electric field equations become,

$$\frac{E^{n+1}_{x,i+1/2,j} - E^n_{x,i+1/2,j}}{\Delta t} = -\frac{j_{x,i+1/2,j}}{\varepsilon_0} + c^2\frac{B^{n+1/2}_{z,i+1/2,j+1/2} - B^{n+1/2}_{z,i+1/2,j-1/2}}{\Delta y}, \tag{15.18}$$

and,

$$\frac{E^{n+1}_{y,i,j+1/2} - E^{n}_{y,i,j+1/2}}{\Delta t} = -\frac{j_{y,i,j+1/2}}{\varepsilon_0} + c^2 \frac{B^{n+1/2}_{z,i+1/2,j+1/2} - B^{n+1/2}_{z,i-1/2,j+1/2}}{\Delta x}. \quad (15.19)$$

One issue that requires care, is that of ensuring that $\nabla.\mathbf{B} = 0$ and $\nabla.\mathbf{E} = \rho/\varepsilon_0$. Provided that $\nabla.\mathbf{B} = 0$ initially, this difference scheme will ensure that it is maintained. Ensuring that Gauss' Law is upheld requires that the $\mathbf{j}$ and ρ used satisfy the continuity equation. This particular problem bears some similarities to ensuring that $\nabla.\mathbf{B} = 0$ in MHD codes. One solution is to employ *divergence cleaning*, i.e. an extra solve is carried out, but divergence cleaning is not a particularly satisfactory solution. The problem can be eliminated by computing $\mathbf{j}$ in a fully charge conserving fashion (first done by Buneman/Morse and Nielson), although this has implications for the noise properties of the EM fields.

In a 1D PIC code some simplifications can be made. The electric field component that coincides with the single spatial direction (which we assume to be E_x) can be determined by integrating Gauss' Law,

$$\frac{\partial E_x}{\partial x} = \frac{Zen_i(x) - en_e(x)}{\varepsilon_0}. \quad (15.20)$$

The transverse field components can then be updated by noting that, for the $E_y - B_z$ case, if we define,

$$F^{\pm} = \frac{E_y \pm B_z}{2}, \quad (15.21)$$

then $F^{\pm}$ are advected at c in the right/left directions respectively with a source term of $-j_y/2$. Therefore if the time step is set to $\Delta x/c$ then the update can be performed via,

$$F^{\pm,n+1}_{j} = F^{\pm,n}_{j\mp1} - \frac{\Delta t}{4}\left(j^{-}_{y,j\mp1} + j^{+}_{y,j}\right), \quad (15.22)$$

where the $-$ superscript on j_y denotes the current density obtained by interpolating the particles onto the grid at t, and the $+$ superscript denotes the current density obtained by interpolating at $t + \Delta t$. This ensures that the current density is centred in space and time.

15.6 Boundary Conditions and Lasers

In the previous sections we have covered all of the core components of a PIC code, but have thus far omitted any discussion of how the boundaries of the computational domain are handled. As PIC codes are grid-based, and the grids must be finite in extent, deciding on appropriate boundary conditions is unavoidable. There are two issues to be decided: (i) What happens to a macroparticle on reaching the grid boundary?, and (ii) What are the boundary conditions on the EM fields?

The typical choices that are made for handling macroparticles at the boundaries include:

- Periodic: Macroparticles that leave through one boundary, reappear at the other with unchanged momentum.
- Reflective: Macroparticles undergo specular reflection at boundaries.
- Open/Absorbing: Macroparticles are lost on passing across boundaries.

Each choice then has implications for the boundary conditions on the EM fields, e.g. if the boundaries are periodic for the macroparticles then the same should apply for the EM fields. In 1D codes the domains can often be made so large on modern computers that they will not affect the system by any appreciable amount. Periodic boundary conditions are easy to implement. So are boundaries that are reflective for the particles. In this case, $E_x = 0$ at these reflective boundaries.

In 2D PIC codes, the two boundary conditions that are used most often are periodic and open boundaries. In many studies, the boundaries are periodic in the direction transverse to the laser and open in the direction parallel to the laser. This arrangement is useful as long as no perturbations reach the transverse boundaries (and propagate back to the centre), as it is often used to simulate systems that have no real periodicity in the transverse direction at all. In many problems, however, it is hard to avoid scattered electromagnetic waves reaching the transverse boundary with a reasonably sized computational domain. In computational electromagnetics, a solution to this problem is to use 'Perfectly Matched Layers' at the transverse boundaries which will absorb these EM waves rather than scatter them back into the box. Perfectly Matched Layers have been implemented in some codes.

Laser pulses can be incorporated into PIC codes in two ways. If there is a large enough vacuum region available, a laser pulse could, in principle, simply be "loaded" onto the grid as an initial condition. In 2D and 3D, this is not an 'affordable' solution so laser pulses are instead launched into the grid at the boundaries.

15.7 Initialisation

The final physics-based core component of a PIC code is the algorithm used to initialise the macroparticles. Laser-plasma researchers are often interested in the relatively simple case of a neutral, field-free, stationary plasma. It is straightforward to assign the correct weight to each macroparticle to ensure that the correct particle density is achieved in each spatial cell. What is slightly non-standard is the initialisation of the distribution function in momentum. Any distribution function can be obtained from a set of uniformly distributed random numbers by the inversion of the cumulative distribution function, i.e. calculate,

$$D(\mathbf{p}) = \frac{\int_0^{\mathbf{p}} f(\mathbf{p}) d^3\mathbf{p}}{\int_0^{\infty} f(\mathbf{p}) d^3\mathbf{p}}, \tag{15.23}$$

then by equating $D(\mathbf{p})$ to a set of uniformly distributed random numbers, one will obtain a set of $\mathbf{p}$s that follow the specified distribution function.

This method of initialising the macroparticles has sometimes been found to be undesirable on the grounds that it is highly noisy. An alternative is to initially distribute the macroparticles in a proscribed order, which is completely without the use of random numbers. This is sometimes referred to as a 'quiet start'.

15.8 Key Issues with PIC Codes

It should be apparent by now that the PIC method is not without it potential pit-falls. This is not to say that the method is fundamentally flawed — it is very powerful — but to point out that the method is prone to error if abused. The first point to make concerns the number of macroparticles, and especially the number per cell. The number of macroparticles per cell (N_{pc}) determines how well the distribution function can be represented in phase space. Clearly one macroparticle in a cell can represent no more than a cell in a hydrocode, but beyond this it rapidly becomes difficult to determine *a priori* whether a given number of particles per cell will be sufficient for a given problem. Often this problem needs to be addressed in an 'experimental' manner, in order to determine whether or not convergence has been attained. Related to this is the problem of noise in PIC codes. PIC codes suffer from shot noise in the gridded quantities that scales as $\propto 1/\sqrt{N_{pc}}$. This is not physical and can often obscure physical phenomena that one is trying to observe.

Even if one has a reasonably high N_{pc} one is not guaranteed to be free of non-physical behaviour. PIC codes can exhibit non-physical instabilities, particularly when the Debye length is shorter than the spatial cell size and the time step is greater than the plasma period. These problems have been examined very thoroughly using formal mathematical theories. Here we will by-pass the formal theory and present results where possible. Firstly, we will look at the leap-frog integrator and how it handles the case where a single macroparticle becomes a simple harmonic oscillator. The difference equations that span two time steps from $t-\Delta t$ to $t+\Delta t$ are:

$$\frac{x^{n+1}-x^{n}}{\Delta t} = v^{n+1/2}, \tag{15.24}$$

$$\frac{v^{n+1/2}-v^{n-1/2}}{\Delta t} = -\omega_0^2 x^n, \tag{15.25}$$

$$\frac{x^{n}-x^{n-1}}{\Delta t} = v^{n-1/2}. \tag{15.26}$$

These three equations can be combined into a single equation for the xs alone. If one now looks for solutions of the form $x^n = Ae^{i\omega t}$, then one obtains the following expression,

$$\sin\left(\frac{\omega t}{2}\right) = \pm\frac{\omega_0 \Delta t}{2}. \tag{15.27}$$

This immediately tells us numerical instability can occur if $\omega_0 \Delta t > 2$, and we will only obtain good accuracy (i.e. $\omega \approx \omega_0$ if $\omega_0 \Delta t/2 \ll 1$. This tells us that the highest natural frequency of the system, i.e. the plasma frequency, needs to be well resolved in time. In interactions with underdense plasmas, where the laser frequency becomes the highest frequency of the system, the same consideration applies.

Secondly we will look at the need to resolve the Debye length. This is somewhat more subtle, as it involves non-physical modes (*aliases*) which can destabilise the plasma. For 1D warm plasmas, the electrostatic dielectric functions can be derived for both the physical case and the PIC case. One finds that the two will only coincide if $\Delta x \ll \lambda_D$. If one doesn't start a PIC simulation of a warm plasma with the Debye length resolved then non-physical electrostatic modes will artificially heat the plasma until the Debye length is resolved. Using smoother functions for macroparticle spatial profile can mitigate this *numerical heating* but not stop it altogether. Note that numerical heating is consistent with our previous comments on energy-momentum conservation in PIC codes, i.e. momentum-conserving PIC algorithms do not conserve energy exactly.

We shall conclude this section by saying the following: PIC codes are excellent tools for understanding real physics, but one must be aware of their limitations and take great care not to abuse them!

15.9 Extending the PIC Method

In the previous sections the reader will have seen that, despite certain limitations, PIC codes are robust and based on relatively simple algorithms. Their simplicity and robustness has lead a number of researchers to try to extend the method to much more than just fully ionised, collisionless plasmas. Hybrid PIC codes are covered in the next section, so here we will discuss a number of other developments of the PIC method:

- **Collisions**: Collisions can be included through Monte-Carlo algorithms. Such methods normally involve randomly pairing macroparticles within each spatial cell, and performing binary Coulomb collisions. It can be shown that, in the limit of large N_{pc}, this is equivalent to the Fokker-Planck collision operator. The additional operations do increase the run-time. Whether or not physically correct transport coefficients are reproduced for a viable N_{pc} needs to be checked.
- **Ionisation**: With an appropriate knowledge of ionisation cross-sections, one can also use Monte-Carlo methods to include electron and ion impact ionisation processes. Field ionisation can similarly be included. How well momentum space is represented by a given N_{pc} is equally important here too.
- **Radiation Reaction**: When charged particles are accelerated they emit radiation. If this acceleration becomes extremely strong then the radiation emission exerts

a non-negligible back-reaction on the charged particle. It is thought that this will become important at the extreme intensities that will be realised by future laser systems ($\approx 10^{23}$Wcm^{-2}).

- **High Energy Photons and QED processes**: At extreme laser intensities ($\gg 10^{23}$Wcm^{-2}) it is possible that very high energy photons (generated by violent acceleration of electrons) will disintegrate into electron-positron pairs in the ultra-strong EM fields. Current research aims to include both photon and pair production into PIC codes via Monte Carlo methods.

15.10 Hybrid Codes

There is considerable interest in subjects such as fast electron transport, and fast ignition ICF, where a small population of super-thermal (a.k.a. *fast*) electrons with very long mean free paths propagate through a dense, collisional, and relatively cool plasma. This is a very demanding problem due to the great disparity in length and time scales. Collisionless PIC codes cannot model the resistivity of the background plasma, and thus can't capture effects due to the finite resistivity of the background plasma (especially magnetic field generation).

One model that has been used to study fast electron transport is the *hybrid code*. The term 'hybrid code' is used in a number of different areas of plasma physics with different meanings. In this context it means that the background plasma is treated as a fluid, and the fast electrons are treated kinetically. Hybrid codes make a number of assumptions that one needs to be aware of:

- **Small Fast Population**: Most importantly, hybrid codes make the assumption that the fast population is much smaller than the background population everywhere, i.e. $n_f \ll n_b$, even though the current density of the fast need not be negligible. Without this, it would be impossible to treat the background as a distinct, quasi-neutral plasma.
- **Validity of Fluid Background**: Hybrid codes also assume that, even if the fasts were absent, that a fluid description of the background plasma is valid on the time and space scales of interest. Since a number of properties of the fluid background will be prescribed, e.g. resistivity, specific heat capacity, etc., all of these models need to be valid and accurate as well.
- **Current Balance**: Provided that the fast electrons are a small population there are strong electrostatic and magnetostatic arguments for current balance holding, i.e.

$$\mathbf{j}_f + \mathbf{j}_b = \frac{\nabla \times \mathbf{B}}{\mu_0}, \tag{15.28}$$

where the 'f' subscript denotes the fast electrons and the 'b' subscript denotes the background electrons. Current balance will only break down if j_f varies strongly over the background electron's collisionless skin depth. If the fast population is indeed very small then this should not occur. The displacement current is also neglected.

- **Ohm's Law**: Hybrid codes assume an Ohm's law to determine the electric field. The simplest one that is typically used being,

$$\mathbf{E} = \eta \mathbf{j}_b, \tag{15.29}$$

which, on assuming current balance and neglecting $\nabla \times \mathbf{B}$ becomes,

$$\mathbf{E} = -\eta \mathbf{j}_f. \tag{15.30}$$

- **Resistive Magnetic Field Generation**: Although not an assumption *per se*, the use of Ohm's Law to determine the electric field, must reduce the induction equation. For the simple Ohm's law this will result in,

$$\frac{\partial \mathbf{B}}{\partial t} = \nabla \times \left(\eta \mathbf{j}_f\right), \tag{15.31}$$

which expresses the resistive generation of magnetic field around a fast electron beam.

These assumptions greatly relax the resolution requirements for these problems, since the cold and dense background has very small spatial and temporal scales that make fully kinetic simulation unfeasible. They also greatly simplify the algorithm, as the reduced Maxwell's equations as much easier to solve. Fast electrons are simply introduced onto the grid in a 'heating zone' according to some prescription that models fast electron generation by the laser interaction. The fast electron macroparticles will also undergo scattering and drag by the background via Monte-Carlo algorithms. Apart from these modifications the same PIC algorithms are employed.

15.11 Summary

At the most basic level, the PIC method is simple and intuitive. The numerical methods that it employs are fairly transparent, as well as being straightforward to understand and program. This can be deceptive however. PIC codes are also subject to a range of non-physical behaviour, which includes statistical noise, non-physical instabilities, non-conservation, and numerical heating. One simply cannot expect to achieve good quality kinetic simulation by simple means without 'paying the piper'!! Having said this, with due care and attention they are excellent numerical tools, that can provide great insight into complex kinetic systems. Apart from urging the interested reader to try writing his or her own 1D PIC code, I also advise him or her to read *Plasma Physics Via Computer Simulation* by Birdsall and Langdon [1]. Other books to look at include [2] and [3].

References

1. C.K. Birdsall, A.B. Langdon, *Plasma Physics via Computer Simulation* (Taylor and Francis, New York, 1991)
2. R.W. Hockney, J.W. Eastwood, *Computer Simulation Using Particles* (Taylor and Francis, New York, 1989)
3. J. Büchner, C.T. Dum, M. Scholer, *Space Plasma Simulation* (Springer, New York, 2003)

Chapter 16
Diagnostics of Laser-Plasma Interactions

David Neely and Tim Goldsack

Abstract The field of laser-plasma diagnostics is large, and is certainly too big to be covered fully in a short article. Here, we concentrate on a number of diagnostics which are of relevance to two areas of significant current interest, namely (1) relativistic laser-plasma interactions, and (2) hohlraum-driven inertial confinement fusion.

16.1 Diagnostics of Relativistic Intensity Interactions

In the high intensity region of laser plasma interactions, significant numbers of ions with energies > 10 MeV/nucleon can be produced if the conditions are correct. The fraction of energy carried away can readily approach 10 % of the initial incident laser energy and there are predictions of significantly higher efficiencies being possible as the interaction intensity is increased. The diagnosis of these particles of ion beams is an important development in laser plasma interactions field which has made significant advances over the last decade and the remainder of this section will examine some of the main techniques currently being employed. Descriptions and reviews of the wider field of plasma diagnostics may be found in references [1–7].

D. Neely (✉)
Central Laser Facility, STFC, Rutherford Appleton Laboratory, Didcot OX11 0QX, UK
e-mail: david.neely@stfc.ac.uk

T. Goldsack
AWE Aldermaston, Reading RG7 4PR, UK
e-mail: tim.goldsack@awe.co.uk

P. McKenna et al. (eds.), *Laser-Plasma Interactions and Applications*,
Scottish Graduate Series, DOI 10.1007/978-3-319-00038-1_16,

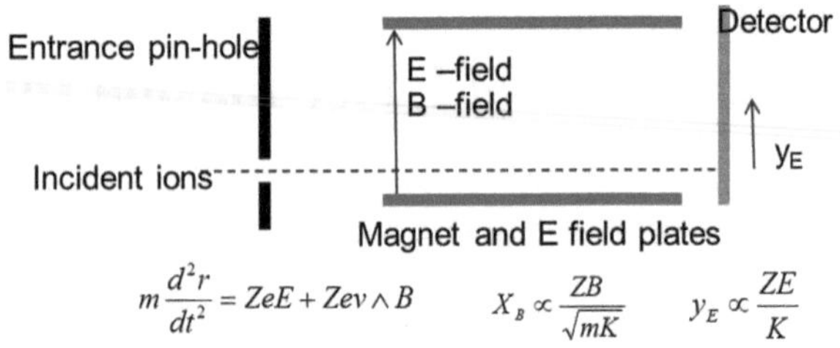

Fig. 16.1 Schematic plan view of a Thomson parabola spectrometer showing how the deflection due to the electric and magnetic fields can be used to separate the distribution of ions with different velocities and charge to mass ratios

16.1.1 The Thomson Parabola Ion Spectrometer

The classical Thomson parabola spectrometer used in laser plasma experiments as shown in Fig. 16.1 consists of parallel electric E and magnetic B fields through which ions are deflected according to their velocity v and charge Ze to mass m ratio. These instruments can be readily operated for ions with energies from keV to GeV with a suitable selection of field strength, geometry and detector. A pin hole placed before the plates samples a small fraction of the ions escaping from the target. The resolution of the instrument is primarily determined by the ratio of the geometrical projection of the ions through the entrance pin-hole to the dispersion due to the electric and magnetic fields and is typically in the region $\Delta E/E \sim 10^2 - 10^3$. Although mutual space charge effects, where, the ions travelling together self repel and the beam effectively blows-up within the instrument can be important for lower energy lighter ions, it is typically not a significant issue even for sub ps bunches of MeV energy protons. The spatial resolution of the detector can also be important if it is inadequate to resolve the geometrical projection of the entrance pin-hole.

Many variants of the basic Thomson parabola spectrometer design exist. In Fig. 16.2 a wedged electric field plate configuration is employed to generate a greater deflection for a given applied voltage [8]. In this instrument, a scintillator screen is coupled to an EMCCD to provide instantaneous readout of the ion spectra. To enable a greater dynamic range of operation, which in this case was limited by the dynamic range of the EMCCD, a double pin-hole is used at the entrance slit which generates two tracks for each charge to mass ratio track. With a ratio of $\sim \times 1{,}000$ in area between the two pin-holes this extends the dynamic range of the instrument. For high charge to mass ions (i.e. protons), the tracks from the two pin holes can be easily separated. However, for heavier ions where the change in charge to mass between adjacent ions is much less the tracks can merge and be difficult to separate.

16.1.2 Particle Detectors

Particle detectors such as CR-39 have been extensively used for charged particle detection from laser produced plasmas [9] as they are simple to use and are virtually

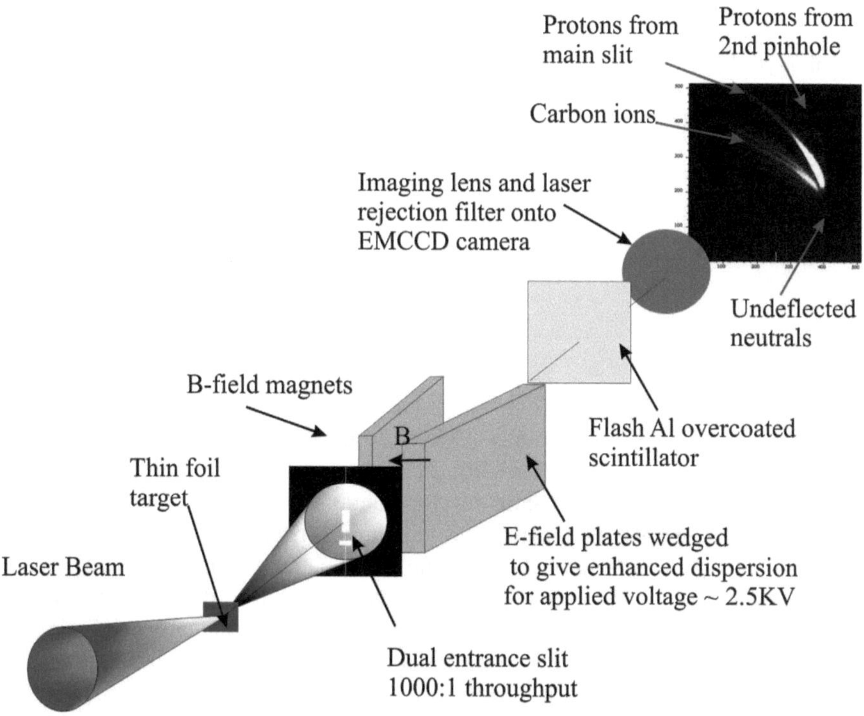

Fig. 16.2 A modified Thomson parabola spectrometer schematic where the electric field plates are wedged to provide greater dispersion for a given applied voltage

immune to radiation and electrons. As incident particles pass through the CR-39 (allyl diglycol carbonate plastic), polymer bonds are damaged due to the high rate of energy deposition (dE/dx), in the long chains of the CR-39. When the plastic is then processed in hot NaOH, the shorter polymers are preferentially dissolved and a pit forms at the surface of the plastic which grows in size with etching/development time. The CR-39 is then scanned, typically using an optical microscope (with $\sim 0.5\,\mu$m resolution) and the size and location of the pits are then identified. For individual protons in the 0.1–4 MeV region the CR-39 has a QE of 1 and every proton which is incident can be detected. An issue arises, when multiple particles are incident very close together. If the distance between the locations where the particles hit the CR-39 is smaller than the diameter of the pit, it becomes difficult to separate the individual pits. In this case, the optimal pit size after development is just resolvable by the optical microscope ($> 0.25\,\mu$m). As the pit size depends on the development time and the deposited energy at the surface dE/dx then conducting multiple development and scanning cycles enables a much higher density of particles to be identified. In Fig. 16.3 an example is given of the particle spectrum recovered using a Thomson Parabola ion spectrometer with CR-39 as the detecting media. For the shortest development time used of 30 min, the

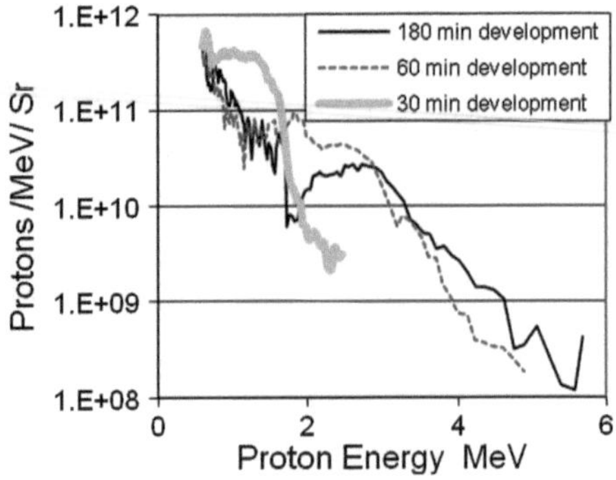

Fig. 16.3 Proton spectrum detected using CR-39 in a Thomson parabola spectrometer using different development times for the CR-39

particles where the Bragg peak is within the layer dissolved are readily detected. However, clearly, particles with $E > 2$ MeV energy are not measurable. During two subsequent developments the detection of protons with energies up to 5 MeV is possible. However, for the 180-min development, the pit sizes for the lower energy protons now overlaps significantly and their number appears to artificially drop. The final particle spectrum is therefore the combination of the maximum particle densities detected for the multiple developments. As well as detecting ions where the Bragg peak is close to the front surface, if the ions are sufficiently energetic, pits are also observed when the Bragg peak coincides with the rear surface, typically ~ 10 MeV for protons when the CR-39 is 1 mm thick.

To measure the flux within a laser driven ion beam, detection using CR-39 and direct counting of the individual ions is only practical when the total number of ions is $< 10^6$. For experiments using drive lasers of > 0.1 J energy, which can readily generate $> 10^9$ ions per shot, sampling using multiple spectrometers to characterise the distribution and then integrating under the profile is routinely used. If the CR-39 is combined with a filter pack it can be used to measure where the edge of the ion beam is, as a function of energy and this can be combined with at least two spectrometer measurements to characterise the distribution. Numerical integration under the interpolated spectral and angular profiles can then be carried out to give a good estimate the total energy contained within the beam [10].

In many experiments Radio Chromic Film (RCF) stacks as shown in Fig. 16.4 are now routinely used to characterise the ion beam distribution [11]. The ion beam is incident upon a stack of films, with the peak sensitivity of each layer corresponding to the Bragg stopping distance of the ions. The angular and spectral profile of the beam can then be deconvolved from the deposited dose. The emission area of the ions can also be diagnosed by imposing structure on the rear surface of the target foil which can be observed on the RCF stack as shown in Fig. 16.4(b).

Typical responses of Radio Chromic Films are in the 5–20 kGy region before saturation effects begin to set-in. The films are normally optically scanned to give a grey-scale image which can be converted to dose using a measured calibration curve [11]. It has been found that scanning at red wavelengths increases the sensitivity at lower doses [12]. In many flat-bed scanners, the optical density

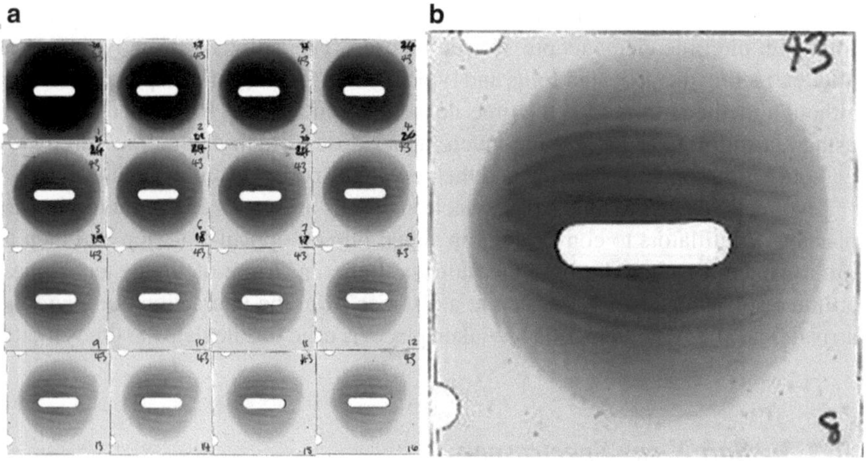

Fig. 16.4 (**a**) Example of a the exposures obtained on an RCF stack in the 4–30 MeV region. (**b**) Enlarged view showing target structure mapped onto the beam which can be used to characterise source size

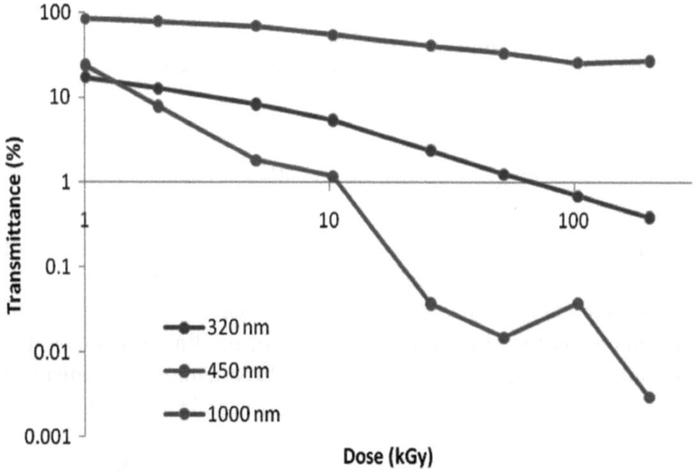

Fig. 16.5 Radio Chromic Film calibration curve at 1,000, 450 and 320 nm wavelengths showing good response up to doses on 200 KGy at 320 nm

resolution is limited by the noise in the optical sensor and scanner electronics. By combining the information from the red, green and blue channels the dynamic range can be routinely increased by $\sim \times 10$. The available dynamic range of RCF can also be increased up to doses of 200 kGy using UV scanning, as illustrated in Fig. 16.5.

Nuclear activation techniques can also be used to measure the ion beam angular and energy distribution [13–15]. In an analogous method to RCF stacks, foils can be used in place of the film and the induced activity measured to characterise the beam. Ion induced reactions with different cross-sections within a single element or foils with multiple elements present can be used. The activity is usually measured after extracting the activated foil/s and typically involves significant post processing although methods to reduce the time needed are underway [16]. An advantage of using activation over RCF is the much higher doses which can be measured [17] without saturation issues arising and the very high spatial resolution achievable.

Two methods employing active detection of ions are to use either micro-channel plates or scintillators to convert the ion signal into an optical emission and then to use a camera or photodiode to measure the light. In 'Time of flight' techniques, a time resolved detector is situated at a distance from the interaction point and the arrival time after the interaction has taken place, is used to calculate the ion velocity.

16.1.3 Soft X-ray Spectroscopy

Spectroscopy of the soft x-ray region (1–50 nm) has played an important part in the analysis and understanding of coronal plasma physics and radiation transport in laser driven plasmas. There are many significant sources of broadening for line emission coming from a given atomic transition in a laser generated plasma. The primary mechanisms are Doppler broadening, opacity, Zeeman and Stark/pressure broadening. The Doppler broadened width $\Delta\lambda_D$ for an ion of temperature T_i is given by:

$$\Delta\lambda_D = 2\lambda\left(\frac{2kT_i \ln 2}{mc^2}\right) = 7.7 \times 10^{-5}\lambda\left(\frac{T_i\,(eV)}{A\,(amu)}\right)^{0.5}$$

where k is the Boltzmann constant, mc^2 the rest mass of the ion and A the atomic weight. The Doppler width for a Ge emission line at 10 nm where the for ions are at a temperature in the range $200 < T_i < 1{,}000$ eV is $\Delta\lambda_D = 1.5 - 2.7 \times 10^{-3}$ nm. The other mechanisms will tend to add to the width of the line or reshape it. However, Doppler broadening is generally the dominant broadening mechanism for typical coronal plasma conditions and to resolve it requires an instrument with a resolving power of at least $\lambda/\Delta\lambda \sim 5{,}000$.

As well as being used for understanding the physical mechanisms at play in a plasma, recent work has also examined laser driven plasmas as secondary sources for a wide variety of applications such as lithography, biological imaging and applications and materials science studies. Plasma based soft x-ray lasers have relied on spectroscopic techniques in their development. From the demonstration of the first saturated soft x-ray laser [18], through beam divergence control [19], to identification of transitions with inner shell holes, medium resolution survey instruments have been the primary instrument of choice. In the field of high harmonic emission from laser driven gaseous targets [20] and solids [21] and in the detection of new generation mechanism, medium resolution instruments have played a pivotal role.

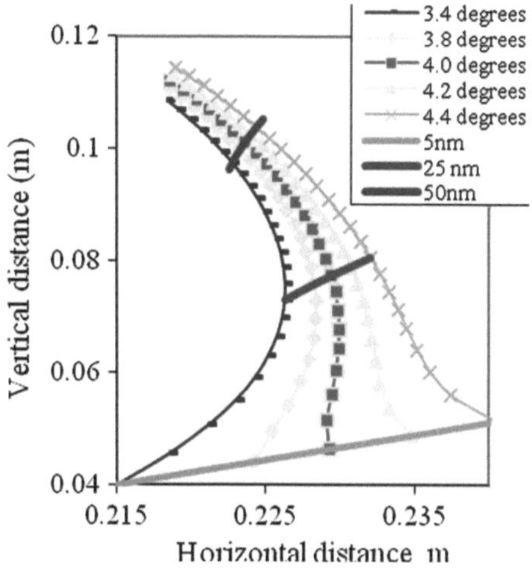

Fig. 16.6 Focal planes for a 1,200 l/mm flat-field spectrometer at different glancing angles with a source at 0.82 m

Each physical measurement or application has different requirements and in general, three types of spectrometer have been developed, (i) high resolution instruments capable of resolving the emission line profile ($\delta\lambda/\lambda > 10^4$), (ii) survey instruments capable of identifying individual lines and measuring the emission across a large fraction of the soft x-ray range few ($\delta\lambda/\lambda \sim$ few $\times 10^2$) and low resolution broad band instruments giving the total energy or power emitted. As the real part of the refractive index for soft x-rays does not deviate significantly from unity, diffraction from gratings or transmission through binary gratings is normally used to disperse the radiation rather than refraction.

Many of the early soft x-ray spectrometers were based on glancing angle diffraction from a curved grating in a Roland circle geometry. The resolving power of an ideal instrument is given by $\delta\lambda/\lambda \sim \rho l$ where ρ is the grating line density and l the illuminated length of grating. The Roland Circle geometry can deliver near diffraction limited performance giving high resolution. However, in this geometry, the soft x-rays are focused onto a curved focal plane where the rays come at a small glancing angle onto the detector. This requires very high accuracy in initial setting up. Coupling over a large wavelength range onto the flat plane, typical of most CCD's and MCP's is difficult and results in a limited in-focus spectral range. An alternative geometry adopted by many laboratories is to use a 'flat-field' spectrometer. In this instrument, the line density of the grating ρ is changed along the length of the grating in a precisely controlled manner to cause additional focusing which effectively rotates the focal plane to be nearly parallel to the grating normal.

In Fig. 16.6, the focal planes for a for a 1,200 l/mm flat-field grating with a point x-ray source located along the x axis at -820 mm and the grating centred at the origin are shown for five different grating glancing angles. The 'flat' focal region of the spectrum can be seen across the 5–25 nm spectral region for the case of 4° glancing angle in this situation.

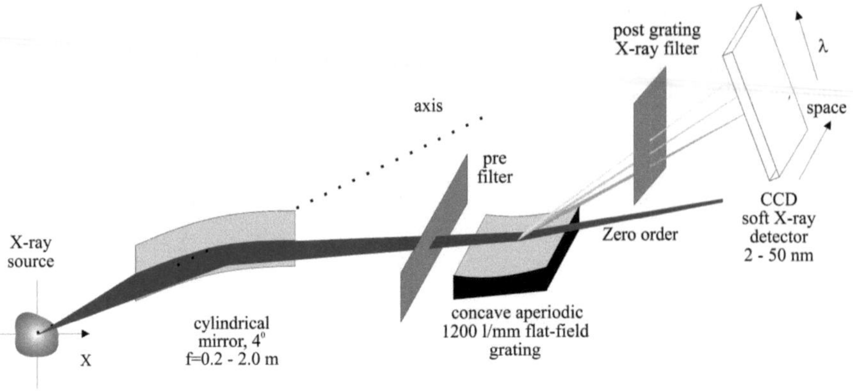

Fig. 16.7 Schematic layout for a soft x-ray flat-field spectrometer

As the total distance that x-rays travel from the entrance slit to the detector is almost constant, curved mirrors in the plane orthogonal to the dispersion have also been coupled to this grating. If the geometry is arranged as shown in Fig. 16.7 so that the source is effectively focused in the orthogonal plane onto the detector plane, then a significant enhancement of sensitivity can be obtained. In this imaging geometry, the spectrometer can be used to provide 1D spatial (in the Z direction as shown in Fig. 16.7) as well as spectral information. If the spectrometer is used in this 'slitless' mode, the size of the source in the y direction is imaged onto the detector by the grating and for mm scale or larger plasmas can significantly reduce the spectral resolving power. However, this effect can be utilised in some cases to create a quasi image of larger plasmas where the spectral and spatial information are both convolved. By careful analysis of the detected signal in multiple orders, deconvolution can be performed in cases where the spectral features are well separated as is usually the case from lower Z emitters in this spectral region.

If a streak camera is used as the detector [22] it is possible to readily obtain ps resolution of the duration of soft x-ray emission. However, the dynamic range of current streak cameras operating with ps resolution is typically $\sim \times 10$.

16.2 Inertial Confinement Fusion Diagnostics

Inertial confinement fusion is achieved by the compression of capsules of DT ice and gas to ultra-high pressures and temperatures by illuminating them with multiple laser-beams ('direct drive') or the x-rays produced inside a hohlraum when that is heated by laser light ('indirect drive'). The latter approach is viewed as likely to be more successful, and is the approach adopted at the National Ignition Facility at LLNL, USA. Target design involves an optimisation of maximising the fuel compression whilst minimising the internal energy imparted to the fuel (i.e. the fuel

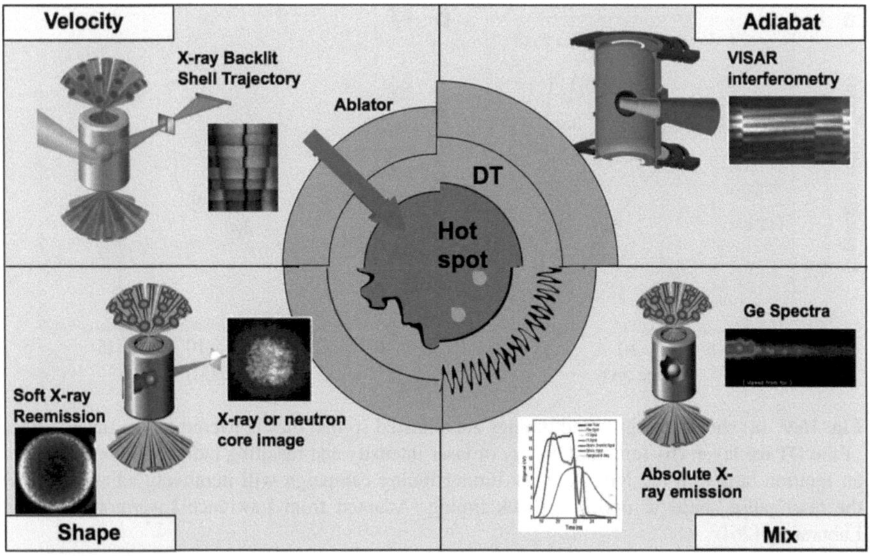

Fig. 16.8 The key implosion parameters which must be controlled if ignition is to be achieved (Figure courtesy of National Ignition Facility, LLNL)

must be kept on a low adiabat). The usual approach is to approximate isentropic compression by employing a sequence of three shock-waves of increasing strength, followed by a fourth, stronger, shock that drives the compression. In order to achieve indirect-drive ignition with the relatively small amount of laser energy available on the NIF, a number of criteria must be met (see Fig. 16.8). The implosion must be sufficiently fast (i.e. the shell must reach a high-enough velocity) and symmetric, the fuel must remain on a low adiabat during the compression, and the amount of hydrodynamic mix must be small. Further details may be found elsewhere (e.g. [23, 24] and references therein).

The diagnostics used to measure each of these parameters are now discussed.

16.2.1 Shell Velocity

The shell velocity is governed by the hohlraum performance, and in particular by the shape and timing of the shocks driven into the ablator by x-rays arising from the interaction of the incoming laser light with the hohlraum wall and – to some extent – with the capsule. Figure 16.9 shows the concept. As the hohlraum is approximately a black-body, it is the temperature reached by the hohlraum which is usually quoted (flux $\propto T^4$), even though the diagnostic (Dante) which is usually employed to measure the temperature actually measures the radiated flux. A description of some early work on the Nation Ignition Facility is given by Meezan et al. [25].

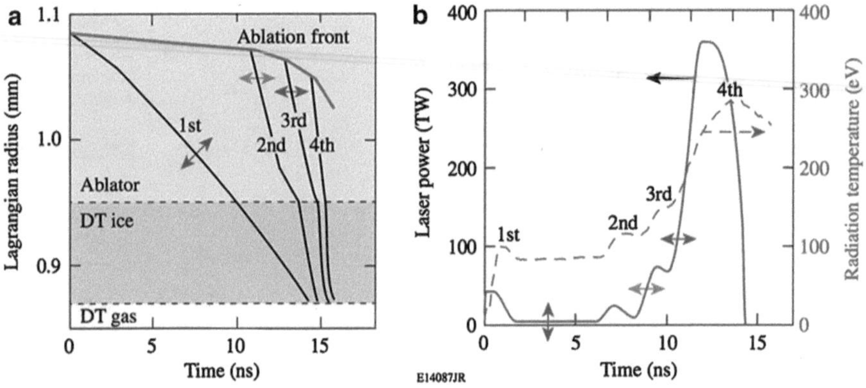

Fig. 16.9 (**a**) shock strengths and timings are adjusted so that they converge on the inner surface of the DT ice layer. (**b**) Temporal history of laser intensity and resulting radiation temperature for an ignition target on the NIF. A shock-timing tuning campaign will iteratively adjust (*arrows*) the laser pulse shape to optimise shock timing (Adapted from Lawrence Livermore National Laboratory [26])

16.2.2 Dante – A Hohlraum Temperature Diagnostic

Dante is an absolutely-calibrated, multi-channel, time-resolved x-ray spectrometer [27]. In France it is known as the DMX [28]. Each channel comprises an x-ray sensitive vacuum photo-diode, with a filter and perhaps an x-ray mirror to define the channel response. Figure 16.10 shows the concept [27].

There are two Dantes on NIF; the one nearest the equatorial plane is shown in Fig. 16.11. The equator is in the horizontal plane at the NIF because the hohlraum axis is vertical and so the beams enter the hohlraum from above and below.

The x-ray diode used on ORION is shown in Fig. 16.12. Photons enter the detector from the right and pass through the grid anode, which is an etched nickel grid (transmission ~ 80), and strike the photocathode, liberating photo-electrons. The outer connector is a $50\,\Omega$ bias cable that maintains a positive bias voltage (typically 1 keV) on the anode grid; the photo-electrons are thus attracted to the anode, inducing an image charge on the anode as they move across the gap. A positive pulse is transmitted along the cathode stalk and propagates to the output cable.

In the ORION Dante there are ten channels, each designed to cover part of the x-ray spectrum through the use of different photo-cathodes, filters and mirrors (Table 16.1).

Spectral coverage is shown in Fig. 16.13 where it can be seen that the device is sensitive only up to around 5 keV. This is appropriate, given that the peak of the black-body spectrum is at around three-times its temperature (around 1 keV for a 300 eV hohlraum, which is the highest temperature that ORION is likely to produce), and that the gold M-band lines are around 2.5 keV.

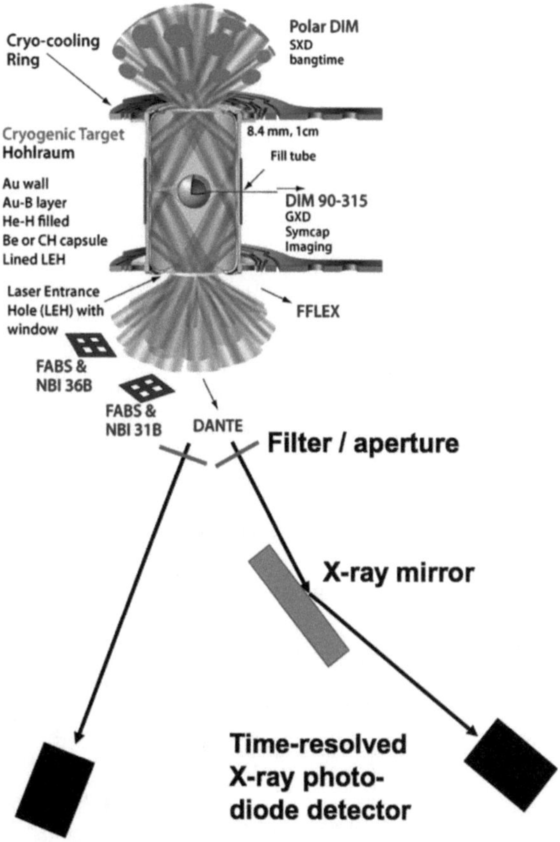

Fig. 16.10 Dante concept. Multiple filtered x-ray diodes (some with mirrors) record time-resolved measurements of x-ray flux in different regions of the x-ray spectrum (Hohlraum image courtesy National Ignition Facility, LLNL)

16.2.3 VISAR – A Shock-Breakout Timing Diagnostic

A VISAR Velocity Interferometer System for Any Reflector (Velocity Interferometer System for Any Reflector) enables changes in velocities to be measured by using the Doppler shift of a laser beam diffusely reflected from a moving target. Figure 16.14 shows the principle.

Figure 16.15 shows how this can be employed inside a capsule to measure the arrival time of the shocks on the inside wall of the DT ice layer and the velocity imparted to it.

Figure 16.16 shows typical VISAR data, showing (a) fringes and (b) the inferred velocities. This information is useful also in determining how much the capsule has been heated by the passage of the shocks.

The ORION laser at AWE will be equipped with a VISAR system, but this is not yet operational.

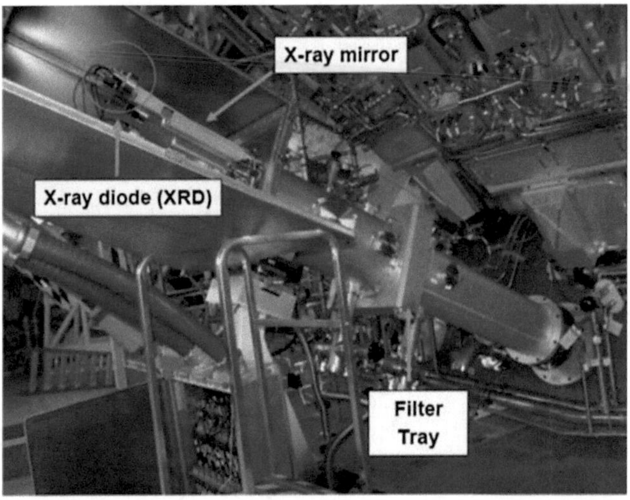

Fig. 16.11 The equatorial NIF Dante (Image courtesy National Ignition Facility, LLNL)

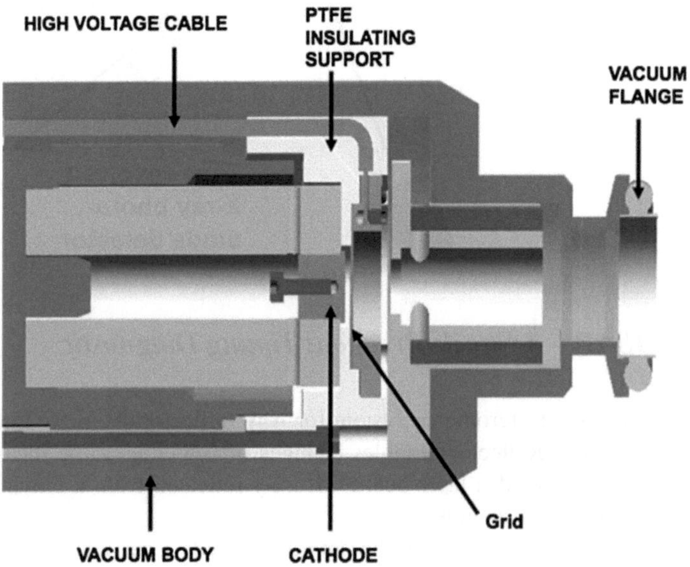

Fig. 16.12 ORION Dante photo-diode

16.2.4 Hohlraum Performance

It might be thought that decreasing the volume of the hohlraum would increase its temperature for a given amount of laser energy shone into the hohlraum. This is true for 'large' hohlraums, but as the size is reduced the plasma ablated from the inside

Table 16.1 ORION Dante channel details

Channel #	Filters (μm) #1	Filters (μm) #2	Filters (μm) #3	Mirrors	Cathodes	E(peak)	ΔE	E/ΔE
1	0.75 Al			C - 7.0°	Cr	20	18	1.1
2	0.75 Al	0.8 Be		C - 7.0°	Cr	62	24	2.6
3	0.2/0.4 B/Lexan	0.2/0.4 B/Lexan		C - 5.0°	Al	161	66	2.4
4	2 Lexan	2 Lexan		Be - 3.5°	Al	276	64	4.3
5	1 V			C - 2.5°	Ni	505	58	8.7
6	0.6 Cu	0.6 Cu			Cr	937	145	6.5
7	0.65 Zn	0.65 Zn			Ni	1026	170	6
8	5.5 Mg	5.5 Mg			Ni	1294	267	4.8
9	5 Al	5 Al			Al	1549	379	4.1
10	5 Pyn	1.5 Fe	0.65 Cr		Al	2989	1495	2

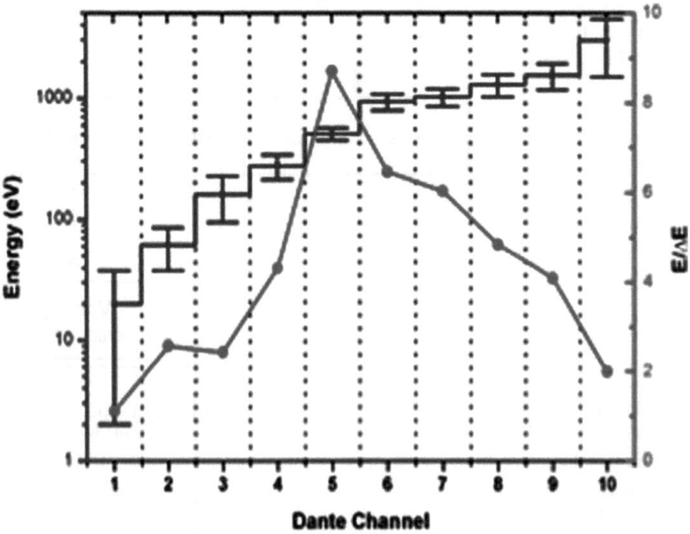

Fig. 16.13 The spectral coverage of the ORION Dante

walls of the hohlraum can move into the laser path, leading to plasma instability growth and the generation of energetic electrons and back-scattered light through the two-plasmon decay and stimulated Brillouin and Raman backscatter. To characterise the energetic electron numbers and energies so generated it is usual to use x-ray spectrometers which operate at higher photon energy than the Dante, and which are usually time-integrating. ORION employs the 'filter-fluorescer' for this. Backscattered light is usually measured in terms of that which falls within the laser's focusing lens(es), and that which falls just outside the lens(es), the latter being known as near-backscatter imaging (NBI). Each of these diagnostics is now discussed.

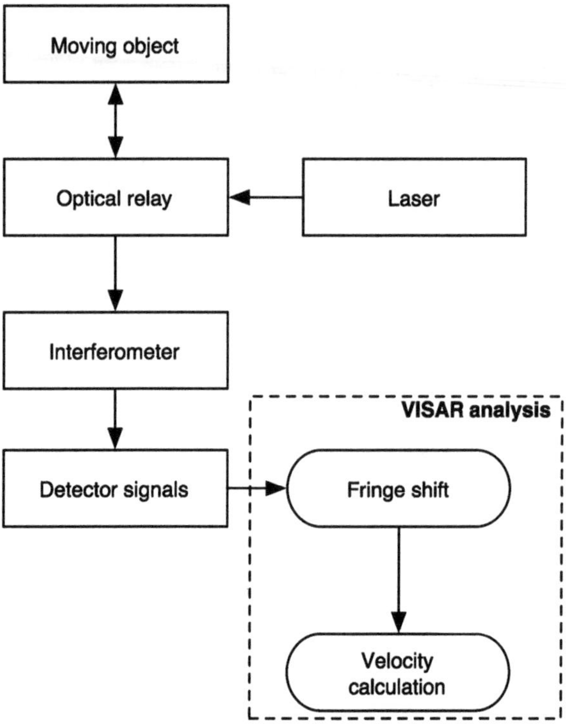

Fig. 16.14 Principle of VISAR. Coherent light illuminates the object of interest. An optical relay directs light toward the object and collects the reflected radiation. Reflected light is sent to an interferometer, producing an output containing the input signal and a time delayed version of the input signal. The output is sensed with fast optical detectors and analysed to infer the object's motion (Figure from Dolan [29])

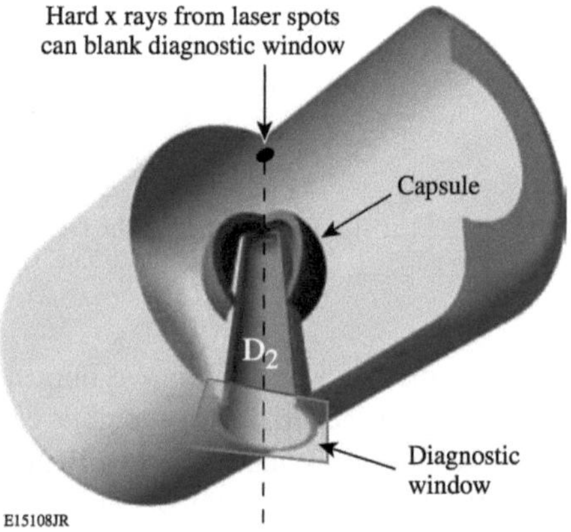

Fig. 16.15 Shock-timing tuning experiments use ignition-style targets that have a re-entrant cone in the capsule. The capsule and cone are filled with liquid deuterium. Optical diagnostics probe the inside of the capsule through the window and aperture in the cone (Adapted from Lawrence Livermore National Laboratory [26])

16.2.5 The Filter-Fluorescer (FFLEX)

The filter-fluorescer (FFLEX) is used to record absolute, time-integrated hard x-ray spectra, from around 20 keV to around 100 keV, using a number of separate channels (eight on ORION and NIF) each of which contains a pre-filter, a fluorescer, and a post-filter. A block diagram of the ORION FFLEX is shown in Fig. 16.17.

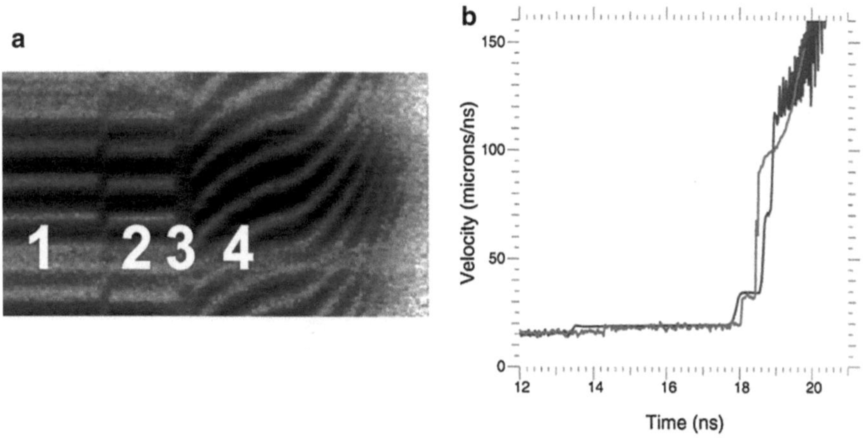

Fig. 16.16 (**a**) Typical VISAR trace showing fringes measured on inside surface of NIF capsule, and (**b**) the inferred velocities (Images courtesy of National Ignition Facility, LLNL)

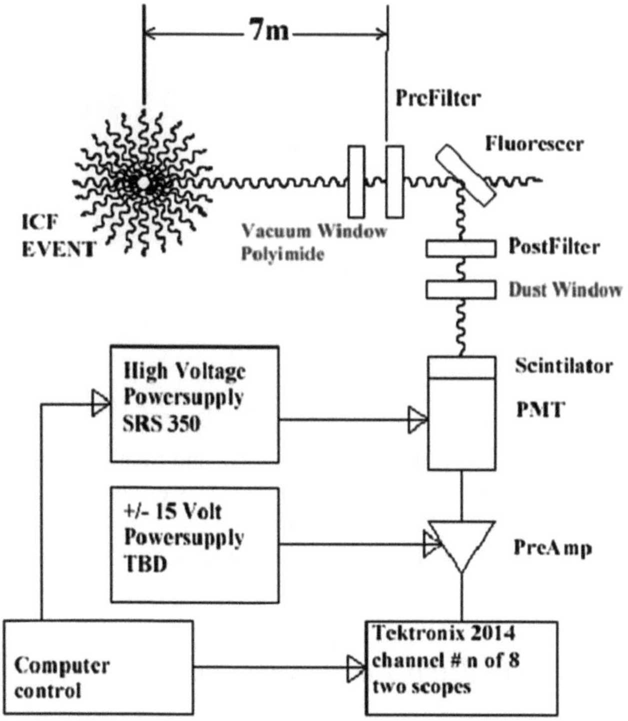

Fig. 16.17 Block diagram of one of the eight channels of the FFLEX used on ORION

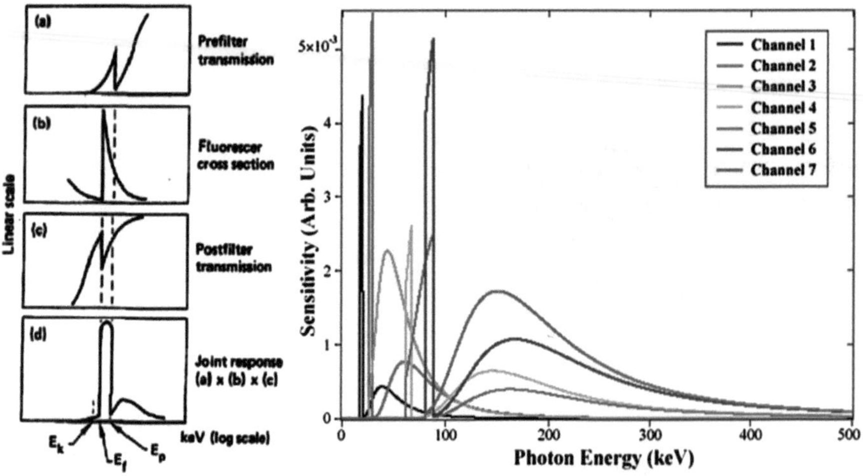

Fig. 16.18 Channel responses for ORION FFLEX

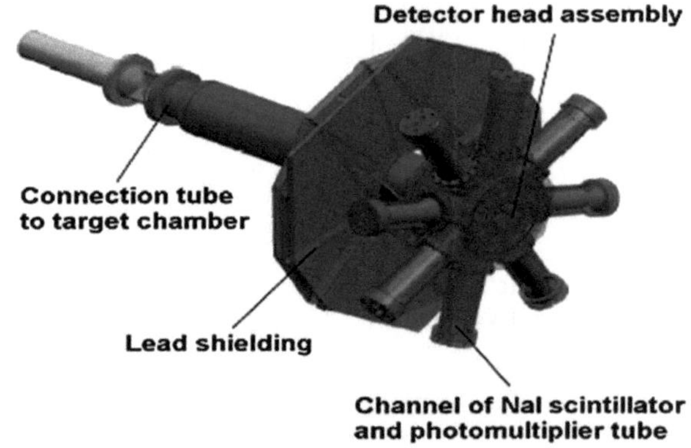

Fig. 16.19 FFLEX as deployed on ORION

By appropriate selection of pre- and post-filters and fluorescers varying channel responses can be defined, as shown in Fig. 16.18. It is usually used to characterise the x-rays generated by so-called hot-electrons generated inside hohlraums.

A CAD drawing of the FFLEX as deployed on ORION is shown in Fig. 16.19.

16.2.6 The Apache High-Energy Spectrometer

The Apache high-energy spectrometer is another absolute, time-integrated x-ray spectrometer, but sensitive to $\sim$ 100 to $\sim$ 2 MeV x-rays. The ORION device has eight channels. It is typically used for hot electron temperature measurements for short-pulse laser-target interactions at $10^{18} - 10^{21}\,\mathrm{Wcm}^{-2}$ intensities. The

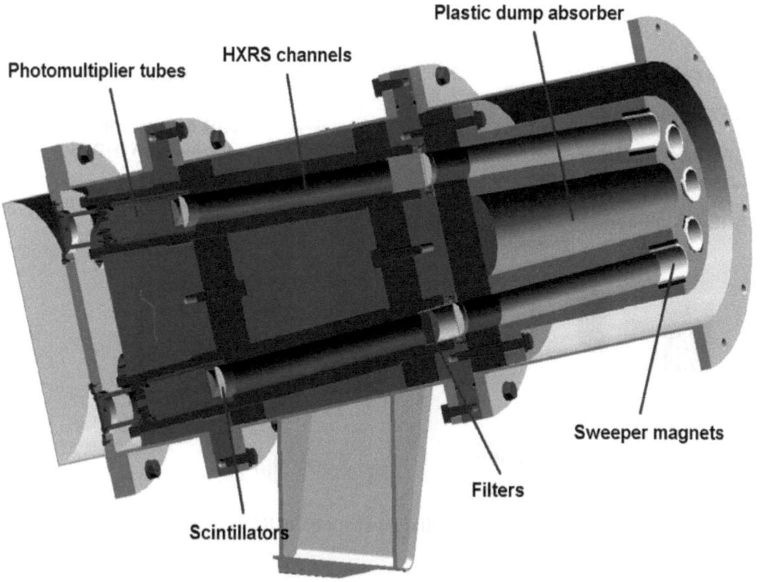

Fig. 16.20 The ORION Apache-like diagnostic has eight channels, each consisting of a filter and a scintillator / photomultiplier

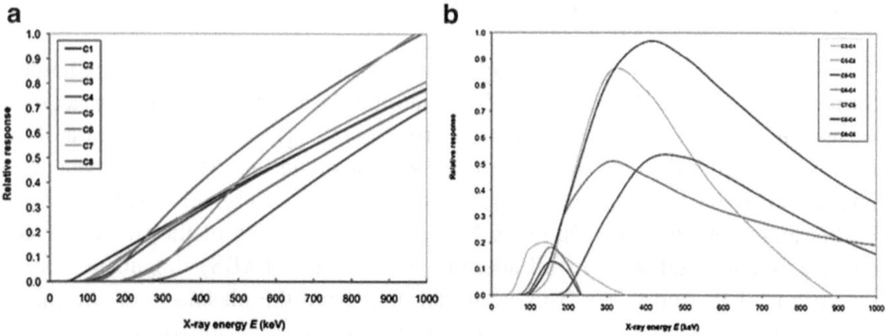

Fig. 16.21 Channel responses (**a**) and spectral definitions achieved by differencing channels (**b**)

ORION device has 1.6 or 17 ns temporal resolution (i.e. two scintillator types) to allow discrimination against charged particles and neutrons. It is shown in Fig. 16.20.

Channel responses are defined in a similar way to those of the FFLEX by choice of filter and scintillator. By differencing channels a degree of spectral sensitivity can be obtained (Fig. 16.21).

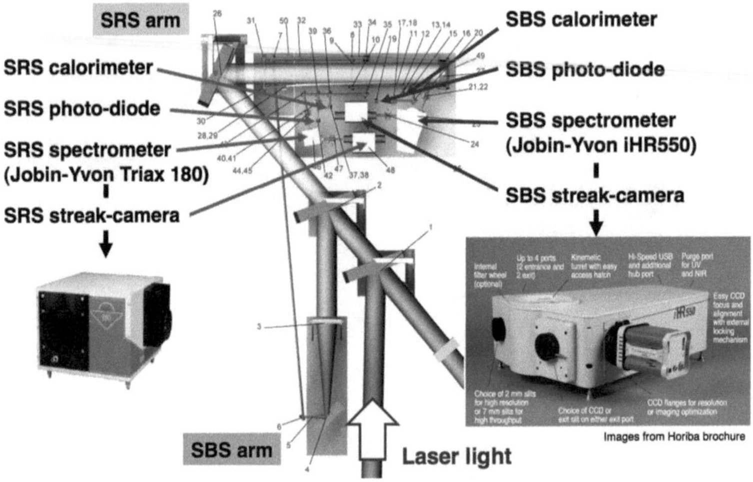

Fig. 16.22 The full-aperture backscatter diagnostics on ORION

16.2.7 Backscattered Light Diagnostics

16.2.7.1 Full-Aperture Backscatter Diagnostics

The back-scatter signal can generally be attributed to two main processes, namely Stimulated Brillouin Scattering (SBS), where the laser interacts with ion acoustic waves, and which typically occupies a narrow wavelength band in the region of the laser wavelength (351 nm), and Stimulated Raman Scattering (SRS), where the laser interacts with electron plasma waves, and which occupies a broad wavelength band in the range 350–700 nm. On ORION one beam from each of the two five-beam clusters is equipped with a full-aperture back-scatter (FABS) diagnostic station (beam-lines LP5 and LP6): back-scattered light from the plasma is re-collimated by the main focussing lens, back through the final turning mirror, and in to the back-scatter station. The diagnostic records SBS and SRS signals independently. It is shown in Fig. 16.22. Each of the two short-pulse (~ 0.5 ps) beams on ORION is also equipped with a FABS capability, though only for SRS as the growth-rate for SBS is expected to be significantly longer than the pulse-length on these beams. On these short-pulse arms the SRS light detected is that which has leaked through the final focusing parabolae.

16.2.7.2 Near-Backscatter Imaging

Near-backscatter imaging (NBI) provides information on the light back-scattered just outside the main focussing optics. On ORION it is used to measures the amount of near-backscattered light in the SBS and SRS spectral bands. Spectralon plates

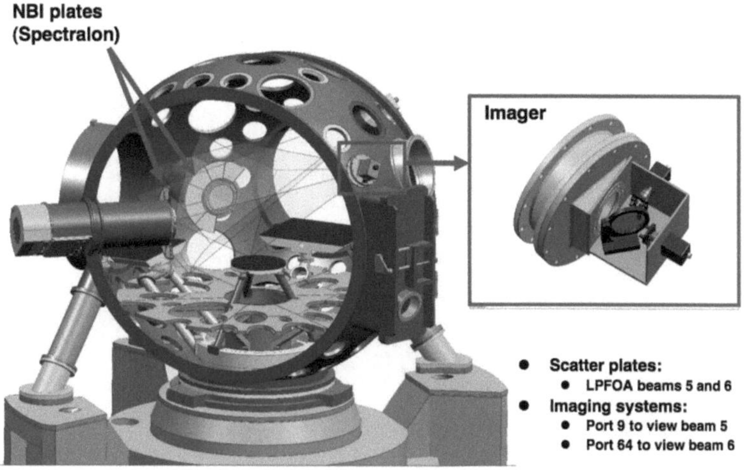

Fig. 16.23 The near-backscatter imaging stations on ORION. Spectralon plates placed around (some of) the focusing optics are viewed by cameras on the opposite side of the target chamber wall to give time-integrated 2D profiles

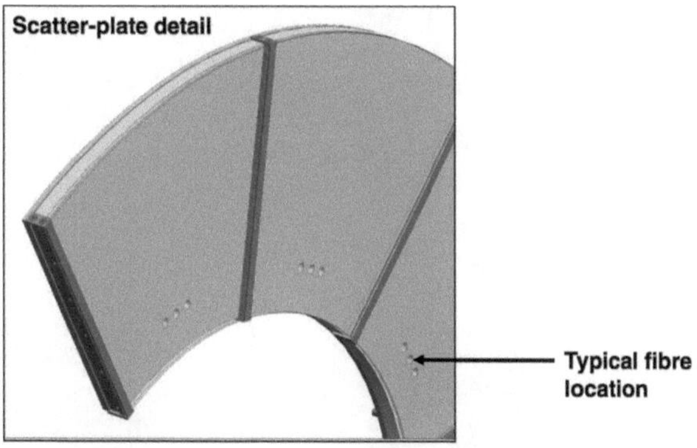

Fig. 16.24 For time-resolved near-backscatter measurements on ORION, light is sampled at the surface of the scatter plate and transported to photodiodes and a fibre spectrometer via optical fibres (*SRS* 460HP single-mode fibres; *SBS* Graded-index fibres; multi-mode fibres for time-integrated signal transportation)

placed around (some of) the focusing optics are viewed by cameras on the opposite side of the target chamber wall to give time-integrated 2D profiles (Fig. 16.23). The diagnostics are on the same long-pulse beams as the FABS.

Time-resolved (time resolution $\sim$ 150 ps) measurements are made in certain locations by fibres inserted into the NBI plates (Fig. 16.24).

16.2.8 Shape: The Gated X-ray Imager

The Gated X-ray Imager, or GXI, is a work-horse diagnostic. It typically consists of a number of separately gated strip-lines which have been deposited on the front of a microchannel plate. The x-ray signal is imaged onto the MCP by a series of pinholes. Fig. 16.25 shows the concept.

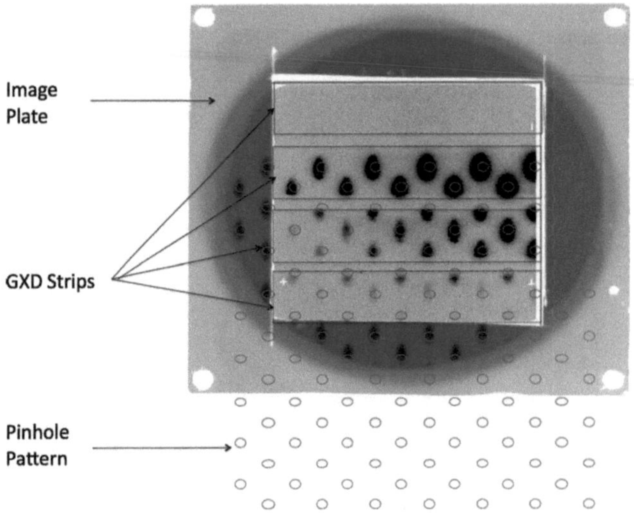

Fig. 16.25 A composite example of the MCP image surrounded by an image plate image and an overlay of the pinhole pattern (From Kyrala et al. [30])

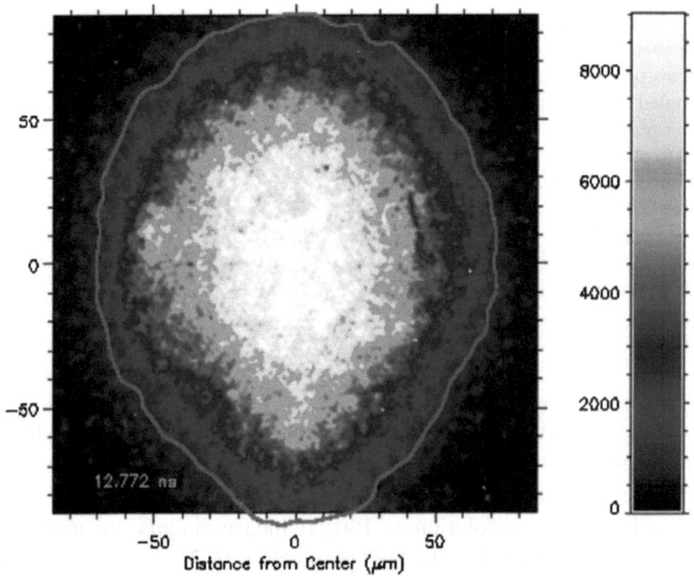

Fig. 16.26 A typical image from the NIF GXD (From Kyrala et al. [30])

As the gate pulse propagates across the surface of the MCP the gain of the MCP increases and then decreases, allowing a short-duration image to be recorded from each pinhole. Fig. 16.26 shows an example from reference [30]. More detail may be found in references [30, 31].

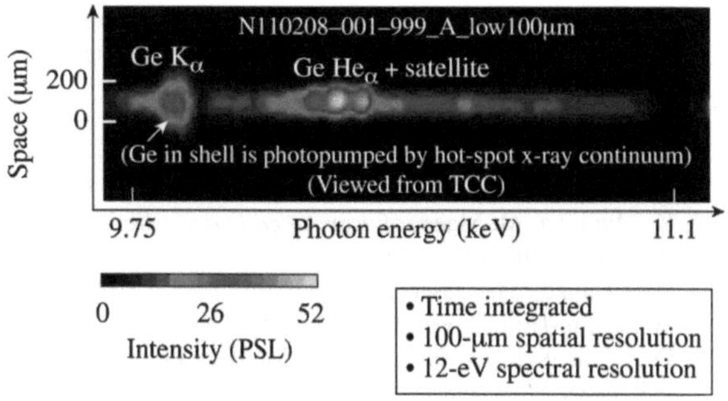

Fig. 16.27 Time-integrated spectrum showing the presence of germanium in the hot-spot (From Regan et al. [34])

16.2.9 Mix

Spectroscopy is a powerful diagnostic tool [32] which may be used to diagnose mix, a phenomenon which is likely to be a significant influence on the performance of ignition capsules [33]. In recent experiments reported by Regan et al. [34] the ablator was doped with germanium to minimise pre-heat of the ablator closest to the DT ice caused by Au M-band emission from the hohlraum x-ray drive. The K-shell line emission (Fig. 16.27) from the ionised germanium that has penetrated into the hot spot provides an experimental signature of hot-spot mix. Analysis of such spectra is often undertaken with the aids of codes such as FLY [35], FLYCHK [36] or SPECT3D [37].

16.3 Summary

The field of laser-plasma diagnostics is a large one, and is certainly too big to be covered comprehensively in a short article such as this. We chose therefore to highlight diagnostics relevant to two areas of significant current interest, namely the interaction of ultra-short laser pulses with matter and hohlraum-driven ICF. In both areas development is rapid, as the number of short-pulse experimental facilities around the world continues to increase, and the significant technical difficulties associated with ICF continue to drive the development of novel diagnostics with high spatial, temporal and spectral resolution. It is hoped that this chapter has whetted the appetite of the reader to explore further some of the recent exciting developments made in this field.

References

1. I.H. Hutchinson, *Principles of Plasma Diagnostics*, 2nd edn. (Cambridge University Press, Cambridge, 2005)
2. A.A. Ovsyannikov, M.F. Zhukov (eds.), *Plasma Diagnostics* (Cambridge University Press, Cambridge, 2000)

3. H.J. Kunze, Plasma physics: confinement, transport and collective effects. in *Lecture Notes in Physics*, vol. 670, ed. by A. Dinklage, T. Klinger, G. Marx, L. Schweikhard (Springer, Berlin/Heidelberg, 2005), pp. 349–373
4. R.H. Huddlestone, S.L. Leonard, *Plasma Diagnostic Techniques* (Academic Press, New York/London, 1965)
5. W. Lochte-Holtgreven, *Plasma Diagnostics* (North-Holland, Amsterdam, 1968)
6. K. Muraoka, M. Maeda, *Laser Aided Diagnostics of Plasmas and Gases* (Institute of Physics Publishing, Bristol, 2001)
7. H.J. Kunze, The laser as a tool for plasma diagnostics. in *Plasma Diagnostics*, ed. by W. Lochte-Holtgreven (North Holland, Amsterdam, 1968)
8. D.C. Carroll et al., Nucl. Instrum. Methods Phys. Res. A **620**(1), 23–27 (2010)
9. A.P. Fews, D.L. Henshaw, Nucl. Instrum. Methods Phys. Res. **197**(2–3), 517–529 (1982)
10. D. Neely et al., Appl. Phys. Lett. **89**, 021502 (2006)
11. F. Nürnberg et al., Rev. Sci. Instrum. **80**, 033301 (2009)
12. D.S. Hey et al., Rev. Sci. Instrum. **79**, 053501 (2008)
13. P. McKenna et al., Phys. Rev. Lett. **91**, 075006 (2003)
14. P. McKenna et al., Phys. Rev. E **70**, 036405 (2004)
15. M. Roth, J. Instrum. **6**, R09001 (2011)
16. R.J. Clarke et al., Appl. Phys. Lett. **89**, 141117 (2006)
17. R.J. Clarke et al., Nucl. Instrum. Methods Phys. Res. A **585**(3), 117–120 (2008)
18. A. Carillon et al., Phys. Rev. Lett. **68**, 2917–2920 (1992)
19. R. Kodama et al., Phys. Rev. Lett. **73**, 3215–3218 (1994)
20. D. Neely et al., AIP Conf. Proc. **426**, 479–484 (1997)
21. P.A. Norreys et al., Phys. Rev. Lett. **76**, 1832–1835 (1996)
22. M.H. Edwards et al., Proc. SPIE **6703**, 67030L (2007)
23. J.D. Lindl, Phys. Plasmas **2**, 3933 (1995)
24. S. Nakai, H. Takake, Rep. Prog. Phys. **59**, 1071–1131 (1996)
25. N.B. Meezan et al., Phys. Plasmas **17**, 056304 (2010)
26. Lawrence Livermore National Laboratory, Quarterly Report **117** (2008)
27. E.L. Dewald et al., Rev. Sci. Instrum. **75**, 3759 (2004)
28. J.L. Bourgade et al., Rev. Sci. Instrum. **72**, 1173 (2001)
29. D.H. Dolan, Sandia National Laboratories Report-2006, 1950 (2006)
30. G.A. Kyrala et al., Rev. Sci. Instrum. **81**, 10E316 (2010)
31. J.A. Oertel et al., Rev. Sci. Instrum. **77**, 10E308 (2006)
32. H.R. Griem, *Principles of Plasma Spectroscopy*, Cambridge Monographs on Plasma Physics 2 (Cambridge University Press, Cambridge, 2005)
33. B.A. Hammel et al., High Energy Density Phys. **6**, 171–178 (2010)
34. S.P. Regan et al., Phys. Plasmas **19**, 056307 (2012)
35. R.W. Lee, J.T. Larsen, J. Quant. Spectrosc. Radiat. Transf. **56**, 535–556 (1996)
36. H.-K. Chung et al., High Energy Density Phys. **1**, 3–12 (2005)
37. J.J. MacFarlane et al., High Energy Density Phys. **3**, 181–190 (2007)

Chapter 17
Microtargetry for High Power Lasers

Martin Tolley and Chris Spindloe

Abstract Microtargetry for high power lasers (HPLs) offers considerable challenges and opportunities at the cutting edge of the application of microtechnology production techniques. In this chapter microtarget production issues are discussed particularly in the context of the mass production of such components which has become one of the major challenges in delivering targets for High Power Laser (HPL) systems and will become essential in the near future as lasers move to application based systems. The challenges of microtarget placement are also discussed.

17.1 What Is a Microtarget and What Are the General Challenges in Fabricating Them?

17.1.1 Introduction

Microtargets have a very broad range of designs and materials combinations but typically individual targets have a scale size of less than a few millimetres with an accuracy of better than two micrometres. An individual microtarget often consists of several components which need to be precisely aligned and assembled, often with reference to a number of other targets in an experimental cluster. The surface finish requirements of individual components can be of the order of 50 nm. An additional class of microtargets of significant interest comprises mounted foils ranging from a few nanometres in thickness to tens of micrometres made from a

M. Tolley (✉) • C. Spindloe
Target Fabrication Group, Central Laser Facility, STFC Rutherford Appleton Laboratory, Chilton, Oxfordshire, UK
e-mail: martin.tolley@stfc.ac.uk; christopher.spindloe@stfc.ac.uk

P. McKenna et al. (eds.), *Laser-Plasma Interactions and Applications*, Scottish Graduate Series, DOI 10.1007/978-3-319-00038-1_17,

variety of materials such as carbon, plastic or gold. Some microtargets need to be used at cryogenic temperatures often incorporating materials which require special handling procedures, for example tritium and beryllium.

17.1.2 Microtarget Production

The production of such complex objects requires the integrated deployment of a wide range of techniques, for example, ultra precision micromachining, complex thin film production, microassembly, lithography, low density materials production and Micro-Electro-Mechanical Systems (MEMS) techniques amongst many others. Such technologies often need to be deployed at the limits of their current capability. Throughout the production process suitable characterisation is necessary, again, sometimes requiring novel solutions. A broad range of challenges often become apparent during the design or processing steps typically centred on scaling issues which arise in microtechnology.

Microtarget fabrication is a specialised application within the general fields of microengineering and microfabrication. It requires the integrated combination of a wide-ranging group of microtechnologies. It should be noted that many processes require modification to be effective in the micro-realm and that capability is one of the key knowledge areas within microtarget fabrication. Within the UK there are two internationally recognised centres of expertise in microtarget fabrication, namely those at the Atomic Weapons Establishment (AWE) and at the Rutherford Appleton Laboratory, Central Laser Facility (RAL, CLF). Also within the UK there are many world-class centres of expertise in a wide range of microtechnologies. When integrated the UK capability is a potent force in microtarget mass production with capabilities that are comparable to anywhere in the world.

Traditionally microtargets were produced in low volume with precision being the main driver rather than time or, to a lesser extent, cost. Not surprisingly from the early days of high power lasers microtarget production tended to begin with precision microcomponent manufacture followed by microassembly all verified by precision characterisation. Generally microcomponents were made using micromachining, chemical techniques or thin film coating. Microassembly was either performed by hand, using specially designed jigs or using microassembly stations. A range of characterisation techniques and equipment were used augmented by specialist development of specific techniques, such as optical shadowgraphs for the characterisation of plastic shells.

The fabrication and the assembly of such precise components is a skill that is developed over a number of years. At the sizes of approximately 10^{-4}–10^{-5}m which is the typical size of microtarget components gravity is no longer the major force that affects an object. Fearing [1] states that for parts with sizes of less than a millimetre (mass less than 10^{-6} kg) the gravitational and inertial forces upon an object become insignificant when manipulating it and forces such as electrostatic or surface tension dominate, see Fig. 17.1. When fabricating parts it is often the case

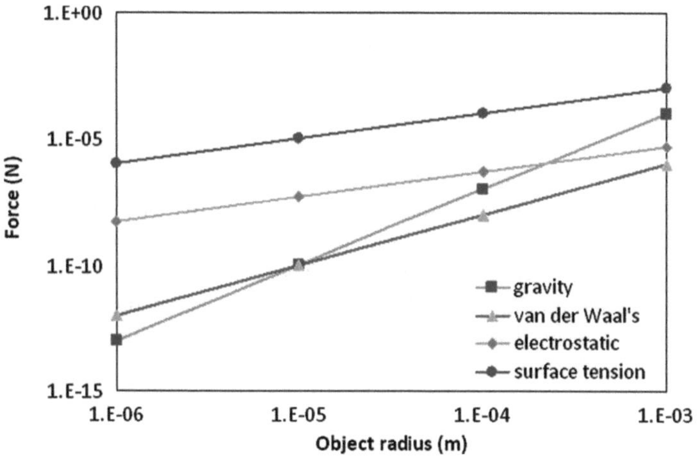

Fig. 17.1 The attractive force between a sphere and a plane [1]

that small components have a tendency to act in a strange manner and assembly of these components is a very challenging activity. Things stick where you don't want them to and don't stick where you want them to go!

17.1.3 Inertial Fusion Energy Targets

A specialist area of microtarget production is the manufacture of targets for Inertial Fusion Energy (IFE). The requirements for such targets are that very high volumes of precise targets need to be fabricated using technologies that are at the cutting edge of current capabilities. The volume production of microtargets has two distinct challenges: making microtargets with sufficient precision and making microtargets at the required production rate. The first of these challenges has, to some extent, already been demonstrated by the production of microtargets for National Ignition Facility (NIF) in the US and generally there is a long heritage of precision microtarget production. The mass production challenge is largely new for microtarget fabrication although many manufacturing sectors do produce precision components and assemblies at rates comparable to those required for IFE. A few microtarget fabrication facilities have been actively pursuing high repetition rate microtarget production. RAL has successful demonstrated a suite of techniques for microtarget batch production to support the repetition rate of the Astra Gemini laser (one shot every 20 s). General Atomics have also pursued a range of microtarget mass production development programmes.

In one program to deliver IFE, (HiPER), there will be two generic approaches to microtarget mass production: scaling up known microtarget production processes to high production rates and applying established mass production processes to microtargets. A strong example of the latter is the use of wafer-based MEMS production techniques which has been demonstrated at RAL. There are multiple opportunities for IP generation within both approaches and this claim is substantiated by the extent and expanding international MEMS industry ($\sim$ £10B in 2011). Furthermore, there will be novel challenges posed by the need to move microtargets within the IFE microtarget production facility. Production process equipment already exists running at the rates required but compatibility with; maintaining efficacy of precision components, a (partially) cryogenic environment and tritium will have to be demonstrated. There is potential cross-over with the medical device and pharmaceutical sectors with their increasing need for specialist environmental control. The production of IFE targets will be dealt with in more detail later in the chapter.

17.2 Microtarget Fabrication Technologies

There are a wide range of complementary technologies that are needed to produce micro-targets. To design, develop and fabricate such targets an understanding of all of the capabilities available is advantageous to best determine the manufacturing process that will deliver targets within the specifications required for each individual experimental campaign. Some of these technologies are summarised below, however this is not an exhaustive list and does not go into the full detail of each technology.

17.2.1 Precision Micromachining

To produce components that are of the order of a mm in size and that have tolerances that are of a few microns requires specialised equipment and highly trained engineers. Target components have in the past been fabricated in low numbers using specialist lathes and multi-axis manual machines that have been fitted with precise stages and microscopes to see the part while in processing. While this can give a high degree of accuracy and surface finish, it is a time consuming process and to develop the skills to manufacture these components can require years of training. Tooling for such components can be expensive and sometimes has to be bespoke for each component. In fact the majority of the development and skill in manufacturing targets is in the understanding of the tool manufacture and the set-up of the components on the machine before fabrication even begins.

In recent years high precision Computer Numerical Control (CNC) machines have been able to produce components that are comparable with the ones previously manually produced. These automated machines can fabricate a wide range of components and allow the use of Computer Aided Design (CAD) /Computer Aided Manufacturing (CAM) production. Typically standard CNC machines can achieve accuracies of 5 μm.

More recently work has been carried out to manufacture components to a higher degree of accuracy than this. High precision Kern CNC machines can produce components with accuracies of $\sim$ 1 μm and with the correct tooling can achieve surface finishes of $\sim$ 0.25 μm roughness. This combined with batch production technology that is available on the machines can allow for large numbers (50–100) of complex targets to be produced in one machining run [2]. This technology can be extended to run up to 1,000 targets in one process, although the running time increases with the number of components.

It should also be considered that when machining to a micron tolerance level that there are a number of other factors that will affect the outcome of the production process other than the specified tolerances of the machine. Factors such as temperature fluctuations and vibrations can expand, contract or vibrate components by a number of microns therefore not allowing you to reach the theoretical limit of the machine. Tool wear throughout a process can cause components to be considerably larger at the end of the run than at the start and reduce surface finishes. Work piece holding is also an important factor. To machine to within a micron, remove the part from the machine and measure and then to continue to machine the part. It is essential that the placement accuracy of the work piece holder is to within a micron.

Other technologies that have not been mentioned but are also useful in the production of targets are electro-discharge machining (EDM) and diamond point turning for ultra high precision surfaces (less than 50 nm Ra). Also multi-axis machining to allow complex 3D geometries has not been mentioned but this technology can allow for targets to be produced with geometries not possible on standard three axis machines.

17.2.2 Thin Film Coating

Thin-film coating is a well understood process with Physical Vapour Deposition (PVD) technologies such as sputter coating being used for over 150 years [3].

In the field of target fabrication, thin film coating is used to manufacture either films of materials that are used as targets, or a coating deposited onto existing components as a thin-film layer to enhance or give the component a feature that is important experimentally.

There are many processes that are suitable for the production of thin film material some of which will be discussed however there are dozens of variations that can be utilised depending on the parameters that are required for the film. The processes can be classified as physical (such as evaporative) or as chemical such as Plasma Enhanced Chemical Vapour Deposition (PECVD).

17.2.2.1 Physical Vapour Deposition (PVD)

This is a term for the varieties of vacuum deposition that form a thin film by the condensation of a vaporised form of the material onto various surfaces (e.g. onto semiconductor wafers, glass slides). The coating method involves purely physical processes such as high temperature vacuum evaporation or plasma sputter bombardment rather than involving a chemical reaction at the surface to be coated. The three most commonly used processes for PVD thin film deposition fall into two distinct classes.

Evaporation - Where a hot source material evaporates and then condenses on a surface [4]. This process takes place through the following steps: (1) the vapour is produced by heating a material until it sublimes, (2) the vapour is transported from the source to the substrate and (3) the vapour condenses to form a solid film on the substrate. As this process takes place in a vacuum and the material has a long mean free path, this process is ideal for coating through masks. In thermal evaporation a source material is heated using an electric filament, usually a boat or coil holding the material. Alternatively, electron beam evaporation uses an electron beam to heat source material that is held in a crucible. There are many factors to consider when evaporating materials, such as the vapour pressure, source container interactions and substrate, all influencing the uniformity and thickness which make a process that seems relatively simple a complex operation.

Sputtering – in its simplest form is knocking an atom out of the surface of a target of coating material [5]. Sputter deposition uses the ejected atoms, under the right circumstances, to build up a coating on a substrate. Usually this coating is up to a micron in thickness. However some systems can coat to much larger thicknesses. Magnetron sputtering uses high strength electric and magnetic fields to confine electrons close to the 'target', a sputtering gas (argon) is ionised and these ions bombard the 'target' ejecting the source atoms. The sputtered atoms are uncharged and can fly from the target to a substrate. Limitations of this process include the fact that the process has a lower mean free path and therefore is less suited to coating though masks. Advantages are that a wider range of materials can be coated and have better adhesion to a substrate. Charge build up on insulating targets can be avoided by use of RF sputtering.

17.2.2.2 Chemical Vapour Deposition (CVD)

This is the process of the deposition of a solid onto a (heated) substrate from a chemical reaction in the vapour phase [6]. The process is very versatile and can produce coatings, powders and fibres. In a typical CVD process, the wafer (substrate) is exposed to one or more volatile precursors, which react and/or decompose on the substrate surface to produce the desired deposit. Frequently volatile by-products are also produced which are removed by gas flow through the reaction chamber.

In CVD, because generally pressures are used above the molecular flow region regime, it is not restricted to line of sight deposition [6] in contrast to evaporative PVD methods and to some extent sputtering. Complex 3D structures and deep holes with an aspect ratio of 10 : 1 can be filled.

Recent years have seen the increase in the use of CVD to produce diamond-like-carbon thin films for ion production experiments. These films can be produced a few nm thick and are extremely strong when compared to other films of similar thickness. Their strength allows for new regions of physics to be explored in which ultra thin films are irradiated with high intensity lasers [7].

17.2.2.3 Plastic Thin Film Coating

Plastic thin film coatings have been produced for laser experiments for a number of years. Typical materials that are used are formvar, polyethylene and polystyrene. These can be made using dip or spin coating methods in which the plastic is dissolved in an appropriate solvent and then the substrate is dipped in the plastic solution or the solution is 'spun' onto the substrate. The thickness of the material produced is dependent on the concentration of the plastic dissolved in the solvent and also on the speed that the substrate is dipped or spun.

Parylene (a trade name for poly-para-xylylenes) is a CVD deposited plastic that forms an ultra-thin, pinhole-free polymer coating [8]. The basic member of the Parylene series is Parylene N (C_8H_8) and forms a completely clear highly crystalline material. Other members include Parylene C and D that have a substitution of one or two chlorine atoms respectively. Parylene is vapour deposited at room temperature and forms highly conformal coatings being able to coat complexly shaped components and also being able to coat through gaps of $\sim$ 10 μm. Parylene's excellent stability and strength make it a useful material for high power laser experiments. Coatings can be produced from $\sim$50 nm up to 10's microns.

17.2.3 Low Density Materials (i.e. Foam and Aerogel Production)

Low density materials (foams and aerogels) are useful in high powers laser experiments because they allow a target or its components to be comprised of materials that are of lower density than the bulk (precursor) material. For example, foams are useful as buffers for preventing hydrodynamic instabilities in ICF experiments [9], for studying shock propagation through low density materials [10], and for studying the dynamics of laser produced shocks [11]. Foams and silica aerogels are produced using chemical techniques.

Foam targets are typically produced by UV polymerisation of a monomer (tri-functional acrylate TMPTA) dissolved in a solvent with small amounts of

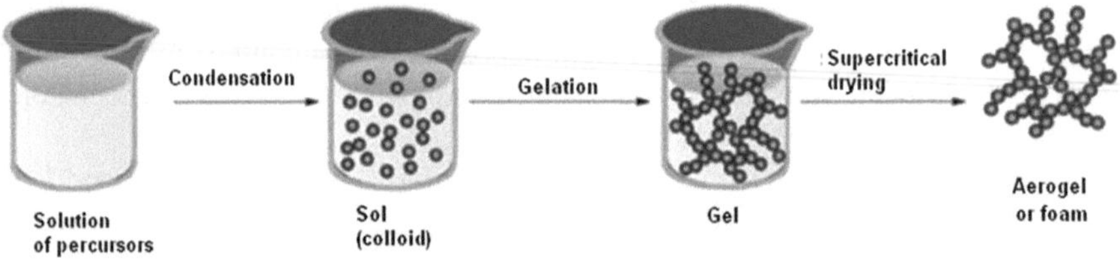

Fig. 17.2 The production of aerogels

initiator (benzil). The polymerised gel is placed in methanol for solvent exchange [12]. When the solvent is exchanged the wet foams are then dried in a critical point dryer using carbon dioxide as the drying fluid. By increasing the temperature and pressure of a liquid above its critical temperature (T_C) and critical pressure (P_C) it becomes a supercritical fluid. The critical point represents the highest temperature and pressure at which the substance can exist as a vapour and liquid in equilibrium. As the critical temperature and pressure is approached, the properties of the gas and liquid phases approach one another, resulting in only one phase at the critical point: a homogeneous supercritical fluid. It expands to fill its container like a gas but with a density similar to that of a liquid. There is no surface tension in a supercritical fluid because there is no liquid/gas phase boundary. By changing the pressure and temperature of the fluid the properties can be 'tuned' to be more liquid or more gas like. By dropping the pressure the fluid becomes a gas that leaves the delicate three-dimensional network without damaging it. The foam is dried in this way because if the liquid was allowed to evaporate the capillary forces would collapse the structure.

Aerogel targets are low density silica based materials that are derived from a gel. As shown in Fig. 17.2 a sol is a stable suspension of colloidal solid particles or polymers in a liquid. A gel is a porous, three-dimensional, continuous solid network surrounding a continuous liquid phase and therefore replacing the liquid with gas produces an aerogel.

Silica aerogels are typically prepared by hydrolysing a tetrafunctional alkoxide precursor (TMOS) employing a base as catalyst (ammonia). Subsequent condensation reactions of the silanol groups produce siloxane bonds (Si–O–Si). These reactions produce anhydrous silica, alcohol and water. During these chemical processes, the 'sol' gradually becomes a gel. Ageing of the gel allows further condensation, dissolution and precipitation within the solid and liquid phases changing its structure and properties. Longer ageing produces stronger and clearer gels. The gel is then dried using the same process as for foams in a critical point dryer.

Acrylic foams can be produced in a range of densities from 2 mg/cm^3 to almost solid density with pore sizes of $\sim 1\,\mu$m. Aerogel density ranges from 1.88 mg/cm^3 to almost solid density with a range or pore sizes (To put this in context the density of dry air is 1.2 mg/cm^3 (at 20°C, 1 atm)).

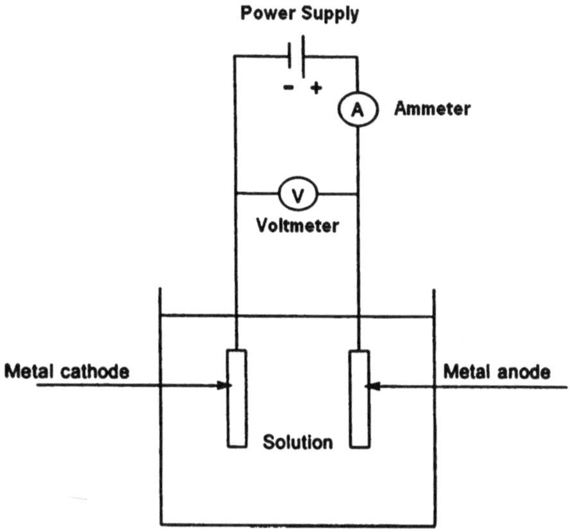

Fig. 17.3 Schematic for electroplating

17.2.4 Electroplating

Electroplating (also called electrodeposition) is the process of producing a coating on a prepared conductive surface by the action of an electric current. The deposition of a metallic coating onto an object is achieved by placing a negative charge on the object to be coated and immersing it into a solution which contains a salt of the metal to be deposited. In other words the object to be plated is made the cathode of an electrolytic cell (Fig. 17.3). The metallic ions of the salt carry a positive charge and are thus attracted to the object. The metal ions receive electrons and are reduced to metallic form at the interface between the solution and the cathode plating onto its surface [13, 14].

Electroplating is used in target fabrication as a means of depositing thick layers of materials to become parts of, or indeed to become the actual target. Electroplating can be used to coat intricate and complex forms with relatively high degrees of accuracy and can coat a wide range of thicknesses up to $\sim 40\,\mu m$. Electroplating has been used to produce cone and hohlraum targets for a number of years [2, 15, 16], for both single and batch production of targets. When combined with high precision micromachining it can be a powerful tool in the capabilities available to manufacture targets. There are a number of materials that can be electroplated, with the most commonly used in target fabrication being gold and copper, however, there are requirements for palladium and lead among others. Varying specific operating parameters can give a variable film surface morphology from a fine grained to a nodular appearance and there are significant challenges in the plating of small precision microparts to ensure that smooth and defect free coatings are achieved. However the benefits are that the costs are significantly reduced when compared to thin film coating to the same thickness and for some materials thin film coating to higher thicknesses is not achievable.

There are also a wide range of other techniques that can be used for electroplating with different power supplies (pulsed plating, etc.). For example, a large number of factors affect coating quality such as surface cleanliness, pH of the solution, temperature and agitation of the sample to name a few that cannot be addressed in the limited scope of this chapter but which have distinct effects on the final product and need to be taken into account.

17.2.5 Micro-Assembly

There is a whole area of science and engineering that is dedicated to the manipulation of small objects. Indeed there are micro-robots that can manipulate thin samples of material that have been machined to sizes of sub 100 μm using a focused ion beam in a scanning electron microscope and pick and place them onto sample holders. There are also many automated systems that can assemble larger components (a few cm) at high repetition rates with high degrees of accuracy. However in the size range for laser micro targets there are few (if any) systems that have the flexibility to manipulate and adaptability of a human hand for assembly of components. The reason for this is that although systems have the ability to pick and place components, they do not have long enough travel, have the ability to change the geometry or the flexibility that is provided by a human. Re-programming complex assembly stations could take to order of a few weeks and when targets need to be manufactured quickly this is not a viable option.

Therefore in some cases, particularly where there is considerable time pressure to ensure experimental delivery, it is acceptable, and at times preferable, to prepare microtarget components 'by hand'. Most simple targets are assembled by hand using fine (surgical) tools such as tweezers, fine paint brushes and scalpels. For example proprietary foils might be cut to size and then subsequently glued to a fibre which is glued on the top of a post. As a very general guiding principle involuntary hand jitter is in the region of 25 μm which gives some indication of an upper bound of potential manual assembly accuracy.

When the accuracy of the human hand is not to the required level there are ways of utilising bespoke jig design for the more complex 3D targets to ensure high repeatability of results and ultra precise assembly. From a pragmatic, but very useful, example being appropriately stacked microscope slides to a precision machined and extensively designed multi-axis target positioner to hold components in place under a microscope and manipulate them while before gluing and characterisation.

High quality binocular optical microscopy is the main interface for target assembly giving the target fabricator a perspective on depth as they assemble components and with the best quality optics possible when working with components that have features that are sometimes of the size of a tenth of a micron.

When working on an experimental campaign such as the National Ignition Campaign (NIC) on the National Ignition Facility (NIF) in the US where target geometry is usually fixed and almost all component assembly is using the same operations it is possible and preferable to design and build complex manipulators [17].

The challenge that faces the target fabrication community is taking the lessons learnt for the manipulator robots that currently exist and developing ways of making them flexible and affordable.

17.2.6 Target Characterisation

It is essential that any target that is fabricated is characterised to ensure that the parameters that are important for the experiment are measured and verified before the target is issued. The inherent nature of a HPL interaction ensures that no post shot characterisation is possible (because the target is destroyed) and, therefore, no characterisation is available to the target fabricator after the laser shot. Generally, characterisation of a high power laser target is as important as any fabrication processes.

There are a vast suite of characterisation methods available that can be applied to individual targets. Techniques range from quick measurements of target dimensions to more complex measurements of elemental composition, surface roughness, crystal structure or grain size. All of the methods have limitations which will not be dealt with in detail here; however the measurement of a target with a number of different instruments will give a reliable measurement or analysis of its properties and provide significant data to inform the experimental user of the details of the produced component.

Typical complementary instruments in a target fabrication laboratory include

- Optical Microscopy for size measurement.
- Scanning Electron Microscopy (SEM) including elemental analysis, backscattered electron detection and 3D reconstruction for elemental and topographical analysis.
- Atomic Force Microscopy for small scale roughness measurement.
- White Light Interferometry for surface roughness measurement.
- Confocal Materials Microscopy for real colour imaging and surface characterisation.
- Touch Probe measurements for data on film thicknesses using step heights.
- Co-ordinate Measuring Microscopes for part measurement and reporting.

Added to this the target fabricator might make use of x-ray tomography, x-ray diffraction, TEM, focused-ion-beam techniques and many others.

A simple question would be to measure a machined profile for form and roughness. But what do we mean by roughness? Any surface can be described by three terms, the form, the waviness and the roughness. Filtering of measured data

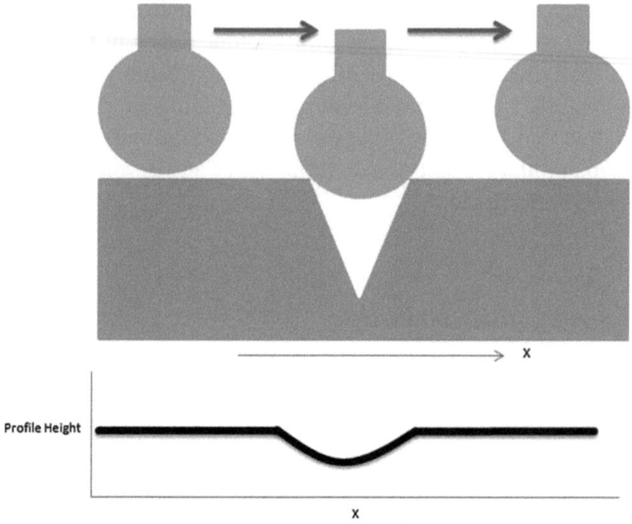

Fig. 17.4 The effect of tip size on measurement

can be used to remove form and give roughness and waviness values and this is a well established technique for mesoscopic objects. The measurement of 3D surface is detailed by NPL [18]. However the challenge for laser targets is that they are so small that there is almost no form or waviness to remove. Take as an example, a set of machined grooves in a 100 μm × 100 μm square target that have a wavelength of 10 μm and amplitude of 1 μm. The periodic features in the sample can be treated as roughness or waviness or form and, without care, filtered out accordingly so that the sample could potentially be measured to be whatever value you want it to be.

A second example is the measurement of a small groove in a sample. It is very easy to measure something that is not actually there. A 10 μm wide 2 μm deep groove can clearly not be measured by a touch probe that has a 5 μm diameter as the tip of the probe will never be able to follow the profile of the surface as shown in Fig. 17.4. Although for many geometries, shape features smaller than the radius can be inferred.

Conversely if you use an optical measurement reflections in the groove may give you a height measurement that is false and give you a virtual 'peak' in the middle of the groove as in Fig. 17.5.

It is therefore crucial to define which parameters are important when characterising a target. For example, surface roughness in terms of R_a (small scale surface roughness) or R_t (the maximum distance between the highest peak and the lowest groove over a distance). Specifically, if you need a smooth sample with no scratches R_t is useful. The sample line-out in Fig. 17.6 has a low R_a and seems smooth, but the high R_t value shows there are scratches on the sample … you might just hit one of these with your laser!

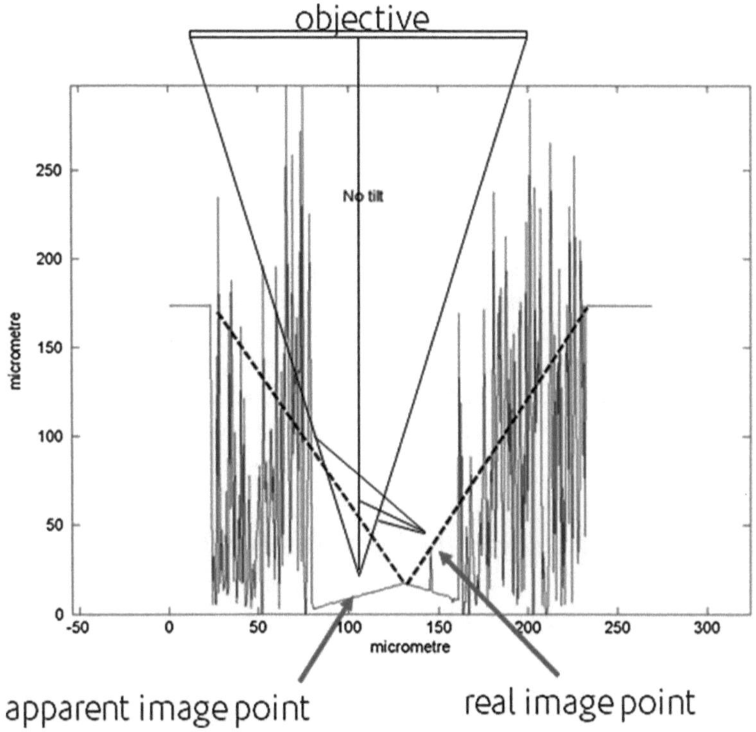

Fig. 17.5 The effect of reflection in the groove

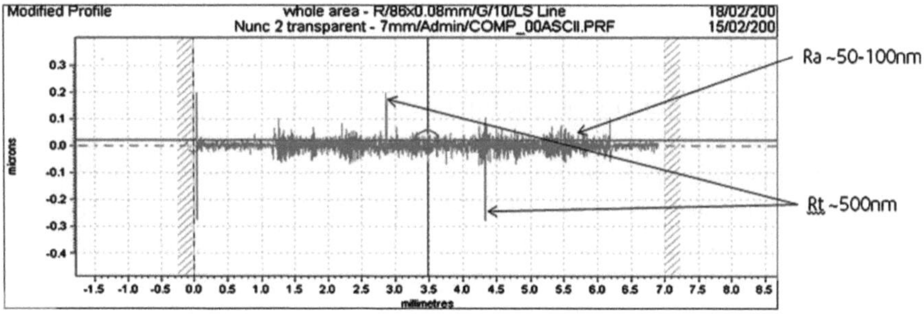

Fig. 17.6 An example of the difference between R_a and R_t

17.3 Microtargetry for High Repetition Rate Lasers

The previous sections have dealt with the technologies that have produced microtargets for laser systems that have shot rates at the maximum of ten per day. Targets can be made quickly by hand for simple geometries or with good project planning even complex 3D targets can be made in high enough numbers to deliver to an experimental campaign that might take 100–200 shots. These techniques were very labour intensive yielding a high value product. However, the recent developments in high power laser technology have produced laser systems that can fire a rates of a shot/minute (Astra Gemini) and in the future systems such as ELI and HiPER will run at rates of Hz or kHz. For the first time, they are bringing to microtargetry the challenges of both medium volume and mass production. Clearly the current technologies are not scalable for delivery to these systems and new technologies will need to be developed [19].

These developments mean that future markets for microtargets seem to be emerging, specifically (1) experimental facilities requiring thousands off (for statistical data gathering) and (2) applications facilities. Examples of the latter are (i) LIBRA [20], requiring sophisticated target delivery techniques with possible application in cancer therapy, and (ii) HiPER [21] requiring prolonged operation at, say, 10 Hz, for commercial laser-driven fusion electricity generation.

17.3.1 Microtechnologies for High Number Production

As previously discussed in this paper anticipated developments in the field of high power laser systems will give increasing peak energies, increasing intensities and higher repetition rates. The most significant impact for microtargetry will almost certainly arise from the increase in repetition rate (with increases of target production and delivery numbers of three to four orders of magnitude anticipated over the next few years). Early experience in the Central Laser Facility at the Rutherford Appleton Laboratory with fielding high repetition rate experiments on the GEMINI laser [22] and work on the HiPER project [21] has already given valuable insight into several important microtargetry issues and some of the main lessons learnt are summarised in the following sections. Perhaps the single most important observation is the necessity to integrate microtarget production solutions with microtarget delivery solutions.

Several techniques are well established for high repetition rate targetry most notably; gas jets, tape drives and droplet generators. However, the techniques have (differing) limitations, most notably target geometry and complexity. Sophisticated 2D and 3D targets in many designs have been requested for high repetition rate experiments. Total redesign of some microtargets has been possible to enable precision machining from solid thereby removing many microassembly steps reducing manual intervention from half a day to 10 min.

17.3.1.1 Thin and Ultra-Thin Targets

Thin foils can be coated onto substrates treated with releasing agents. Such foils can then be floated off onto the surface of water and subsequently lifted from the surface onto a suitable mount, typically having through holes. It has been possible

Fig. 17.7 An array of AFI cone mandrels made by the CLF

to develop a mount that enabled simultaneous mounting over an array (for example 10×10) of holes giving multiple thin film targets. Some of the challenge with these techniques is to ensure that the foils are sufficiently flat. Complex multilayer foil target arrays have been similarly prepared. It has also been possible to produce ultrathin foils (thinner than 50 nm) in a range of materials (with the thinnest, although not in arrays, of 2.5 nm carbon). Diamond-like-carbon targets exhibit more strength at thicknesses of a few nm and so are more successful for use in array targets when shooting nm thickness foils. At these thicknesses, shock damage from the laser pulse and debris from neighbouring targets can cause target damage before the shot and so shielding between targets in arrays is a consideration when planning such experiments. This is a relatively simple way of producing simple 2D targets. However it is limited to flat foils. With some more complex mount designs it is possible to produce double foils, but these mounts are difficult to prepare in high numbers.

17.3.1.2 Ultra-Precision CNC Milling

As detailed in Sect. 17.2 there have been developments in precision CNC machining that make it possible to produce high aspect ratio 3D microparts with submicron accuracy. The technique has been developed at RAL to produce microparts (for example AFI cones and hohlraums as in Fig. 17.7) in batches with high yield. As detailed earlier in the chapter there is considerable capital investment needed in the machine, the tooling and the staff training as well as the environmental considerations when producing such challenging parts. However is has been possible to produce batch of 50–100 [2] components in repeatable runs that have allowed statistical studies to be carried out on targets such as cones that were not previously possible.

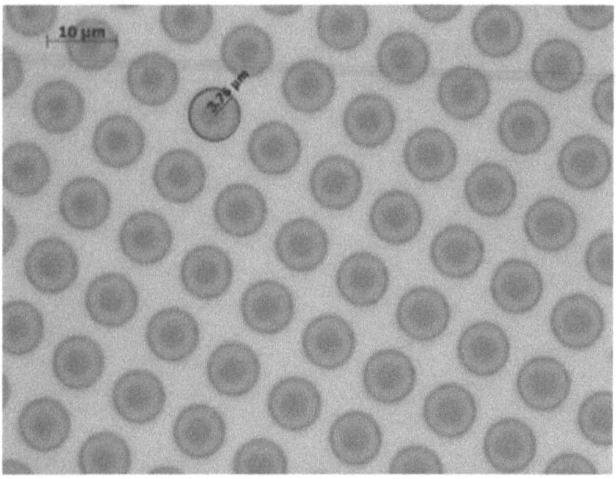

Fig. 17.8 Plastic disc targets (on Si wafer); 7.5 µm diameter, 1 µm thick

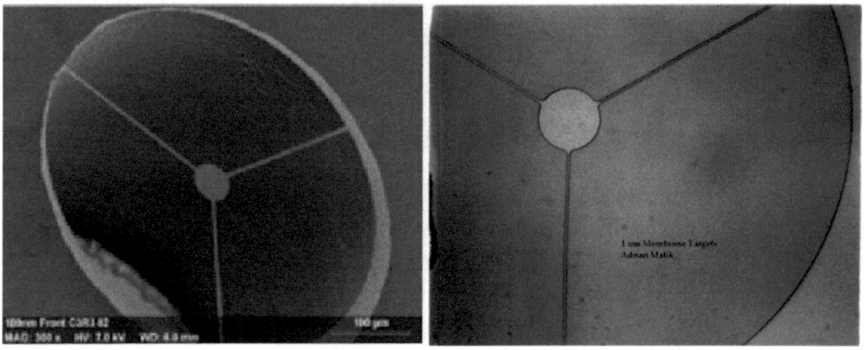

Fig. 17.9 Thin film targets of silicon nitride (32 µm diameter, 50 nm thick)

17.3.1.3 Wafer-based and MEMS Techniques

Micro-Electro-Mechanical Systems (MEMS) fabrication can be used to mass produce target components with very high accuracies and high number. This technology uses silicon wafer based techniques to mass produce geometries that would not be possible with other techniques. At RAL a series of 2D and 3D targets have been fabricated. Simple targets such as disks of plastic of a few microns in diameter (Fig. 17.8) can be made extremely simply (although picking them up is more challenging), 2D objects such as Si pillars can be fabricated using deep reactive ion etching and a particularly good example is the silicon nitride membrane targets (Fig. 17.9) shot at RAL by Strangio et al in 2006 [23].

Although significant initial expense can be required for MEMS manufacturing the technique does give the possibility of producing (large numbers of) microtargets or microcomponents which it is not possible to produce using other techniques. Additionally, if large numbers are required then significant cost savings have been demonstrated.

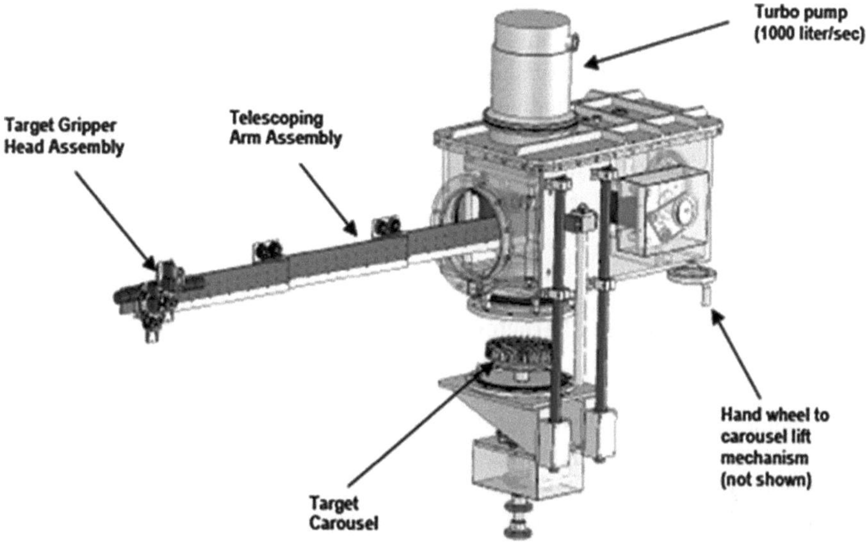

Fig. 17.10 High precision high repetition rate target inserter showing inserter arm (extended) and carousel (General Atomics and the Central Laser Facility)

17.3.1.4 High Rep-rate Microtarget Placement in Chamber

With the introduction of high repetition rates high accuracy microtarget placement at matching rates has become a significant technical challenge. Target wheels have been used in many facilities but have significant limitations if several tens of shots are required without breaking vacuum. In a joint collaboration between General Atomics and CLF an Inserter system (see Fig. 17.10) was developed to meet the requirement. The system comprises a linear arm equipped with grippers that can pick up targets mounted on special carriers and then place them on a hexapod (for accurate positioning) in the interaction chamber. Positional accuracies of a few microns can be achieved. Target carriers are individually identified using a machine readable (2D) bar code.

Future projects, such as HiPER, will almost certainly require the development of injector technology in which (cryogenic) microtargets will be injected with high accuracy into an interaction chamber reaching their shot position without mechanical support.

17.3.1.5 Established Microtarget Characterisation Techniques

Historically as mentioned in Sect. 17.2 a complimentary range of characterisation equipment is used in targetry which has developed without the need for high throughput or rapid data taking. It is worth noting that a very large amount of characterisation time (easily over half) is spent during the development of production processes and for many, especially novel, target types relatively little time is required for quality assurance on finished product, not least due to a quality 'gating' process throughout individual target production. Typical characterisation processes include; high resolution optical microscopy, tactile thin film measurement, scanning electron microscopy and white light interferometry. For more accurate measurements, for example sphere mapping, atomic force microscopy is used.

17.3.1.6 High Rep-rate Characterisation and Quality Assurance

Several methods of automated metrology already exist in various sectors of industry, for example automated CMMs (coordinate measuring machines). Some of the methods have been directly applied to mass-produced microtarget component metrology. Challenges have particularly arisen for 3D microtargets and this may require the introduction of automation to techniques such as confocal microscopy. One possibility arising from suitable automated metrology is to store the information as a 3D spatial image (for example). This gives rise to a large amount of metrology data (of the order of 1 GB per target) but there are significant advantages from both reducing the amount of non-automated metrology and also introducing the possibility of *post hoc* characterisation. A sophisticated data management system has been developed to record the characterisation data for each individual target and synchronise it with the specific individual shot data. Such high levels of control of individual target data are highly amenable to quality management systems and several microtarget fabrication facilities have already introduced ISO9001.2008, which would be an obvious international standard to use.

17.3.1.7 High Rep-rate Logistics and Methodology for Targetry

Target production and placement may become a limiting factor in future high rep rate experiments. Accurate insertion/injection of targets is a significant challenge and the solutions are intimately related to microtarget design and production. Also insertion mechanisms introduce further experimental complexity (and possibly extra characterisation). Due to microtarget production times (i) there can be a significant number of un-shot targets at the end of a high repetition rate experiment if it does not run smoothly and (ii) the ability to make target modifications in response to ongoing experimental data is significantly reduced. Large amounts of metrology data are produced which needs careful control, especially in synchronising with other (shot and experimental diagnostic) data streams. To enable the targetry activities new production and characterisation techniques will be required. For efficient experimental delivery there is a necessity for early and detailed planning of targetry.

17.4 Targetry for IFE

The HiPER project is moving into an R&D phase with an increasingly clear vision of the stages required to demonstrate inertial fusion energy (IFE) as a power source. One of the major technical challenges will be to demonstrate the production and delivery to chamber of microtargets. The project baseline targets and targetry-relevant system requirements are reviewed. An update is given of the current status of the HiPER targetry workpackage summarising the coordinated range of progress which has been made within the project's preparatory phase. A forward strategy is then presented in the context of the targetry technology development plan. The full delivery plan is complex and only its essential structure will be presented in this paper focussing primarily on mass production issues and risk reduction. General technical issues of significance for targetry are also discussed.

17.4.1 Introduction

HiPER will be a European laser-driven fusion demonstration reactor facility in which Inertial Fusion Energy (IFE) can be studied [21]. The main goal of HiPER is to study the technology and physics of laser-driven fusion as a basis for commercial IFE reactors.

At the beginning of the preparatory phase of the project the major technical challenges for HiPER targetry were identified and subsequently allocated as the foci of work for individual groups. Throughout the preparatory phase, group activities were progressively integrated feeding into a coordinated, forward-looking strategy forming the targetry sections of the HiPER Business Case. This article, published at a time which is at the end of the preparatory phase and just as HiPER moves into its next phase, has two main and interwoven themes: to review the targetry work accomplished so far and also to look forward and indicate how the targetry activity can be developed to support the needs of HiPER to the point of IFE demonstration.

A considerable amount of (published) work has been conducted over the past 50 years in microtargetry covering a broad sweep of microtechnologies which have been integrated. Microtargets probably represent some of the most complex and demanding micro-objects ever constructed, for example the cryogenic targets for NIF and LMJ. However, the challenges of mass production for microtargetry are only beginning to be explored, although it should be immediately noted that high fidelity microproduction is already an everyday capability in many commercial sectors, primarily the MEMS market. Initial experience with microtarget mass production has shown that the challenge of target placement is intimately related to microtarget production and the two activities need to be approached in an integrated way from the onset.

(Note: Throughout this article the term 'microtargetry' includes both microtarget fabrication and microtarget placement. 'Placement' means placing the microtarget

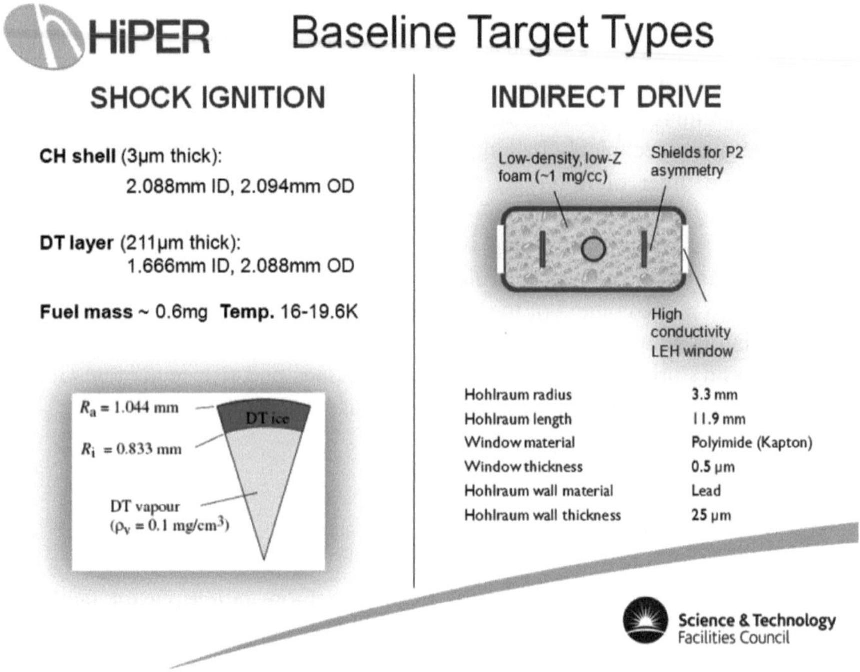

Fig. 17.11 HiPER baseline target designs

with sufficient spatial and angular accuracy in a position for shooting. 'Insertion' means holding the microtarget in position using a physical support during a shot. 'Injection' means firing the microtarget into position using an injector and the microtarget does not have a physical support but is shot while in unsupported motion.)

17.4.2 HiPER Targetry Requirements

17.4.2.1 Baseline Target Designs

There are currently two baseline target designs for HiPER: Shock Ignition and Indirect Drive (Fig. 17.11). Throughout most of the Preparatory phase an Advanced Fast Ignition (AFI) cone + shell target design was also studied within the targetry workpackage. Many important techniques were learnt from work on AFI targets which can be generalised to other target designs and components, most notably hohlraum cans.

It was understood from the project onset that the baseline designs would almost certainly be different to those of the final targets; indeed it might be impossible to actually make the baseline targets. However, it was imperative to have the baseline designs to enable targetry work to begin. Additionally, it was agreed at the beginning that throughout the entire HiPER project there would be an ongoing iterative process between the targetry, modelling, facility design and experimental activities to mutually refine the designs at specified stages. This reflects the nature of HiPER as being an end-directed project to build an IFE power station rather than a physics driven project. Specifically, for example, target production capabilities may well require a loosening of target specifications, particularly if economic viability of a power plant is a major project driver.

17.4.2.2 Repetition Rate

It is stated in the shot rate document that there will be three modes of operation initially in HiPER:

- Senario 1: 100 shots in a burst no yield shots (HDT or similar targets)
- Senario 2: 100 shots in a burst including 2 consecutive yield shots
- Senario 3: 100 shots in a burst including 5 equally spaced yield shots (A burst is a continual stream of targets running at 5–10 Hz.)

Additionally for HiPER the possibility of providing a burst of 100 yield and 1,000 (mixed) shot bursts will be considered under scenarios 4 and 5.

For HiPER, to demonstrate power production capability, the targetry production and injection requirements are to operate at 10 Hz continuously for several days with all shots potentially being full yield shots.

To run an IFE reactor in continuous operation at 10 Hz will require $\sim 900{,}000$ targets a day, and to run at 16 Hz will require $\sim 1{,}500{,}000$ targets a day. This clearly has extensive implications for mass production of components and assemblies.

A key issue in HiPER will be to assess the targetry requirements for IFE high gain high repetition rate scale-up. Such targets will have the same general features as single shot targets but the emphasis will be on demonstrating high number scale-up capabilities and new processes for production will almost certainly have to be developed. Target mass-production, injection and tracking are key demonstrators for proving laser-driven fusion as a realistic option for commercial energy production.

17.4.2.3 Particular Targetry Challenges

As previously mentioned early identification was made of the challenges from a microtarget fabrication perspective. There are many significant challenges that are associated with high gain IFE targets but some which received particular attention in the project follow. (1) Targets will almost certainly have a thin-walled microballoon component with an internal layer of deuterium/tritium (ice). (2) For some targets the layer may be carried on foam (particularly to remove the need for layering). (3) Indirect drive targets will require the production of hohlraums that may be made of lead.

17.4.2.4 Survivability

Target survival is defined as being the retention of the capability of a target to undergo fusion at the appropriate point in its lifecycle. From its manufacture until its ignition during the fusion event within the target chamber a target must retain its fundamental properties by being able to accommodate all perturbations and environmental changes to which it is exposed. The life of a target for HiPER can be considered in eight key stages these being; production, storage, transport to injector loader, injector loading, injection, separation from sabot, steering and exposure to the chamber environmental conditions.

17.4.2.5 Responding to the Requirements

In the Preparatory phase specific areas of concern within the requirements were identified and the major aspects of the work performed, usually by individual partners, is summarised in Sect. 17.4.3. The approach to scaling up targetry to IFE rates, directed by the requirements, is given in Sect. 17.4.4.

17.4.3 Preparatory Phase Work

17.4.3.1 Partners' Work Areas

During the Preparatory phase individual partners worked in particular areas of targetry. They are: CEA (France), cryogenic single shot targets and modification of the LMJ inserter to shoot them; LPI (Russia), cryogenic fuelled shell rapid layering; General Atomics (US), injection, tracking and engagement; UPM (Spain), advanced target materials; TUD (Germany) medium repetition rate cryogenic targetry; STFC (UK) microtarget mass production.

Good and increasingly integrated progress was made in many areas. Some specific aspects are highlighted in the following sections.

17.4.3.2 CEA Single Shot Targetry and AFI Targets

CEA have produced AFI targets (Fig. 17.12) by the production of thin walled micro shells and the insertion of a gold cone. The shells are produced using droplet orifice techniques to fabricate Poly(AlphaMethylStyrene) (PAMS) mandrels which are then coated with glow discharge polymer.

The GDP targets are laser drilled to produce openings for the cone and the fill tube and then the targets are assembled. Capillary tubes have been fabricated to fill the shells with helium and leak tests have been carried out at cryogenic temperature.

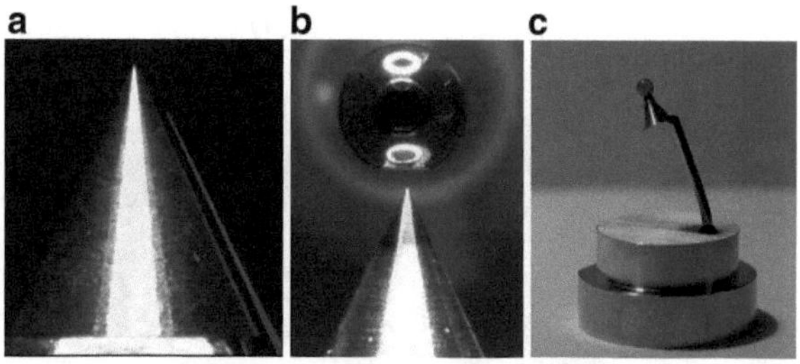

Fig. 17.12 From left to right: (**a**) AFI Cone; (**b**) assembly of the cone shell; (**c**) the final target [24]

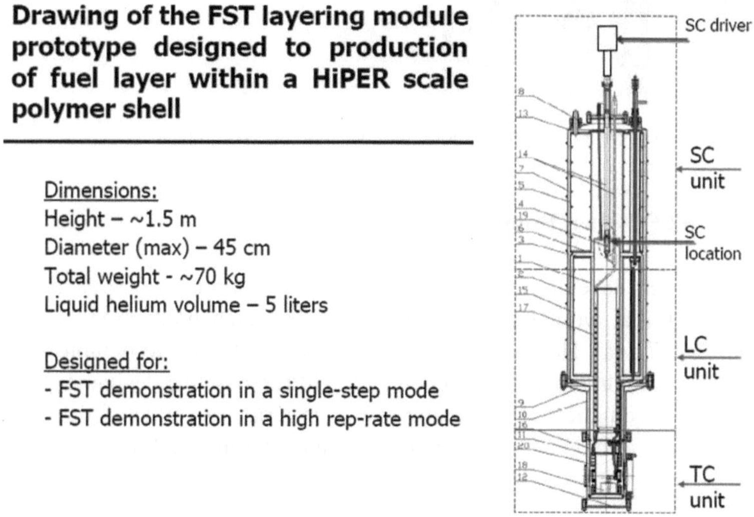

Fig. 17.13 FST Layering prototype for HiPER facility (designed by researchers at the Lebedev Physical Institute)

Importantly the trials demonstrated that it is possible to form a glue joint that is sufficiently leak tight at 77 K (The leak rate was less than 1×10^{-8} mbar.L/s) [24].

17.4.3.3 Cryogenic Shell Rapid Layering

The Lebedev Physical Institute has proposed the Free Standing Target (FST) technology for filling and layering targets for HiPER [25]. Test models for the layering module have been designed (Fig. 17.13), and layering units for undersize shells have been built and demonstrated to produce layered shells, the layering time being of the order 10s.

The Lebedev Physical Institute has also proposed designs for a full scale target delivery system for the HiPER facility running in burst mode (100 shots) with sabots carrying the target through an electromagnetic injector to the target chamber.

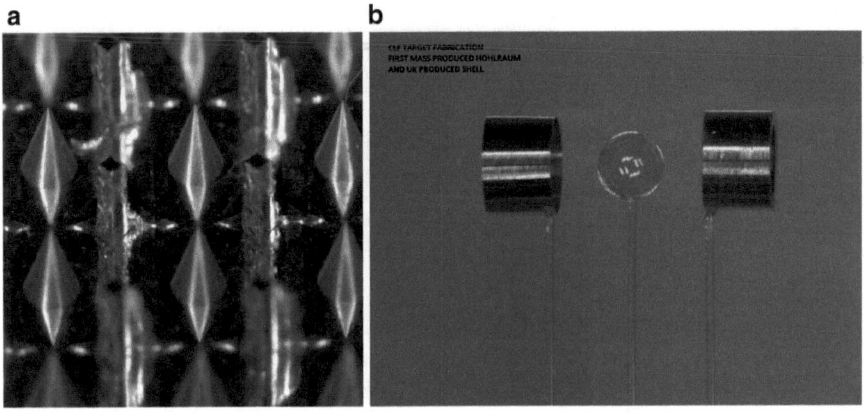

Fig. 17.14 (**a**) AFI cone mandrel; (**b**) hohlraum batch produced components

17.4.3.4 STFC Mass Production

STFC has been developing the capability to mass produce components for IFE targets [2]. Initial work had focused around the batch production of AFI cone targets (Fig. 17.14a). However, this work has been extended to demonstrate batch production of hohlraum components for the HiPER baseline design (Fig. 17.14b). Initial trails have indicated promising results in terms of surface finish and yield. Future work to develop complex geometries and other materials are in progress.

17.4.4 Moving Forwards and the Business Case

During the later stages of the preparatory phase a coordinated review was taken of the targetry work in the light of the requirements (which had been defined by then) for subsequent phases of HiPER and IFE. The review then formed the basis for targetry section of the business case.

17.4.4.1 The Business Case

In its simplest form the targetry section of the Business Case reviews the current Technology Readiness Levels (TRLs) of all aspects of targetry and also gives a detailed Technology Development Plan (TDP). This is done in a way that shows risk reduction. The main purpose is to demonstrate a credible way to develop targetry to meet the Requirements of HiPER.

Because targetry spans many technologies the targetry section of the Business Case is complex and only crucial aspects are given in this section. The particular nature of targetry enables separate technologies to be developed almost independently early in the R&D phase and progressively integrate them throughout 4a and then fully in 4b. This allows significant de-risking during the R&D phase as well as more flexible funding opportunities.

17.4.4.2 Shell Production

There are three potential techniques suitable for shell production for HiPER; (1) wet chemistry/thin film coating, (2) Atomic Layer Deposition (ALD), and (3) dielectrophoretics/microfluidics.

The introduction of innovative production techniques, particularly those which are only recently of sufficient maturity to be applicable to shell production, may offer opportunities for cost-effective scale up.

It is immediately noted that the different baseline target designs require significantly different amounts of post-production processing for shells. Most notably cone + shell targets require the placement of a re-entrant cone into the shell. Furthermore there are potentially modifications to shells that may result from the mode of filling.

In all cases shell sphericity and roughness requirements are stringent. Specifications may vary between target types.

17.4.4.3 Non-shell component Production

Current capability is for batch sizes of 50 using established ultra precision micromachining and coating techniques. However, the established process is not realistically scalable to HiPER 4b. Several currently maturing technologies have the potential for producing the high numbers of ultra precision microcomponents which will be required. Hot pressing and Metal Injection Moulding (MIM) techniques are potentially appropriate for HiPER and IFE.

17.4.4.4 Micro-Assembly

This task is the assembly of targets consisting of more than one component (assuming that it cannot be manufactured already assembled). Currently this is, at best, a semi-automated task but for volume manufacture under the levels of accuracy required increasing levels of automation will be necessary. R&D projects in robotic microassembly are underway. This includes pick and place, adhesive application (and cure) and real-time optical recognition within a production environment. The robots have a number of axes of freedom and can be combined with other assembly machines to form an integrated microassembly solution. With sub-micron accuracy and computer control software it is possible to pick and place parts ready for glue application and curing. This programme will increasingly support high rep rate target production and the initial stages of the project are showing considerable scope for addressing the many technical challenges.

Alternatively, high volume production/assembly may be based on wafer-based fabrication techniques.

17.4.4.5 Target Fill and Layering

Target manufacture, fill and layering techniques vary widely between the baseline target types. There are, however, two main modes for filling: injection filling (in which the fuel mixture is injected into the shell through a microhole bored in the wall, typically via an attached ultra small bore fill tube) and permeation filling (in which the fuel is forced to diffuse through the shell wall using elevated external pressure).

Generally, however, and cutting across target types, layering, if used, will probably be performed in one of three regimes; (1) small batch processing of, say, 1–100 (or possibly 1,000) targets at a time using a LMJ-scaled layering chamber which is of particular relevance for HiPER. (2) Large batch processing of, say, 10,000–1,000,000 targets at a time using, for example, a fluidised bed technique for application on HiPER. (3) Continuous production, running at 10 Hz possibly with parallel production units, based on, for example, microencapsulation or microfluidic technology, again for HiPER.

17.4.4.6 Characterisation

During manufacture and processing shells can currently only be characterised slowly, particularly for the outer roughness (which is data of great significance for assessing target viability). Looking ahead to IFE the characterisation needs to be done quickly, possibly on a statistical basis, and feed back to the continuously optimise the target production line.

Historically shell parameters have been used which are particularly suitable for physics modelling, however, for a production environment, especially HiPER 4b, there is significant scope for choosing parameters which are rapidly applicable in a high throughput microproduction environment.

Throughout production and, if it occurs, storage, fuelled targets experience tritium decay/heating issues that may affect the final target ignition viability. Characterisation will need to be deployed to assess potentially deleterious changes. Characterisation equipment will be working in a challenging environment (cryogenic temperatures, elevated radiation) and also have to analyse targets in a way that does not affect the measurements.

17.4.4.7 Other Technologies

It is clearly understood that tritium handling procedures will be a major part of the targetry activities. However, the procedures are mature and understood well.

17.4.4.8 Cross-cutting Technologies and Evolution

Shell production processes, target types and fill techniques are highly inter-related issues and cannot be considered in isolation from each other.

Within the evolution of the full HiPER programme it is inevitable that there will be refinements and innovations in target design. Specifically, for example, foam inner layers, radially graded materials and ultra hard materials have been discussed. External (metal) coatings have also been discussed, primarily as IR reflection coatings whilst the target transits the interaction chamber.

17.5 Pragmatics of Progress

The modelling, experimental, targetry and chamber design work within HiPER are very strongly inter-related. Specifically, the progressive refinement of the (base line) target design(s) will be achieved through controlled, iterative interactions. This will be a two way process: target designs will be refined in response to results from both modelling and experiments and at the same time R&D work performed within targetry will progressively establish the range of target designs which can be practically produced.

As a specific example the target design is influenced by (a range of) factors such as the method of filling and the injection velocity. If the targets are injected at high speed, then there is less black body radiation absorption from the chamber walls; therefore there is less or no need for coating which gives suitability for permeation filling. If the injection velocity is low (leading to a higher positional accuracy and less demand on the tracking system) then the target will be heated by the black body radiation from the chamber and a coating may be required to reflect the radiation and the coating may necessitate a fill tube for shell filling.

17.6 The HiPER Target Fabrication Facility

Target production requires the following:

A specifically designed building including:

1. A cryogenic target production area where liquefied tritium/deuterium can be added to targets within a glovebox environment on stable foundations (local decoupled foundations) similar to semi-conductor manufacture
2. Nuclear ventilation including tritium recovery processes to minimise the discharge of tritium to atmosphere
3. Target characterisation equipment to demonstrate compliance with finalised target specifications
4. Cryogenic target storage
5. Cryogenic target transfer to a magazine (or similar) suitable for interfacing with the target injector

The remit of the HiPER development facility may be satisfied through manual target production which is currently demonstrable as a technology however, there are requirements to be satisfied during the lifetime of HiPER to reduce the cost per target significantly and further, to produce targets at a rate commensurate with the operation of a reactor. This can only be achieved through the automation of the target stream. This therefore imposes two further requirements on the HiPER project under the assumptions that a reactor will be operating at 10 Hz and 24 h per day during phase 4b:

1. The HiPER project shall provide the technology basis to demonstrate yield target production at a rate of one million targets of an appropriate quality per day in a scalable fashion.
2. The HiPER project shall demonstrate a cost per target under the above regime of less than 1 euro per target to be achievable (this being within a factor of ten of commercial power production cost requirements).

Targets within a reactor environment would also have to be filled with tritium, predominantly bred within the reactor blanket. Deuterium fuel will need to be refined from (sea) water in sufficient quantity to provide the D fuel and/or recycled from that unburned within the reactor. This imposes one further need on target production to enable the future reactor:

1. The HiPER project phase 4a shall demonstrate the viability of the process for recovery and if necessary, refining of tritium from blanket material in suitable quantities and at a suitable cost to provide a continuous fuel source for the reactor in a scalable fashion.

17.7 Conclusions

Current high power laser microtarget production capabilities offer significant challenges for the integrated application of a range of microtechnologies including characterisation. The latest high repetition rate high power lasers which are currently coming on-line offer wide-ranging opportunities for microtechnology to meet medium volume and mass production requirements. Ultraprecision micromachn-ing and wafer-based manufacture have been successfully deployed as an initial approach. High repetition rate target placement gives another range of opportunities for microtechnology solutions.

In summary, high power laser microtargetry is an excellent example of applied microtechnology offering a wide range of challenges and rewards.

Acknowledgements The authors would like to thank a wide range of collaborators whose work has contributed significantly to this article. In particular all current and past staff on the Target Fabrication Laboratory at the Central Laser Facility, RAL, specifically Donna Wyatt, Sam Serra, and Ian East; all current and past staff of the Target Fabrication Laboratory at AWE, and Wigen Nazarov at the University of St Andrews.

References

1. R.S. Fearing, IROS, vol. 2. International Conference on Intelligent Robots and Systems-Volume 2 (1995), p. 2212
2. C. Spindloe et al., Fusion Sci. Technol. **59**(1), 221 (2011)
3. D.M. Hoffman, B. Singh, J.H. Thomas, *Handbook of Vacuum Science and Technology* (Academic Press, New York, 1998), p. 612
4. K. Seshan, *Handbook of Thin Film Deposition Processes and Techniques – Principles, Methods, Equipment and Applications*, 2nd edn. (William Andrew Publishing/Noyes, Norwich, 2002), p. 14
5. B. Chapman, *Glow Discharge Processes* (New York, Wiley, 1980), p. 185
6. H.O. Pierson, *Handbook of Chemical Vapor Deposition* (Elsevier, Burlington, 1999)
7. C.A.J. Palmer et al., Central Laser Facility Annual Report (2008/2009), p. 29
8. Specialty Coating Systems, www.scscoatings.com
9. M. Dunne et al., Phys. Rev. Lett. **75**, 21 (1995)
10. D. Hoarty et al., Phys. Rev. Lett. **78**, 17 (1997)
11. A. Benuzzi et al., Phys. Plasmas **5**, 2827 (1998)
12. J.W. Falconer et al., J. Vac. Sci. Technol. A **8**(2), 968 (1989)
13. Electrochemistry Encyclopedia, http://electrochem.cwru.edu/encycl/art-e01-electroplat.htm
14. L.J. Durney, *Electroplating Engineering Handbook* (Chapman and Hall, London, 1996)
15. M. Beardsley et al., Central Laser Facility Annual Report (2005/2006), pp. 221–222
16. S. Serra et al., Central Laser Facility Annual Report (2010/2011), p. 58
17. R.C. Montesanti et al., Fusion Sci. Technol. **59**(1), 70 (2011)
18. R.K. Leach et al., Measurement Good Practice Guide, No. 108, National Physical Laboratory (2008)
19. L.A. Gizzi et al., (2nd ICUIL) AIP Conf. Proc. **1209**, 134–143 (2010)
20. www.qub.ac.uk/sites/LIBRA/
21. www.hiper-laser.org/
22. www.clf.stfc.ac.uk/Facilities/Astra/12254.aspx
23. C. Strangio et al., Proceedings 29 ECLIM (2006), pp. 384–390
24. F.B. Said et al., Fusion Sci. Technol. **59**(1), 234 (2011)
25. E. Koresheva et al., Laser and Particle Beams **23**, 563–571 (2005)

Index

P. McKenna et al. (eds.), *Laser-Plasma Interactions and Applications*,
Scottish Graduate Series, DOI 10.1007/978-3-319-00038-1,

MIX
Papier aus verantwortungsvollen Quellen
Paper from responsible sources
FSC® C105338

If you have any concerns about our products,
you can contact us on
ProductSafety@springernature.com

In case Publisher is established outside the EU,
the EU authorized representative is:
Springer Nature Customer Service Center GmbH
Europaplatz 3, 69115 Heidelberg, Germany

Printed by Libri Plureos GmbH
in Hamburg, Germany